The Biofuels Handbook

The Biofuels Handbook

Edited by Damian Price

SYRAWOOD
PUBLISHING HOUSE

New York

Published by Syrawood Publishing House,
750 Third Avenue, 9th Floor,
New York, NY 10017, USA
www.syrawoodpublishinghouse.com

The Biofuels Handbook
Edited by Damian Price

International Standard Book Number: 978-1-68286-678-8 (Hardback)

Cataloging-in-Publication Data

The biofuels handbook / edited by Damian Price.
    p. cm.
Includes bibliographical references and index.
ISBN 978-1-68286-678-8
1. Biomass energy. 2. Fuel. 3. Energy conversion. I. Price, Damian.
TP339 .B56 2019
662.88--dc23

TABLE OF CONTENTS

PREFACE

This book has been a concerted effort by a group of academicians, researchers and scientists, who have contributed their research works for the realization of the book. This book has materialized in the wake of emerging advancements and innovations in this field. Therefore, the need of the hour was to compile all the required researches and disseminate the knowledge to a broad spectrum of people comprising of students, researchers and specialists of the field.

Biofuel is a source of energy manufactured from renewable organic materials like agricultural or human waste, decayed forest waste, etc. through intensive biological processes. Some examples of biofuels are bioethanol, biodiesel, biomass, etc. Research in biofuels cover investigations into better biofuels derived from Jatropha, fungi, animal gut bacteria, etc. Chapters compiled in this book present numerous researches in a comprehensive form in order to equip the reader with extensive knowledge related to this field of study. Topics covered herein present upcoming theories and concepts while also presenting the practical applications. As this field is emerging at a fast pace, this book will help engineers, ecologists, environmentalists, academicians and students associated with the field of biofuels and energy production.

At the end of the preface, I would like to thank the authors for their brilliant chapters and the publisher for guiding us all-through the making of the book till its final stage. Also, I would like to thank my family for providing the support and encouragement throughout my academic career and research projects.

Editor

Comparison of four glycosyl residue composition methods for effectiveness in detecting sugars from cell walls of dicot and grass tissues

Ajaya K. Biswal[1,2,3], Li Tan[1,2,3], Melani A. Atmodjo[1,2,3], Jaclyn DeMartini[3,4,5], Ivana Gelineo-Albersheim[2,3], Kimberly Hunt[2,3,6], Ian M. Black[2], Sushree S. Mohanty[2,3], David Ryno[2,3], Charles E. Wyman[3,4] and Debra Mohnen[1,2,3]*

Abstract

Background: The effective use of plant biomass for biofuel and bioproduct production requires a comprehensive glycosyl residue composition analysis to understand the different cell wall polysaccharides present in the different biomass sources. Here we compared four methods side-by-side for their ability to measure the neutral and acidic sugar composition of cell walls from herbaceous, grass, and woody model plants and bioenergy feedstocks.

Results: Arabidopsis, *Populus*, rice, and switchgrass leaf cell walls, as well as cell walls from *Populus* wood, rice stems, and switchgrass tillers, were analyzed by (1) gas chromatography–mass spectrometry (GC–MS) of alditol acetates combined with a total uronic acid assay; (2) carbodiimide reduction of uronic acids followed by GC–MS of alditol acetates; (3) GC–MS of trimethylsilyl (TMS) derivatives; and (4) high-pressure, anion-exchange chromatography (HPAEC). All four methods gave comparable abundance ranking of the seven neutral sugars, and three of the methods were able to quantify unique acidic sugars. The TMS, HPAEC, and carbodiimide methods provided comparable quantitative results for the specific neutral and acidic sugar content of the biomass, with the TMS method providing slightly greater yield of specific acidic sugars and high total sugar yields. The alditol acetate method, while providing comparable information on the major neutral sugars, did not provide the requisite quantitative information on the specific acidic sugars in plant biomass. Thus, the alditol acetate method is the least informative of the four methods.

Conclusions: This work provides a side-by-side comparison of the efficacy of four different established glycosyl residue composition analysis methods in the analysis of the glycosyl residue composition of cell walls from both dicot (Arabidopsis and *Populus*) and grass (rice and switchgrass) species. Both primary wall-enriched leaf tissues and secondary wall-enriched wood/stem tissues were analyzed for mol% and mass yield of the non-cellulosic sugars. The TMS, HPAEC, and carbodiimide methods were shown to provide comparable quantitative data on the nine neutral and acidic sugars present in all plant cell walls.

Keywords: Cell wall, Sugar composition, Feedstock, Secondary cell wall, Biofuel, Uronic acid

Background

Cell walls constitute the bulk of plant biomass, a major renewable resource for biofuel and biomaterial production [1, 2]. The use of plants as a source for bioproducts [3] is expected to increase as fossil fuel supplies decrease, mitigation strategies for climate change intensify [4], and the world population increases [5]. Plant cell walls are a matrix of complex non-cellulosic polymers (hemicelluloses and pectins), cellulose microfibrils and fibers, proteins and proteoglycans, and in tissues

*Correspondence: dmohnen@ccrc.uga.edu
[2] Complex Carbohydrate Research Center, University of Georgia, 315 Riverbend Rd., Athens, GA 30602-4712, USA
Full list of author information is available at the end of the article

with secondary walls, the phenolic polymer lignin [6]. The choice of plant species for production of specific bioproducts is influenced by the quality and quantity of the cell wall polymers [7]. Although all plant cell walls contain the same general types of polymers, the specific amounts of the different polymers and their unique glycosyl residue content and linkages vary in different types of plants (e.g., woody versus herbaceous dicots versus grasses) and in different tissues and cell types. Since the different cell wall polymers have unique physical–chemical properties, their suitability as a resource for specific bioproducts also varies, underscoring the need for cell wall analysis methods that can detect critical differences in different biomass sources.

A full analysis of cell wall structure requires the use of detailed, time-consuming, and often expensive analytical methods [8]. However, an initial assessment of the polymer content of biomass samples can be obtained by analysis of the glycosyl residue composition. Multiple methods exist to measure the sugar content of plant cell walls; however, these methods have not been compared side-by-side for effectiveness in analyzing the same tissues from multiple plant species. Here we compare the four most common sugar composition methods for their ability to reproducibly quantify the greatest number of different types of sugars present in cell walls of dicot and grass species. The goal was to provide researchers a reference source for selecting a preferred sugar analysis method for comparison of cell walls from different species, cell types, and/or walls from native versus mutant/transgenic/variant plants.

The two most common plant cell wall sugar composition analysis methods are the alditol acetate (AA) and the trimethylsilyl (TMS) methods [9]. The AA method involves hydrolysis of monosaccharides from alcohol insoluble residues (AIR) and their reduction to alditols using sodium borohydride, followed by acetylation with acetic anhydride to volatilize them for gas chromatography and mass spectrometry (GC–MS) (Fig. 1A) [10]. The AA method has been used to study the sugar composition of many plant species, including Arabidopsis [11], Italian ryegrass [10], potatoes [12], barley [13], tobacco [14], Populus [15], rice [16], and switchgrass [17]. The limitation of this method is that it does not measure acidic sugars. In contrast, the TMS method involves sequential methanolysis and trimethylsilylation of the hydrolyzed sugars to yield TMS-methyl glycosides (Fig. 1C), enabling detection of both neutral and acidic sugars. The TMS method has been used to study sugar composition in a variety of plant species including Arabidopsis [18], rice [19], carrots and apples [20], and Populus [21].

Uronic acids (UAs) are ubiquitous acidic sugars in plant cell wall non-cellulosic polysaccharides, including galacturonic acid (GalA) in pectins and glucuronic acid (GlcA) in hemicellulosic xylans. UAs are abundant in dicot primary walls, of lesser abundance in dicot secondary walls, and of low abundance in grass walls. Many researchers have thus used the AA method to analyze the cell walls of grasses and dicot secondary walls. However, this results in an underestimation, or total lack of recognition, of the presence of UAs in such biomass, as well as the risk of not identifying UA-containing matrix polysaccharides, such as pectin and glucuronoxylan that have been shown to impact biomass recalcitrance [15, 21–23]. Thus, the glycosyl residue composition analysis methods that detect both neutral and acidic sugars, such as the TMS method, are preferable for the most complete analyses. However, despite its advantage over the AA method, the TMS method is not without drawbacks. TMS derivatization of methyl glycosides results in multiple anomeric forms of the monosaccharide derivatives, yielding multiple peaks for each sugar that can be difficult to distinguish and quantitate [20, 24].

Another method to analyze both neutral and acidic sugars is the carbodiimide method, which entails reduction of UAs to their respective neutral sugars with subsequent analysis by the AA method (Fig. 1B) [25]. Specifically, the carboxyl groups of UAs in un-degraded polymeric material are activated with a water-soluble carbodiimide and reduced with sodium borodeuteride to yield 6,6-dideuterio sugars. The UAs are quantified as the increased amount of their respective neutral sugars in a pre-reduced compared to un-reduced sample. This method has been used to study the cell wall sugar composition of apple [26] and maize [25].

Liquid chromatography-based methods are also available that detect both neutral and acidic sugars in hydrolyzed cell wall samples. High-pressure, anion-exchange chromatography (HPAEC) coupled with electrochemical detection (ECD) allows for direct analysis of monosaccharides and oligosaccharides without derivatization or labeling. It uses high pH (pH 12–13) to partially deprotonate the sugar hydroxyl groups, yielding sugar anions that can be separated on anion-exchange columns [27, 28]. This method has been used to analyze cell walls from multiple plant species including Arabidopsis [29], wheat [30], potato [31], rice [32], and switchgrass [17]. HPAEC, however, has the disadvantage that it is not readily adaptable to mass spectrometry for confirmation of sugar identity.

Here we compare four different sugar composition analysis methods (AA, carbodiimide, TMS, and HPAEC) for their ability to quantify the sugar composition of cell walls from leaves of Arabidopsis, Populus, rice, and switchgrass. Our objective was to identify quantitative, reliable, and facile methods for analysis of the glycosyl

Comparison of four glycosyl residue composition methods for effectiveness in detecting...

3

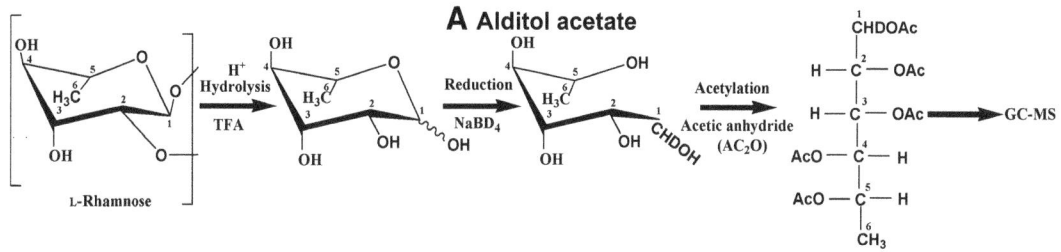

A Alditol acetate

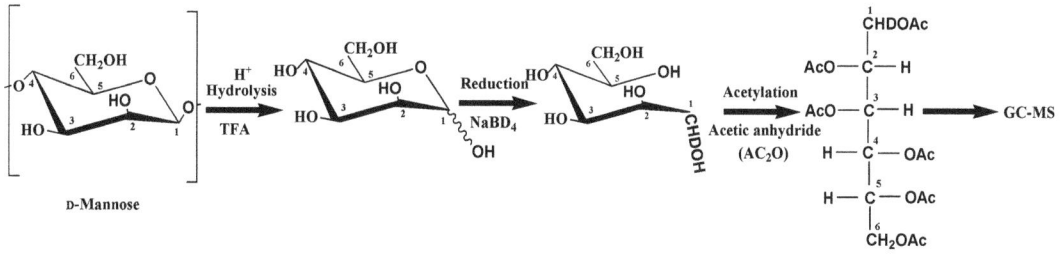

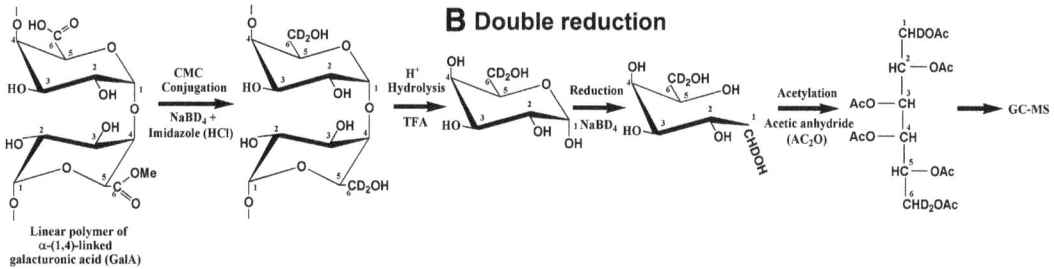

B Double reduction

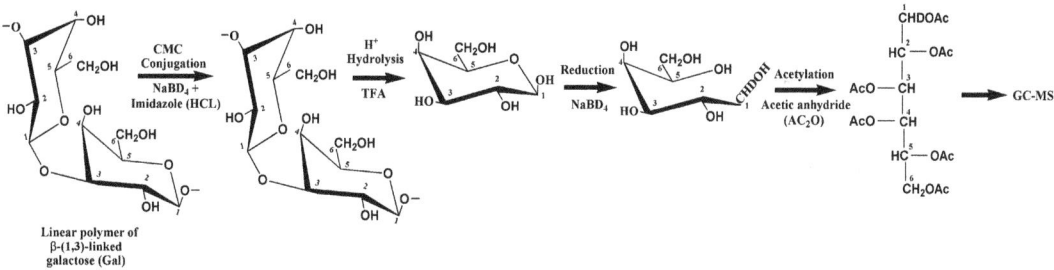

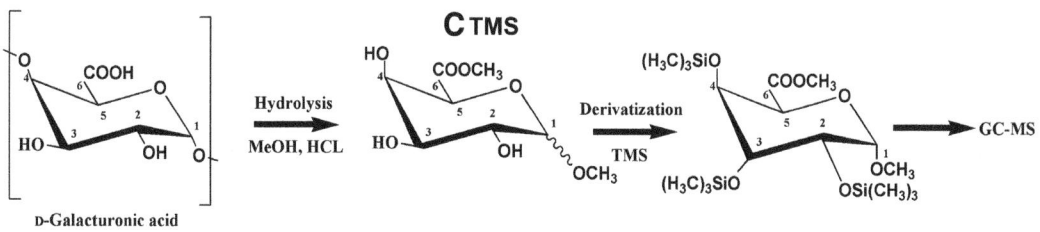

C TMS

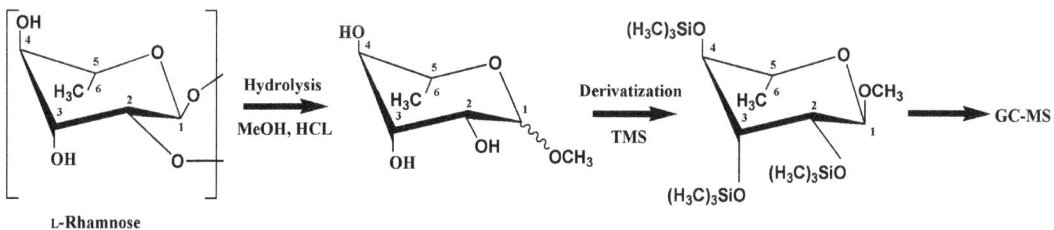

(See figure on previous page.)
Fig. 1 Schematic overview of the derivatization of sugar residues by the **A** alditol acetate, **B** carbodiimide, and **C** TMS plant cell wall glycosyl residue composition analysis methods. Schematic depicts analysis of the designated terminal and internal sugars of the indicated plant cell wall polysaccharides: **A** terminal Rha and internal Man residues, **B** internal GalA residues of homogalacturonan and Gal residues in β-1,3-linked galactan, **C** internal GalA and terminal Rha residues. Cyclic and linear sugars are depicted as Haworth and Fischer projections, respectively

residue composition of plant cell walls. Such information is essential to understand plant cell wall structure/function relationships and cell wall structures associated with biomass recalcitrance and/or bioproduct quality. To the best of our knowledge, this is the first side-by-side comparison of the different analytical methods using the same tissue sources from multiple dicot and grass species. To ensure that the results from the analysis of leaves are applicable to other types of biomass tissues (e.g., stems), we also compared the performance of the four methods in the analysis of cell walls from *Populus* wood, rice stem, and switchgrass tillers. We conclude that the TMS, HPAEC, and carbodiimide methods are the preferred methods to obtain quantitative and reproducible sugar composition data on the major neutral and acidic sugars present in all dicot and grass biomass.

Methods
Plant material and growth conditions
Arabidopsis wild-type (WT) [*Arabidopsis thaliana* (L.) Heynh. var. Columbia S6000] plants were grown essentially as described [33]. Briefly, sterilized seeds were sown on media plates containing half-strength Murashige and Skoog basal salts (Sigma-Aldrich Corp., St. Louis, MO) and 5.5 g/L plant agar (Research Products International Corp., Mount Prospect, IL) with pH adjusted to 5.7 prior to autoclaving. Seed-containing plates were kept in a growth chamber with 60% relative humidity, 150 μmol photons/m^2/s light, and photoperiod cycle of light for 14 h at 19 °C and dark for 10 h at 15 °C. Following germination, 10-day-old seedlings were transferred to soil and grown to maturity in a growth chamber under the same growth conditions as above. Fertilizer (Peters 20/20/20 with micronutrients) was applied once a week or as needed.

Populus deltoides Bartr. ex Marsh. clone WV94 plants were obtained from ArborGen Inc. (Ridgeville, SC) as plantlets generated in vitro from petiole explants via callus. Rooted plantlets grown for 4–6 weeks in magenta boxes were cleaned with running water to remove media and charcoal, and transplanted into soil in 3.8-L (1 gallon) pots. Growth conditions and plant maintenance were as previously described [21] and summarized below. The soil was a mix of 1 bag (2.8 cubic feet) of Fafards 3B Soil mix (GroSouth Inc, Atlanta, GA), 250 mL osmocote, 84 mL bone meal, 84 mL gypsum, and 42 mL dolomite/limestone. Plants were grown in the greenhouse for 9 months under a 16-h light/8 h dark cycle at 25–32 °C with constant misting, and fertilized weekly with Peters 20-10-20 (nitrogen-phosphorus-potassium; GroSouth Inc, Atlanta, GA).

Switchgrass (*Panicum virgatum* L.) var. Alamo II genotype ST1 [34] was grown for 2 months as seedlings in 3.8-L (1 gallon) pots followed by transfer to 11.4-L (3 gallon) pots and a further 6 weeks of growth. Growing medium was a soil mixture consisting of two 2.8-cubic-feet bags of Fafards 3B (GroSouth Inc, Atlanta, GA), one 2.8-cubic-feet of River Bottom Sand (Redland Sand, Watkinsville, GA), and 118 mL of Osmocote Plus granular fertilizer (18-9-12 minors, 8–9 month release). After planting, plants were fertilized once a week with 440 ppm Jack's Peat Lite Special 20-10-20 (nitrogen-phosphorus-potassium; GroSouth Inc, Atlanta, GA).

Rice (*Oryza sativa* L.) seeds var. IAC 165 obtained from the USDA National Plant Germplasm System were grown in 1.9-L (1/2 gallon) pots for 2 weeks. The seedlings were then transferred to 11.4-L (3 gallon) pots and grown in a greenhouse under a 16-h light/8 h dark cycle at 25–32 °C. Growing medium was the same soil mixture as described above for switchgrass. At the time of planting, plants were fertilized with 1.2-mL (1/4 teaspoon) Sprint 330 Iron Chelate and 3.75 g Jack's Peat Lite Special 20-10-20 per 2 L water. After planting, plants were fertilized once a week with 440 ppm of the Peat Lite Special 20-10-20.

Isolation of plant samples and preparation of cell walls as alcohol insoluble residues (AIR)
Leaf samples were harvested from 5-week-old Arabidopsis, 8-week-old rice, and 10-week-old switchgrass and *Populus*, ground to a fine powder using liquid nitrogen, and stored at −80 °C until use. Biomass samples were isolated as follows: rice stem from 3-month-old plants, switchgrass whole tillers harvested at the R1 stage [35], and *Populus* wood from 9-month-old plants [21]. Harvested biomass samples were air dried completely and milled to a 20-mesh (0.85 mm) particle size using a Wiley Mini-Mill (model number: 3383L10, Thomas Scientific). For *Populus* wood, the bottom 6 cm of stem measured from the soil surface was collected from 9-month-old plants, the bark peeled using a razor, the remaining stem air dried, and the pith removed using a hand drill prior to milling. AIR was prepared from the ground tissue/biomass powder and destarched prior to analysis as described [33].

Comparison of four glycosyl residue composition methods for effectiveness in detecting...

5

Glycosyl residue composition analysis by the alditol acetate (AA) derivatization method

The neutral sugar composition of AIR was analyzed by the AA derivatization method [36] with modification. Briefly, 100–500 µg AIR was incubated in 0.2–1 mL 2 M trifluoroacetic acid (TFA, Thermo Fisher Scientific, Waltham, MA) at 121 °C for 2 h, followed by reduction with 200–300 µL of 10 mg/mL $NaBD_4$ in 1 M ammonium hydroxide for at least 2 h to overnight at room temperature (RT). The borodeuteride solution was neutralized by adding 3–4 drops glacial acetic acid, and dried down twice with 200 µL methanol:acetic acid (9:1 [v/v]) and thrice with 200 µL anhydrous methanol under a stream of air. The samples were incubated with 250 µL acetic anhydride and 250 µL concentrated TFA for 10 min at 50 °C and dried down with 20–30 drops of isopropanol under a stream of air. To the dried samples, 1 mL of 0.2 M sodium carbonate and 1 mL of methylene chloride (Sigma) was added, the samples vortexed, and the upper aqueous layer removed. The bottom organic layer containing AA derivatives of hydrolyzed sugars was washed thrice with 1 mL deionized water (ddH_2O), transferred to a clean tube, dried down, and resuspended in ~100 µL methylene chloride. The samples (~1 µL) were injected using the splitless injection mode and helium as carrier gas onto an SP-2330 Supelco column (30 m × 0.25 mm, 0.25 µm film thickness) connected to a Hewlett–Packard chromatograph (5890) coupled to a mass spectrometer for GC–MS analysis. AA derivatives were separated using the following temperature gradient: 80 °C for 2 min, 80–170 °C at 30 °C/min, 170–240 °C at 4 °C/min, and 240 °C for 20 min, and were ionized by electron impact at 70 eV. A sample of the GC profile of alditol acetate derivatives of the neutral sugar standards is provided in Additional file 1. GC peak areas were used to determine the response factors for each sugar relative to the internal standard myo-inositol, and subsequently used to determine the amount of sugars in the wall samples [8].

Colorimetric determination of uronic acid (UA) content

The UA content of AIR samples was determined using a modification of the methods of Blumenkrantz and Asboe-Hansen [37], Filisetti-Cozzi and Carpita [38], and van den Hoogen et al. [39] as described below. The hydrolysis of AIR samples was performed independently with either sulfuric acid (H_2SO_4) or TFA [40]. Briefly, 0.4 mg AIR was suspended in 0.4 mL ddH_2O and mixed thoroughly with 40 µL of 4 M sulfamic acid–potassium sulfamate (pH 1.6). The sample was subsequently hydrolyzed with 2.4 mL of 12.5 mM sodium tetraborate in either concentrated H_2SO_4 or 2 M TFA, with incubation for 20 min at 100 °C for H_2SO_4 hydrolysis or for 2 h at 120 °C for TFA hydrolysis. The reaction mixture was

cooled immediately, and mixed with 80 µL of 0.15% (w/v) m-hydroxybiphenyl in 0.5% (w/v) NaOH by vortexing. After 5–10 min, the pink color that developed was measured as absorbance at 540 nm using a microtiter plate reader and the UA content was estimated by comparison to a standard curve of GalA (Sigma-Aldrich, St. Louis, MO) as illustrated in Additional file 2.

Glycosyl residue composition analysis by uronic acid reduction using the carbodiimide method

Glycosyl residue composition was analyzed by initial activation of UAs in an underivatized sample using the carbodiimide method [41] followed by reduction with $NaBD_4$ to convert UAs to their respective 6,6-dideuterio sugars [25, 42, 43]. The AIR sample (10 mg) was suspended in 3 mL 0.033 M sodium acetate pH 4.6. With continuous stirring, 250 mg of CMC [N-cyclohexyl-N'-(2-morpholinoethyl) carbodiimide] methyl-p-toluene sulfonate (Sigma) powder was added, and the pH kept at 4.8 by dropwise addition of 1 M HCl for 2 h. The mixture was chilled on ice, mixed with 1 mL ice cold 4 M imidazole–HCl pH 7.0, and immediately 300 mg $NaBD_4$ was added to the suspension with continuous stirring on ice for 1 h. Excess borodeuteride was afterwards destroyed by dropwise addition of glacial acetic acid. The sample was dialyzed against running ddH_2O for at least 36 h, frozen, lyophilized [26], and one mg of the lyophilized material subjected to sugar composition analysis by the AA method as described above. The amount of the UAs, GalA, and GlcA was calculated as the increase in the amount of galactose (Gal) and glucose (Glc), respectively, compared to the amount measured in un-reduced samples analyzed directly using the AA method [42].

Glycosyl residue composition analysis by GC–MS of TMS-derivatized methyl glycosides

Glycosyl residue composition of AIR was determined by GC–MS of per-O-trimethylsilyl (TMS) derivatives of monosaccharide methyl glycosides produced by acidic methanolysis as previously described [8, 44]. AIR (100–300 µg) was aliquoted into individual tubes, supplemented with 20 µg inositol as internal standard, and lyophilized. The dry samples were hydrolyzed for 18 h at 80 °C in 1 M methanolic-HCl (Supelco, St. Louis, MO), cooled to RT, evaporated under a stream of air, and dried twice more with anhydrous methanol. The released glycosyl residues were derivatized with 200 µL TriSil Reagent (Thermo Fisher Scientific, Waltham, MA) at 80 °C for 20 min. Cooled samples were evaporated under a stream of air, resuspended in 3 mL hexane, and filtered through packed glass wool. Dried samples were resuspended in 150 µL hexane and 1 µL sample injected using helium as carrier gas onto a Supelco EC-1 fused silica capillary column (30 m × 0.25 mm ID) on an Agilent 7890A gas chromatograph interfaced to a

5975C mass spectrometer. The temperature gradient was: 80 °C for 2 min, 80–140 °C at 20 °C/min, 140–200 °C at 2 °C/min, and 200–250 °C at 30 °C/min. A sample of the GC profile of the TMS-derivatized sugar standards is provided in Additional file 3. The response factor for each sugar was determined from the GC peak area (or total peak areas of multiple peaks if multiple derivatives were formed) of each sugar standard relative to the internal standard myo-inositol, and the value subsequently used to calculate the amount of each sugar in the wall samples [8].

Glycosyl residue composition analysis by HPAEC

AIR (100 µg) was refluxed in 400 µL 2 M TFA at 120 °C for 1 h [31, 45] and the resulting solution dried under a stream of air with addition of isopropanol. The dried residue was dissolved in 200 µL ddH_2O and the solution centrifuged for 5 min. The supernatant was diluted 1:3 with ddH_2O and 50 µL of the diluted supernatant injected into a Dionex ICS-3000 HPLC system (Dionex, Sunnyvale, CA) for monosaccharide analysis by high pH anion-exchange chromatography with electrochemical detection in the carbohydrate mode. The buffers used were A—nanopure water, B—200 mM NaOH, and C—1 M NaOAc. Two programs were used to detect different monosaccharides. Program 1 was used to quantify fucose (Fuc), rhamnose (Rha), arabinose (Ara), Gal, Glc, GalA, and GlcA using a Dionex PA20 column (3 × 150 mm) at a 0.5 mL/min flow rate (see Additional file 4A for a sample chromatogram). The column was equilibrated at 1% buffer B for 30 min prior to each separation. The gradient was: 0 min 1% buffer B, 0.1 min 10% buffer B, 2 min 10% buffer B, 4 min 1% buffer B, 15 min 0% buffer B, 25 min 5% buffer B and 10% buffer C, 30 min 5% buffer B and 50% buffer C, and 35 min 1% buffer B. Program 2 was used to quantify xylose (Xyl) and mannose (Man) (see Additional file 4B) using a Dionex PA1 column (4 × 150 mm) and a flow rate of 1 mL/min. The column was equilibrated with 1% of buffer B for 30 min, the sample eluted isocratically at 1% buffer B for 40 min, and the column regenerated with 100% buffer B for 5 min. For both programs, a standard mixture containing known concentrations of different sugars was used to plot concentration-peak area standard curves. The amount of each monosaccharide was calculated from the standard equations based on the corresponding peak area as registered by ECD.

Results

Our goal was to determine the preferred sugar composition analysis method(s) for use across dicot and monocot grass species, which have been shown to have characteristically distinct cell wall compositions and structures [46–48]. *Populus* and switchgrass were chosen as bioenergy crops and Arabidopsis and rice as model plants for

the comparison. Leaves were used as the initial target tissue for this study for three reasons. First, leaves are comparable organs between these two groups. In contrast, for example, *Populus* wood has a different tissue structure and composition compared to rice and switchgrass tillers. Secondly, leaves are a major biomass resource. For example, leaves comprise a significant proportion (25–44%) of switchgrass biomass [49]. Thirdly, leaves are composed of both primary and secondary walls, and thus contain the majority of the different types of cell wall polysaccharides.

We first compared the four different glycosyl residue composition analysis methods for their ability to measure the nine major neutral and acidic sugars in leaf AIR from Arabidopsis, *Populus*, rice, and switchgrass. We present the data in both the relative yield (mol%) (Fig. 2; Additional file 5) and the mass yield (µg sugar/mg AIR) (Table 1; Figs. 3, 4). Mol% data provide information on the relative molar proportions of the different sugar residues, which are indicative of the relative amounts of different non-cellulosic wall polymers in different wall samples. As such, mol% data provide a facile means to compare sugar compositions of different cell wall samples, even when the total amount of polysaccharides in the walls differs as can be the case, for example, in mutant versus wild-type samples [50]. Mass yield data provide information on the actual measurable amounts of the different sugars present in cell walls from different plant samples, and thus, are indicative of the effectiveness of the methods in quantifying both major and minor sugars. Finally, we further compared the efficacy of the four methods in the analysis of AIR from *Populus* wood, rice stem, and switchgrass tiller, which represent biomass from secondary cell-wall-enriched biofuel feedstock tissues (Table 2; Figs. 5, 6, 7; Additional file 6).

Glycosyl residue composition analysis by the alditol acetate (AA) method

The glycosyl residue compositions of AIR from leaves of Arabidopsis, *Populus,* rice, and switchgrass were measured by production of AA derivatives followed by GC–MS. The method allows the detection of neutral sugars, but not acidic sugars. Using the AA derivatization method, Gal (28 mol%), Xyl (22 mol%), and Ara (18 mol%) were identified as the predominant non-cellulosic sugars in Arabidopsis leaf cell walls (Fig. 2; Additional file 5). Arabidopsis AIR also had substantial amounts of Rha (12 mol%) and Glc (12 mol%), and lesser amounts of Man (5 mol%) and Fuc (3 mol%). The major sugars in *Populus* leaf AIR were Ara (36 mol%), Xyl (25 mol%), and Gal (18 mol%). These leaf AIR sugar compositions of Arabidopsis and *Populus* with large amounts of Gal and Ara (predominant sugars in pectin) are consistent with the dicot pectin-rich, Type

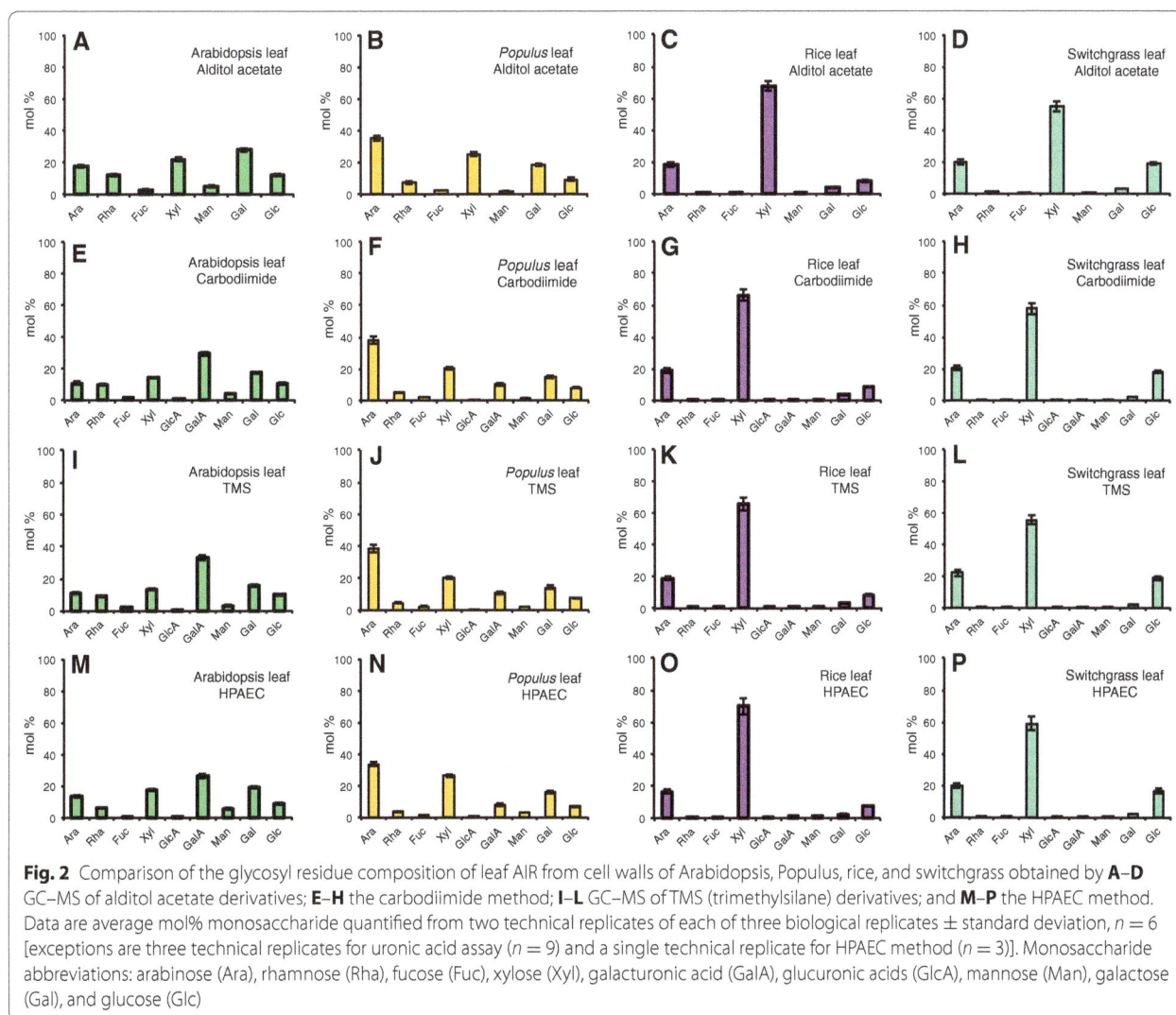

Fig. 2 Comparison of the glycosyl residue composition of leaf AIR from cell walls of Arabidopsis, Populus, rice, and switchgrass obtained by **A–D** GC–MS of alditol acetate derivatives; **E–H** the carbodiimide method; **I–L** GC–MS of TMS (trimethylsilane) derivatives; and **M–P** the HPAEC method. Data are average mol% monosaccharide quantified from two technical replicates of each of three biological replicates ± standard deviation, $n = 6$ [exceptions are three technical replicates for uronic acid assay ($n = 9$) and a single technical replicate for HPAEC method ($n = 3$)]. Monosaccharide abbreviations: arabinose (Ara), rhamnose (Rha), fucose (Fuc), xylose (Xyl), galacturonic acid (GalA), glucuronic acids (GlcA), mannose (Man), galactose (Gal), and glucose (Glc)

I primary walls [6]. The high level of Xyl is likely from xyloglucan and, to a lesser extent, xylan [6, 51]. In contrast, rice, switchgrass, and other Poales and commelinid monocots have Type II walls that contain arabinoxylans and β1,3:β1,4 mixed-linkage glucans as the predominant hemicellulosic polysaccharides, and have significantly less pectin [46, 52, 53]. This was confirmed by the AA analysis data (Fig. 2; Additional file 5), which identified Xyl, Ara, and Glc as the predominant sugars in leaf AIR from rice (68, 18, and 8 mol%, respectively) and switchgrass (55, 20, and 19 mol%, respectively), as well as significantly lower amounts of Rha. Leaf walls of both grasses had similar relative amounts of the different sugars, with Xyl as the main non-cellulosic sugar. The majority of Xyl in grass walls arises from arabinoxylan, with a smaller amount from xyloglucan [13, 54]. Since the contribution of Ara from pectin is very small in grasses, the Ara was derived largely from arabinoxylan [54].

Measurement of uronic acid (UA) content of plant biomass
Since the AA method detects only neutral sugars, it was necessary to use an independent method to quantify the amount of acidic sugars in the AIR samples. We analyzed the total UA content of AIR from leaves of Arabidopsis, *Populus*, rice, and switchgrass using a method that combines sulfamate and biphenyl reagents to yield a pink-colored product representative of the UA content. The simultaneous use of sulfamate and biphenyl reagent reduces the brown color, which can develop from neutral sugars and interfere with detection of UAs [37, 38, 55]. Since the hydrolysis procedure used to release monosaccharides from the polymers can affect total sugar yield, here we compared two different hydrolysis methods for the UA analyses. Figure 3A shows that with sulfuric acid hydrolysis, the UA content was 160 and 124 µg/mg for leaf AIR from Arabidopsis and *Populus*, respectively, consistent with dicot Type I primary cell walls that are relatively

Table 1 Comparison of different glycosyl residue analysis methods for µg monosaccharide quantified per mg of AIR from four different leaf samples

Leaf-method	µg glycosyl residue/mg leaf AIR											
	Ara	Rha	Fuc	Xyl	GlcA	GalA	Man	Gal	Glc	Total neutral	Total acidic	Total
Arabidopsis												
Alditol acetate	21.8 ± 0.8	13.3 ± 0.7	3.5 ± 0.3	23.9 ± 1.3	–	–	5.2 ± 0.3	38.6 ± 2.3	14.0 ± 0.6	120.3		120.3
Uronic acid*					108.6 ± 3.2						108.6	108.6
Alditol acetate + uronic acid**	21.8 ± 0.8[a]	13.3 ± 0.7[a]	3.5 ± 0.3[b]	23.9 ± 1.3[a]	108.6 ± 3.2		5.2 ± 0.3[b]	38.6 ± 2.3[a]	14.0 ± 0.6[a]	120.3[a]	108.6[c]	228.9[a]
Carbodiimide	27.4 ± 1.8[b]	23.5 ± 1.5[c]	2.7 ± 0.02[a]	32.7 ± 1.4[b]	1.7 ± 0.01	64.8 ± 4.1[b]	4.8 ± 0.3[a]	46.6 ± 2.1[c]	26.1 ± 1.3[b]	163.8[b]	66.5[a]	230.2[b]
TMS	29.2 ± 2.2[c]	22.1 ± 1.9[b]	3.5 ± 0.03[b]	32.2 ± 1.1[b]	1.9 ± 0.02	72.5 ± 3.3[b]	5.2 ± 0.5[a]	42.3 ± 2.6[b]	24.3 ± 2.2[b]	158.8[b]	74.4[b]	233.1[b]
HPAEC	27.4 ± 1.4[b]	20.0 ± 1.0[b]	3.3 ± 0.02[b]	35.5 ± 1.5[c]	1.8 ± 0.01	62.4 ± 3.9[a]	8.9 ± 0.7[c]	41.0 ± 2.8[b]	23.4 ± 1.6[b]	159.5[b]	64.2[a]	223.6[a]
Populus												
Alditol acetate	55.2 ± 3.8	13.6 ± 0.7	5.3 ± 0.2	41.5 ± 2.6	–	–	3.9 ± 0.4	35.7 ± 2.5	20.5 ± 0.8	175.7		175.7
Uronic acid*					91.4 ± 4.1						91.4	91.4
Alditol acetate + uronic acid**	55.2 ± 3.8[a]	13.6 ± 0.7	5.36 ± 0.2	41.5 ± 2.6[a]	91.4 ± 4.1		3.9 ± 0.4[a]	35.7 ± 2.5[a]	20.5 ± 0.8	175.7[a]	91.4[c]	267.1[a]
Carbodiimide	98.1 ± 5.1[b]	11.2 ± 0.7	5.1 ± 0.2	51.5 ± 2.9[b]	0.9 ± 0.01	29.5 ± 3.1[b]	3.6 ± 0.3[b]	44.6 ± 2.9[b]	21.2 ± 1.3	235.3[b]	30.4[a]	265.7[a]
TMS	99.2 ± 4.9[b]	12.4 ± 0.9	5.9 ± 0.4	54.1 ± 3.2[b]	1.0 ± 0.01	32.8 ± 2.9[b]	5.1 ± 0.4[b]	42.3 ± 2.1[c]	20.2 ± 1.8	239.2[b]	33.8[b]	272.8[b]
HPAEC	95.7 ± 6.3[b]	10.6 ± 0.8	4.2 ± 0.3	60.3 ± 4.1[c]	0.7 ± 0.01	25.6 ± 2.6[a]	8.6 ± 0.6[c]	34.3 ± 1.8[a]	19.7 ± 1.4	233.4[b]	26.3[a]	259.7[a]
Rice												
Alditol acetate	64.1 ± 3.9	2.8 ± 0.1	0.7 ± 0.02	220.3 ± 8.9	–	–	1.3 ± 0.01	16.5 ± 1.1	31.1 ± 2.4	336.8		336.8
Uronic acid*					12.1 ± 0.9						12.1	12.1
Alditol acetate + uronic acid**	64.1 ± 3.9[a]	2.8 ± 0.1	0.7 ± 0.02	220.3 ± 8.9[b]	12.1 ± 0.9		1.3 ± 0.01[a]	16.5 ± 1.1[b]	31.1 ± 2.4[b]	336.8[a]	12.1[c]	348.9[c]
Carbodiimide	70.3 ± 3.6[b]	3.1 ± 0.3	0.5 ± 0.01	212.3 ± 9.3[a]	1.1 ± 0.20	4.3 ± 0.4[b]	1.2 ± 0.02[a]	11.6 ± 0.8[a]	34.3 ± 2.8[b]	333.3[a]	5.4[a]	338.5[a]
TMS	73.4 ± 4.2[b]	2.8 ± 0.3	0.4 ± 0.01	214.5 ± 11.3[a]	1.2 ± 0.02	4.9 ± 0.4[b]	2.0 ± 0.01[a]	11.4 ± 0.9[b]	33.5 ± 1.9[b]	338.0[b]	6.1[b]	343.9[b]
HPAEC	75.3 ± 5.8[c]	2.9 ± 0.2	0.8 ± 0.02	228.9 ± 14.3[c]	0.6 ± 0.01	3.6 ± 0.3[a]	3.1 ± 0.04[a]	10.8 ± 0.6[a]	25.6 ± 2.1[a]	347.2[b]	4.4[a]	351.6[c]
Switchgrass												
Alditol acetate	71.3 ± 2.4	2.5 ± 0.3	1.0 ± 0.02	281.0 ± 13.3	–	–	1.4 ± 0.02	15.8 ± 1.6	50.2 ± 4.1	423.2		423.2
Uronic acid*					4.5 ± 0.4						4.5	4.5
Alditol acetate + uronic acid**	71.3 ± 2.4[a]	2.5 ± 0.3	1.0 ± 0.02[b]	281.0 ± 13.3[a]	4.5 ± 0.4		1.4 ± 0.02[a]	15.8 ± 1.6[c]	50.2 ± 4.1[c]	423.2[a]	4.5	427.7[a]
Carbodiimide	75.8 ± 3.3[a]	2.5 ± 0.3	0.6 ± 0.01[b]	290.8 ± 12.7[b]	0.8 ± 0.01	3.6 ± 0.2[a]	2.0 ± 0.02[b]	12.6 ± 0.7[b]	43.7 ± 3.9[b]	428.0[b]	4.4	432.3[a]
TMS	80.3 ± 5.1[b]	2.7 ± 0.4	0.8 ± 0.01[b]	295.3 ± 13.8[b]	0.9 ± 0.01	4.1 ± 0.4[b]	1.0 ± 0.01[a]	11.9 ± 1.0[b]	45.3 ± 3.5[b]	437.3[b]	5.0	442.3[b]
HPAEC	84.0 ± 3.7[b]	2.0 ± 0.2	0.6 ± 0.01[a]	299.3 ± 15.8[b]	0.9 ± 0.01	3.0 ± 0.3[a]	2.3 ± 0.02[b]	9.0 ± 2.8[a]	36.0 ± 2.8[b]	433.2[b]	3.9	437.1[b]

Data are means of three biological replicates ± standard deviation, with two technical replicates for alditol acetate, carbodiimide, and TMS methods ($n = 6$), three technical replicates for uronic acid assay ($n = 9$), and a single technical replicate for HPAEC method ($n = 3$). Different letters in superscript following the values indicate significant differences between the amounts ("a" represents the lowest amount) of a particular (or total) sugar residue in the cell walls of a species across different methods (and not between different types of sugars within that leaf sample analyzed using a particular method). Statistics are one-way analysis of variance (ANOVA) followed by Tukey's multiple comparison tests with significant P value <0.05

* Uronic acid assay data shown here are from analysis using TFA hydrolysis

** Alditol acetate + uronic acid data shown here are the combined data of the respective individual assays

Comparison of four glycosyl residue composition methods for effectiveness in detecting...

9

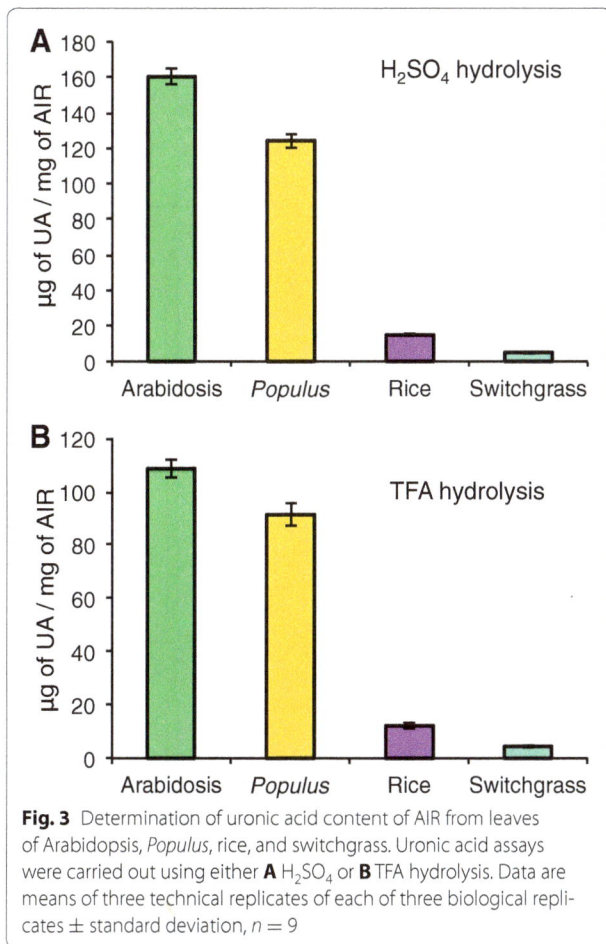

Fig. 3 Determination of uronic acid content of AIR from leaves of Arabidopsis, *Populus*, rice, and switchgrass. Uronic acid assays were carried out using either **A** H_2SO_4 or **B** TFA hydrolysis. Data are means of three technical replicates of each of three biological replicates ± standard deviation, $n = 9$

rich in GalA-containing pectin and in agreement with previously published UA content of AIR from Arabidopsis leaves [56]. As expected for low-pectin-content grass cell walls, the UA content of rice and switchgrass leaf AIR hydrolyzed with sulfuric acid was significantly lower than in the dicots, being 15 and 5 μg/mg AIR, respectively. Compared to sulfuric acid hydrolysis, the TFA hydrolysis yielded lower UA content for all four species, being 109, 91, 12, and 4.5 μg/mg for Arabidopsis, *Populus*, rice, and switchgrass leaf AIR, respectively (Fig. 3B).

Glycosyl residue composition analysis by the carbodiimide method

Both neutral and acidic sugars can be detected by the carbodiimide method. The carboxylic acid moieties of UAs are activated by a water-soluble carbodiimide to form products that can be reduced to primary alcohols [41] (Fig. 1B). Using this method, GalA (30 mol%), Gal (18 mol%), and Xyl (15 mol%) were the major non-cellulosic monosaccharides detected in Arabidopsis leaf (Fig. 2; Additional file 5), with moderate amounts

of Ara (11 mol%), Glc (11 mol%), and Rha (10 mol%). In *Populus* leaf AIR, Ara (38 mol%), Xyl (21 mol%), and Gal (15 mol%) were the major monosaccharides (Fig. 2; Additional file 5), followed by GalA (10 mol%) and Glc (8 mol%). Similar to the trends observed in the AA data above, the carbodiimide method also detected Xyl, Ara, and Glc as the predominant sugars in leaf AIR of rice (66, 19, and 9 mol%, respectively) and switchgrass (58, 20, and 18 mol%, respectively) (Fig. 2; Additional file 5).

Glycosyl residue composition analysis by the trimethylsilyl (TMS) method

In the TMS method, leaf AIR is hydrolyzed in the presence of methanol to generate methyl glycosides, which are subsequently derivatized with TMS and the resulting TMS ethers separated and identified by GC–MS (Fig. 1C). The TMS analysis identified the most abundant sugar in Arabidopsis leaf AIR as GalA (33 mol%), other major sugars being Gal (16 mol%), Xyl (14 mol%), and lesser amounts of Ara (11 mol%), Glc (10 mol %), Rha (9 mol%), Man (3 mol%), Fuc (2 mol%), and GlcA (1 mol%) (Fig. 2; Additional file 5). The most abundant monosaccharide in *Populus* leaf AIR was Ara (39 mol%) with other major sugars being Xyl (20 mol%), Gal (14 mol%), GalA (11 mol%), and Glc (7 mol%) (Fig. 2; Additional file 5). TMS analysis of rice leaf AIR identified more than 60% of total sugar content as Xyl (66 mol%) and Ara (19 mol%), consistent with the high arabinoxylan content of grass Type II cell walls (Fig. 2; Additional file 5). A considerable amount of Glc (9 mol%) was also present in rice leaf AIR along with measurable amounts of the UAs GalA and GlcA. In switchgrass, Xyl was the most abundant (56 mol%) sugar followed by Ara (22 mol%) and Glc (19 mol%) (Fig. 2; Additional file 5). A smaller amount of Gal (2 mol%) and trace amounts of GalA and GlcA were also detected in switchgrass leaf AIR.

Glycosyl residue composition analysis by the HPAEC method

Leaf AIR from Arabidopsis, *Populus*, rice, and switchgrass was hydrolyzed with TFA and the resulting monosaccharides were separated and quantified by HPAEC. HPAEC composition analysis of Arabidopsis leaf AIR detected GalA (26 mol%), Gal (19 mol%), and Xyl (18 mol%) as the predominant non-cellulosic cell wall sugars (Fig. 2; Additional file 5). Trace amounts of GlcA and Fuc were also detected. HPAEC analysis of *Populus* leaf AIR indicated a large Ara content (34 mol%) with other major sugars being Xyl (26 mol%) and Gal (16 mol%). Measurable amounts of GalA (8 mol%) and Rha (4 mol%) were also present. The HPAEC data for rice and switchgrass leaf AIR (Fig. 2; Additional files 5) revealed Xyl, Ara, and Glc as the predominant sugars (70, 17, and 8 mol%, respectively, in rice; 60, 20, and 17 mol%, respectively, in switchgrass).

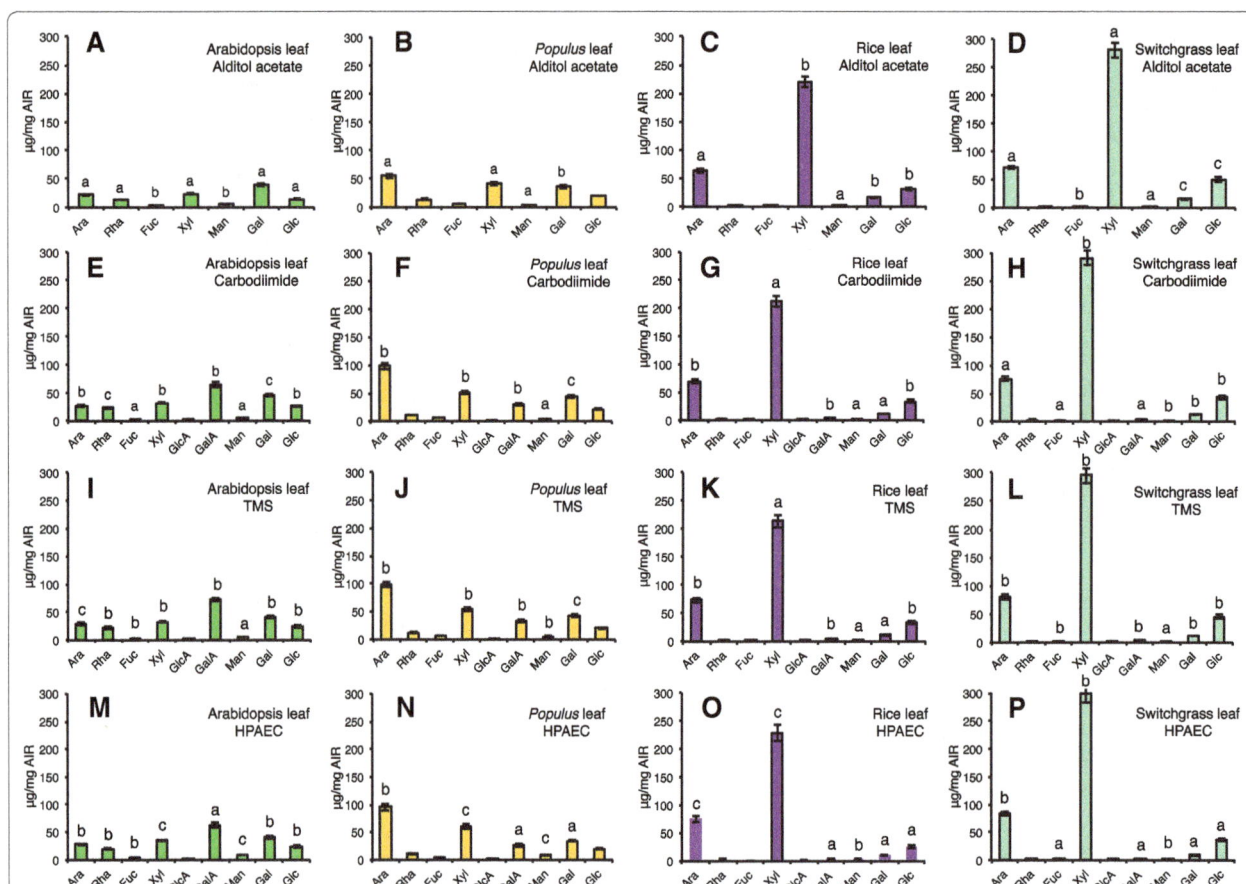

Fig. 4 Comparison of the glycosyl residue composition of leaf AIR from cell walls of Arabidopsis, *Populus*, rice, and switchgrass obtained by **A–D** GC–MS of alditol acetate derivatives; **E–H** the carbodiimide method; **I–L** GC–MS of TMS (trimethylsilane) derivatives; and **M–P** the HPAEC method. Data are average μg monosaccharide quantified per mg of leaf AIR from two technical replicates of each of three biological replicates ± standard deviation, $n = 6$ [exceptions are three technical replicates for uronic acid assay ($n - 9$) and a single technical replicate for HPAEC method ($n = 3$)]. Monosaccharide abbreviations: arabinose (Ara), rhamnose (Rha), fucose (Fuc), xylose (Xyl), galacturonic acid (GalA), glucuronic acids (GlcA), mannose (Man), galactose (Gal), and glucose (Glc). *Different letters* indicate significant differences between the amounts ("a" represents the lowest amount) of a particular sugar residue in the cell walls of a species across different methods (and not between different types of sugars within that biomass sample analyzed using a particular method). Statistics are one-way analysis of variance (ANOVA) followed by Tukey's multiple comparison tests with significant P value <0.05

Comparison of the glycosyl residue composition analysis methods for analysis of leaf biomass

A comparison of the glycosyl residue compositions obtained from leaf AIR from the four different plant sources using the four different analysis methods can be made based on both the relative mol% yield (Fig. 2; Additional file 5) and the μg sugar/mg AIR mass yield (Table 1; Figs. 3, 4) of sugars. An overview of the data showed that the carbodiimide, TMS, and HPAEC methods were able to detect the most common nine neutral and acidic sugars, while the AA method detected the most common seven neutral sugars.

For a more in depth analysis, we first compared the four methods for their ability to detect neutral sugars. All four methods gave the same relative abundance order for the neutral sugars present in leaf AIR from each of the

four plant species, based on both the relative (mol%) and the mass (μg/mg AIR) sugar yields (Figs. 2, 4; Table 1; Additional file 5). A minor exception was the HPAEC method which gave reversed orders, compared to the other three methods, for the three least abundant neutral sugars (Rha, Man, and Fuc) in the majority of the samples (Table 1). It is noteworthy, however, that the alditol acetate method often gave the lowest neutral sugar measurement, especially in the dicot samples, as apparent from the total neutral sugar mass yields and from some of the individual sugar (particularly Ara, Xyl, Glc, and Rha) mass yields (Fig. 4; Table 1).

The greatest mass yield of total acidic sugars from leaf AIR samples was obtained using the UA method for Arabidopsis, *Populus,* and rice, and with the TMS method for switchgrass (Table 1). However, the TMS

Table 2 Comparison of different glycosyl residue analysis methods for µg monosaccharide quantified per mg of AIR from three different biofuel feedstock biomass samples from three different species

Biomass-method	µg glycosyl residue/mg biomass AIR											
	Ara	Rha	Fuc	Xyl	GlcA	GalA	Man	Gal	Glc	Total neutral	Total acidic	Total
Populus wood												
Alditol acetate	6.7 ± 0.6	3.0 ± 0.3	0.4 ± 0.01	175.3 ± 12.3	-	-	12.1 ± 0.9	9.4 ± 0.8	20.6 ± 1.5	227.5		227.4
Uronic acid*	-	-	-	-	37.5 ± 1.3		-	-	-		37.5	
Alditol acetate + uronic acid**	6.7 ± 0.6[a]	3.0 ± 0.3[a]	0.4 ± 0.01	175.3 ± 12.3[a]	37.5 ± 1.3	-	12.1 ± 0.9[a]	9.4 ± 0.8[a]	20.6 ± 1.5	227.5[a]	37.5[b]	264.9[a]
Carbodiimide	9.0 ± 0.5[b]	4.0 ± 0.5[a]	0.5 ± 0.01	190.1 ± 10.2[c]	1.8 ± 0.1	16.4 ± 0.9	13.0 ± 1.0[a]	12.0 ± 0.8[b]	22.5 ± 1.8	251.1[b]	18.2[a]	269.3[a]
TMS	9.3 ± 0.6[b]	4.2 ± 0.4[a]	0.4 ± 0.01	197.6 ± 7.9[c]	2.0 ± 0.2	18.1 ± 1.1	13.2 ± 0.9[a]	10.7 ± 0.6[a]	20.9 ± 1.3	256.3[b]	20.1[a]	276.4[b]
HPAEC	11.8 ± 0.8[c]	5.0 ± 0.6[b]	0.5 ± 0.01	185.9 ± 8.5[b]	2.9 ± 0.2	17.9 ± 1.3	14.6 ± 0.8[b]	12.3 ± 0.9[b]	21.3 ± 1.9	251.4[b]	20.8[a]	272.2[b]
Rice stem												
Alditol acetate	38.9 ± 3.1	3.0 ± 0.3	1.0 ± 0.02	223.6 ± 9.9	-	-	2.0 ± 0.2	15.1 ± 1.3	29.3 ± 3.1	312.9		312.9
Uronic acid*	-	-	-	-	11.9 ± 0.8		-	-	-		11.9	
Alditol acetate + uronic acid**	38.9 ± 3.1[b]	3.0 ± 0.3[b]	1.0 ± 0.02[b]	223.6 ± 9.9[a]	11.9 ± 0.8	-	2.0 ± 0.2	15.1 ± 1.3	29.3 ± 3.1	312.9[a]	11.9[b]	324.8[a]
Carbodiimide	41.3 ± 3.8[b]	2.3 ± 0.3[b]	0.4 ± 0.01[a]	225.6 ± 16.2[b]	0.9 ± 0.01	8.0 ± 0.6[b]	2.0 ± 0.2	16.8 ± 0.9	31.3 ± 2.9	319.7[b]	8.9[a]	328.6[a]
TMS	39.5 ± 2.7[b]	2.0 ± 0.2[a]	0.8 ± 0.01[b]	230.3 ± 10.3[b]	1.0 ± 0.01	8.9 ± 0.5[b]	1.5 ± 0.1	15.8 ± 1.4	30.8 ± 2.6	320.7[b]	9.9[a]	330.6[a]
HPAEC	35.9 ± 3.4[a]	2.5 ± 0.2[b]	0.8 ± 0.01[b]	242.7 ± 14.7[c]	1.9 ± 0.02	6.5 ± 0.5[a]	2.9 ± 0.3	14.6 ± 0.7	32.8 ± 2.2	332.2[c]	8.4[a]	340.5[b]
Switchgrass tiller												
Alditol acetate	52.0 ± 2.1	3.3 ± 0.4	0.7 ± 0.01	265.2 ± 9.6	-	-	1.8 ± 0.02	16.5 ± 2.0	22.5 ± 2.1	362.0		361.9
Uronic acid*	-	-	-	-	4.7 ± 0.4		-	-	-		4.7	
Alditol acetate + uronic acid**	52.0 ± 2.1	3.3 ± 0.4[c]	0.7 ± 0.01	265.2 ± 9.6[a]	4.7 ± 0.4	-	1.8 ± 0.02[b]	16.5 ± 2.0	22.5 ± 2.1	362.0[a]	4.7	366.6[a]
Carbodiimide	50.3 ± 2.6	2.3 ± 0.2[b]	0.4 ± 0.01	270.4 ± 12.7[b]	0.7 ± 0.01	3.0 ± 0.4[a]	1.4 ± 0.02[a]	14.3 ± 0.8[b]	22.4 ± 1.9	361.5[a]	3.7	365.1[a]
TMS	52.7 ± 3.2	2.6 ± 0.2[b]	0.3 ± 0.01	278.8 ± 17.3[c]	0.8 ± 0.01	3.4 ± 0.4[b]	1.0 ± 0.01[a]	13.1 ± 1.1[a]	22.0 ± 2.3	370.5[b]	4.2	374.7[b]
HPAEC	53.6 ± 3.7	1.5 ± 0.1[a]	0.5 ± 0.01	270.6 ± 14.3[b]	0.9 ± 0.01	2.9 ± 0.2[a]	2.1 ± 0.01[c]	15.9 ± 0.9[b]	20.9 ± 2.6	365.1[a]	3.8	368.9[a]

Data are means of three biological replicates ± standard deviation, with two technical replicates for alditol acetate, carbodiimide, and TMS methods (n = 6), three technical replicates for uronic acid assay (n = 9), and a single technical replicate for HPAEC method (n = 3). Different letters in superscript following the values indicate significant differences between the amounts ("a" represents the lowest amount) of a particular (or total) sugar residue in the cell walls of a species across different methods (and not between different types of sugars within that biomass sample analyzed using a particular method). Statistics are one-way analysis of variance (ANOVA) followed by Tukey's multiple comparison tests with significant *P* value <0.05

* Uronic acid assay data shown here are from analysis using TFA hydrolysis

** Alditol acetate + uronic acid data shown here are the combined data of the respective individual assays

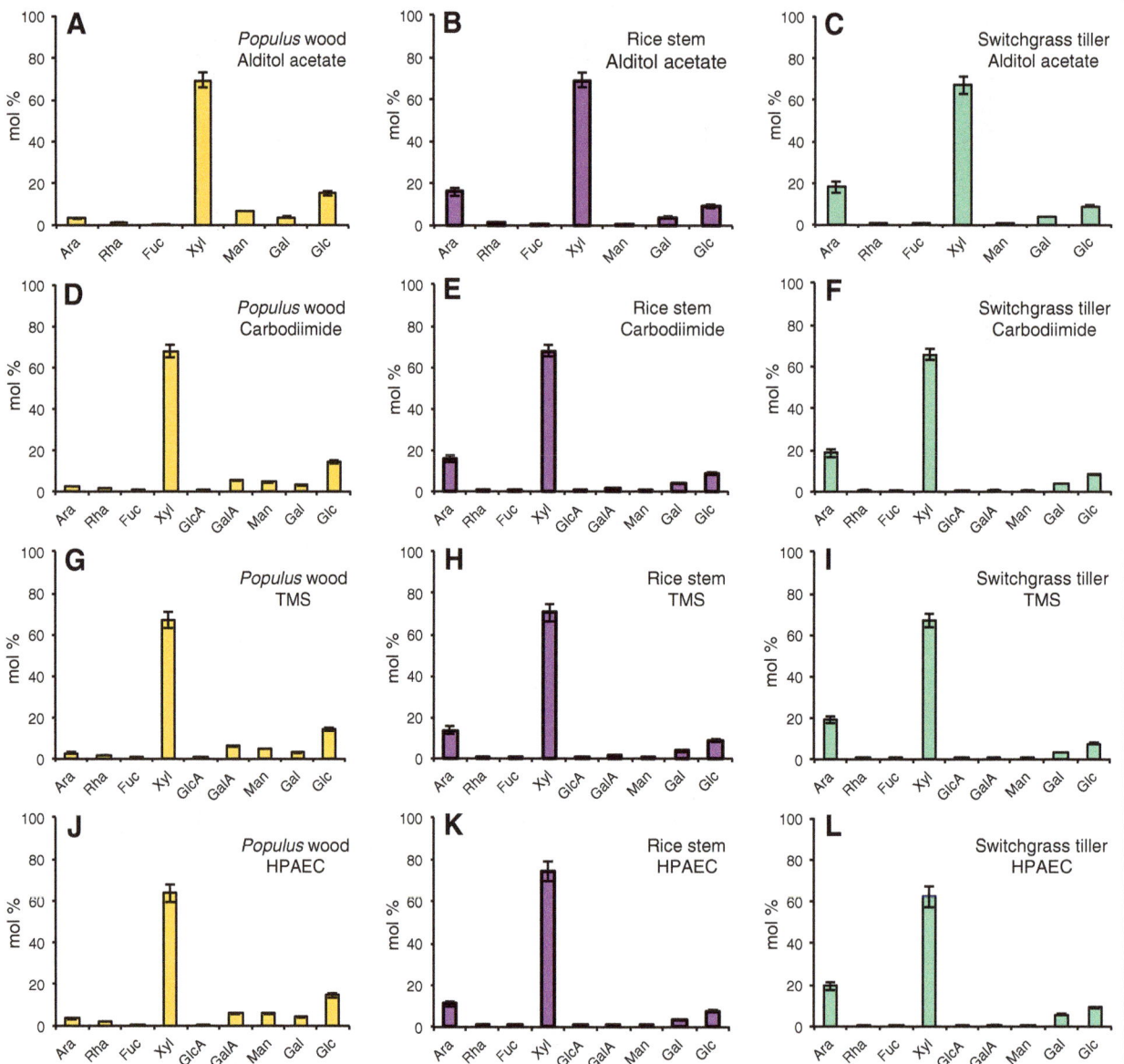

Fig. 5 Comparison of the glycosyl residue composition of AIR from cell walls of Populus wood, rice stem, and switchgrass tiller biomass obtained by **A–C** GC–MS of alditol acetate derivatives; **D–F** the carbodiimide method; **G–I** GC–MS of TMS (trimethylsilane) derivatives; and **J–L** the HPAEC method. Data are average mol% monosaccharide quantified from two technical replicates of each of three biological replicates ± standard deviation, $n = 6$ [exceptions are three technical replicates for uronic acid assay ($n = 9$) and a single technical replicate for HPAEC method ($n = 3$)]. Monosaccharide abbreviations: arabinose (Ara), rhamnose (Rha), fucose (Fuc), xylose (Xyl), galacturonic acid (GalA), glucuronic acids (GlcA), mannose (Man), galactose (Gal), and glucose (Glc)

method provided the greatest μg/mg yield of specific acidic sugars (i.e., GlcA and GalA) for all leaf samples (Table 1). The greatest total sugar yield (neutral + acidic sugars) from leaf AIR was obtained using the TMS method for Arabidopsis, *Populus* and switchgrass, and using the HPAEC method for rice (Table 1). Interestingly, the total sugar yield measured from the same amount of starting AIR was much greater from both monocot grasses (~1.3–3.5 times greater) than from the dicots, regardless of the analysis method used (Table 1).

Comparison of the glycosyl residue composition analysis methods for analysis of wood and stem biomass

To evaluate the efficacy of the four methods for analysis of the sugar composition of biomass biofuel feedstock rich in secondary walls (e.g., stems), we analyzed AIR

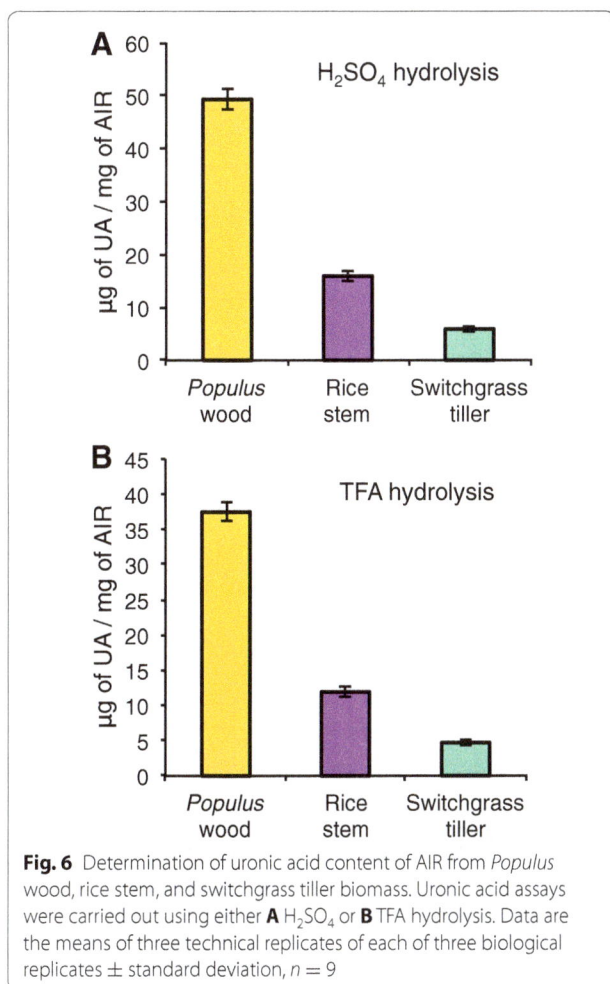

Fig. 6 Determination of uronic acid content of AIR from *Populus* wood, rice stem, and switchgrass tiller biomass. Uronic acid assays were carried out using either **A** H_2SO_4 or **B** TFA hydrolysis. Data are the means of three technical replicates of each of three biological replicates ± standard deviation, $n = 9$

from *Populus* wood, rice stems, and switchgrass tillers (Table 2; Figs. 5, 6, 7; Additional file 6). The results showed several trends similar to those obtained with the leaf samples. For example, (1) all four methods gave the same relative abundance order of the different neutral sugars based on both mol% and mass yield (µg/mg AIR), again with the exception of reversed orders of Man and Rha abundance using the HPAEC method. (2) As observed in the analysis of leaf AIR, the AA method provided the lowest total mass yield of neutral sugars in the dicot tissue sample, *Populus* wood, compared to the other three methods (Table 2). This trend, however, was again not so obvious for the grass biomass. (3) The greatest amount of total acidic sugars from all feedstock AIR samples was obtained using the UA assay. (4) The total sugar yield from the grass tissues (319–340 and 362–375 µg/mg AIR from rice stem and switchgrass tiller, respectively) was greater than from than from the dicot *Populus* wood (227–276 µg/mg AIR), again a trend similar to that obtained with the leaf samples.

As expected for dicot secondary wall-enriched samples, the Xyl content of *Populus* wood (Fig. 7) was substantially greater than from *Populus* leaves (Fig. 4), due to the abundance of xylan in secondary walls. In contrast, the most measurable change observed in the grass stem and tiller samples (Fig. 7) was a marked decrease in the Ara content compared to the leaf samples (Fig. 4).

Discussion

Plant cell walls comprise the bulk of plant biomass. The demand for biofuels and bioproducts has spurred research to identify biomass sources with desirable properties and to improve the quality and/or quantity of such biomass. Such studies require comparison of the sugar composition of biomass from different species and from different tissues. Although most biomass feedstocks consist predominantly of cellulose, xylan, and lignin, increasing evidence shows that even seemingly minor components of the biomass (e.g., pectin) can significantly impact wall structure, plant growth, yield, and biomass recalcitrance [15, 21, 22, 57]. Thus, sensitive, accurate, and preferably high-throughput analytical method(s) are needed to identify and quantify the different major and minor neutral and acidic sugars that constitute the non-cellulosic polysaccharides of plant cell walls.

Here we assessed four different methods, i.e., AA–UA assay, carbodiimide, TMS, and HPAEC, for their ability to quantitatively measure the sugar composition of non-cellulosic polysaccharides in cell walls from four different species representing dicots and grasses. Since the hydrolysis conditions used in these methods do not appreciably hydrolyze cellulose, the results are indicative of non-cellulosic sugar content. The four methods were compared for their ability to detect and quantify the nine most common monosaccharides present in plant cell walls, the yield of sugar detected by each method, and the ease and practicality of use of each method (summarized in Additional file 7).

All four methods were able to detect and quantify the seven major neutral sugars in both leaf and stem biomass samples from the different species, even at relatively low amounts. All four methods also gave the same mol% and µg sugar/mg AIR abundance ranking of the neutral sugars, with a minor exception of the HPAEC method for which the abundance ranking of the less abundant sugars Rha, Man, and Fuc was often reversed compared to the other methods. However, only three of the methods, the carbodiimide, TMS and HPAEC, were able provide a comparable quantitative and qualitative evaluation of the nine major neutral and acidic sugars present in all plant biomass.

The AA method is the most commonly used method for sugar composition analysis of plant biomass, likely

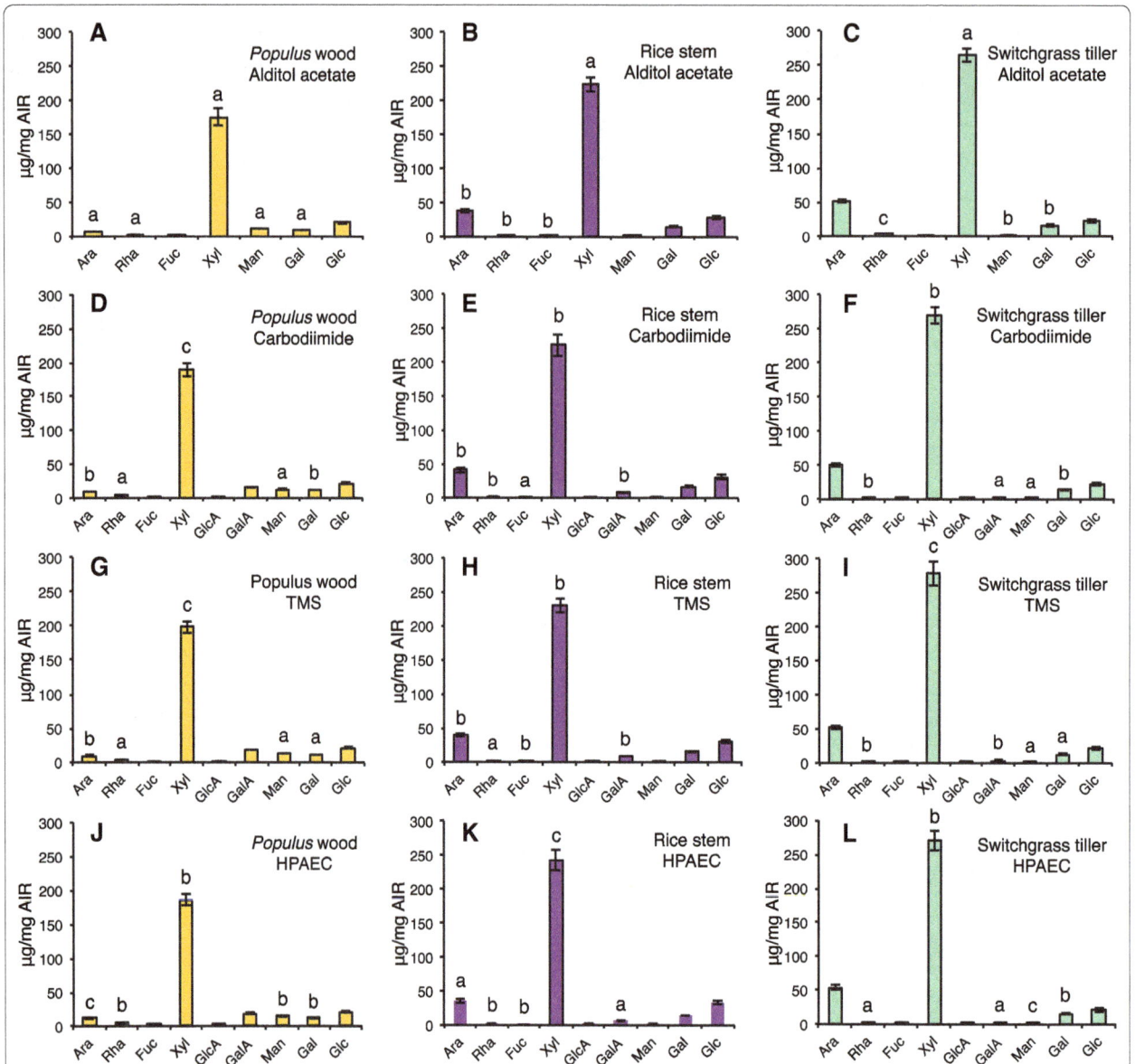

Fig. 7 Comparison of the glycosyl residue composition of biomass AIR from cell walls of Populus wood, rice stem, and switchgrass tiller obtained by **A–C** GC–MS of alditol acetate derivatives; **D–F** the carbodiimide method; **G–I** GC–MS of TMS (trimethylsilane) derivatives; and **J–L** the HPAEC method. Data are average µg monosaccharide quantified per mg of leaf AIR from two technical replicates of each of three biological replicates ± standard deviation, $n = 6$ [exceptions are three technical replicates for uronic acid assay ($n = 9$) and a single technical replicate for HPAEC method ($n = 3$)]. Monosaccharide abbreviations: arabinose (Ara), rhamnose (Rha), fucose (Fuc), xylose (Xyl), galacturonic acid (GalA), glucuronic acids (GlcA), mannose (Man), galactose (Gal), and glucose (Glc). *Different letters* indicate significant differences between the amounts ("a" represents the lowest amount) of a particular sugar residue in the cell walls of a species across different methods (and not between different types of sugars within that biomass sample analyzed using a particular method). Statistics are one-way analysis of variance (ANOVA) followed by Tukey's multiple comparison tests with significant P value <0.05

due to the relatively simple GC chromatograms produced which have single peaks for each sugar, making quantification easier [12]. However, the inability of the AA method to provide quantitative data for the specific acidic sugars (i.e., GalA and GlcA) is a major limitation when measuring the composition of plant cell wall

biomass, since these sugars are critical components in the pectic and hemicellulosic polymers. The UA assay is often carried out in conjunction with the AA method to complement the results of the AA method and provide a measure of the total acidic sugar content of the tissue. However, the UA assay does not provide information

Comparison of four glycosyl residue composition methods for effectiveness in detecting...

15

about the amounts of the individual acidic sugars, GlcA and GalA. In this study, we compared the use of TFA, a solvent used in the AA method, versus the more typical sulfuric acid, to hydrolyze AIR samples for the UA analysis. The yield of UA was lower using TFA hydrolysis compared to hydrolysis by sulfuric acid [11, 58, 59]. However, even with TFA hydrolysis, the UA assay still generally provided the greatest total UA values compared to the other three methods, particularly from samples with high pectin content such as in the dicot samples (Tables 1, 2). For example, inspection of µg sugar/mg AIR data for Arabidopsis leaf (Table 1) indicates that acidic sugars (GlcA + GalA) account for 29, 32, and 29% of the biomass based on analyses using the carbodiimide, TMS and HPAEC methods, respectively, but rather 47% of the biomass based on AA and UA assays. Thus, the comparative results presented here show that the amount of total UAs measured using the AA–UA methods may not be comparable to the amount of GalA + GlcA detected using the carbodiimide, TMS and HPAEC methods. Furthermore, although the acidic sugar yield was high using the UA assay, it does not differentiate between GalA and GlcA which is necessary to study specific wall components such as pectin and glucuronoxylan, respectively. Thus, the AA–UA assay method does not provide complete sugar composition information for plant biomass, and it yields a different relative amount of acidic versus neutral sugars compared to the other three methods.

The carbodiimide method takes advantage of the simplicity of AA chromatographic profiles by reducing the UAs to their neutral sugars prior to the AA procedure, thus enabling detection of GalA and GlcA in addition to the neutral sugars. Its drawbacks, however, include the time-consuming and laborious steps required to modify the UAs, which added up to three additional days of experimental time on top of that needed for the AA part of the procedure, and the greater amount of starting AIR needed (e.g., 10 mg versus 100–400 µg) (see "Methods"; Additional file 7). Moreover, to quantify the UAs, the amounts of GalA and GlcA are indirectly determined by comparing the Gal and Glc peaks obtained from the AA and the carbodiimide methods [42], requiring a sample to be measured in parallel by both methods. Thus, twice the number of samples need to be processed (compared to the AA method) when using the carbodiimide method.

The TMS method requires the simplest sample preparation compared to the other GC–MS-based methods. In our hands, this method yielded the highest overall amounts of sugar for Arabidopsis, *Populus* and switchgrass and comparable amounts for rice, compared to the other methods. It also detected the greatest amount of GalA and GlcA in AIR samples from most samples compared to the carbodiimide and HPAEC methods. The major difficulty with the TMS method is the interpretation of the GC profiles. TMS derivatization of methyl glycosides results in derivatives of both the α- and β-anomeric configurations as well as the pyranose and furanose ring forms of each sugar, yielding multiple peaks for each sugar in the chromatogram. This can be managed, however, by comparison of the sample chromatograms with chromatograms of respective sugar standards and confirmation of peak identity from the MS spectra. Beyond routine plant cell wall sugar composition analysis, the TMS method also allows detection of amino sugars [60], other unusual sugars (e.g., 2-*O*-methylxylose, 2-*O*-methylfucose, acetic acid, Kdo, Dha) [20], and fatty acids [61], making it a versatile analytical method.

In the HPAEC method, hydrolyzed sugars are analyzed directly by liquid chromatography with electrochemical detection, without the need for a time-consuming and sometimes incomplete derivatization step. The HPAEC method clearly required the least amount of time for sample preparation compared to the other methods. For this study, we chose to perform the chromatography using two different columns/gradients to enable accurate detection and quantification of all nine monosaccharides, with the downside that a longer analysis time was required per sample. Other HPAEC gradient schemes that allowed separation of the nine sugars using one column in a single run have been reported [31, 45], which may reduce the analysis time considerably. However, in our hands these methods did not provide sufficient base line separation for Xyl and Man, especially for cell wall samples that are rich in xylan and/or xyloglucan. Another drawback of the HPAEC method is that it is not readily compatible with MS to allow confirmation of peak identity, a critical limitation since HPAEC retention time alone is (sometimes) not sufficient to conclusively identify a compound. The sugar peaks could indeed be collected, but would require further treatment, e.g., to remove salts and reduce sample volumes before being subjected for MS verification. Such steps would add labor and time factors to the analysis.

With the above differences noted, all four methods tested enabled general conclusions regarding the cell wall content of the biomass to be made. For example, all four methods indicated significant differences in the glycosyl residue composition of leaf AIR from the grasses switchgrass and rice (Type II cell walls) compared to the dicots Arabidopsis and *Populus* (Type I cell walls). The latter were relatively richer in UAs, particularly GalA, which is consistent with the higher pectin content and the former were richer in Xyl and Glc, which is consistent with the greater xylan content and presence of mixed-linkage glucans in grass primary walls. The analysis of *Populus* wood, rice stem, and switchgrass tiller using the four methods provided sugar composition results consistent with tissues enriched in secondary walls (Table 2; Figs. 5, 6; Additional file 6). For example, these methods identified a greater amount of Xyl and a reduced amount of GalA, Rha, Ara, and Gal in *Populus* wood (Table 2) compared to *Populus* leaf (Table 1), consistent with the higher glucuronoxylan and lower pectin content in *Populus* secondary walls compared to primary walls. The results also yielded some unexpected findings. Overall, all four methods detected a greater amount of total sugar from the same amount of starting AIR from grasses compared to dicots, regardless of whether the tissues were enriched in primary or secondary walls (compare total sugar values in Tables 1, 2). This result suggests that the non-cellulosic polysaccharides may be present in greater amounts in grasses than in dicots [46]. Alternatively, it is possible that the non-cellulosic polysaccharides are held less tightly in the walls of grasses than in dicots. For example, intrinsic differences in the cell wall structure and/or architecture of these two different phylogenetic groups of plants, such as distinct cross-linking between wall components and different overall wall structural features, could account for the observation. This phenomenon warrants further study.

Although all four methods provided generally comparable sugar compositions, the results indicate that use of the AA + UA method alone to analyze plant biomass has limitations compared to the other methods. The results also make clear that the choice of glycosyl residue composition analysis method is critical to obtain a complete set of neutral + acidic sugar composition data for the analysis of cell wall polymers in biomass and for detailed mechanistic interpretation of the results. According to the American Society for Testing and Materials [9], the AA and TMS methods are the most accurate for analysis of sugars in plant biomass. Based on our comparison, we found that the TMS method gave a slightly greater yield of the majority of sugars, including acidic sugars, in plant biomass (Tables 1, 2), although the carbodiimide and HPAEC methods also provided highly comparable results. In summary, this study provides a basis for selecting a sugar analysis method that is commensurate with the experimental goals. We recommend that the TMS, HPAEC, or carbodiimide methods be used when the goals include detailed mechanistic interpretations regarding plant cell wall (biomass) structure.

Additional files

Additional file 1. Gas chromatographic (GC) profile of the derivatized sugar standards in the alditol acetate (AA) method. The standard mixture consists of 0.5 µg of each sugar (in bold), supplemented with myo-inositol (0.2 µg, in bold) as an internal standard. Note that ribose (in brackets) is also included in the chromatogram shown. Derivatized sugars are separated on a SP-2330 Supelco column (30 m × 0.25 mm, 0.25 µm film thickness) connected to a Hewlett–Packard chromatograph (5890) using helium as the carrier gas with an oven temperature program as described in the "Methods".

Additional file 2. An example of the standard curve used in the uronic acid assay.

Additional file 3. Gas chromatographic (GC) profiles of the derivatized sugar standards in the trimethylsilyl (TMS) method. The standard mixtures 1 and 2 consist of the nine monosaccharides (each 0.5 µg, shown in bold): arabinose (Ara), rhamnose (Rha), fucose (Fuc), xylose (Xyl), mannose (Man), galactose (Gal), glucose (Glc), galacturonic acid (GalA), and glucuronic acid (GlcA), supplemented with myo-inositol (Inos, 0.2 µg, in bold) as an internal standard. Also included in the chromatogram shown are (in parentheses) ribose (Rib), N-acetylmannosamine (ManNac), N-acetylglucosamine (GlcNAc), and N-acetylgalactosamine (GalNAc). The derivatized sugars are separated on a Supelco EC-1 fused silica capillary column (30 m × 0.25 mm ID) on an Agilent 7890A gas chromatograph using helium as the carrier gas with temperature gradient as described in the "Methods".

Additional file 4. Chromatographic profiles of the sugar standards in the HPAEC method. As outlined in "Methods" section, the HPAEC analyses were carried out in two separate runs using two different programs (i.e. different columns and gradients) for each sample. In bold are the sugars quantified using the respective program. (A) Program 1 was used to quantify the amounts of fucose (Fuc), rhamnose (Rha), arabinose (Ara), galactose (Gal), glucose (Glc), galacturonic acid (GalA), and glucuronic acid (GlcA) on a Dionex PA20 column eluted using a NaOH/NaOAc gradient. (B) Program 2 was used to quantify the amounts of xylose (Xyl) and mannose (Man), which eluted as one peak in program 1, on a Dionex PA1 column eluted isocratically using 2 mM NaOH.

Additional file 5. Comparison of the mol% of the different types of sugars in leaf AIR from Arabidopsis, *Populus*, rice and switchgrass using the four different glycosyl residue composition analysis methods.

Additional file 6. Comparison of the mol% of the different types of sugars in AIR from *Populus* wood, rice stem and switchgrass tiller biomass using the four different glycosyl residue composition analysis methods.

Additional file 7. Comparison of the four glycosyl residue composition analysis methods.

Abbreviations
AA: alditol acetate; AIR: alcohol insoluble residue; Ara: arabinose; ddH₂O: deionized water; ECD: electrochemical detection; Fuc: fucose; Gal: galactose; GalA: galacturonic acid; GC–MS: gas chromatography–mass spectrometry; Glc: glucose; GlcA: glucuronic acid; HPAEC: high-pressure, anion-exchange chromatography; Man: mannose; Rha: rhamnose; RT: room temperature; TFA: trifluoroacetic acid; TMS: trimethylsilyl; UA: uronic acid; Xyl: xylose; WT: wild type.

Authors' contributions
AKB and DM planned and designed the research. AKB, LT, JD, IG-M, KH, IMB, SSM, MA, and DR performed research. AKB, LT, JD, MA, CEW, and DM analyzed data. AKB, MA, JD, CEW, and DM wrote the manuscript. All authors read and approved the final manuscript.

Author details

[1] Department of Biochemistry and Molecular Biology, University of Georgia, Athens, GA 30602, USA. [2] Complex Carbohydrate Research Center, University of Georgia, 315 Riverbend Rd., Athens, GA 30602-4712, USA. [3] DOE-BioEnergy Science Center (BESC), Oak Ridge 37831, TN, USA. [4] Center for Environmental Research and Technology (CE-CERT) and Department of Chemical and Environmental Engineering, University of California Riverside, Riverside 92507, CA, USA. [5] Present Address: DuPont Industrial Biosciences, Palo Alto, CA 94304, USA. [6] Present Address: South Georgia State College, Douglas, GA 31533, USA.

Acknowledgements

We thank L. Scott Forsberg for constructive comments and suggestions on the manuscript, and Kristen A. Engle for help with statistics.

Competing interests

The authors declare that they have no competing interests.

Funding

This work was supported by BioEnergy Science Center grant DE-PS02-06ER64304 and partially funded by the Department of Energy Center Grant DE-SC0015662. The BioEnergy Science Center is a US Department of Energy Bioenergy Research Center supported by the Office of Biological and Environmental Research in the Department of Energy's Office of Science.

References

1. Himmel ME, Ding SY, Johnson DK, Adney WS, Nimlos MR, Brady JW, et al. Biomass recalcitrance: engineering plants and enzymes for biofuels production. Science. 2007;315:804–7.
2. Laser M, Lynd LR. Comparative efficiency and driving range of light- and heavy-duty vehicles powered with biomass energy stored in liquid fuels or batteries. Proc Natl Acad Sci USA. 2014;111:3360–4.
3. US-DOE. Breaking the biological barriers to cellulosic ethanol: a joint research agenda. In: Secondary breaking the biological barriers to cellulosic ethanol: a joint research agenda. U.S. Department of Energy Office of Science. 2006. http://www.genomicscience.energy.gov/biofuels/. Accessed 26 Aug 2016.
4. Edenhofer O, Pichs-Madruga R, Sokona Y, Farahani E, Kadner S, Seyboth K, et al (eds.). IPCC, 2014: climate change 2014: mitigation of climate change. In: Contribution of working group III to the fifth assessment report of the intergovernmental panel on climate change. Cambridge, United Kingdom and New York, NY, USA; 2014.
5. Abel GJ, Barakat B, Kc S, Lutz W. Meeting the sustainable development goals leads to lower world population growth. Proc Natl Acad Sci USA. 2016;113:14294–9.
6. Albersheim P, Darvill A, Roberts K, Sederoff R, Staehelin A. Plant cell walls. 1st ed. New York: Garland Science; 2011.
7. Keijsers ER, Yilmaz G, van Dam JE. The cellulose resource matrix. Carbohydr Polym. 2013;93:9–21.
8. York WS, Darvill AG, McNeill M, Stevenson TT, Albersheim P. Isolation and characterization of plant cell walls and cell wall components. Methods Enzymol. 1985;118:3–40.
9. ASTM-E1821-01. Standard test method for determination of carbohydrates in biomass by gas chromatography. ASTM International, West Conshohocken, PA. 2007. www.astm.org. Accessed 14 Mar 2016.
10. Blakeney AB, Harris PJ, Henry RJ, Stone BA. A simple and rapid preparation of alditol acetates for monosaccharide analysis. Carbohydr Res. 1983;113:291–9.
11. Fujikura U, Elsaesser L, Breuninger H, Sanchez-Rodriguez C, Ivakov A, Laux T, et al. Atkinesin-13A modulates cell-wall synthesis and cell expansion in *Arabidopsis thaliana* via the THESEUS1 pathway. PLoS Genet. 2014;10:e1004627.
12. Brunton NP, Gormley TR, Murray B. Use of the alditol acetate derivatization for the analysis of reducing sugars in potato tubers. Food Chem. 2007;104:398–402.
13. Gibeaut DM, Pauly M, Bacic A, Fincher GB. Changes in cell wall polysaccharides in developing barley (*Hordeum vulgare*) coleoptiles. Planta. 2005;221:729–38.
14. Sims IM, Munro SL, Currie G, Craik D, Bacic A. Structural characterisation of xyloglucan secreted by suspension-cultured cells of *Nicotiana plumbaginifolia*. Carbohydr Res. 1996;293:147–72.
15. Biswal AK, Soeno K, Gandla ML, Immerzeel P, Pattathil S, Lucenius J, et al. Aspen pectate lyase *Ptx*PL1-27 mobilizes matrix polysaccharides from woody tissues and improves saccharification yield. Biotechnol Biofuels. 2014;7:11.
16. Wang X, Cheng Z, Zhao Z, Gan L, Qin R, Zhou K, et al. *BRITTLE SHEATH1* encoding OsCYP96B4 is involved in secondary cell wall formation in rice. Plant Cell Rep. 2016;35:745–55.
17. Mazumder K, York WS. Structural analysis of arabinoxylans isolated from ball-milled switchgrass biomass. Carbohydr Res. 2010;345:2183–93.
18. Tan L, Eberhard S, Pattathil S, Warder C, Glushka J, Yuan C, et al. An Arabidopsis cell wall proteoglycan consists of pectin and arabinoxylan covalently linked to an arabinogalactan protein. Plant Cell. 2013;25:270–87.
19. Li M, Xiong G, Li R, Cui J, Tang D, Zhang B, et al. Rice cellulose synthase-like D4 is essential for normal cell-wall biosynthesis and plant growth. Plant J. 2009;60:1055–69.
20. Doco T, O'Neill MA, Pellerin P. Determination of the neutral and acidic glycosyl-residue compositions of plant polysaccharides by GC-EI-MS analysis of the trimethylsilyl methyl glycoside derivatives. Carbohyd r Polym. 2001;46:249–59.
21. Biswal AK, Hao Z, Pattathil S, Yang X, Winkeler K, Collins C, et al. Down-regulation of *GAUT12* in *Populus deltoides* by RNA silencing results in reduced recalcitrance, increased growth and reduced xylan and pectin in a woody biofuel feedstock. Biotechnol Biofuels. 2015;8:41.
22. Lionetti V, Francocci F, Ferrari S, Volpi C, Bellincampi D, Galletti R, et al. Engineering the cell wall by reducing de-methyl-esterified homogalacturonan improves saccharification of plant tissues for bioconversion. Proc Natl Acad Sci USA. 2010;107:616–21.
23. Petersen PD, Lau J, Ebert B, Yang F, Verhertbruggen Y, Kim JS, et al. Engineering of plants with improved properties as biofuels feedstocks by vessel-specific complementation of xylan biosynthesis mutants. Biotechnol Biofuels. 2012;5:84.
24. Burton RA, Gidley MJ, Fincher GB. Heterogeneity in the chemistry, structure and function of plant cell walls. Nat Chem Biol. 2010;6:724–32.
25. Kim JB, Carpita NC. Changes in esterification of the uronic acid groups of cell wall polysaccharides during elongation of maize coleoptiles. Plant Physiol. 1992;98:646–53.
26. Pena MJ, Carpita NC. Loss of highly branched arabinans and debranching of rhamnogalacturonan I accompany loss of firm texture and cell separation during prolonged storage of apple. Plant Physiol. 2004;135:1305–13.
27. Currie HA, Perry CC. Resolution of complex monosaccharide mixtures from plant cell wall isolates by high pH anion exchange chromatography. J Chromatogr A. 2006;1128:90–6.
28. Weitzhandler M, Pohl C, Jandik P, Cheng J, Avdalovic N. CarboPac PA20: a new monosaccharide separator column with electrochemical detection with disposable gold electrodes. J Biochem Biophys Methods. 2004;60:309–17.
29. Draeger C, Fabrice TN, Gineau E, Mouille G, Kuhn BM, Moller I, et al. Arabidopsis leucine-rich repeat extensin (LRX) proteins modify cell wall composition and influence plant growth. BMC Plant Biol. 2015;15:155.
30. Obel N, Porchia AC, Scheller HV. Dynamic changes in cell wall polysaccharides during wheat seedling development. Phytochemistry. 2002;60:603–10.
31. Øbro J, Harholt J, Scheller HV, Orfila C. Rhamnogalacturonan I in *Solanum tuberosum* tubers contains complex arabinogalactan structures. Phytochemistry. 2004;65:1429–38.
32. Chiniquy D, Sharma V, Schultink A, Baidoo EE, Rautengarten C, Cheng K, et al. XAX1 from glycosyltransferase family 61 mediates xylosyltransfer to rice xylan. Proc Natl Acad Sci USA. 2012;109:17117–22.
33. Caffall KH, Pattathil S, Phillips SE, Hahn MG, Mohnen D. *Arabidopsis thaliana* T-DNA mutants implicate *GAUT* genes in the biosynthesis of pectin and xylan in cell walls and seed testa. Mol Plant. 2009;2:1000–14.
34. King ZR, Bray AL, Lafayette PR, Parrott WA. Biolistic transformation of elite genotypes of switchgrass (*Panicum virgatum* L.). Plant Cell Rep. 2014;33:313–22.
35. Moore KJ, Moser LE, Vogel KP, Waller SS, Johnson BE, Pedersen JF. Describing and quantifying growth stages of perennial forage grasses. Agron J. 1991;83:1073–7.
36. Albersheim P, Nevins DJ, English PD, Karr A. A method for the analysis of

sugars in plant cell-wall polysaccharides by gas-liquid chromatography. Carbohydr Res. 1967;5:340–5.

37. Blumenkrantz N, Asboe-Hansen G. New method for quantitative determination of uronic acids. Anal Biochem. 1973;54:484–9.

38. Filisetti-Cozzi TM, Carpita NC. Measurement of uronic acids without interference from neutral sugars. Anal Biochem. 1991;197:157–62.

39. van den Hoogen BM, van Weeren PR, Lopes-Cardozo M, van Golde LM, Barneveld A, van de Lest CH. A microtiter plate assay for the determination of uronic acids. Anal Biochem. 1998;257:107–11.

40. Wikiera A, Mika M, Starzynska-Janiszewska A, Stodolak B. Development of complete hydrolysis of pectins from apple pomace. Food Chem. 2015;172:675–80.

41. Taylor RL, Conrad HE. Stoichiometric depolymerization of polyuronides and glycosaminoglycuronans to monosaccharides following reduction of their carbodiimide-activated carboxyl groups. Biochemistry. 1972;11:1383–8.

42. Carpita NC, McCann MC. Some new methods to study plant polyuronic acids and their esters. In: Townsend RR, Hotchkiss Jr AT, editors. Techniques in glycobiology. New York: Marcel Dekker, Inc.; 1997. p. 595–612.

43. Maness NO, Ryan JD, Mort AJ. Determination of the degree of methyl esterification of pectins in small samples by selective reduction of esterified galacturonic acid to galactose. Anal Biochem. 1990;185:346–52.

44. Merkle RK, Poppe I. Carbohydrate composition analysis of glycoconjugates by gas-liquid chromatography/mass spectrometry. Methods Enzymol. 1994;230:1–15.

45. Gardner SL, Burrell MM, Fry SC. Screening of *Arabidopsis thaliana* stems for variation in cell wall polysaccharides. Phytochemistry. 2002;60:241–54.

46. Vogel J. Unique aspects of the grass cell wall. Curr Opin Plant Biol. 2008;11:301–7.

47. Pauly M, Keegstra K. Cell-wall carbohydrates and their modification as a resource for biofuels. Plant J. 2008;54:559–68.

48. Carpita NC, Gibeaut DM. Structural models of primary cell walls in flowering plants: consistency of molecular structure with the physical properties of the walls during growth. Plant J. 1993;3:1–30.

49. Mann DGJ, Labbé N, Sykes RW, Gracom K, Kline L, Swamidoss IM, et al. Rapid assessment of lignin content and structure in switchgrass (*Panicum virgatum* L.) grown under different environmental conditions. Bioenergy Res. 2009;2:246–56.

50. Cavalier DM, Lerouxel O, Neumetzler L, Yamauchi K, Reinecke A, Freshour G, et al. Disrupting two *Arabidopsis thaliana* xylosyltransferase genes results in plants deficient in xyloglucan, a major primary cell wall component. Plant Cell. 2008;20:1519–37.

51. Mortimer JC, Faria-Blanc N, Yu X, Tryfona T, Sorieul M, Ng YZ, et al. An unusual xylan in Arabidopsis primary cell walls is synthesised by GUX3, IRX9L, IRX10L and IRX14. Plant J. 2015;83:413–26.

52. Carpita NC. Structure and biogenesis of the cell walls of grasses. Annu Rev Plant Physiol Plant Mol Biol. 1996;47:445–76.

53. Fincher GB. Revolutionary times in our understanding of cell wall biosynthesis and remodeling in the grasses. Plant Physiol. 2009;149:27–37.

54. Carpita NC, Defernez M, Findlay K, Wells B, Shoue DA, Catchpole G, et al. Cell wall architecture of the elongating maize coleoptile. Plant Physiol. 2001;127:551–65.

55. Galambos JT. The reaction of carbazole with carbohydrates. I. Effect of borate and sulfamate on the carbazole color of sugars. Anal Biochem. 1967;19:119–32.

56. Bethke G, Thao A, Xiong G, Li B, Soltis NE, Hatsugai N, et al. Pectin biosynthesis is critical for cell wall integrity and immunity in *Arabidopsis thaliana*. Plant Cell. 2016;28:537–56.

57. Francocci F, Bastianelli E, Lionetti V, Ferrari S, De Lorenzo G, Bellincampi D, et al. Analysis of pectin mutants and natural accessions of Arabidopsis highlights the impact of de-methyl-esterified homogalacturonan on tissue saccharification. Biotechnol Biofuels. 2013;6:163.

58. Pettolino FA, Walsh C, Fincher GB, Bacic A. Determining the polysaccharide composition of plant cell walls. Nat Protoc. 2012;7:1590–607.

59. Sjöström E, Westermark U. Chemical composition of wood and pulps: basic constituents and their distribution. In: Sjöström E, Alén R, editors. Analytical methods in wood chemistry, pulping, and papermaking. Berlin: Springer; 1999. p. 1–19.

60. Villas-Boas SG, Smart KF, Sivakumaran S, Lane GA. Alkylation or silylation for analysis of amino and non-amino organic acids by GC–MS? Metabolites. 2011;1:3–20.

61. Rohloff J. Analysis of phenolic and cyclic compounds in plants using derivatization techniques in combination with GC–MS-based metabolite profiling. Molecules. 2015;20:3431–62.

2

Lignin valorization: lignin nanoparticles as high-value bio-additive for multifunctional nanocomposites

Dong Tian[1,2,3], Jinguang Hu[2,3*], Jie Bao[2], Richard P. Chandra[3], Jack N. Saddler[3] and Canhui Lu[1*]

Abstract

Background: Although conversion of low value but high-volume lignin by-product to its usable form is one of the determinant factors for building an economically feasible integrated lignocellulose biorefinery, it has been challenged by its structural complexity and inhomogeneity. We and others have shown that uniform lignin nanoparticles can be produced from a wide range of technical lignins, despite the varied lignocellulosic biomass and the pretreatment methods/conditions applied. This value-added nanostructure lignin enriched with multifunctional groups can be a promising versatile material platform for various downstream utilizations especially in the emerging nanocomposite fields.

Results: Inspired by the story of successful production and application of nanocellulose biopolymer, two types of uniform lignin nanoparticles (LNPs) were prepared through self-assembling of deep eutectic solvent (DES) and ethanol-organosolv extracted technical lignins derived from a two-stage fractionation pretreatment approach, respectively. Both LPNs exhibited sphere morphology with unique core–shell nanostructure, where the DES–LNPs showed a more uniform particle size distribution. When incorporated into the traditional polymeric matrix such as poly(vinyl alcohol), these LPN products displayed great potential to formulate a transparent nanocomposite film with additional UV-shielding efficacy (reached ~80% at 400 nm with 4 wt% of LNPs) and antioxidant functionalities (reached ~160 μm mol Trolox g^{-1} with 4 wt% of LNPs). At the same time, the abundant phenolic hydroxyl groups on the shell of LNPs also provided good interfacial adhesion with PVA matrix through the formation of hydrogen bonding network, which further improved the mechanical and thermal performances of the fabricated LNPs/PVA nanocomposite films.

Conclusions: Both LNPs are excellent candidates for producing multifunctional polymer nanocomposites using facile technical route. The prepared transparent and flexible LNPs/PVA composite films with high UV-shielding efficacy, antioxidant activity, and biocompatibility are promising in the advanced packaging field, which potentially provides an additional high-value lignin product stream to the lignocellulose biorefinery. This study could open the door for the production and application of novel LNPs in the nascent bioeconomy.

Keywords: Lignin nanoparticles, UV-shielding, Antioxidant, Polymer nanocomposite, Biorefinery

*Correspondence: jinguang@mail.ubc.ca; canhuilu@scu.edu.cn
[1] State Key Laboratory of Polymer Materials Engineering, Polymer Research Institute of Sichuan University, Chengdu 610065, China
[2] State Key Laboratory of Bioreactor Engineering, East China University of Science and Technology, 130 Meilong Road, Shanghai 200237, China
Full list of author information is available at the end of the article

Background

Conversion of major components (cellulose, hemicellulose, and lignin) of lignocellulosic biomass into usable platform is essential in the integrated biorefinery concept, which simplifies the subsequent production of fuels, chemicals, and materials [1, 2]. Although deconstructing of biomass carbohydrates into a valuable hexose/pentose sugar platform through biochemical approaches has been succeed for many years, the efficient utilization of lignin component has still been challenging [3, 4]. Both the traditional pulp and paper industry and the emerging cellulosic ethanol plant have been liberating a huge pile of technical lignins from lignocellulosic biomass; however, most of these lignins have been directly burned as the industrial "waste" for energy generation [4]. The high-value utilization of lignin via hydrocarbon fuel or aromatic polymer precursor production is attractive but still challenged by its structural complexity and inhomogeneity [5]. Recently, we and others have shown that the uniform lignin nanoparticles (LNPs) could be produced from a wide range of lignin by-products, regardless of their varied chemical structures [6–11]. These LNPs hold huge potential for downstream valorization due to their unique morphology and abundant multifunctional groups.

The nature of lignin is highly branched, three-dimensional polymer derived from three phenylpropane units (monolignols), namely, guaiacyl (G, conniferyl alcohol), syringyl (S, sinapyl alcohol), and p-hydroxyphenyl (H, p-coumaryl alcohol). When the prevalent solution-based self-assembly micellization process is employed to produce LNPs from the amphiphilic lignin fragments, the hydrophobic part of lignin (phenylpropanoid units) aggregates to form the micelle core in the solution, while the hydrophilic part of lignin (mainly phenolic and aliphatic hydroxyl groups) forms the micelle shell, simultaneously [6, 9]. Thus, the obtained LNPs exhibit unique core–shell nanostructure with abundant phenolic hydroxyl groups exposed on the shell of the LNPs. The greatly improved availability of phenolic hydroxyl groups on the shell of LNPs allows them to disperse well and stable in aqueous solution even for several months [7].

The production and application of functional polymer-based nanocomposites present new market opportunities for various bio-additives [12]. Although traditional inorganic nanomaterials could effectively endow the polymer nanocomposites with additional functionalities such as conductivity, antibacterial activity, flame resistance, etc., unexpected environmental and/or health problems occurs due to their poor biodegradability and biocompatibility [13]. In addition, some inorganic nanomaterials could also induce serious polymer matrix degradation. For example, when prevalent titanium dioxide and zinc oxide nanoparticles are used as UV-absorber additives, they catalyze the cleavage of polymer macromolecular chain due to their intrinsic photocatalytic activity [14–17]. Alternatively, LNPs which are produced from natural lignocellulosic biomass might be promising alternatives to those inorganic nanomaterials for producing functional polymer composites. Considering the outstanding UV shielding and antioxidant properties of the phenolic hydroxyl groups [18], LNPs might be suitable for producing functional protective nanocomposites by introducing functionalities to the polymer matrix while overcoming the above disadvantages from inorganic nanomaterials.

In the work reported here, we assessed the technical feasibility of valorizing lignin through producing LNPs/polymer nanocomposite films with both UV-shielding and antioxidant functionalities. Two technical lignins, DES and organosolv lignin isolated from steam pretreated hardwood poplar using a deep eutectic solvent (DES, an emerging solvent for biomass fractionation) and traditional ethanol organosolv, respectively [2, 19], were initially upgraded to their usable form of LNPs using prevalent micellization process. Then, the prepared LNPs were incorporated into the testing polymer poly(vinyl alcohol) (PVA), a biodegradable synthetic polymer material with wide commercial applications, to produce nanocomposite films via the facile solution-cast method. The overall performance of the resulting nanocomposite films including UV-shielding efficacy, antioxidant activity, and mechanical strength was systematically assessed. The possible interactions between two LNPs and PVA matrix were also comparatively evaluated. Results showed that both LNPs were great candidates for producing high-value functional polymer nanocomposites, while the organosolv LNPs with higher amount of phenolic hydroxyl groups exposed on the nanoparticle shell (assessed by quantitative ^{31}P NMR) exhibited better overall performances than the DES LNPs [20]. We hope that the work reported here could open the door for the production and application of a wide range of novel LNP-based polymer nanocomposites.

Results and discussion
Synthesis and characterization of lignin nanoparticles

To achieve full utilization of lignocellulosic biomass and easy integration of LNPs production into current biorefinery concept, a two-step pretreatment strategy, mild steam pretreatment followed by solvent extraction, was employed to produce DES and organosolv technical lignins from raw hardwood poplar while facilitating the conversion of cellulose/hemicellulose component to hexose/pentose sugar platform according to the previous reports [2, 21] (for the details of the fractionation pretreatment, see Additional file 1: Figure S1). The purity of the two technical lignins was higher than 98% according to HPLC analysis reported previously [9]. When these two technical lignins were dissolved in dimethylsulfoxide (2 mg mL^{-1}) and subjected to micellization using dialysis, uniform lignin nanoparticle dispersions (referred as DLNPs and OLNPs, respectively) were obtained. Scan electron microscopy (SEM) and atomic force microscopy (AFM, Additional file 1: Figure S2) images showed that both lignin nanoparticles products had sphere morphological structure, while DLNPs had a more uniform particle size distribution (Fig. 1a). The core–shell structure of the two lignin nanoparticles was confirmed by high-resolution transmission electron microscopy (TEM) images (Additional file 1: Figure S3). The dark black color of the sphere particles indicates the core, while the grey color around it indicates the shell. The shell thickness was about 10–20 nm. When the dynamic properties of the two lignin nanoparticle dispersions were further analyzed by dynamic light scattering (DLS), the DLNPs gave an average particle size of 195 nm with a polydispersity index (PDI) of 0.08, while the OLNPs exhibited a similar average particle size (197 nm) but indeed a much higher PDI (0.17) (Fig. 1a). The zeta-potential value (also measured by DLS) of the lignin nanoparticle dispersion was −37.5 and −35.8 mV for DLNPs and OLNPs, respectively, which indicated a relative high stability of these two lignin nanoparticles in water [7]. The uniform particle size, regular-sphere structure, and high stability of these two lignin nanoparticles indicated that they might be promising candidates for the production of nanocomposite films with PVA polymer.

UV shielding, antioxidant, and mechanical performance of the lignin nanoparticles/PVA composite films

As expected, the lignin nanoparticles/PVA composites were easily prepared by a simple solution-cast method due to the good properties of these lignin nanoparticles as mentioned above (Fig. 1a). The influence of the content of lignin nanoparticles (0–4 wt% lignin nanoparticles based on the dry weight of PVA) on the UV shielding and antioxidant performances of the resulting composite films was first assessed. The related mechanism of UV

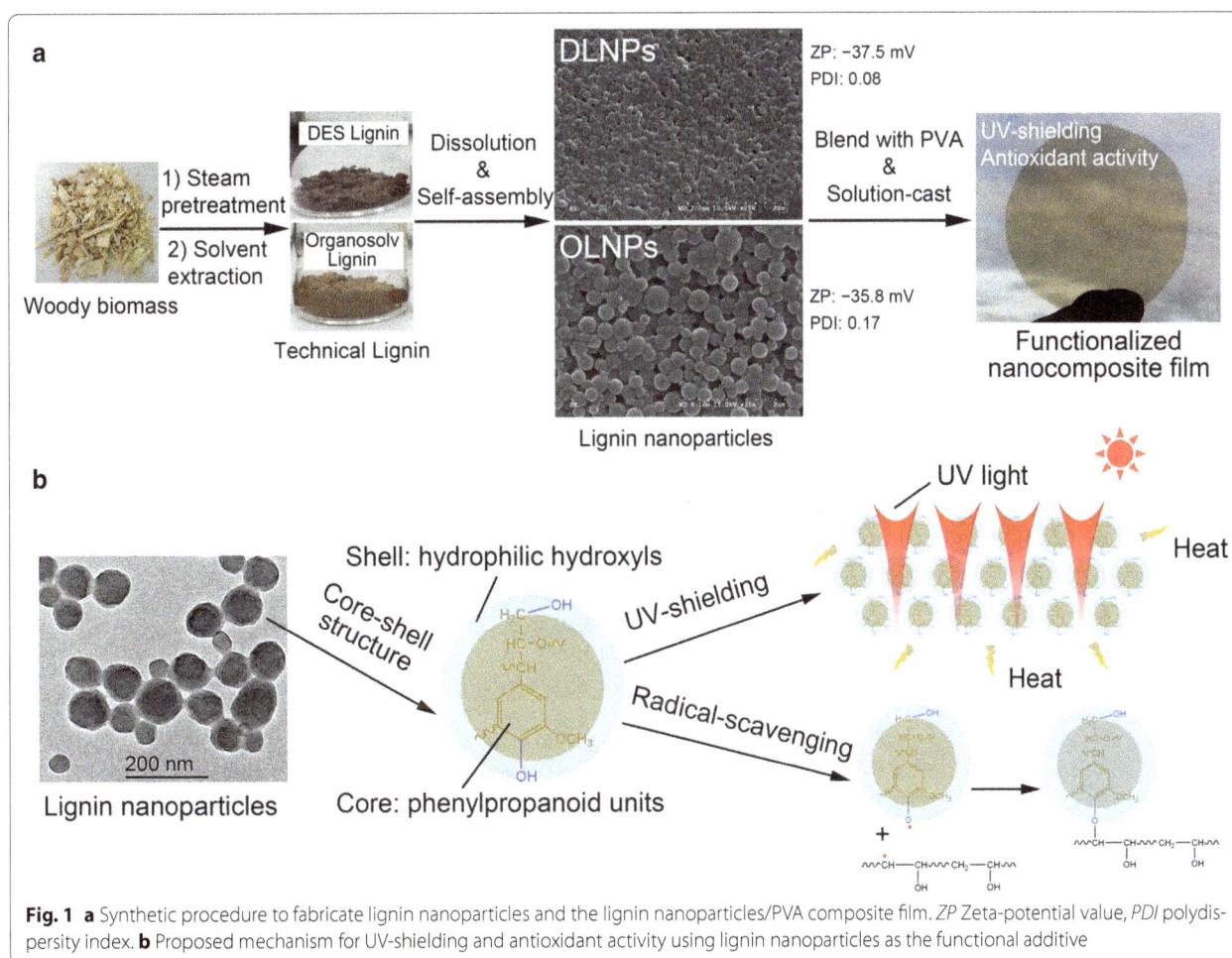

Fig. 1 **a** Synthetic procedure to fabricate lignin nanoparticles and the lignin nanoparticles/PVA composite film. *ZP* Zeta-potential value, *PDI* polydispersity index. **b** Proposed mechanism for UV-shielding and antioxidant activity using lignin nanoparticles as the functional additive

shielding and antioxidant is proposed in Fig. 1b according to the previous reports [13, 15], and LNPs are suggested to block the ultraviolet light by absorbing its photon energy and further converting it to heat with the corresponding hydrophilic chromophores (mainly phenolic hydroxyl, carbonyl, and carboxyl groups) exposed on the particle shell. Then, the generated heat is gradually released out of the nanocomposite films without causing PVA degradation, while these phenolic hydroxyl groups could also easily quench active radicals through an electron transfer process [22, 23]. When all the prepared films were exposed to UV–Visible light with a wavelength from 200 to 800 nm (Fig. 2), it was apparent

that the nanocomposite films could efficiently block the ultraviolet lights especially for UVB (280–315 nm) even with a low lignin nanoparticles content (0.5 wt%), while the neat PVA film (control) was almost transparent for all the testing ultraviolet lights (Fig. 2). Although further increase of lignin nanoparticles contents from 0.5 to 4 wt% resulted in an obvious improvement of the shielding efficacy for both UVB (280–315 nm) and UVA (315–400 nm), it slightly sacrificed the visible light transparency simultaneously (Fig. 2). In general, the OLNPs/PVA film exhibited higher shielding efficacy compared to the DLNPs/PVA film at the same lignin nanoparticles content, and it was also worth noting that the shielding

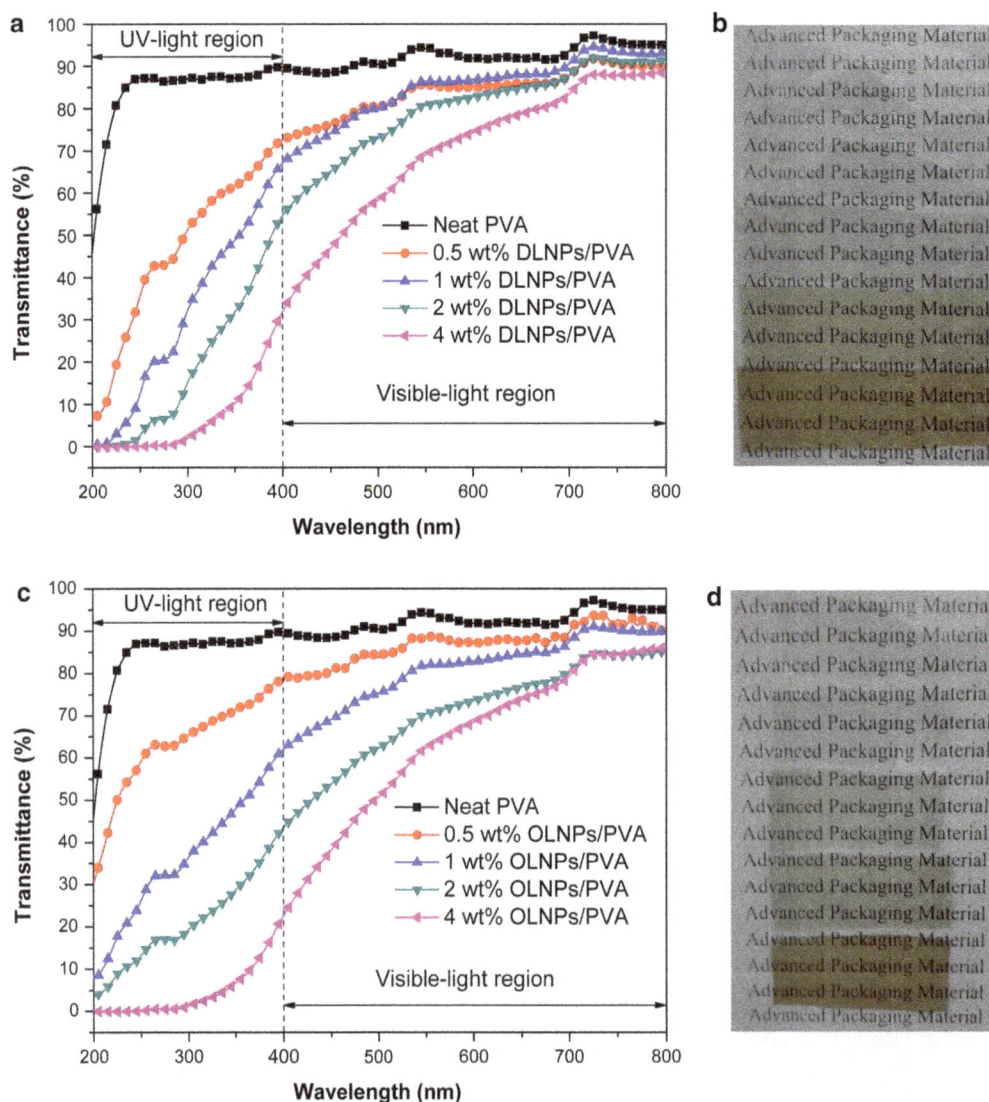

Fig. 2 UV–Vis light transmittance spectra and digital photographs of a and b DLNPs/PVA and c and d OLNPs/PVA composite films with 0–4 wt% lignin nanoparticles. The digital photographs show the high optical transparency of the lignin nanoparticles/PVA composite films (from top to bottom, the content of lignin nanoparticles in the film was increasing from 0 to 4 wt%)

efficacy of DLNPs/PVA and OLNPs/PVA nanocomposite films could reach nearly 100% for UVB with only 4% (w/w) lignin nanoparticles content (Fig. 2a, c). All these results indicated that the addition of lignin nanoparticle to the PVA film could provide an efficient UV block capacity without influencing its visible light transparency [11].

Further translating the UV–Visible transmittance spectra into Tauc's plot with the frequency dependent absorption coefficient according to the previous reports provided additional information about the band structure and optical energy bandgap (E_g) of the composite films [15, 24]. The principle of evaluating the optical property of the composites using E_g is that the photons with energy higher than the band-gap energy will be absorbed by the corresponding molecules in the composite [24]. When the E_g value of each composite film was calculated, it decreased with the increase of lignin nanoparticle content (3.81 − 2.49 eV for DLNPs/PVA and 3.54 − 1.96 eV for OLNPs/PVA, for the details of E_g calculation, see Additional file 1: Figure S4), indicating that more ultraviolet lights with wider wavelength range could be blocked. The E_g values also provided a fair compare of the overall optical performance among various UV-absorbing materials. The results showed that the UV-shielding efficacy of these two types of lignin nanoparticles was comparable to prevalent metal oxide nanoparticles and other emerging biopolymer nanoparticles such as polydopamine and melanin [13, 15, 25, 26].

We further assessed the influence of lignin nanoparticles on the antioxidant properties of the prepared nanocomposite films using prevalent Trolox equivalent antioxidant capacity (TEAC) measurement, which employed stable 1,1-diphenyl-2-picrylhydrazyl (DPPH) as the testing radicals and Trolox as the internal standard to evaluate the radical-scavenging ability of the composites according to the previous reports (Fig. 3, and results were expressed as μmol Trolox per gram of composite film) [27]. As expected, the antioxidant activities of the films gradually increased with the increasing content of lignin nanoparticles, while OLNPs/PVA film exhibited higher antioxidant activity than DLNPs/PVA at the same lignin nanoparticles content, likely due to more available phenolic hydroxyl groups in the OLNPs [23, 28]. The antioxidant activity of the film was nearly zero for PVA, but dramatically increased to 129 (DLNPs/PVA) and 157 μm mol Trolox per gram composite (OLNPs/PVA) after incorporation of 4 wt% lignin nanoparticles. Previous reports have shown that the TEAC value of pure lignin and other natural phenolic compounds was ~500 μm mol Trolox g^{-1} [22, 27]. However, the work reported here showed that

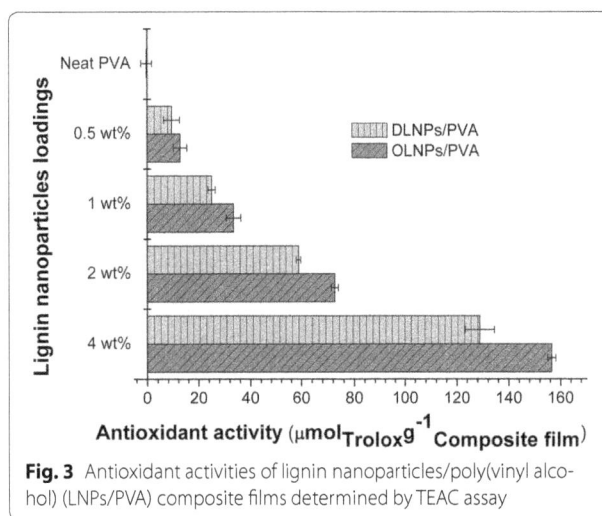

Fig. 3 Antioxidant activities of lignin nanoparticles/poly(vinyl alcohol) (LNPs/PVA) composite films determined by TEAC assay

although the content of lignin nanoparticles in the composite films was only 4 wt%, their TEAC values could reach ~150 μm mol Trolox g^{-1} despite of different testing solvents employed. These results indicated that the prepared lignin nanoparticles had rather high radical-scavenging ability.

Although bulk lignin could be directly blended with a polymer matrix using thermal extrusion/injection method to produce a composite, the poor interfacial binding between bulk lignin and the polymer matrix usually resulted in the deterioration of its mechanical performance [18]. The nanoeffects of lignin nanoparticles including increased surface area and good dispersion state potentially enhance their compatibility with polymer matrix; thus, the resulting composite might exhibit a better mechanical performance [8, 11]. When the mechanical properties of the above nanocomposites were further checked, there was indeed an increase instead of a deterioration of the tensile strength, indicating a certain reinforcement effect of lignin nanoparticles. The tensile strength of the composite films increased from 50 to ~55 MPa for DLNPs/PVA and to ~60 MPa for OLNPs/PVA while only slightly compromising its elongation performance simultaneously (Fig. 4). It seemed that apart from the intrinsic properties of lignin nanoparticles themselves, their interfacial adhesion and dispersion state in the PVA matrix also played an important role in determining the overall performances of the nanocomposites. Therefore, we next assessed the interactions between lignin nanoparticles and PVA matrix. As the composite films containing 4 wt% lignin nanoparticles had the best UV shielding, antioxidant, and mechanical performances, we selected them as the testing samples for the subsequent analysis.

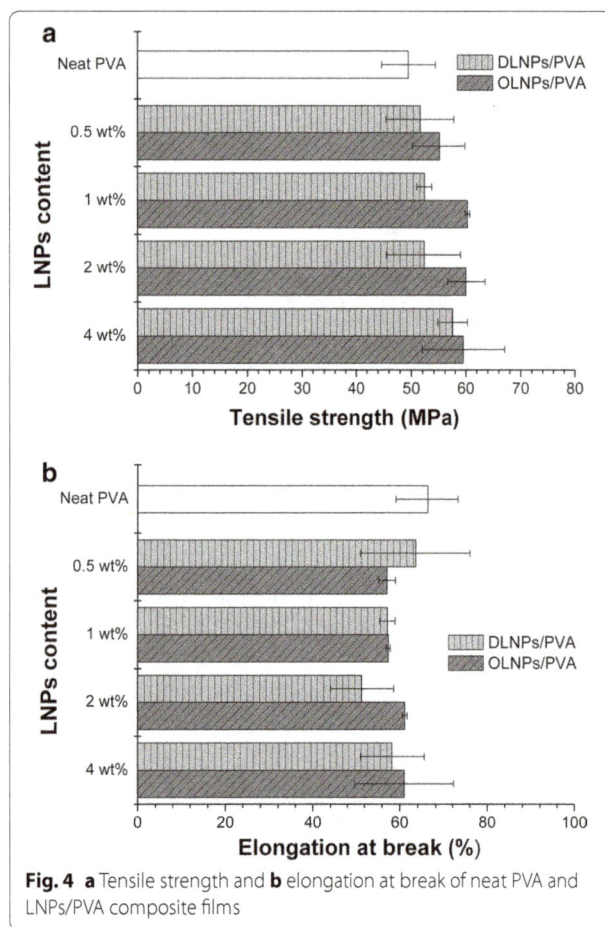

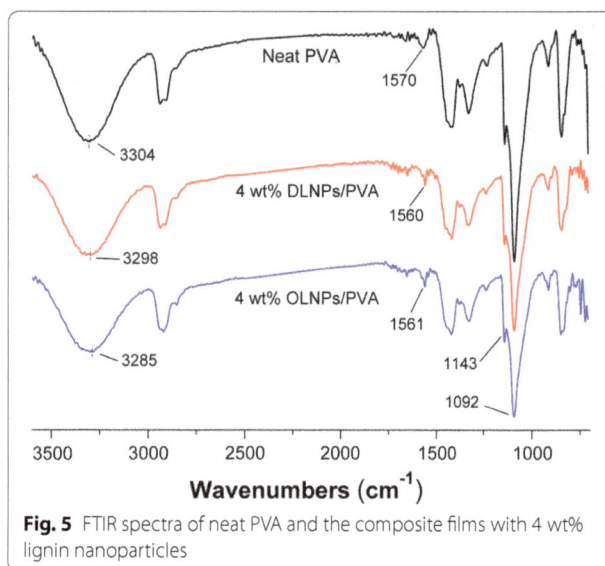

Fig. 5 FTIR spectra of neat PVA and the composite films with 4 wt% lignin nanoparticles

Fig. 4 **a** Tensile strength and **b** elongation at break of neat PVA and LNPs/PVA composite films

Interactions between lignin nanoparticles and PVA matrix

As mentioned earlier, the shell of the lignin nanoparticles was mainly composed of hydrophilic hydroxyl groups, which could potentially form strong interactions with the hydroxyl groups of PVA [6]. When Fourier transform infrared spectroscopy (FTIR) analysis was conducted on the selected samples to confirm this hypothesis, it was shown that the –OH stretching band for the neat PVA shifted from 3304 cm^{-1} to a lower wavenumber of 3298 cm^{-1} (DLNPs/PVA) and 3285 cm^{-1} (OLNPs/PVA), respectively, upon incorporating with 4 wt% lignin nanoparticles, indicating that hydrogen bonds were formed between the PVA and the lignin nanoparticles (Fig. 5) [29]. In addition, the FTIR spectrum of neat PVA exhibited a strong stretching vibrational band of carbon–carbon double bonds at 1570 cm^{-1}, which was likely resulted by the radical-induced degradation of PVA macromolecular chains, since the PVA solution was prepared under heating at an open atmosphere (Fig. 5) [30]. However, this absorption peak became much narrower and also shifted to lower wavenumbers (1560 cm^{-1} for DLNPs/PVA and 1561 cm^{-1} for OLNPs/PVA, respectively) when lignin nanoparticles were added, indicating that

lignin nanoparticles could stabilize the PVA macromolecular structure through trapping the generated free radicals [15]. The carbon–carbon double bonds in the composite films might be negatively charged by lignin nanoparticles; thus, their FTIR spectra exhibited such a wavenumber shift according to the previous reports [15, 30]. The above analysis suggested that the lignin nanoparticles were able to interact with the PVA macromolecular chains through hydrogen bonding and radical-scavenging reactions.

The good interfacial adhesion between lignin nanoparticles and PVA matrix was further evidenced by transmission electron microscopy (TEM) observations, as shown in Fig. 6. Due to the good dispersion state of the lignin nanoparticles in water, the resulting nanocomposite films exhibited well-defined sea-island structure, where lignin nanoparticles were dispersed at nanoscale without aggregation. The morphological structure of the lignin nanoparticles was slightly changed after incorporation into the PVA matrix, but still remained their sphere-like shape (Figs. 1, 6), suggesting that the shell of lignin nanoparticles (composed of hydroxyl groups) indeed interacted with PVA as hypothesized earlier. These strong interactions were likely the driving force in improving the mechanical performance of the composite films. In addition, the good dispersion state of lignin nanoparticles was also responsible for the excellent UV shielding and antioxidant performance of the composite films.

Influence of lignin nanoparticles on the crystalline and thermal properties of the composite films

It is acknowledged that the nanofiller itself and its interactions with the polymer matrix can affect the crystalline and thermal properties of the resulting composites,

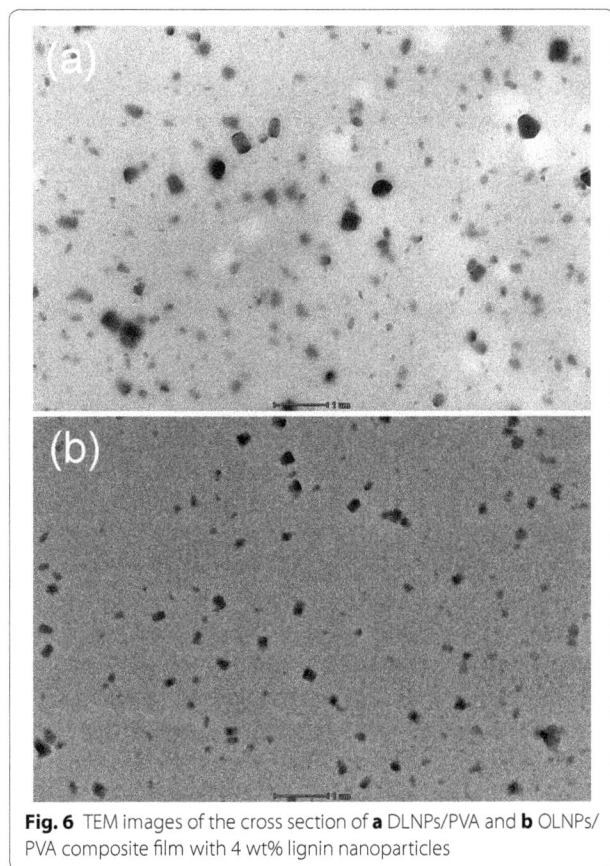

Fig. 6 TEM images of the cross section of **a** DLNPs/PVA and **b** OLNPs/PVA composite film with 4 wt% lignin nanoparticles

which also play an important role in determining the downstream processability and usability of the composites [13]. Thus, crystalline and thermal analysis were subsequently carried out with the selected sample films using prevalent techniques [differential scanning calorimetry (DSC), X-ray diffraction (XRD), and thermogravimetric analysis (TGA)] and the crucial results are summarized in Table 1 (for detail information of the results, see Additional file 1: Figures S5–S7). It was shown that the melting point (T_m) of the composite films was unchanged compared to the neat PVA film, but the degree of crystallinity (X_c) was decreased from 21.7 to 19.2% for DLNPs/PVA and to 15.6% for OLNPs/PVA, respectively. For neat PVA, the macromolecular chains regularly stacked together to form crystalline regions, where the extensively existed

intra- and intermolecular hydrogen bonding network could further enhance its crystallization [31], whereas incorporation of the amorphous lignin nanoparticles disrupted this hydrogen bonding network and new hydrogen bonds were formed between lignin nanoparticles and PVA matrix (Fig. 5), leading to less crystalline regions formed in the composite films (Table 1) [29]. In contrast to other biopolymer nanoparticles that they could facilitate the crystallization process through the nucleation effect [13], lignin nanoparticles showed limited contribution to the PVA crystal formation and growth, even though they were compatible with PVA matrix. As evidenced by XRD analysis, the crystalline structure of the composites was nearly the same as that of neat PVA film (Additional file 1: Figure S6).

TGA results show that the thermal stability of the composite films was slightly improved when lignin nanoparticles were incorporated. Neat PVA film exhibited an initial decomposition temperature (T_i) of 247 °C and a maximum decomposition temperature (T_p) of 270 °C, respectively, corresponding to the elimination of side hydroxyl groups and the partial chain-scission process, where a considerable amount of free radicals was generated [32]. It was likely that the incorporated lignin nanoparticles could trap these radicals and, therefore, retard the decomposition of the composites. As shown in Table 1, both DLNPs/PVA and OLNPs/PVA composites exhibited higher T_i and T_p. Such thermal stability improvement for other PVA-based nanocomposites has also been reported, which contain similar phenolic compound nanofillers such as melanin, polydopamine, and Kraft nanolignin particles prepared through high shear homogenization [10, 13, 33].

Phenolic substructures of the lignin nanoparticles

It was interesting that although DLNPs and OLNPs were derived from the same biomass substrate, both of which exhibited similar particle size (about 200 nm) and sphere morphology, OLNPs showed a better performance in enhancing the overall protective, mechanical, and thermal properties of the prepared composite films according to the above results. We notice that the phenolic hydroxyls within the lignin nanoparticles are the main responsible functional groups that influence the overall properties of

Table 1 Crystalline and thermal properties of neat PVA and lignin nanoparticles/PVA composite films

Sample	T_m (°C)	ΔH_m (J g^{-1})	X_c (%)	τ (nm)	T_i (°C)	T_p (°C)
Neat PVA	229	34.9	21.7	3.3	247	270
4 wt% DLNPs/PVA	229	29.7	19.2	3.3	253	274
4 wt% OLNPs/PVA	228	24.1	15.6	3.6	254	278

T_m melting point, ΔH_m the heat of fusion, X_c degree of crystallinity, τ crystal size, and T_i and T_p initial and maximum decomposition temperature, respectively

the nanocomposites, which encouraged us to further look at the phenolic substructures of these two lignin nanoparticles using emerging quantitative ^{31}P NMR technic according to previously reported procedures [20]. In this method, an internal standard and the lignin sample are suitably phosphitylated with the phosphorous reagent, and then, all the phosphorus-tagged hydroxyl groups belonging to lignin including phenolic, aliphatic, and carboxylic hydroxyls could be readily quantified by ^{31}P NMR spectroscopy. When the contents and locations of the hydroxyl groups in these two lignin nanoparticles were determined and compared (Table 2, for the ^{31}P NMR spectra, see Additional file 1: Figure S8), it was apparent that OLNPs indeed had a higher amount of total phenolic hydroxyl groups (3.37 mmol g^{-1}) compared to DLPNs (2.44 mmol g^{-1}) as expected before. Organosolv extraction employing a much aggressive extraction solvent (ethanol–water) with sulphuric acid as the catalyst tended to extensively cleavage the β–O–4′ linkages in the biomass lignin thus forms a large amount of free phenolic hydroxyl groups [23]. Meanwhile, the acid-catalyzed condensation between the aromatic active sites and the generated free radicals resulted in an increased content of syringyl hydroxyl groups (2.03 mmol g^{-1}) [2]. On the contrary, DES lignin extracted by lactic acid–betaine solvent system at milder conditions likely underwent less fragmentation and condensation; therefore, the resulting DLNPs showed less content of total phenolic hydroxyl groups and condensed aromatics (Table 2) [21]. To conclude, the higher content of total phenolic hydroxyl groups in the lignin nanoparticles not only enabled the resulting composite film with higher UV-shielding and radical-scavenging ability, but also provided stronger interactions between the lignin nanoparticles and the PVA matrix.

Conclusions

Lignin nanoparticles are good candidates for next-generation functional nanocomposites. In addition to blocking ultraviolet light and scavenging free radicals, the enriched phenolic hydroxyl groups on the shell of the

Table 2 Contents (mmol g^{-1}) and locations of hydroxyl groups in these two lignin nanoparticles as determined by quantitative ^{31}P NMR spectroscopy

Sample	DLNPs	OLNPs
Aliphatic–OH	2.40	1.04
Syringyl–OH	1.31	2.03
Guaiacyl–OH	0.90	1.20
p-hydroxyphenyl–OH	0.23	0.15
Carboxylic–OH	0.18	0.09
Total phenolic–OH	2.44	3.37

lignin nanoparticles also provide good interfacial adhesion with poly(vinyl alcohol). It is also shown that the phenolic substructures of the lignin nanoparticles, which are determined by the employed extraction/pretreatment technique routes, significantly influence the overall properties of the downstream nanocomposite films. Further conducting a hydrophobic modification on these lignin nanoparticles potentially enables them to be compatible with nonpolar polymer matrix such as polyethylene and polypropylene; therefore, their applications could be greatly extended.

Experimental
Fabrication and characterization of lignin nanoparticles

A mild steam pretreatment was conducted with raw hardwood poplar to pre-extract hemicellulose while facilitating the subsequent lignin extraction according to method previously reported [34]. The DES formulated by a certain amount of lactic acid and betaine was selected as the extraction solvent, since it showed quite high lignin solubility among all the assessed DESs as reported previously [21]. DES lignin extraction was carried out at atmospheric pressure on a hot plate equipped with a digital controller and magnetic stirring. One gram of steam pretreated poplar biomass (dry matter) and 20 g of DES were transferred to a 100 mL conical flask and heated at 130 °C for 3 h with continuous stirring. The reaction mixture was then cooled to about 80 °C, after which acetone/water mixture (50/50 by volume) was added to wash the DES and lignin away from the obtained cellulose-rich pulp by vacuum filtration. DES lignin was precipitated by evaporating of acetone from the washes and washed with distilled water three times. Organosolv lignin was obtained using ethanol/water (50/50 by wt, 1% H_2SO_4 as the catalyst) as the extraction solvent with the optimized conditions according to our previous report [2]. 100 g (on a dry matter basis) of steam pretreated poplar biomass was cooked at 170 °C for 1 h at a liquid-to-solid ratio of 7:1 using a four-vessel (2 L each) rotating digester (Aurora Products, Savona, BC, Canada). At the end of the extraction period, the vessel was cooled to room temperature in a water bath. Then, the liquid fraction was collected by vacuum filtration. Organosolv lignin was precipitated by adding ten times of hot water to the filtrate, and then, the collected lignin was washed with distilled water. Both these two obtained technical lignins were dried in a vacuum oven for further use.

The prevalent dialysis method was employed in this study to fabricate lignin nanoparticles. Briefly, 400 mg of the technical lignin was dissolved in 200 mL of dimethylsulfoxide [DMSO, its Hildebrand solubility parameter (δ value) was close to various types of lignin] to form a homogenous solution [35]. Then, the resulting lignin solution was introduced into a dialysis tube (Spectra/Por® 2 Standard RC Dry Dialysis Tubing, 12–14 kDa, Spectrum Labs, USA). The dialysis was conducted in excess of tap water (periodically replaced) and stopped until no DMSO trace was checked in the waste water. Finally, the obtained lignin nanoparticle dispersion was stored in the refrigerator (4 °C) for further characterization after adjusting its concentration to 4 mg mL^{-1} by evaporating excess water.

Scanning electron microscopy (SEM) images of the lignin nanoparticles were taken using a JEOL JSM-5600 SEM (Japan) with the freeze-dried sample. Prior to imaging, the sample was sputter-coated with Pd–Au alloy to build up the charge on the surface.

The particle size, particle size distribution, and zeta potential of the lignin nanoparticle dispersions were measured with a Malvern Zetasizer Nano-ZS90 Instrument.

Preparation and characterization of lignin nanoparticles/PVA composite films

The composite films were prepared using facile solution-cast method. First, 10 g of PVA (Product No. 563900, Sigma-Aldrich) was dissolved in 190 g of water with heating (90 °C) for 2 h to make a 5 wt% PVA solution. Then, the required amount of lignin nanoparticle dispersion and PVA solution was mixed and sonicated for 1 h. After that, the homogeneous mixture was degased and poured onto a polished glass plate. The composite films were obtained by evaporating the water from the gel-like films at room temperature and dried at 60 °C in the oven. The resulting composite films were quite uniform with an average thickness of about 30 μm.

The UV-shielding performance and optical transparency of the composite films were measured on a UV–Visible spectrophotometer. The free-radical-scavenging activity was evaluated using TEAC assay according to the method previously reported with some modifications [22, 27]. Briefly, 7.4 mg of Diphenyl-1-picrylhydrazyl (DPPH, Product No. D9132, Sigma-Aldrich) was dissolved in 100 mL of methanol to obtain an absorbance of 1.8 at 520 nm. Each composite sample (30 mg) was dissolved in 5 mL of DMSO and stirred for 3 h. Then, 1 ml of the fresh DPPH solution was mixed with 0.2 μL of the sample solution and incubated for 1 h at 30 °C. The absorption of the reacted mixture was immediately measured at 520 nm. Trolox (Product No. 238813, Sigma-Aldrich) solutions in DMSO at various concentrations (0.1–1 mmol L^{-1}) were used for calibration. The results were expressed as μmol equivalents of Trolox per gram of the composite film.

Mechanical properties were measured using Instron 5567 Universal Testing Machine at a crosshead speed of 50 mm min^{-1}. Five specimens of each sample were tested and the averaged results were presented.

Fourier transform infrared (FTIR) spectra were conducted by a Nicolet 560 spectrophotometer (USA). Transmission electron microscopy (TEM) was performed using a transmission electron microscope (JEOL JEM-100CX, Japan). The sample was prepared by ultrathin section before imaging. Differential scanning calorimetry (DSC) analysis was conducted on a NETZSH 204 DSC differential scanning calorimeter under a flowing N$_2$ with the heating scan from ambient temperature to 250 °C at a heating rate of 10 °C min^{-1}. The degree of crystallinity of PVA was calculated based on the following equation [31]:

$$X_c = \frac{\Delta H_m}{w \Delta H_{mo}}$$

where w is the weight fraction of PVA matrix in the composite film, ΔH_m is the heat of fusion of the composite, and ΔH_{mo} is the heat of fusion of 100% crystalline PVA (161 J g^{-1}). X-ray diffraction (XRD) patterns were collected on a Philips Analytical X'Pert X-diffractometer (Philips Co., Netherlands), using Cu–Ka radiation ($\lambda = 0.1540$ nm) at an accelerating voltage of 40 kV and the current of 40 mA. The data were collected from $2\theta = 5$–$60°$ with a step interval of $0.03°$. The crystallite size was calculated using the Scherrer equation [31]:

$$\tau = \frac{K\lambda}{\beta \cos\theta}$$

where K is a constant (0.94), β is the full-width at half-maximum in radians and θ is the position of the peak (half of the plotted 2θ value).

Thermal stability was measured on a TG209 F1 instrument (NETZSCH Co., Germany). About 5–8 mg of the composite sample was heated in a platinum crucible from room temperature to 600 °C at a heating rate of 20 °C min^{-1} under nitrogen atmosphere.

Determine the content of hydroxyl groups of lignin nanoparticles using ^{31}P NMR spectroscopy

^{31}P NMR analysis was performed following the reported procedure [27]. An accurately weighed amount of lignin (20 mg) was dissolved in 500 μL of an anhydrous pyridine and deuterated chloroform mixture (1.6:1, v/v) with stirring. The anhydrous pyridine was purified by Soxlet's extraction and dewatered with molecular sieve prior to use. 100 μL of cyclohexanol (10.85 mg mL^{-1} in anhydrous pyridine and deuterated chloroform 1.6:1, v/v) and 100 μL of chromium(III) acetylacetonate solution (5 mg mL^{-1} in anhydrous pyridine and deuterated chloroform 1.6:1, v/v) were further added as an internal standard and relaxation reagent, respectively. The mixture was reacted with 100 μL of phosphitylating reagent (2-chloro-4,4,5,5-tetramethyl-1,3,2-dioxaphospholane, TMDP, Product No. 447536, Sigma-Aldrich) and transferred into a 5 mm

NMR tube. The NMR spectra of freshly prepared samples were acquired immediately at room temperature on a Bruker AV II 600 MHz spectrometer equipped with a QNP cryoprobe. Chemical shifts were calibrated relative to the phosphitylation product of TMDP with water (sample moisture), which gave a sharp and stable signal at 132.2 ppm.

Additional file

Additional file 1: Figure S1. Material balance of each crucial step during the two-stage fractionation pretreatment approach. The tailored two-stage pretreatment could greatly enhance the enzymatic hydrolyzability of cellulose fraction while producing a usable lignin fraction for further valorization. **Figure S2.** AFM images of (a) and (b) DLNPs and (c) and (d) OLNPs. **Figure S3.** High-resolution TEM images of (a) DLNPs and (b) OLNPs. **Figure S4.** Translation of the UV–Visible transmittance spectra into Tauc's plots to calculate the optical energy bandgap (E_g) of each nanocomposite film (a) DLNPs/PVA, (b) OLNPs/PVA. **Figure S5.** Differential scanning calorimetry (DSC) curves of heating scans for neat PVA and 4 wt% lignin nanoparticles/PVA composite films. **Figure S6.** X-ray diffraction (XRD) patterns of neat PVA and 4 wt% lignin nanoparticles/PVA composite films. **Figure S7.** Thermal gravity (TG) and Differential thermal gravity (DTG) curves of neat PVA and 4 wt% lignin nanoparticles/PVA composite films. **Figure S8.** Quantitative ^{31}P NMR spectra of these two lignin nanoparticles tagged with the phosphorous reagent using cyclohexanol as internal standard

Abbreviations

DES: deep eutectic solvent; LNPs: lignin nanoparticles; DLNPs: deep eutectic solvent lignin nanoparticles; OLNPs: organosolv lignin nanoparticles; UV: ultraviolet light; PVA: poly(vinyl alcohol); DPPH: 1,1-diphenyl-2-picrylhydrazyl; TEAC: trolox equivalent antioxidant capacity; XRD: X-ray diffraction patterns; CrI: crystallinity index; FTIR: Fourier transform infrared spectroscopy; TEM: transmission electron microscopy; SEM: scanning electron microscopy; DMSO: dimethylsulfoxide; DSC: differential scanning calorimetry; TGA: thermogravimetric analysis; HPLC: high-performance liquid chromatography

Authors' contributions

All authors (DT, JH, JB, RPC, JNS, and CL) contributed jointly to all aspects of the work reported in the manuscript. DT performed most of the experiments and data analysis at Sichuan University. All authors read and approved the final manuscript.

Author details

[1] State Key Laboratory of Polymer Materials Engineering, Polymer Research Institute of Sichuan University, Chengdu 610065, China. [2] State Key Laboratory of Bioreactor Engineering, East China University of Science and Technology, 130 Meilong Road, Shanghai 200237, China. [3] Forest Products Biotechnology/Bioenergy Group, Department of Wood Science, Faculty of Forestry, University of British Columbia, 2424 Main Mall, Vancouver, BC V6T 1Z4, Canada.

Acknowledgements

The authors thank Dr. Andrew Lewis at Simon Fraser University (British Columbia, Canada) for the technical help of NMR tests and the Analytical and Testing Center at Sichuan University for imaging tests.

Competing interests

The authors declare that they have no competing interests.

Funding

This work was supported by Chinese Scholarship Council (Grant No. 201406240173); Natural Science Foundation of China (No. 51473100); State Key Laboratory of Polymer Materials Engineering (Grant No. sklpme2015-2-02), and Open Funding Project of the State Key Laboratory of Bioreactor Engineering.

References

1. Zhang YHP. Reviving the carbohydrate economy via multi-product ligno-cellulose biorefineries. J Ind Microbiol Biotechnol. 2008;35:367–75.
2. Panagiotopoulos IA, Chandra RP, Saddler JN. A two-stage pretreatment approach to maximise sugar yield and enhance reactive lignin recovery from poplar wood chips. Bioresour Technol. 2013;130:570–7.
3. Alvira P, Tomás-Pejó E, Ballesteros M, Negro MJ. Pretreatment technologies for an efficient bioethanol production process based on enzymatic hydrolysis: a review. Bioresour Technol. 2010;101:4851–61.
4. Bruijnincx PCA, Rinaldi R, Weckhuysen BM. Unlocking the potential of a sleeping giant: lignins as sustainable raw materials for renewable fuels, chemicals and materials. Green Chem. 2015;17:4860–1.
5. Liu W-J, Jiang H, Yu H-Q. Thermochemical conversion of lignin to functional materials: a review and future directions. Green Chem. 2015;17:4888–907.
6. Qian Y, Deng Y, Qiu X, Li H, Yang D. Formation of uniform colloidal spheres from lignin, a renewable resource recovered from pulping spent liquor. Green Chem. 2014;16:2156–63.
7. Lievonen M, Valle-Delgado JJ, Mattinen M-L, Hult E-L, Lintinen K, Kostiainen MA, et al. Simple process for lignin nanoparticle preparation. Green Chem. 2016;18:1416–22.
8. Zhao W, Simmons B, Singh S, Ragauskas A, Cheng G. From lignin association to nano-/micro-particle preparation: extracting higher value of lignin. Green Chem. 2016;18:5693–700.
9. Tian D, Hu J, Chandra RP, Saddler JN, Lu C. Valorizing recalcitrant cellulolytic enzyme lignin via lignin nanoparticles fabrication in an integrated biorefinery. ACS Sustain Chem Eng. 2017;5:2702–10.
10. Nair SS, Sharma S, Pu Y, Sun Q, Pan S, Zhu JY, et al. High shear homogenization of lignin to nanolignin and thermal stability of nanolignin-polyvinyl alcohol blends. ChemSusChem. 2014;7:3513–20.
11. Yang W, Owczarek JS, Fortunati E, Kozanecki M, Mazzaglia A, Balestra GM, et al. Antioxidant and antibacterial lignin nanoparticles in polyvinyl alcohol/chitosan films for active packaging. Ind Crops Prod. 2016;94:800–11.
12. Markarian J. Biopolymers present new market opportunities for additives in packaging. Plast Addit Compd. 2008;10:22–5.
13. Xiong SQ, Wang Y, Yu JR, Chen L, Zhu J, Hu ZM. Polydopamine particles for next-generation multifunctional biocomposites. J Mater Chem A. 2014;2:7578–87.
14. Ren J, Wang S, Gao C, Chen X, Li W, Peng F. TiO2-containing PVA/xylan composite films with enhanced mechanical properties, high hydrophobicity and UV shielding performance. Cellulose. 2015;22:593–602.
15. Wang Y, Li T, Ma P, Bai H, Xie Y, Chen M, et al. Simultaneous enhancements of UV-shielding properties and photostability of poly(vinyl alcohol) via incorporation of sepia eumelanin. ACS Sustain Chem Eng. 2016;4:2252–8.
16. Tu Y, Zhou L, Jin YZ, Gao C, Ye ZZ, Yang YF, et al. Transparent and flexible thin films of ZnO-polystyrene nanocomposite for UV-shielding applications. J Mater Chem. 2010;20:1594–9.
17. Han C, Wang F, Gao C, Liu P, Ding Y, Zhang S, et al. Transparent epoxy–ZnO/CdS nanocomposites with tunable UV and blue light-shielding capabilities. J Mater Chem C. 2015;3:5065–72.
18. Kai D, Tan MJ, Chee PL, Chua YK, Yap YL, Loh XJ. Towards lignin-based functional materials in a sustainable world. Green Chem. 2016;18:1175–200.
19. Procentese A, Johnson E, Orr V, Garruto Campanile A, Wood JA, Marzocchella A, et al. Deep eutectic solvent pretreatment and subsequent saccharification of corncob. Bioresour Technol. 2015;192:31–6.
20. Pu Y, Cao S, Ragauskas AJ. Application of quantitative 31P NMR in biomass lignin and biofuel precursors characterization. Energy Environ Sci. 2011;4:3154–66.
21. Francisco M, van den Bruinhorst A, Kroon MC. New natural and renewable low transition temperature mixtures (LTTMs): screening as solvents for lignocellulosic biomass processing. Green Chem. 2012;14:2153–7.
22. Delgado-Andrade C, Rufián-Henares JA, Morales FJ. Assessing the antioxidant activity of melanoidins from coffee brews by different antioxidant methods. J Agric Food Chem. 2005;53:7832–6.
23. Pan X, Kadla JF, Ehara K, Gilkes N, Saddler JN. Organosolv ethanol lignin from hybrid poplar as a radical scavenger: relationship between lignin structure, extraction conditions, and antioxidant activity. J Agric Food Chem. 2006;54:5806–13.

24. Ravindrachary V, Nayak SP, Dutta D, Pujari PK. Free volume related fluorescent behavior in electron beam irradiated chalcone doped PVA. Polym Degrad Stab. 2011;96:1676–86.

25. Wang Y, Wang Z, Ma P, Bai H, Dong W, Xie Y, et al. Strong nanocomposite reinforcement effects in poly (vinyl alcohol) with melanin nanoparticles. RSC Adv. 2015;5:72691–8.

26. Xie S, Zhao J, Zhang B, Wang Z, Ma H, Yu C, et al. Graphene oxide transparent hybrid film and its ultraviolet shielding property. ACS Appl Mater Interfaces. 2015;7:17558–64.

27. Sun SL, Wen JL, Ma MG, Sun RC, Jones GL. Structural features and antioxidant activities of lignins from steam-exploded bamboo (*Phyllostachys pubescens*). J Agric Food Chem. 2014;62:5939–47.

28. Sadeghifar H, Argyropoulos DS. Correlations of the antioxidant properties of softwood kraft lignin fractions with the thermal stability of its blends with polyethylene. ACS Sustain Chem Eng. 2015;3:349–56.

29. Kubo S, Kadla JF. The formation of strong intermolecular interactions in immiscible blends of poly(vinyl alcohol) (PVA) and lignin. Biomacromolecules. 2003;4:561–7.

30. Ali ZI, Ali FA, Hosam AM. Effect of electron beam irradiation on the structural properties of PVA/V2O5 xerogel. Spectrochim Acta-Part A Mol Biomol Spectrosc. 2009;72:868–75.

31. Sun X, Lu C, Liu Y, Zhang W, Zhang X. Melt-processed poly(vinyl alcohol) composites filled with microcrystalline cellulose from waste cotton fabrics. Carbohydr Polym. 2014;101:642–9.

32. Pandey S, Pandey SK, Parashar V, Mehrotra GK, Pandey AC. Ag/PVA nanocomposites: optical and thermal dimensions. J Mater Chem. 2011;21:17154–9.

33. Dong W, Wang Y, Huang C, Xiang S, Ma P, Ni Z, et al. Enhanced thermal stability of poly(vinyl alcohol) in presence of melanin. J Therm Anal Calorim. 2014;115:1661–8.

34. Hu J, Pribowo A, Saddler J. Oxidative cleavage of some cellulosic substrates by auxiliary activity (AA) Family 9 enzymes influences the adsorption/desorption of hydrolytic cellulase enzymes. Green Chem. 2016;18:6329–36.

35. Sannigrahi P, Ragauskas AJ. Fundamentals of biomass pretreatment by fractionation. In: Wyman CE, editor. Aqueous pretreatment of plant biomass for biological and chemical conversion to fuels and chemicals. London: Wiley; 2013. p. 201–22.

3

Engineering *Shewanella oneidensis* enables xylose-fed microbial fuel cell

Feng Li[1,2] [iD], Yuanxiu Li[1,2], Liming Sun[3], Xiaofei Li[1,2], Changji Yin[1,2], Xingjuan An[1,2], Xiaoli Chen[1,2], Yao Tian[1,2] and Hao Song[1,2]*

Abstract

Background: The microbial fuel cell (MFC) is a green and sustainable technology for electricity energy harvest from biomass, in which exoelectrogens use metabolism and extracellular electron transfer pathways for the conversion of chemical energy into electricity. However, *Shewanella oneidensis* MR-1, one of the most well-known exoelectrogens, could not use xylose (a key pentose derived from hydrolysis of lignocellulosic biomass) for cell growth and power generation, which limited greatly its practical applications.

Results: Herein, to enable *S. oneidensis* to directly utilize xylose as the sole carbon source for bioelectricity production in MFCs, we used synthetic biology strategies to successfully construct four genetically engineered *S. oneidensis* (namely XE, GE, XS, and GS) by assembling one of the xylose transporters (from *Candida intermedia* and *Clostridium acetobutylicum*) with one of intracellular xylose metabolic pathways (the isomerase pathway from *Escherichia coli* and the oxidoreductase pathway from *Scheffersomyces stipites*), respectively. We found that among these engineered *S. oneidensis* strains, the strain GS (i.e. harbouring *Gxf1* gene encoding the xylose facilitator from *C. intermedi*, and *XYL1*, *XYL2*, and *XKS1* genes encoding the xylose oxidoreductase pathway from *S. stipites*) was able to generate the highest power density, enabling a maximum electricity power density of 2.1 ± 0.1 mW/m^2.

Conclusion: To the best of our knowledge, this was the first report on the rationally designed *Shewanella* that could use xylose as the sole carbon source and electron donor to produce electricity. The synthetic biology strategies developed in this study could be further extended to rationally engineer other exoelectrogens for lignocellulosic biomass utilization to generate electricity power.

Keywords: Microbial fuel cell, Synthetic biology, Xylose, *Shewanella oneidensis* MR-1

Background

Bio-electrochemical systems enabled many practical applications in environments and energy fields [1–7], including microbial fuel cell (MFC) for simultaneous organic wastes treatment and electricity harvest [8–12], microbial electrolysis cells for hydrogen production [13–16], and microbial electrosynthesis for production of valuable chemicals from CO_2 bioreduction [17–22]. Many mono-, di-saccharides as well as complex carbohydrates like starch and organics in wastewater and marine sediment have been used in MFCs for the production of electricity [8, 23, 24]. Xylose, one of primary ingredients from hydrolysis of lignocellulosic biomass, is the second most abundant carbohydrate after glucose in nature [25–27]. Conversion of xylose to electricity energy using MFC would thus provide a sustainable and green energy, which received increased attention in recent few years [24, 28–30]. However, xylose is hard to be effectively utilized by many microorganisms due to slow utilization rate and inefficient metabolic pathways of xylose [26, 31–35].

Shewanella oneidensis, one of the most well established metal-reducing exoelectrogens [36, 37], is capable of conducting extracellular electrons transfer (EET) through its metal-reducing (Mtr) pathway [38–42], being extensively studied for the optimization of MFC performance

*Correspondence: hsong@tju.edu.cn
[1] Key Laboratory of Systems Bioengineering (Ministry of Education), School of Chemical Engineering and Technology, Tianjin University, Tianjin 300072, China
Full list of author information is available at the end of the article

[40, 41, 43–47], MFC-based logic gate [48–50], bioremediation of toxic metals [51], etc., in recent decade. However, the wild-type (WT) *S. oneidensis* could only use three- (or two-) carbon substrates (e.g. lactate, pyruvate and acetate) as their carbon and energy sources, with an exception of N-acetyl-glucosamine (NAG) as a high-carbon carbohydrate [45, 52, 53], while common pentoses or hexoses (e.g. xylose and glucose), the most abundant composition of biomass, could not be utilized by the WT *S. oneidensis* owing to its incomplete sugar utilization pathways [36, 54, 55]. Such defect enormously restricted the wide applications of *S. oneidensis*.

Recently, several strategies were developed to use xylose for electricity generation in *Shewanella*-inoculated MFCs. Firstly, an adaptive evolution approach was developed to activate an otherwise silent xylose metabolic pathway, i.e. oxidoreductase pathway in the WT *S. oneidensis*, thus generating a *S. oneidensis* mutant XM1 that could metabolize xylose as the sole carbon and energy source [56]. Secondly, microbial consortia including fermenters and exoelectrogens were developed to accomplish xylose-powered MFCs, in which the engineered *Escherichia coli* played as a fermenter to metabolize xylose for the synthesis of metabolites such as lactate and formate to feed the *S. oneidensis* as the carbon source and electron donor, thus enabling an indirect utilization of xylose by *S. oneidensis* for bioelectricity production [24].

Herein, we used synthetic biology strategy to rationally engineer *S. oneidensis* that could use xylose as the sole carbon source and electron donor for electricity generation in MFCs. To enable *S. oneidensis* to be able to use xylose, the xylose transporters (i.e. glucose/xylose facilitator encoded by gene *Gxf1* from *Candida intermedia* [57, 58] and D-xylose-proton symporter encoded by gene *xylT* from *Clostridium acetobutylicum* [59]), synthetic isomerase pathway (including the genes *xylA* and *xylB* from *E. coli* [60]), and oxidoreductase (including from *Scheffersomyces stipites* [61]) pathway for xylose metabolism were heterologously expressed in *S. oneidensis* in a combinatorial way. Thus, four recombinant *S. oneidensis* strains were synthesized (see Fig. 1). Xylose-fed MFCs experiments proved that these engineered *S. oneidensis* MR-1 strains were conferred with the ability of utilizing xylose to produce electricity, and the engineered *S. oneidensis* strain GS provided the highest electricity generation. Compared with the *S. oneidensis* strain XM1 previously evolved by an adaptive evolution strategy [56], our rationally engineered *S. oneidensis* strains GS and XS (bearing the oxidoreductase pathway from *S. stipites*) showed a higher xylose consumption rate and a superior growth rate. In addition, the relative higher electricity generation by the GS strain than other engineered strains can be attributed to the higher intracellular riboflavin level and reducing equivalents in

the GS. To the best of our knowledge, this was the first report on the rationally designed *Shewanella* that gained the expanded metabolic capability of using xylose as sole carbon source and electron donor to produce electricity.

Results and discussion
Engineered xylose-utilizing *S. oneidensis* strain via synthetic biology strategies

A few xylose metabolic pathways in microorganisms were found, including the oxidoreductase, isomerase, and Weimberg–Dahms pathways [56]. For example, *E. coli* is a robust and well-studied xylose scavenger [56, 62], which could metabolize xylose by the isomerase pathway; however, *S. stipites* [56, 61] could utilize oxidoreductase pathway for the metabolism of xylose. In the xylose isomerase pathway of *E. coli* [56, 60, 63], xylose isomerase encoded by the gene *xylA* converts xylose to xylulose, which is then phosphorylated by xylulokinase encoded by the gene *xylB* to xylulose 5-phosphate (X-5-P), and then enters the pentose phosphate pathway (see Fig. 1). In the oxidoreductase pathway of *S. stipites* [26], NAD(P)H-dependent xylose reductase encoded by the gene *XYL1* converts intracellular xylose to xylitol, which is then oxidized to xylulose by xylitol dehydrogenase (XDH) encoded by the gene *XYL2*. Xylulose is then phosphorylated by xylulokinase encoded by the gene *XKS1* to xylulose 5-phosphate (X-5-P), which enters the pentose phosphate pathway, similar to the isomerase pathway (see Fig. 1).

To facilitate convenient and fast multigene assembly in *S. oneidensis*, a Biobrick compatible vector named pYYDT including an IPTG-inducible promoter PlacIq-lacIq-Ptac was well developed in our laboratory (Additional file 1: Figure S1B) [64]. Furthermore, to avoid the codon usage bias and prevent blocked translation due to shortage of tRNAs for rare codons between *S. oneidensis* and other bacteria, in vitro chemical synthesis of codon-optimized genes instead of direct cloning from other bacteria was used. The xylose metabolic pathway was then assembled by several routines of Biobrick ligation steps of the relevant genes. With the combinations of the two xylose transporters and the two xylose-utilizing metabolic pathways (the isomerase and the oxidoreductase pathways), four recombinant *S. oneidensis* strains harbouring engineered gene assembly (plasmid) for enhanced xylose transport and metabolism were synthesized, respectively, which were XE (including the gene *xylT* for xylose symporter, *xylA* and *xylB* for the xylose isomerase pathway), GE (including *Gxf1* for the xylose facilitator, *xylA* and *xylB* for the xylose isomerase pathway), XS (including *xylT* for xylose symporter, and *XYL1*, *XYL2*, and *XKS1* for the xylose oxidoreductase pathway), and GS (including *Gxf1* for the xylose facilitator, and *XYL1*, *XYL2*, and *XKS1* for the xylose oxidoreductase pathway).

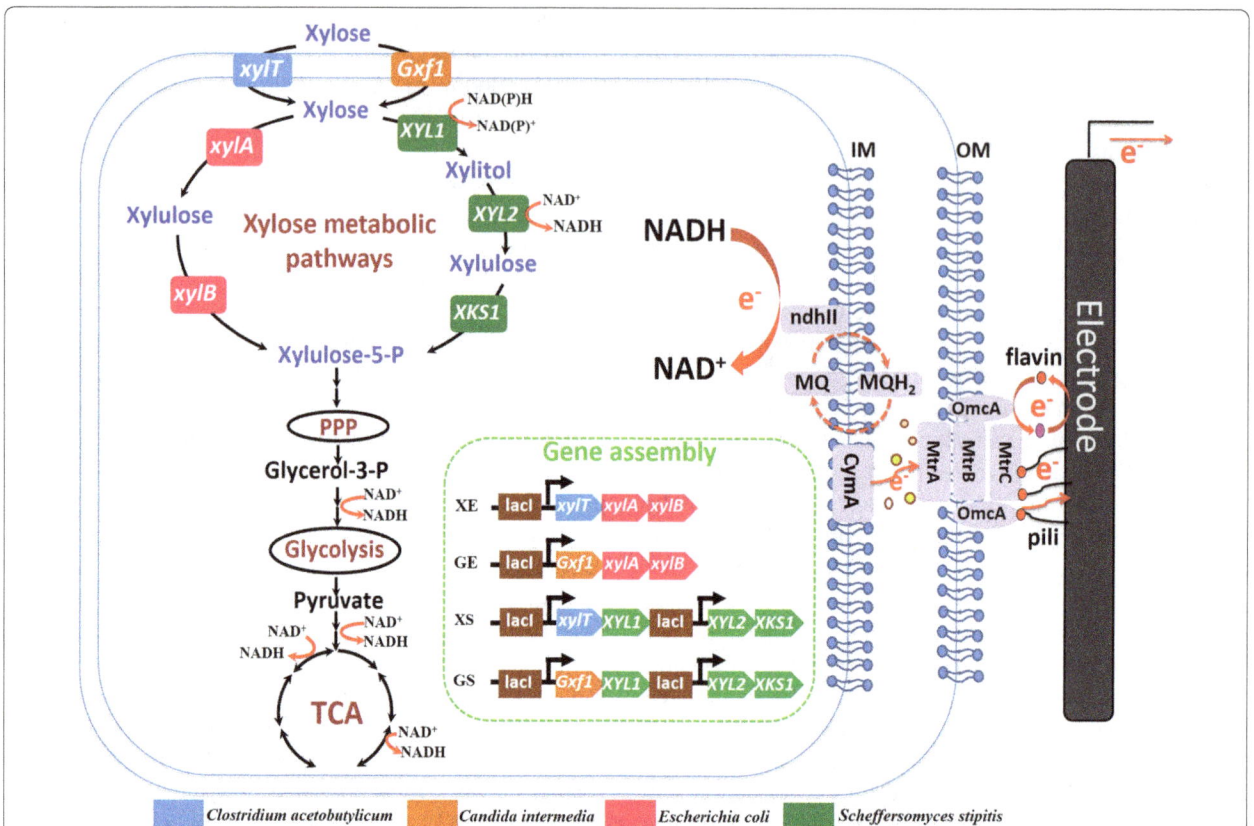

Fig. 1 Synthetic biology strategies for the construction of four recombinant *S. oneidensis* strains (namely XE, GE, XS, and GS) to enable xylose utilization and electricity generation of *S. oneidensis*. Xylose transporter genes included *xylT* (the gene encoding D-xylose-proton symporter) from *Clostridium acetobutylicum* and *Gxf1* (the gene encoding glucose/xylose facilitator 1) from *Candida intermedia*. The xylose isomerase pathway included *xylA* (the gene encoding xylose isomerase) and *xylB* (the gene encoding xylulokinase) from *E. coli*. The oxidoreductase pathway included *XYL1* (the gene encoding D-xylose reductase), *XYL2* (the gene encoding xylitol dehydrogenase), and *XKS1* (the gene encoding D-xylulokinase) from *Scheffersomyces stipites*. Four gene assemblies (plasmids), namely XE, GE, XS, and GS (as shown in the *green-dash square*) were synthesized for the enhanced xylose transport and metabolism, which transformed into *S. oneidensis*, respectively, to construct four recombinant *S. oneidensis* strains

Evaluation of xylose utilization and cell growth of the recombinant *S. oneidensis*

The cell growth and xylose consumption by the wild-type (WT, harbouring the pYYDT empty vector) and four genetically engineered *S. oneidensis* strains (i.e. harbouring XE, GE, XS, and GS, respectively) were evaluated in SBM supplemented with 5 mM xylose as the sole carbon source.

Under aerobic conditions, the WT *S. oneidensis* strain showed almost no growth and xylose consumption, while the four engineered *S. oneidensis* strains showed a superior growth over the WT *S. oneidensis* strain. In addition, the growth rate of the engineered strains XS and GS (harbouring the oxidoreductase pathway) was faster than that of the strains XE and GE (harbouring the isomerase pathway) (Fig. 2a). The engineered strain XS and GS consumed xylose at a rate of ~28.1 and ~35.2 μM/h, which was faster than that of the engineered strains XE and GE (~11.2 and ~20.3 μM/h) (Fig. 2b). Thus, the rate of xylose

consumption of these engineered strains was in good agreement with that of the growth rate, respectively.

The anaerobic respiratory capabilities of the WT and the recombinant *S. oneidensis* were also determined under anaerobic conditions with xylose as the sole electron donor and fumarate as the electron acceptor. Similar to the aerobic conditions, the four genetically engineered strains grew faster than that of the WT strain. The recombinant strains XS and GS (harbouring the oxidoreductase pathway) consumed xylose at a rate of ~14.8, and ~17.2 μM/h, respectively, which had a faster xylose consumption rate than that of the strains XE and GE (harbouring the isomerase pathway, ~6.3 and ~9.7 μM/h, respectively) (Fig. 2c, d). Furthermore, the engineered strain GS could intake xylose faster than XS, which indicated that the glucose/xylose facilitator Gxf1 enabled a higher xylose transportation than that of the D-xylose-proton symporter XylT. It was revealed that sugar uptake via facilitated diffusion by Gxf1 required less energy

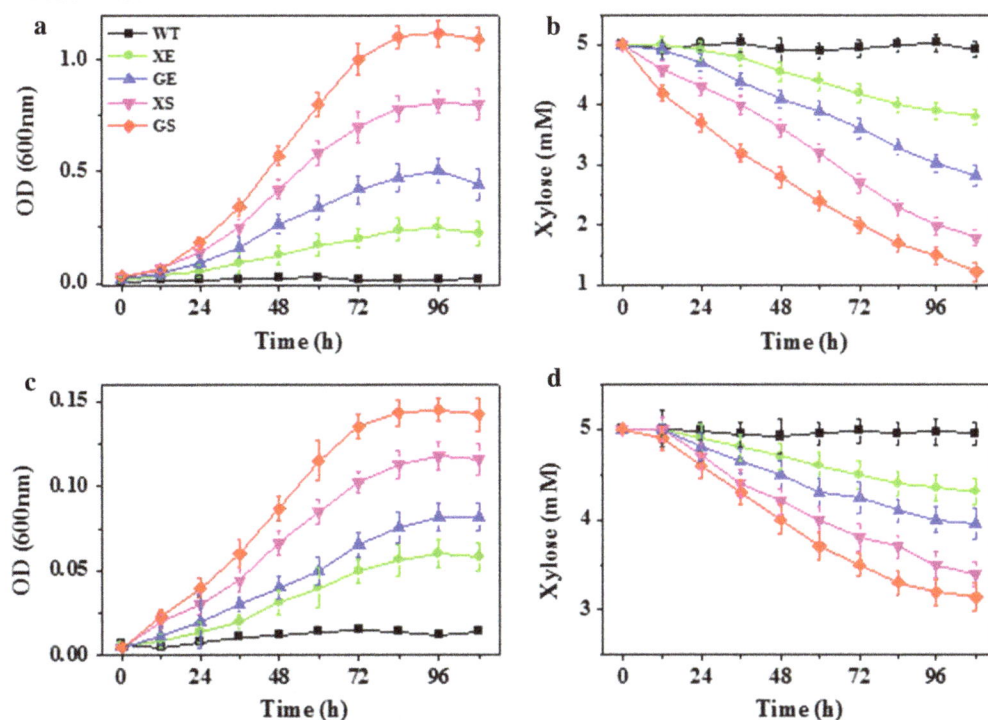

Fig. 2 Growth curves and xylose consumption of the WT and the recombinant *S. oneidensis* strains. **a** Aerobic growth curve (OD_{600} ~ t) in SBM supplemented with 5 mM xylose. **b** Xylose consumption under aerobic conditions. **c** Anaerobic growth curve (OD_{600} ~ t) in SBM supplemented with 5 mM xylose. **d** Xylose consumption under anaerobic conditions. The *error bars* were calculated from triplicate experiments

(ATP) than proton symport XylT, and thus the facilitator protein would probably be more efficient with higher substrate affinity under oxygen-limited or anaerobic conditions where ATP production is restricted in our MFC conditions [58, 65]. In addition, all recombinant *S. oneidensis* strains were able to utilize lactate as the sole carbon source at a rate similar to that of the WT strain, suggesting that the lactate metabolism of *S. oneidensis* was not altered by such engineering efforts (data not shown).

Thus, our results indicated that the introduction of one of the synthetic xylose transporters (the D-xylose-proton symporter from *C. acetobutylicum* and the glucose/xylose facilitator from *C. intermedia*) and one of the metabolic pathways (i.e. the isomerase pathway from *E. coli* and the oxidoreductase pathways from *S. stipites*) could successfully confer *Shewanella* strains with the ability of utilizing xylose as the sole carbon source for the cell growth. Especially, our rationally designed *S. oneidensis* strains XS and GS (bearing the oxidoreductase pathway from *S. stipites*) showed a higher consumption of xylose and a superior growth rate than that of the *S. oneidensis* strain XM1 (that was recently developed through an adaptive evolution strategy) [56]. *Escherichia coli* (the BL21 strain) harbouring those genes related to xylose transport and metabolism exhibited a superior xylose consumption rate

(~455 μM/h), i.e. ~12 times faster than that of the engineered *S. oneidensis* GS (~35.2 μM/h) (Additional file 1: Figure S2). This result indicated that although the engineered *S. oneidensis* was enabled the capability of xylose utilization, there was much room to further improve its xylose consumption rate by synthetic biology endeavours.

MFC performance and bio-electrochemical analyses

MFC was used to examine the extracellular electron transfer and power generation by the engineered *S. oneidensis* MR-1 using xylose as the sole carbon source. The WT and the engineered *S. oneidensis* strains were inoculated into the anodic chamber of MFCs, respectively, with a 2 kΩ external resistor, across which the voltage output was recorded.

Initially, 18 mM lactate was used (as the favourable carbon source of *Shewanella*) to feed the engineered *S. oneidensis* strains in MFCs to verify the capacity of power output of each strain (Fig. 3). After the output voltage decreased to baseline levels (indicating the depletion of lactate), 18 mM xylose was added into the anodic chamber as the carbon source. Obviously, the output voltages of these engineered *S. oneidensis* strains with xylose as the carbon source were lower than those of lactate as the carbon source, because lactate is the favourable carbon source

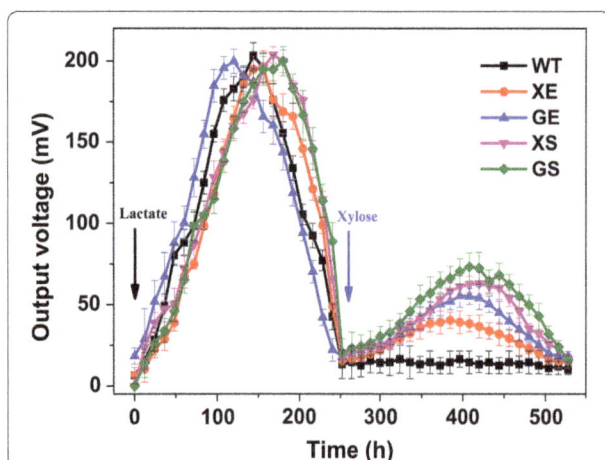

Fig. 3 Output voltage of the WT and four recombinant *S. oneidensis* strains (XE, GE, XS, and GS) with different carbon sources in the anodic chamber of MFCs. The output voltage of the four strains XE, GE, XS, and GS, using lactate and xylose as the carbon source, respectively. 18 mM lactate (the favourable carbon source of *Shewanella*) was added at the initiation of MFC operations (as indicated by the *black arrow*). Upon the depletion of lactate and vanishing of electricity output, 18 mM xylose was added at ~260 h as the carbon source (as indicated by the *blue arrow*) to illustrate the power generation capability of these *S. oneidensis* strains using xylose as the sole carbon source. The *error bars* were calculated from triplicate experiments

for the growth and respiration of *Shewanella* (Fig. 3). When lactate was used as the carbon source, the maximum output voltages could increase to ~205 ± 7.2 mV ($n = 3$) for both the WT and engineered *S. oneidensis* strains. However, the WT *S. oneidensis* strain could barely generate any voltage output when xylose was used as the carbon source, which indicated that the WT *S. oneidensis* could not utilize xylose. Upon genetic programming of the xylose transporter and metabolic pathway into *S. oneidensis*, the recombinant *S. oneidensis* strains, namely XE, GE, XS, and GS, could generate a maximum output voltage of ~40.5 ± 5.1, ~55.5 ± 4.8, ~63.2 ± 6.2, and ~73.4 ± 5.8 mV ($n = 3$), respectively. Furthermore, the multiple cycles of voltage output of these genetically engineered *S. oneidensis* strains showed the stability of power generation in the semi-batch xylose-fed MFCs (Fig. 4a).

We observed that the strains GS and XS harbouring the synthetic oxidoreductase xylose metabolic pathway could generate a higher voltage output than those of the XE and GE harbouring the synthetic isomerase xylose metabolic pathway (Fig. 4a). Bio-electrochemical analyses were further conducted to study the EET efficiency of these rationally engineered strains in MFCs. The cyclic voltammetry (CV) at 1 mV/s was applied to reveal the redox reaction kinetics at the interfaces of bacterial cells and anodes. As shown in Fig. 4b, there were typical redox peaks of flavins in the CV curves starting

from around −0.4 V (vs. Ag/AgCl), which showed that flavins-mediated extracellular electron transfer was the dominating mechanism for bioelectricity production in these strains [64, 66]. The power output curves (output voltage vs. current density) and the polarization curves (power density vs. current density), which were obtained by varying load resistances to show the dependence of voltage and power on the current, helped to further investigate the bioelectricity generation capability of the engineered *S. oneidensis* strains (Fig. 4c). Notably, the dropping slope of the polarization curve obtained from the engineered *S. oneidensis* strain GS (harbouring the xylose facilitator and the xylose oxidoreductase pathway) was smaller than those obtained from the other three engineered *S. oneidensis* stains (i.e. XE, GE, XS), implying that the internal charge transfer resistance of the MFC inoculated with GS was relatively smaller (Fig. 4c). The power density were calculated, which showed that the engineered *S. oneidensis* strain GS obtained a maximum power density of ~2.1 ± 0.1 mW/m^2 ($n = 3$), which was ~0.3, ~0.9, ~1.1 times higher than that of XE, GE, and XS, respectively (Fig. 4c). Previous xylose-fed MFCs generally used sludge, natural or synthetic microbial consortia, the power generation of which were in the range of 6.3–2330 mW/m^2 (as shown in Additional file 2: Table S1), higher than that of our recombinant *S. oneidensis* strain. Thus, future engineering of *Shewanella oneidensis* to enable higher output electricity remained of paramount importance.

Biochemical characterizations showed that the engineered strains GS and XS had a higher utilization efficiency of xylose and higher growth rate, and a more efficient formation of biofilm attached on the anodes (Fig. 5a). Meanwhile, the engineered strains GS and XS could also generate higher intracellular reducing equivalents (i.e. NADH/NAD$^+$, Fig. 5b). Such a high intracellular releasable electron pool (i.e. NADH) had resulted from the oxidative reaction of xylitol to xylulose, mediated by the reduction of NAD$^+$ to NADH in the oxidoreductase pathway [65, 67–69]. Both the efficient biofilm formation on the anodes [24, 70] and higher intracellular reducing equivalents [71, 72] in the engineered *S. oneidensis* strains GS and XS synergistically enabled an enhanced EET efficiency and electricity generation. In addition, an increase in the secretion of riboflavin in the recombinant strains also enabled an increase in the output voltage of MFCs. The increased biosynthesis of riboflavin would be attributed to the biosynthesis of xylulose 5-phosphate (X-5-P) owing to the heterologously introduced xylose metabolism pathway (i.e. the oxidoreductase pathway). X-5-P, as a metabolic product of the oxidoreductase pathway, was converted to ribulose-5-P, a crucial precursor for the biosynthesis of riboflavin,

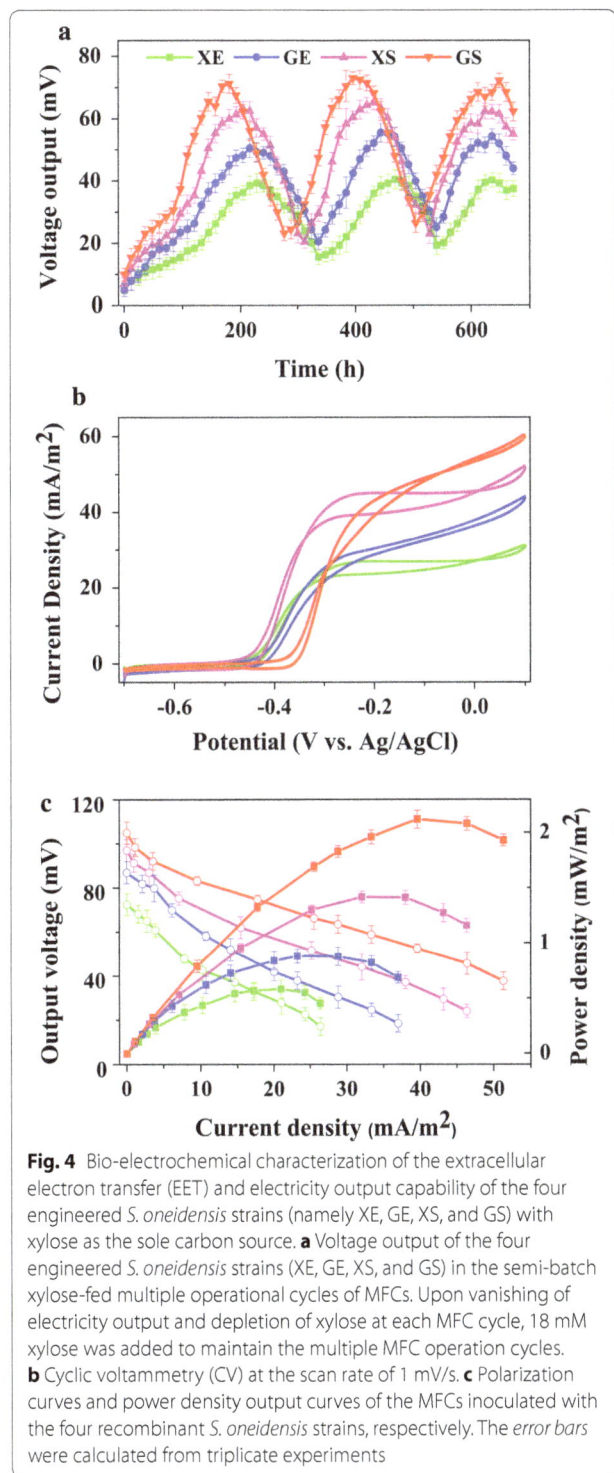

Fig. 4 Bio-electrochemical characterization of the extracellular electron transfer (EET) and electricity output capability of the four engineered *S. oneidensis* strains (namely XE, GE, XS, and GS) with xylose as the sole carbon source. **a** Voltage output of the four engineered *S. oneidensis* strains (XE, GE, XS, and GS) in the semi-batch xylose-fed multiple operational cycles of MFCs. Upon vanishing of electricity output and depletion of xylose at each MFC cycle, 18 mM xylose was added to maintain the multiple MFC operation cycles. **b** Cyclic voltammetry (CV) at the scan rate of 1 mV/s. **c** Polarization curves and power density output curves of the MFCs inoculated with the four recombinant *S. oneidensis* strains, respectively. The *error bars* were calculated from triplicate experiments

Conclusions

To the best of our knowledge, this research is the first to use synthetic biology strategy to rationally engineer *S. oneidensis* MR-1 to enable direct utilization of xylose as the sole carbon source and electron donor for bioelectricity production in MFCs. The efficient xylose metabolic pathways (the isomerase pathway or the oxidoreductase pathway) combined with two different xylose transporters were heterologously expressed in *S. oneidensis* MR-1 to construct four engineered *S. oneidensis* strains (namely XE, GE, XS, and GS), which could successfully utilize xylose under anaerobic and aerobic conditions. These recombinant *S. oneidensis* strains could generate bioelectricity in MFCs with xylose as the sole carbon source and electron donor. The maximum power density of the MFC inoculated with the engineered *S. oneidensis* strain GS (harbouring the xylose facilitator and the xylose oxidoreductase pathway) could reach ~2.1 ± 0.1 mW/m^2. This rationally engineered xylose transport and metabolic pathway significantly expanded the spectrum of carbon source that could be used by *S. oneidensis*. In the foreseeable future, with continuous development of synthetic biology strategies [73–75] to engineer exoelectrogens, a diverse array of organics such as lignocellulosic biomass and recalcitrant wastes may be more efficiently converted to electricity power.

Methods

In vitro gene synthesis

The information and coding sequences of the genes (Additional file 2: Tables S2 and S3) were extracted from the NCBI database and adapted for optimal expression in *S. oneidensis* MR-1 by a Java codon adaption tool (JCAT) in order to prevent blocked translation due to shortage of tRNAs for rare codons [49]. Each gene component was synthesized as a Biobrick [76, 77], and restriction enzyme sites of *Eco*RI, *Xba*I, *Spe*I, and *Sbf*I were avoided in the codon-optimized sequences. The optimized gene sequence was flanked by an upstream prefix (containing *Eco*RI and *Xba*I), a RBS site (BBa_B0034, iGEM) located at 6 bp ahead of the start codon, and a downstream suffix (containing *Spe*I and *Sbf*I) (Additional file 1: Figure S1A). The designed gene sequences were synthesized in vitro, verified by Sanger sequencing (AuGCT, China).

Plasmid construction, transformation, and culture conditions

All plasmid constructions were performed in *E. coli* Trans T1. The *E. coli* strains were cultured in the LB (Luria–Bertani) medium at 37 °C with 200 rpm. The plasmid to be transformed into *S. oneidensis* MR-1 (ATCC 700550) was firstly transformed into the plasmid donor strain *E. coli* WM3064 (auxotroph), and then transferred into *S. oneidensis* by conjugation. Then, 100 µg/ml 2,

by ribulose-phosphate 3-epimerase encoded by the *rpe* gene in the pentose phosphate pathway. Subsequently, ribulose-5-P and guanosine triphosphate (GTP) were converted to riboflavin via the riboflavin biosynthesis pathway (Additional file 1: Figure S3) [24, 36, 53].

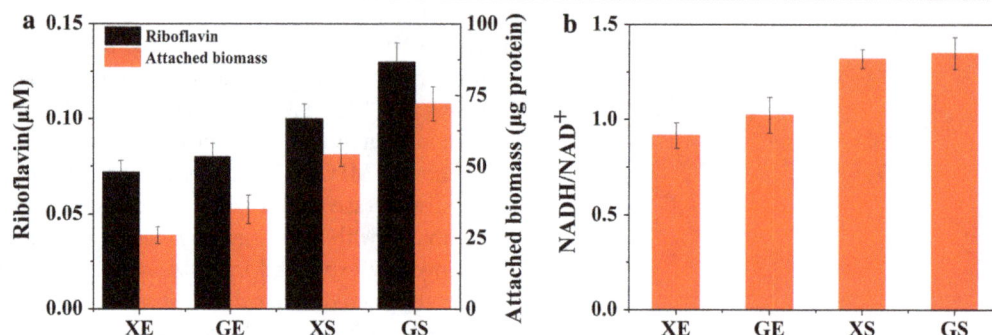

Fig. 5 Biochemical analyses of the four engineered *S. oneidensis* strains harbouring either the synthetic isomerase pathway (XE and GE) or the oxidoreductase pathway (XS and GS), respectively. **a** Riboflavin concentration in the anolytes of MFCs, and the attached biomass of each strain on anode surfaces. **b** Quantitative measurements of the ratio of NADH/NAD$^+$ in these engineered *S. oneidensis* strains in MFCs. All *error bars* were calculated from triplicate experiments

6-diaminopimelic acid (DAP) was added for the growth of *E. coli* WM3064. Whenever needed, 50 μg/ml kanamycin was added in the culture medium for plasmid maintenance. All the strains and plasmids used in this study are listed in Table (Additional file 2: Table S4).

Determination of cell growth and xylose utilization

To determine cell growth and xylose utilization under both aerobic and anaerobic conditions, 0.5 ml of the wild-type (WT) or engineered xylose-utilizing *S. oneidensis* strain culture suspension was inoculated into 15 ml *Shewanella* basal medium (SBM) [53] (Additional file 2: Table S5), supplemented with 5 mM xylose as the electron donor and carbon source in the test tube. When needed, 10 mM sodium fumarate was supplemented as the electron acceptor, which was stoichiometrically sufficient from both theoretical calculations and experimental validations (Additional file 1: Figure S4). The cell cultures were incubated at 30 °C, and samples were withdrawn periodically for the determination of cell density (optical density at 600 nm, i.e. OD_{600}) and xylose consumption. The OD_{600} was measured by an ultraviolet and visible spectrophotometer (TU-1810, Beijing, China).

BES setup

To evaluate the efficiency of extracellular electron transfer (EET), the overnight *Shewanella* culture suspension (1.5 ml) was inoculated into 150 ml fresh LB broth at 30 °C with shaking (200 rpm) till the OD_{600} reached 0.6–0.8. Then, the cells were harvested by centrifugation and washed 3 times with fresh M9 buffer (Additional file 2: Table S6). The cell pellets were subsequently resuspended in 140 ml electrolyte (5% LB broth plus 95% M9 buffer supplemented with 18 mM lactate or xylose). 50 μg/ml kanamycin was added to ensure consistent culture condition. The medium was supplemented with 0.1 mM IPTG as the inducer of the tac promoter. Our

previous experiments proved that IPTG had no effect on the cell physiology and EET of *Shewanella* [70]. The dual-chamber MFCs were used in this study, namely the anodic and cathodic chambers (140 ml working volume) separated by the nafion 117 membrane (DuPont Inc., USA), were the same as those used in the previous study. Carbon cloth was used as the electrodes for both the anode (2.5 cm × 2.5 cm, i.e. the geometric area is 6.25 cm^2) and the cathode (2.5 cm × 3 cm). The cathodic electrolyte consisted of 50 mM $K_3[Fe(CN)_6]$ in 50 mM K_2HPO_4 and 50 mM KH_2PO_4 solution. To measure the voltage generation, a 2 kΩ external resistor was connected into the external circuit of MFCs, and the output voltage (V) across the external loading resistor (R) was measured by a digital multimeter (DT9205A).

Electrochemical analyses

Cyclic voltammetry (CV) was performed in a three-electrode configuration with an Ag/AgCl reference electrode on a CHI 1000C multichannel potentiostat (CH Instrument, Shanghai, China). At the pseudo-steady state of MFCs, the polarization curves were obtained by varying the external resistor. Current density (I) was calculated as $I = V$ (output voltage)$/R$ (external resistance), and power density (P) was calculated as $P = V \times I$. Then, the I and P were normalized to the projected geometric area of the anode to obtain the current density and power density, respectively [78].

Quantification of metabolites

For the quantification of riboflavin, the samples in the MFC supernatant were firstly centrifuged (35,000 rpm for 20 min) and filtered (0.22 μm), and then, the eluted media were detected by a liquid chromatograph-tandem mass spectrometer (LC–MS) (Agilent LCMS-1290-6460) in a positive ion mode using a Waters XBridge C8 column (2.1 × 100 mm; particle size: 3.5 μm). Xylose in the anolytes were analysed using a high-performance liquid chromatography (HPLC) system equipped with a diode array detector. Sulphuric acid (5 mM) was used as the mobile phase flowing at 0.6 ml/min through the Aminex HPX-87H column (Bio-Rad, USA), which was incubated at 50 °C. Signals at 190 nm were used to quantify xylose.

Quantification of intracellular NADH/NAD⁺

Cells (10 ml) were collected by centrifugation (10,000 rpm at 4 °C for 5 min) and immediately re-suspended in 300 μl of 0.2 M HCl (for NAD^+) or 0.2 M NaOH (for NADH). The suspensions were boiled for 7 min, rapidly quenched in an ice bath, and added with 300 μl of 0.1 M NaOH (for NAD^+) or 0.1 M HCl (for NADH). Cell debris was removed by centrifugation at 10,000 rpm for 10 min, and the supernatant was used in a cycling assay to determine the amounts of NAD^+ and NADH [79, 80]. Meanwhile, the cell concentration for the detection of NAD^+ and NADH concentration was detected by plate counts on LB agar.

Measurement of electrode-attached biomass

The electrode was placed in a 50-ml tube containing 5 ml of 0.2 mol/l NaOH, then vortexed for 2 min, and incubated in a water bath to lyse cells at 96 °C for 30 min. The extracts were tested by bicinchoninic acid protein assay kit (Solarbio, China) after being cooled to room temperature.

Additional files

Additional file 1: Figure S1. Construction of synthetic xylose metabolic pathways in *Shewanella oneidensis* MR-1. (A) Schematic of the plasmid with a synthesized functional fragment of genes. The restriction sites EcoRI and XbaI with the ribosome binding site (RBS) are located upstream of each codon-optimized gene sequence, while the restrictions SpeI and PstI are located downstream of the gene. (B) Four plasmid constructs with xylose utilization pathways. To construct the multigene assembly in *S. oneidensis*, a Biobrick compatible expression vector pYYDT was adopted, which was previously constructed in our laboratory. Layout of the four plasmid constructs containing gene components in the xylose pathway examined in this study. **Figure S2.** Xylose consumption rate by *E. coil* (BL21) and by the recombinant *S. oneidensis* strain. The error bars were calculated from triplicate experiments. **Figure S3.** Metabolic pathway of riboflavin synthesis from xylose fermentation in *S. oneidensis*. A synthetic intracellular xylose metabolic pathway, i.e. the oxidoreductase pathway including genes *XYL1*, *XYL2* and *XKS1* from *S. stipites*, is incorporated into *S. oneidensis* MR-1 to enable the direct utilization of xylose. Xylulose 5-phosphate, as a metabolite in the oxidoreductase pathway, was converted to ribulose-5-P by ribulose-phosphate 3-epimerase (encoded by the *rpe* gene) in the pentose phosphate pathway, which was a crucial precursor for the biosynthesis of riboflavin via the riboflavin synthesis pathway. **Figure S4.** Xylose consumption under anaerobic conditions with 10 mM and 50 mM fumarate. The error bars were calculated from triplicate experiments.

Additional file 2: Table S1. Summary of the reported energy output of Xylose-Fed MFCs. **Table S2.** Genes used in this study. **Table S3.** Synthesized sequences of genes in this study. **Table S4.** Strains and plasmids used in this study. **Table S5.** Main constituents for *S. oneidensis* basal medium (SBM). **Table S6.** Main constituents for M9 buffer.

Abbreviations
EET: extracellular electron transfer; MQ: methyl naphthoquinone; IM: inner membrane; OM: outer membrane; CymA: inner membrane tetraheme *c*-type cytochromes; Mtr: metal-reducing; MtrA: periplasmic decaheme; MtrB: β-barrel trans-OM protein; MtrC and OmcA: two OM decaheme *c*-type cytochromes; TCA: tricarboxylic acid cycle; ndhII: NADH dehydrogenase; P: phosphate; PEP: phosphoenolpyruvate; NAD^+: nicotinamide adenine dinucleotide; NADH: reduced nicotinamide adenine dinucleotide; $NADP^+$: nicotinamide adenine dinucleotide phosphate; NADPH: reduced nicotinamide adenine dinucleotide phosphate; ATP: adenosine triphosphate; ADP: adenosine diphosphate; X-5-P: xylulose 5-phosphate; GTP: guanosine triphosphate; IPTG: isopropyl β-D-1-thiogalactopyranoside; DAP: 2, 6-diaminopimelic acid; MFC: microbial fuel cell; OD: optical density; CV: cyclic voltammetry; XDH: xylitol dehydrogenase; NAG: N-acetyl-glucosamine.

Authors' contributions
FL designed the project, performed experiments, analysed data, and drafted the manuscript. YL, LS and XL performed some experiments, collected data, analysed data, and drafted the manuscript; CY, XA, XC and YT provided some reagents, helped design the experiment and drafted the manuscript; HS supervised the project, analysed the data, and critically revised the manuscript. All authors read and approved the final manuscript.

Author details
¹ Key Laboratory of Systems Bioengineering (Ministry of Education), School of Chemical Engineering and Technology, Tianjin University, Tianjin 300072, China. ² SynBio Research Platform, Collaborative Innovation Centre of Chemical Science and Engineering, Tianjin University, Tianjin 300072, China. ³ Petrochemical Research Institute, PetroChina Company Limited, Beijing 102206, People's Republic of China.

Acknowledgements
Not applicable.

Competing interests
The authors declare that they have no competing interests.

Funding
This research was supported by the National Natural Science Foundation of China (NSFC 21376174, 21621004), the National Basic Research Program of China ("973" Program: 2014CB745103), and Tianjin Science & Technology Council (13JCYBJC40700).

References
1. Butti SK, Velvizhi G, Sulonen MLK, Haavisto JM, Oguz Koroglu E, Yusuf Cetinkaya A, Singh S, Arya D, Annie Modestra J, Vamsi Krishna K, Verma A, Ozkaya B, Lakaniemi A-M, Puhakka JA, Venkata Mohan S. Microbial electrochemical technologies with the perspective of harnessing bioenergy: maneuvering towards upscaling. Renew Sust Energ Rev. 2016;53:462–76.
2. Mohan S, Butti S, Amulya K, Dahiya S, Modestra J. Waste biorefinery:a new paradigm for a sustainable bioelectro economy. Trends Biotechnol. 2016;34(11):852–5.
3. Xie X, Ye M, Hsu PC, Liu N, Criddle CS, Cui Y. Microbial battery for efficient energy recovery. Proc Natl Acad Sci. 2013;110(40):15925–30.
4. Wang H, Luo H, Fallgren PH, Jin S, Ren ZJ. Bioelectrochemical system platform for sustainable environmental remediation and energy generation. Biotechnol Adv. 2015;33(3–4):317–34.
5. Wang H, Ren ZJ. A comprehensive review of microbial electrochemical systems as a platform technology. Biotechnol Adv. 2013;31(8):1796–807.
6. Harnisch F, Schroder U. From MFC to MXC: chemical and biological cathodes and their potential for microbial bioelectrochemical systems. Chem Soc Rev. 2010;39(11):4433–48.
7. Kumar A. The ins and outs of microorganism–electrode electron transfer reactions. Nat Rev Chem. 2017;1(3):0024.

8. Chaudhuri S, Lovley D. Electricity generation by direct oxidation of glucose in mediatorless microbial fuel cells. Nat Biotechnol. 2003;21(10):1229–32.
9. Logan BE. Exoelectrogenic bacteria that power microbial fuel cells. Nat Rev Microbiol. 2009;7(5):375–81.
10. Lovley DR. Bug juice: harvesting electricity with microorganisms. Nat Rev Microbiol. 2006;4(7):497–508.
11. Logan BE, Rabaey K. Conversion of wastes into bioelectricity and chemicals by using microbial electrochemical technologies. Science. 2012;337(6095):686–90.
12. Bond DR, Holmes DE, Tender LM, Lovley DR. Electrode-reducing microorganisms that harvest energy from marine sediments. Science. 2002;295(5554):483.
13. Jafary T, Wan RWD, Ghasemi M, Kim BH, Jahim JM, Ismail M, Lim SS. Biocathode in microbial electrolysis cell: present status and future prospects. Renew Sust Energ Rev. 2015;47:23–33.
14. Luo H, Jenkins PE, Zen Z. Concurrent desalination and hydrogen generation using microbial electrolysis and desalination cells. Environ Sci Technol. 2011;45(1):340–4.
15. Mehanna M, Kiely PD, Call DF, Logan BE. Microbial electrodialysis cell for simultaneous water desalination and hydrogen gas production. Environ Sci Technol. 2010;44(24):9578–83.
16. Zhang Y, Angelidaki I. Microbial electrolysis cells turning to be versatile technology: recent advances and future challenges. Water Res. 2014;56(3):11–25.
17. Rabaey K, Rozendal RA. Microbial electrosynthesis-revisiting the electrical route for microbial production. Nat Rev Microbiol. 2010;8(10):706–16.
18. Han L, Opgenorth PH, Wernick DG, Rogers S, Wu TY, Higashide W, Malati P, Huo YX, Cho KM, Liao JC. Integrated electromicrobial conversion of CO_2 to higher alcohols. Science. 2012;335(6076):1596.
19. Sakimoto KK, Wong AB, Yang P. Self-photosensitization of nonphotosynthetic bacteria for solar-to-chemical production. Science. 2016;351(6268):74–7.
20. Liu C, Colon BC, Ziesack M, Silver PA, Nocera DG. Water splitting-biosynthetic system with CO_2 reduction efficiencies exceeding photosynthesis. Science. 2016;352(6290):1210–3.
21. Sadhukhan J, Lloyd JR, Scott K, Premier GC, Yu EH, Curtis T, Head IM. A critical review of integration analysis of microbial electrosynthesis (MES) systems with waste biorefineries for the production of biofuel and chemical from reuse of CO_2. Renew Sust Energ Rev. 2016;56:116–32.
22. Choi O, Sang BI. Extracellular electron transfer from cathode to microbes: application for biofuel production. Biotechnol Biofuels. 2016;9(1):1–14.
23. Lin T, Bai X, Hu Y, Li B, Yuan Y, Song H, Yang Y, Wang J. Synthetic *Saccharomyces cerevisiae-Shewanella oneidensis* consortium enables glucose-fed high-performance microbial fuel cell. AlChE J. 2016.
24. Yang Y, Wu Y, Hu Y, Cao Y, Poh CL, Cao B, Song H. Engineering electrode-attached microbial consortia for high-performance xylose-fed microbial fuel cell. ACS Catal. 2015;5(11):6937–45.
25. Rubin EM. Genomics of cellulosic biofuels. Nature. 2008;454(7206):841–5.
26. Kuhad RC, Gupta R, Khasa YP, Singh A, Zhang YHP. Bioethanol production from pentose sugars: current status and future prospects. Renew Sust Energ Rev. 2011;15(9):4950–62.
27. Jackson S, Nicolson S. Xylose as a nectar sugar: from biochemistry to ecology. Comp Biochem Phys B. 2002;131(4):613–20.
28. Catal T, Li K, Bermek H, Liu H. Electricity production from twelve monosaccharides using microbial fuel cells. J Power Sources. 2008;175(1):196–200.
29. Huang L, Logan BE. Electricity production from xylose in fed-batch and continuous-flow microbial fuel cells. Appl Microbiol Biotechnol. 2008;80(4):655–64.
30. Huang L, Zeng RJ, Angelidaki I. Electricity production from xylose using a mediator-less microbial fuel cell. Bioresour Technol. 2008;99(10):4178–84.
31. Utrilla J, Licona-Cassani C, Marcellin E, Gosset G, Nielsen LK, Martinez A. Engineering and adaptive evolution of escherichia coli, for D-lactate fermentation reveals gatc as a xylose transporter. Metab Eng. 2012;14(5):469–76.
32. Zhou H, Cheng JS, Wang BL, Fink GR, Stephanopoulos G. Xylose isomerase overexpression along with engineering of the pentose phosphate pathway and evolutionary engineering enable rapid xylose utilization and ethanol production by *Saccharomyces cerevisiae*. Metab Eng. 2012;14(6):611–22.
33. Young EM, Comer AD, Huang H, Alper HS. A molecular transporter

34. engineering approach to improving xylose catabolism in *Saccharomyces cerevisiae*. Metab Eng. 2012;14(4):401.
34. Yuan Y, Du J, Zhao H. Customized optimization of metabolic pathways by combinatorial transcriptional engineering. Nucleic Acids Res. 2012;40(18):177–209.
35. Xiao H, Li Z, Jiang Y, Yang Y, Jiang W, Gu Y, Yang S. Metabolic engineering of d-xylose pathway in *Clostridium beijerinckii* to optimize solvent production from xylose mother liquid. Metab Eng. 2012;14(5):569.
36. Fredrickson JK, Romine MF, Beliaev AS, Auchtung JM, Driscoll ME, Gardner TS, Nealson KH, Osterman AL, Pinchuk G, Reed JL, Rodionov DA, Rodrigues JL, Saffarini DA, Serres MH, Spormann AM, Zhulin IB, Tiedje JM. Towards environmental systems biology of *Shewanella*. Nat Rev Microbiol. 2008;6(8):592–603.
37. Kumar R, Singh L, Zularisam AW. Exoelectrogens: recent advances in molecular drivers involved in extracellular electron transfer and strategies used to improve it for microbial fuel cell applications. Renew Sust Energ Rev. 2016;56:1322–36.
38. Coursolle D, Gralnick JA. Modularity of the Mtr respiratory pathway of *Shewanella oneidensis* strain MR-1. Mol Microbiol. 2010;77(4):995–1008.
39. Okamoto A, Nakamura R, Hashimoto K. In-vivo identification of direct electron transfer from *Shewanella oneidensis* MR-1 to electrodes via outer-membrane OmcA–MtrCAB protein complexes. Electrochim Acta. 2011;56(16):5526–31.
40. Hartshorne RS, Reardon CL, Ross D, Nuester J, Clarke TA, Gates AJ, Mills PC, Fredrickson JK, Zachara JM, Shi L, Beliaev AS, Marshall MJ, Tien M, Brantley S, Butt JN, Richardson DJ. Characterization of an electron conduit between bacteria and the extracellular environment. Proc Natl Acad Sci. 2009;106(52):22169–74.
41. Clarke TA, Richardson DJ. Structure of a bacterial cell surface decaheme electron conduit. Proc Natl Acad Sci. 2011;108(23):9384–9.
42. Mao L, Verwoerd WS. Theoretical exploration of optimal metabolic flux distributions for extracellular electron transfer by *Shewanella oneidensis* MR-1. Biotechnol Biofuels. 2014;7(1):1–20.
43. El-Naggar MY, Wanger G, Leung KM, Yuzvinsky TD, Southam G, Yang J, Lau WM, Nealson KH, Gorby YA. Electrical transport along bacterial nanowires from *Shewanella oneidensis* MR-1. Proc Natl Acad Sci. 2010;107(42):18127–31.
44. Okamoto A, Hashimoto K, Nealson KH, Nakamura R. Rate enhancement of bacterial extracellular electron transport involves bound flavin semiquinones. Proc Natl Acad Sci. 2013;110(19):7856–61.
45. Pinchuk GE, Rodionov DA, Yang C, Li X, Osterman AL, Dervyn E, Geydebrekht OV, Reed SB, Romine MF, Collart FR, Scott JH, Fredrickson JK, Beliaev AS. Genomic reconstruction of *Shewanella oneidensis* MR-1 metabolism reveals a previously uncharacterized machinery for lactate utilization. Proc Natl Acad Sci. 2009;106(8):2874–9.
46. Pirbadian S, Barchinger SE, Leung KM, Byun HS, Jangir Y, Bouhenni RA, Reed SB, Romine MF, Saffarini DA, Shi L, Gorby YA, Golbeck JH, El-Naggar MY. *Shewanella oneidensis* MR-1 nanowires are outer membrane and periplasmic extensions of the extracellular electron transport components. Proc Natl Acad Sci. 2014;111(35):12883–8.
47. White GF, Shi Z, Shi L, Wang Z, Dohnalkova AC, Marshall MJ, Fredrickson JK, Zachara JM, Butt JN, Richardson DJ, Clarke TA. Rapid electron exchange between surface-exposed bacterial cytochromes and Fe(III) minerals. Proc Natl Acad Sci. 2013;110(16):6346–51.
48. Hu Y, Wu Y, Mukherjee M, Cao B. A near-infrared light responsive c-di-GMP module-based AND logic gate in *Shewanella oneidensis*. Chem Commun. 2017;53:1646–8.
49. Hu Y, Yang Y, Katz E, Song H. Programming the quorum sensing-based AND gate in *Shewanella oneidensis* for logic gated-microbial fuel cells. Chem Commun. 2015;51(20):4184–7.
50. Li Z, Rosenbaum MA, Venkataraman A, Tam TK, Katz E, Angenent LT. Bacteria-based AND logic gate: a decision-making and self-powered biosensor. Chem Commun. 2011;47(11):3060–2.
51. Ding Y, Peng N, Du Y, Ji L, Cao B. Disruption of putrescine biosynthesis in *Shewanella oneidensis* enhances biofilm cohesiveness and performance in Cr(VI) Immobilization. Appl Environ Microbiol. 2013;80(4):1498–506.
52. Serres MH, Riley M. Genomic analysis of carbon source metabolism of *Shewanella oneidensis* MR-1: predictions versus experiments. J Bacteriol. 2006;188(13):4601–9.
53. Flynn CM, Hunt KA, Gralnick JA, Srienc F. Construction and elementary

mode analysis of a metabolic model for *Shewanella oneidensis* MR-1. Biosystems. 2012;107(2):120–8.

54. Pinchuk GE, Geydebrekht OV, Hill EA, Reed JL, Konopka AE, Beliaev AS, Fredrickson JK. Pyruvate and lactate metabolism by *Shewanella oneidensis* MR-1 under fermentation, oxygen limitation, and fumarate respiration conditions. Appl Environ Microbiol. 2011;77(23):8234–40.

55. Pinchuk GE, Hill EA, Geydebrekht OV, De Ingeniis J, Zhang X, Osterman A, Scott JH, Reed SB, Romine MF, Konopka AE, Beliaev AS, Fredrickson JK, Reed JL. Constraint-based model of *Shewanella oneidensis* MR-1 metabolism: a tool for data analysis and hypothesis generation. PLoS Comput Biol. 2010;6(6):e1000822.

56. Sekar R, Shin HD, DiChristina TJ. Activation of an otherwise silent xylose metabolic pathway in *Shewanella oneidensis*. Appl Environ Microbiol. 2016;82(13):3996–4005.

57. Runquist D, Fonseca C, Radstrom P, Spencer-Martins I, Hahn-Hagerdal B. Expression of the Gxf1 transporter from *Candida intermedia* improves fermentation performance in recombinant xylose-utilizing *Saccharomyces cerevisiae*. Appl Microbiol Biotechnol. 2009;82(1):123–30.

58. Leandro MJ, Goncalves P, Spencer-Martins I. Two glucose/xylose transporter genes from the yeast *Candida intermedia*: first molecular characterization of a yeast xylose-H$^+$ symporter. Biochem J. 2006;395(3):543–9.

59. Gu Y, Ding Y, Ren C, Sun Z, Rodionov DA, Zhang W, Yang S, Yang C, Jiang W. Reconstruction of xylose utilization pathway and regulons in *Firmicutes*. BMC genomics. 2010;11:255.

60. Zhang M, Eddy C, Deanda K, Finkelstein M, Picataggio S. Metabolic Engineering of a pentose metabolism pathway in ethanologenic *Zymomonas mobilis*. Science. 1995;267(5195):240–3.

61. Jeffries TW, Grigoriev IV, Grimwood J, Laplaza JM, Aerts A, Salamov A, Schmutz J, Lindquist E, Dehal P, Shapiro H, Jin YS, Passoth V, Richardson PM. Genome sequence of the lignocellulose-bioconverting and xylose-fermenting yeast *Pichia stipitis*. Nat Biotechnol. 2007;25(3):319–26.

62. Nduko JM, Matsumoto K, Ooi T, Taguchi S. Effectiveness of xylose utilization for high yield production of lactate-enriched P(lactate-co-3-hydroxybutyrate) using a lactate-overproducing strain of *Escherichia coli* and an evolved lactate-polymerizing enzyme. Metab Eng. 2013;15:159–66.

63. Wovcha MG, Steuerwald DL, Brooks KE. Amplification of d-xylose and D-glucose isomerase activities in *Escherichia coli* by gene cloning. Appl Environ Microbiol. 1983;45(4):1402.

64. Yang Y, Ding Y, Hu Y, Cao B, Rice SA, Kjelleberg S, Song H. Enhancing bidirectional electron transfer of *Shewanella oneidensis* by a synthetic flavin pathway. ACS Synth Biol. 2015;4(7):815–23.

65. Jeffries TW. Engineering yeasts for xylose metabolism. Curr Opin Biotech. 2006;17(3):320–6.

66. Baron D, LaBelle E, Coursolle D, Gralnick JA, Bond DR. Electrochemical measurement of electron transfer kinetics by *Shewanella oneidensis* MR-1. J Biol Chem. 2009;284(42):28865–73.

67. Verho R, Londesborough J, Penttila M, Richard P. Engineering redox cofactor regeneration for improved pentose fermentation in *Saccharomyces cerevisiae*. Appl Environ Microbiol. 2003;69(10):5892–7.

68. Kuyper M, Winkler AA, van Dijken JP, Pronk JT. Minimal metabolic engineering of *Saccharomyces cerevisiae* for efficient anaerobic xylose fermentation: a proof of principle. FEMS Yeast Res. 2004;4(6):655–64.

69. Kim SR, Ha SJ, Kong II, Jin YS. High expression of XYL2 coding for xylitol dehydrogenase is necessary for efficient xylose fermentation by engineered *Saccharomyces cerevisiae*. Metab Eng. 2012;14(4):336–43.

70. Liu T, Yu YY, Deng XP, Ng CK, Cao B, Wang JY, Rice SA, Kjelleberg S, Song H. Enhanced *Shewanella* biofilm promotes bioelectricity generation. Biotechnol Bioeng. 2015;112(10):2051–9.

71. Han S, Gao X, Ying H, Zhou CC. NADH gene manipulation for advancing bioelectricity in *Clostridium ljungdahlii* microbial fuel cells. Green Chem. 2016;18(8):2473–8.

72. Yong YC, Yu YY, Yang Y, Li CM, Jiang R, Wang XY, Wang J, Song H. Increasing intracellular releasable electrons dramatically enhances bioelectricity output in microbial fuel cells. Electrochem Commun. 2012;19:13–6.

73. Jensen HM, Cantor CR. Engineering of a synthetic electron conduit in living cells. Proc Natl Acad Sci. 2010;107(45):19213–8.

74. TerAvest MA, Ajo-Franklin CM. Transforming exoelectrogens for biotechnology using synthetic biology. Biotechnol Bioeng. 2016;113(4):687–97.

75. Jensen H, Teravest M, Kokish M, Ajofranklin CM. CymA and exogenous flavins improve extracellular electron transfer and couple it to cell growth in Mtr-expressing *Escherichia coli*. ACS Synth Biol. 2016;5(7):679–88.

76. Tsvetanova B, Peng L, Liang X, Li K, Yang JP, Ho T, Shirley J, Xu L, Potter J, Kudlicki W, Peterson T, Katzen F. Genetic assembly tools for synthetic biology. Methods Enzymol. 2011;498:327–48.

77. Ellis T, Adie T, Baldwin GS. DNA assembly for synthetic biology: from parts to pathways and beyond. Integr Biol (Camb). 2011;3:109–18.

78. Yong YC, Yu YY, Zhang X, Song H. Highly active bidirectional electron transfer by a self-assembled electroactive reduced-graphene-oxide-hybridized biofilm. Angew Chem Int Ed. 2014;53(17):4480–3.

79. Yong XY, Feng J, Chen YL, Shi DY, Xu YS, Zhou J, Wang SY, Xu L, Yong YC, Sun YM, Shi CL, OuYang PK, Zheng T. Enhancement of bioelectricity generation by cofactor manipulation in microbial fuel cell. Biosens Bioelectron. 2014;56:19–25.

80. Bernofsky C, Swan M. An improved cycling assay for nicotinamide adenine dinucleotide. Anal Biochem. 1973;53(2):452–8.

4

The impact of considering land intensification and updated data on biofuels land use change and emissions estimates

Farzad Taheripour, Xin Zhao and Wallace E. Tyner[*]

Abstract

Background: The GTAP model has been used to estimate biofuel policy induced land use changes and consequent GHG emissions for more than a decade. This paper reviews the history of the model and database modifications and improvements that have occurred over that period. In particular, the paper covers in greater detail the move from the 2004 to the 2011 database, and the inclusion of cropland intensification in the modeling structure.

Results: The results show that all the changes in the global economy and agricultural sectors cause biofuels induced land use changes and associated emissions can be quite different using the 2011 database versus 2004. The results also demonstrate the importance of including land intensification in the analysis. The previous versions of GTAP and other similar models assumed that changes in harvested area equal changes in cropland area. However, FAO data demonstrate that it is not correct for several important world regions. The model now includes land intensification, and the resulting land use changes and emission values are lower as would be expected.

Conclusions: Dedicated energy crops are not similar to the first generation feedstocks in the sense that they do not generate the level of market-mediated responses which we have seen in the first-generation feedstocks. The major market-mediated responses are reduced consumption, crop switching, changes in trade, changes in intensification, and forest or pasture conversion. These largely do not apply to dedicated energy corps. The land use emissions for cellulosic feedstocks depend on what we assume in the emissions factor model regarding soil carbon gained or lost in converting land to these feedstocks. We examined this important point for producing bio-gasoline from miscanthus. Much of the literature suggests miscanthus actually sequesters carbon, if grown on the existing active cropland or degraded land. We provide some illustrative estimates for possible assumptions. Finally, it is important to note the importance of the new results for the regulatory process. The current California Air Resources Board carbon scores for corn ethanol and soy biodiesel are 19.8 and 29.1, respectively (done with a model version that includes irrigation). The new model and database carbon scores are 12 and 18, respectively, for corn ethanol and soy biodiesel. Thus, the current estimates values are substantially less than the values currently being used for regulatory purposes.

Keywords: Land use change, Biofuel emissions, Intensive versus extensive margin, GTAP model and database

Background

The GTAP-BIO model has been developed and frequently improved and updated to evaluate biofuels induced land use changes and their consequent emissions [1–7]. The modifications made in this model can be divided into three groups: modifications and updates in the GTAP-BIO database; changes in model parameters; and improvements in the modeling structure. This paper briefly reviews these changes, introduces a set of new modifications into the model and its database, and examines induced land use emissions for several biofuel pathways using the new model and its database.

The previous version of this model uses an old databases (GTAP database version 7) which represents the world economy in 2004. During the past decade, the

*Correspondence: wtyner@purdue.edu
Department of Agricultural Economics, Purdue University, West Lafayette, USA

global economy has changed considerably. In particular, since 2004, major changes occurred in the agricultural and biofuel markets. Recently, a new version of the GTAP database (version 9) which represents the world economy in 2011 has been published. However, as usual, this standard database does not explicitly represent production and consumption of biofuels. We have added biofuels (including traditional biofuels and several advanced cellulosic biofuels) into this database to take the advantages of the newer databases. This allows us to examine the economic and land use consequences of the first- and second-generation biofuels using the updated database.

Several recent publications [8–15] have shown that that land intensification in crop production (in terms of expansion in multiple cropping and/or returning unused cropland to crop production) has increased in several regions across the world. Typically, economic models, including GTAP-BIO, ignore this kind of intensification. Recently, we improved the GTAP-BIO model to take into account land intensification in crop production. We use this model in combination with the new database mentioned above to assess the land use impacts of several biofuel pathways. We compare the results of the new simulations with their corresponding results obtained from the older versions.

Methods
GTAP-BIO database version 9
The standard GTAP databases do not include production, consumption, and trade of biofuels. Taheripour et al. [16] introduced the first generation of biofuels (including grain ethanol, sugarcane ethanol, and biodiesel) into the GTAP standard database version 6, which represented the world economy in 2001 [17]. The early versions of the GTAP-BIO model were built on this database and used in several applications and policy analyses [3, 4, 18–21]. The California Air Resources Board (CARB) developed its first set of ILUC values using this database and early versions of the model [22]. The Argonne National Lab also used the results of this model in developing the early versions of the life cycle analyses (LCA) of biofuels [21, 23].

When the standard GTAP database version 7, which represented the world economy in 2004 was released [24], Taheripour and Tyner [25] introduced first- and second-generation biofuels into this database. Several alternative aggregations of this database have been developed and used in various studies to evaluate the economic and land use impacts of biofuel production and polices [26–31]. CARB has used this database to develop its final ILUC values [32, 33], and Argonne National Lab also used the outcomes obtained from this database in its more recent LCA analyses.

The GTAP-BIO 2004 database in comparison to its 2001 version had several advantages including but not limited to: (1) providing data on cropland pasture for the US and Brazil; (2) disaggregating oilseeds into soybeans, rapeseed, palm, and other oilseeds; (3) disaggregating coarse grains into sorghum and other coarse grains; (4) introducing cellulosic crops and corn stover collection as new activities into the database; (5) disaggregating vegetable oil industry into soybean oil, rapeseed oil, palm oil, and other vegetable oils and fats and their corresponding meal products; (6) dividing the standard food industry of GTAP into two distinct food and feed industries; and (7) covering a wide range of biofuels including ethanol produced from grains, ethanol produced from sugar crops, four types of biodiesel produced from soybean oil, rapeseed oil, palm oil, and other oils and fats, three types of cellulosic ethanol produced form corn stover, switchgrass, and miscanthus and three types of drop-in cellulosic biofuels produced from the corn stover, switchgrass, and miscanthus.

The GTAP-BIO 2004 database with all of the above advantages is now out-of-date. During the past decade, the global economy has changed significantly with major consequences for agricultural and energy markets including biofuels. On one hand, demand for agricultural products has increased across the world at different rates due to growths in income and population. Expansion in biofuel production due to public policies has contributed to the expansion in demand for agricultural products in some regions and at the global scale, as well. On the other hand, the agricultural sector has evolved considerably across the world: crop production and its geographical distribution have changed, the mix of crops produced in most countries has changed, crop yields have improved due to technological progress in many regions, crop production has been negatively affected in some regions due to severe climate conditions, and international trade in agricultural products has changed. Major changes occurred in the livestock industry, as well: demand for meat and meat products has shifted from red meat towards white meat, more biofuels by-products and meals were used in animal feed rations, and land intensification has been extended in the livestock industry. The biofuel industry has grown rapidly across the world and, in particular, in US, Brazil, and EU. Biofuel producers now operate more efficiently than before. Unlike the early 2000s, the biofuel industry is now a mature industry which operates without government subsidies. However, they still benefit from biofuel mandates. The 2004 database misses all these changes and many other changes which occurred in the global economy. Therefore, it becomes necessary to update the GTAP-BIO database.

To accomplish this task, following our earlier work in this area [16, 25, 34], we explicitly introduced biofuels into the latest publicly released version (V9) of the standard GTAP database which represents the world economy in 2011 [35]. This means is that all the steps that we followed to introduce biofuels into the 2001 and 2004 databases had to be repeated for the 2011 GTAP database but using 2011 data for all the biofuels components. Thus, production, consumption, trade, prices, and co-products had to be introduced into the 2011 database. The full description of this task is reported in [36]. Here, we explain the main important aspects of this task.

Data collection

Production and consumption of biofuels for 2011 are taken from the US Energy Information Administration (EIA) website (http://www.eia.gov). The EIA provides data on ethanol and biodiesel produced across the world by country. Harvested area, crop produced, area of forest, pasture, and cropland for 2011 are obtained from the FAOSTAT database http://faostat3.fao.org/home\E; for details, see [37]. Data on vegetable oils and meals produced, consumed, and traded in 2011 were collected by country from the world oil database [38] and used to split the GTAP vegetable oil sector into different types of vegetable oils and meals.

Introducing new non-biofuel sectors into the standard database

As mentioned above in our earlier work [16, 25, 34], we developed a process to further disaggregate coarse grains, oilseeds, vegetable oils, and food sectors of the GTAP original database to additional new sectors to support various biofuel pathways and their links with the agricultural, livestock, food, and feed industries. Using the collected data mentioned in "Data collection" section, we repeated that process for the 2011 database.

In addition, unlike the earlier versions of the GTAP-BIO databases, a blend sector was added to the database to represent a new industry which blends biofuels with traditional fuels. The earlier versions of this database assumed that biofuels are directly used by the refinery sector (as an additive to the traditional fuels) or consumed by households (as substitutes for the traditional fuels). The new blend sector takes the traditional fuels used in transportation and blends them with biofuels. This sector supplies the blended fuels to the transportation sectors and final users.

Introducing biofuel sectors into the standard database

In our earlier work [16, 25, 34], a process was also designed and implemented to introduce biofuels into a standard GTAP database. We followed and improved that process to introduce biofuels into the GTAP database version 9.

This process first determines the original GTAP sectors which biofuels are embedded. Then, data were obtained on monetary values of biofuels produced by country; a proper cost structure for each biofuel pathway; users of biofuels; and feedstock for each biofuel. Finally, it uses these data items and a set of programs to introduce biofuels into the database. As an example, in the standard GTAP database, the US corn ethanol is imbedded in the food sector. Therefore, this sector was divided into food and ethanol sectors. To accomplish this task, we needed to evaluate monetary values of corn ethanol and its by-product (DDGS) produced in the US at 2011 prices. We also needed to determine the cost structure of this industry in the US in 2011, as well. This cost structure should represent the shares of various inputs (including intermediate inputs and primary factors of production) used by the ethanol industry in its total costs in 2011. For the case of US corn ethanol, which represents a well-established industry in 2011, these data items should match with national level information. Hence, as mentioned in the previous section, we collected data from trusted sources to prepare required data for all types of the first generation of biofuels produced across the world in 2011. For the second generation of biofuels (e.g., ethanol produced from switchgrass or miscanthus) which are not produced at commercial level, we rely on the literature to determine their production costs and also their cost structures. For these biofuels, we also need to follow the literature to define new sectors (e.g., miscanthus or switchgrass) and their cost structures to include their feedstock at 2011 prices.

After preparing this information, we used a set of codes and the SplitCom program [39] to insert biofuels into the national input–output tables of the standard database. The SplitCom program allows users to split a particular sector into two or more sectors while maintaining the national SAM tables in balance. To split a particular sector, the program takes the original database (including regional SAM tables) and some additional external data items and then runs the split process. In general, in each split process, the additional external data items are: (1) the name of original sector; (2) the name of new sectors; (3) the cost structure of new sectors; (4) users of the new sectors; (5) share of each user in each new product; and (6) trade flows of new products. See these references for more details [16, 25, 34, 36].

Other important data modifications

In addition to the above modifications, we made several adjustments in the standard GTAP database to match with real-world observations. The major adjustments are:

- Production and sales of US coarse grains are adjusted according to the USDA data. The modified GTAP-

BIO US input–output table shows that 11.3, 26.8, and 61.9% of corn used by livestock industry are consumed by dairy, ruminant, and non-ruminant subsectors, respectively. The corresponding original GTAP figures are about 48, 7, and 45%. We altered the original GTAP figures to match with the USDA data.

- The standard GTAP database underestimates the monetary value of vegetable oils and their meals produced in the US. This is fixed using the world oil database [38]. According to this database which reports vegetable oils and meals produced across the world and using a set of price data for these products obtained from the FAOSTATA, we estimated that the US vegetable oil industry produced about $36.5 billion in 2011. The corresponding GTAP figure was about $25 billion.

- The monetary values of vegetable oils used in non-food uses presented in the input–output tables of some countries were smaller than the monetary values of vegetable oils needed to support their biodiesel production. The input–output tables of these countries were properly modified to solve these inconsistencies.

- Cropland pasture data were added for Canada [39], and proper changes were made in the input–output table of this country. Cropland pasture was updated for the US and Brazil according to the existing data for 2011.

The GTAPADJUST program developed by Horridge [40] and several programs developed by the authors were used to carry out the above changes and adjustments. The GTAPADJUST program allows users to modify elements of the SAM tables while maintaining required balances.

In conclusion, the GTAP-BIO databases for 2004 and 2011 represent the same regional and sectoral aggregation schemes, except for the blend sector which was added to the 2011 database. While these two databases represent the same aggregation schemes, they represent entirely different data content. Finally, it is important to note that a GTAP-BIO database including cellulosic biofuels is labeled GTAP-BIO-ADV. The GTAP-BIO and GTAP-BIO-ADV versions for each year represent the same data contents, but the latter represents the second-generation biofuel pathways with very small production levels.

Database comparison

Here, we briefly compare the new GTAP-BIO database which represents the world economy in 2011 with the 2004 version. See [36] for the full comparison of these two databases. Note that in CGE models, the data for the base year represent all economic data for that year, and, in some circumstances, because of annual variability, the base year may not be completely representative of trends. The impacts of this issue normally are not large, but it is an issue for all CGE models.

Expansion in biofuel production Total biofuel production (including ethanol and biodiesel) has rapidly increased from 8.4 billion gallons (BGs) in 2004 to 29 BGs in 2011 at the global scale, a tremendous growth of 19.4% per year over this time period. In 2004, Brazil, US, and EU were the main biofuel producers. In this year, they were producing about 4, 3.4, and 0.7 BGs biofuels (manly ethanol), respectively. In 2011, about 22.9 BGs of ethanol and 6.2 BGs of biodiesel were produced across the world. The largest ethanol producers including US, Brazil, and EU produced 13.9, 6, and 1.1 BGs of ethanol in 2011. The next three largest ethanol producers were China (with 0.6 BGs), Canada (0.5 BGs), and South America (0.2 BGs). The largest biodiesel producers including EU, US, and South America produced 2.7, 1, and 0.9 BGs of biodiesel in 2011. The next three largest biodiesel producers were Brazil (with 0.7 BGs), Malaysia and Indonesia (0.3 BGs), and South East Asia (0.2 BGs).

Economy-wide comparison Many changes occurred in the global economy. Population increased by about 550.4 million across the world between 2004 and 2011. Major changes occurred in sub-Saharan Africa (144.2 million or 19.6%), India (by 134 million or 12.3%), and Middle East and North Africa (48.6 million or 14.2%). In most developed countries and regions, population has been increased slightly or decreased.

In 2004, EU, US, and Japan had the largest shares in the global production of goods and services (measured with GDP) with 31.5, 28.5, and 11.4% shares, respectively. In 2011, the shares of these regions dropped to 24.6%, 21.7, and 8.3%. Instead, the share of China from global productions of goods and services has increased from 4.6% in 2004 to 10.6% in 2011. As a measure of income, GDP per capital at current prices has increased all across the world in 2004–11. Large changes occurred in China (301%), Brazil (274%), and Russia (236%).

The share of consumption and investment in GDP in 2004 and 2011 are not very different in many regions. However, some regions like China, India, East Asia, Malaysia–Indonesia, and Russia allocated larger shares of their GDP to investment and spend less on consumption in 2011 compared with 2004.

Between 2004 and 2011, in several regions across the world, the shares of agricultural, processed food and feed, biofuels, and energy sectors in GDP increased, but the total share of other goods and services decreased. Some countries experienced differently. For instance, the agricultural share in total output declined in some countries such as Brazil, China, and India. In these countries, agricultural activities experienced rapid growths, but their growth rates were smaller than the growth rates of other economic activities.

At the national level, the shares of domestic and export uses in total value of output of each region have not significantly changed. However, at commodity level, important changes occurred. For example, consider a few examples from the US economy. In 2004, the US exported 32% of its coarse grains to other countries. This figure was about 19% in 2011. That is basically is due to the expansion in domestic use of corn for ethanol production. On the other hand, the US exports of DDGS have increased from 1 million metric tons in 2004 to about 8 million metric tons in 2011. During this time period, the share of exports in total output of soybeans increased from 44 to 53%. As another example, the share of domestic use in total energy produced in the US decreased from 97% in 2004 to 91% in 2011.

The regional GTAP input–output tables represent the cost structure of sectors/industries in each region. The cost structures of the well-established sectors have not significantly changed. However, changes are large for the ethanol and biodiesel sectors. These industries were relatively new in 2004 with large shares for capital and smaller shares for feedstocks. In 2011, these industries became more mature and well established with lower shares for capital and higher shares for feedstock. For example, the share of capital in total costs of ethanol sector dropped from 52.2% in 2004 to 18.5% in 2011. That reflects the fact that emerging sectors use more capital at the early stages of their development paths. When well established, the share of capital usually drops, but the share of intermediate inputs goes up. For example, the share of non-energy intermediate inputs (mainly corn) in total costs of ethanol sector increased from 38.3% in 2004 to 76.1% in 2011. This difference is also due to the higher corn price in 2011 compared with 2004. Notice that the price of corn was exceptionally high in 2011, and therefore, the share of this input in total cost of ethanol was slightly higher in this year. This share has been around 65 to 75% in recent years.

Biophysical data The GTAP-BIO database includes data on land cover, harvested area, and crop production by region. It also represents cropland pasture in a few counties. Here, we examine changes in these variables between 2004 and 2011.

Land cover At the global scale, areas of forest and cropland increased by 7.8 and 17.5 million hectares, respectively, while area of pasture decreased by 41.7 million hectares. This means that at the global scale, the livestock industry in 2011 is using less land directly compared with 2004. At the regional level, the largest expansion in cropland occurred in sub-Saharan Africa (by 15.7 million hectares), and the largest reduction was observed in the US (by 10.5 million hectares).

Harvested area At the global scale, harvested area increased by 94 million hectares between 2004 and 2011. As mentioned earlier in this paper, the area of cropland has increased by 17.5 million hectares during the same time period. Comparing these two figures indicates that the harvested area has grown faster than land cover between 2004 and 2011. This could be due to some combination of reductions in crop failure and idled land and increases in double cropping between 2004 and 2011. The largest expansions in harvested area occurred in sub-Saharan Africa (by 32.5 million hectares), India (by 21.9 million hectares), and China (by 13.7 million hectares). Harvested area decreased in a few regions slightly.

Among crops at the global scale, the largest expansion in harvested area is for oilseeds (by 33.2 million hectares). At the global scale, the smallest increase in harvested area was for wheat. The harvested area of wheat increased only by 3.4 million hectares between 2004 and 2011.

Harvested area decreased in all crop categories in the US, except for coarse grains. The harvested area of coarse grains increased by 2 million hectares. This reflects the need for more corn for ethanol production in the US. In the EU, the harvested area of almost all crops decreased, except for oilseeds. This reflects the need for more oilseeds for biodiesel production in the EU.

Crop production At the global level, production of paddy rice, wheat, coarse grains, oilseeds, and other crops increased by 115.4 million metric tons (MMT), 66.8, 127.7, 178, and 907.3 MMT, repressively, between 2004 and 2011. The per capita production for all of these crop categories also increased by 9, 1.8, 5.5, 18.7, and 52 kg, respectively. Thus, more food is available to consume per person. Of course, some of these crops are consumed for non-food uses (e.g., corn for ethanol or oilseeds for biodiesel), but some of them (like rice and wheat) are basic food crops.

The largest increases in crop production occurred in Brazil (by 368.6 MMT), China (by 325.7 MMT), India (by 305.9 MMT), and sub-Saharan Africa (by 128.2 MMT) between 2004 and 2011. Crop production has fallen (by 68.4 MMT) in Canada. Again, that is basically due to a correction in the GTAP data for Canada as indicated above. In the US only production of coarse grains has increased by 4.2 MMT, while production of other crops has decreased between 2004 and 2011.

Yield Crop yields increased in many regions. At the global scale rice, wheat, coarse grains, oilseeds, and other crop yields increased by 9.7, 8.8, 7.8, 13.8, and

7.2%, respectively, between 2004 and 2011. The largest growth in crop yields occurred in Brazil (ranging from 26 to 38%), India (ranging from 10 to 40%), Russia (ranging from 10 to 35%), and members of the former Soviet Union (ranging from 15 to 40%). In many other regions, yields also increased by large percentages.

In the US, yield has slightly increased for paddy rice, wheat, and other crops, and decreased for coarse grains (by 4%) and soybeans (by 0.2%) between 2004 and 2011. It is important to note that the US corn yield was more than 10 metric tons per hectare in 2004, higher than the normal trend. On the other hand, it was about 9.2 metric tons per hectare in 2011, below the normal trend.[1] Therefore, while corn yield follows an upward trend in the US, our data show a reduction in coarse grain yield between 2004 and 2011.

Cropland pasture Cropland pasture represents a portion of cropland which has been cultivated and used for crop production in the past, but currently is in pasture. The GTAP-BIO 2004 database includes cropland pasture only for US (25 million hectares) and Brazil (23.6 million hectares). The area of cropland pasture in US has dropped to 5.2 million hectares in 2011, according to the US census. Due to the lack of information, we assumed that the area of cropland pasture in Brazil has dropped to 11.8 million hectares in 2011. Finally, with access to new data, about 5.2 million hectares of cropland pasture was added to the database for Canada.

Improvements in GTAP-BIO model

Birur et al. [1] used an improved version of the GTAP-E model [41] and developed the first version of the GTAP-BIO model to analyze the impacts of biofuel production on energy and agricultural markets and to study the market. This early model version was able to trace market-mediated responses due to biofuel production. Responses such as but not limited to: (1) increases in crop prices due to expansion in feedstock demand for biofuel production; (2) reductions in crop demands in non-biofuel uses such as food and feed; (3) changes in the global trade of crops and other agricultural products; (4) expansion in crop supplies across the world; (5) substitution between biofuels and fossil fuels; (6) crop switching as relative prices changed; and (7) competition for limited resources. However, the model was not able to accurately quantify these impacts and was missing several other important market-mediated responses due to several limitations.

The first version of the model did not include biofuel by-products such as Distiller's Dried Grains with Soluble (DDGS) and oilseed meals. Hence, the model was missing the impacts of biofuel production on the livestock industry and animal feed rations. Therefore, it provided misleading results on livestock demand for crops, leading to overestimation of biofuel impacts on demand for crops and land use changes. In addition, the first model did not consider the fact that productivity of new land likely would be lower than the existing cropland. Furthermore, the first model did not include any yield response to higher crop prices. More importantly, it was incapable to trace changes in physical land. Over the past decade, many modifications were introduced to GTAP-BIO to improve its performance and eliminate its initial deficiencies. Golub and Hertel [42] explained some of the early modifications. Here, we briefly outline them and introduce some newer modifications.

Taheripour et al. [3, 4] introduced biofuel by-products in the model and defined a module to take into account substitution between biofuels by-products (such as DDGS and oilseed meals) and feed crops in livestock feed rations. Hertel et al. [20] improved the model to distinguish between productivities of the new and existing croplands. They developed a new land supply system to trace changes in physical land. In addition, they defined a module to better take care of crop yield responses to changes in crop prices and production costs. The impacts of these modifications on the outcomes of the model were substantial, basically leading to lower induced land use changes compared with the initial model.

The three main modifications made by Hertel et al. [20] were significant contributions. However, these authors established their modifications based on some limited real-world observations. First, they assumed that the productivity of new land is about 2/3 of the productivity of existing cropland everywhere across the world. Second, they assumed that the land transformation elasticity among forest, pasture, and cropland equals to 0.2 across the world, and also used a uniform land transformation elasticity of 0.5 to govern allocation of cropland across alternative crops everywhere around the world. Finally, they assumed that crop yield response with respect to changes in profitability of crop production is uniform across regions and crops. They also assumed that crop harvest frequency remains fixed, meaning no expansion in multiple cropping and no conversion of idled cropland to crop production. Many of these limitations were removed over time.

Tyner et al. [23] partially removed the last issue mentioned above by introducing cropland pasture into the model for only US and Brazil, where data were available. Cropland pasture is a particular marginal cropland that usually is used as pasture land but moves to cropland when more cropland is needed. The model developed by these authors and the subsequent work continued to

---

[1] The US corn yields for 2004 and 2011 are obtained from the USDA database.

ignore multiple cropping and assumed idled cropland will remain idle.

Taheripour et al. [5] used a biophysical model (TEM) and estimated a set of extensification parameters which represent productivity of new cropland versus the existing land by region at the spatial resolution of Agro-Ecological Zone. Using a tuning process, Taheripour and Tyner [29] developed a set of land transformation elasticities by region according to recent real-world observations on land use changes across the world. These land transformations elasticities govern land allocation across land cover categories and distribute cropland among crops.

Recently, Taheripour et al. [43] introduced several more important improvements: First, they altered the land use module of the model to take into account intensification in cropland due to multiple cropping and/or returning idled cropland to crop production. They defined a new set of regional intensification parameters and determined their magnitudes according to observed land use changes across the world in recent years. They also altered the assumption that the elasticity of yield improvement with respect to changes in profitability of crops is uniform across regions. Instead, they defined regional yield responses and tuned their magnitudes according to observed regional changes in crop yields.

These model improvements were targeted towards the first-generation biofuels. Taheripour and Tyner [44] developed a special version of the model (called GTAP-BIO-ADV) to examine the economic and land use impacts of the second-generation biofuels. Unlike other versions of the GTAP-BIO model which put all crops in one nest in the land supply tree, the GTAP-BIO-ADV model uses a different land supply tree which puts cropland pasture and dedicated crops (such as miscanthus and switchgrass) in one nest and all other crops in another nest and allows the land to move between the two nests. They used this setup to avoid conversion of food crops to dedicated energy crops to make greater use of cropland pasture (a representative for marginal land) to produce dedicated energy crops. The GTAP-BIO-ADV model was developed prior to the tuning process described above and only includes those model modifications that were available when the model was developed in 2011.

This paper brings all the modifications explained above less than one umbrella and generates a comprehensive model to have the first- and second-generation biofuels in one model. We also match the model with the 2011 GTAP-BIO database introduced in the data section. Then, we examined the land use impacts and the biofuel pathways outlined in the next sections. Henceforth, we refer to this model as GTAP-BIO-ADV11.

The modeling framework used in this paper is based on the latest model introduced by Taheripour et al. [43] which includes all the modification made in the GTAP-BIO model over time including intensification in cropland due to multiple cropping and returning idled cropland to crop production. To do simulations for the second-generation biofuels, we alter the land supply tree of this model according to the land supply tree of the GTAP-BIO-ADV model. The top left and right panels of Fig. 1 represent the land supply trees of the latest version of the GTAP-BIO and GTAP-BIO-ADV models, respectively. The bottom panel of this figure shows the mix of these two panels which we used in this paper. As shown in the bottom panel, the land supply tree of the new model uses two nests to govern changes in land cover and two nests to manage allocation of cropland among crops, including miscanthus and switchgrass. At the lowest level of this tree, available land is allocated between forest and a mix of cropland–pasture. The second level allocates the mix of cropland–pasture to cropland and pasture. Then, at the third level, cropland is divided between the traditional crops (first nest of cropland) and dedicated crops including cropland pasture (second nest of cropland). Finally, at the top level, the first category of land is allocated among the traditional crops, and the second category between miscanthus, switchgrass, and cropland pasture.

The land transformation elasticities used with this specification match the tuned elasticities reported by Taheripour and Tyner [29] for the land cover and allocation of cropland among the traditional crops. For the cropland nest including miscanthus, switchgrass, and cropland pasture, following Taheripour and Tyner [44], we used a relatively large land transformation elasticity to support the idea of producing dedicated crops on marginal cropland and to avoid a major competition between the traditional crops and dedicated energy crops. For the nest between the first and second groups of cropland, we use the same tuned land transformation elasticities which we used in land allocation among the first group of crops (i.e., traditional crops). With this assignment, the new model replicates the results of the old model for the first-generation biofuels.

The modeling framework developed by Taheripour et al. [43] takes into account intensification in cropland due to multiple cropping and/or conversion of unused cropland. These authors introduced a new land intensification factor into the model and tuned it according to the actual recent historical observations. The modeling framework used in this paper adopts the approach developed by these authors. However, it required changes to introduce land intensification in the new model which uses a different land supply structure.

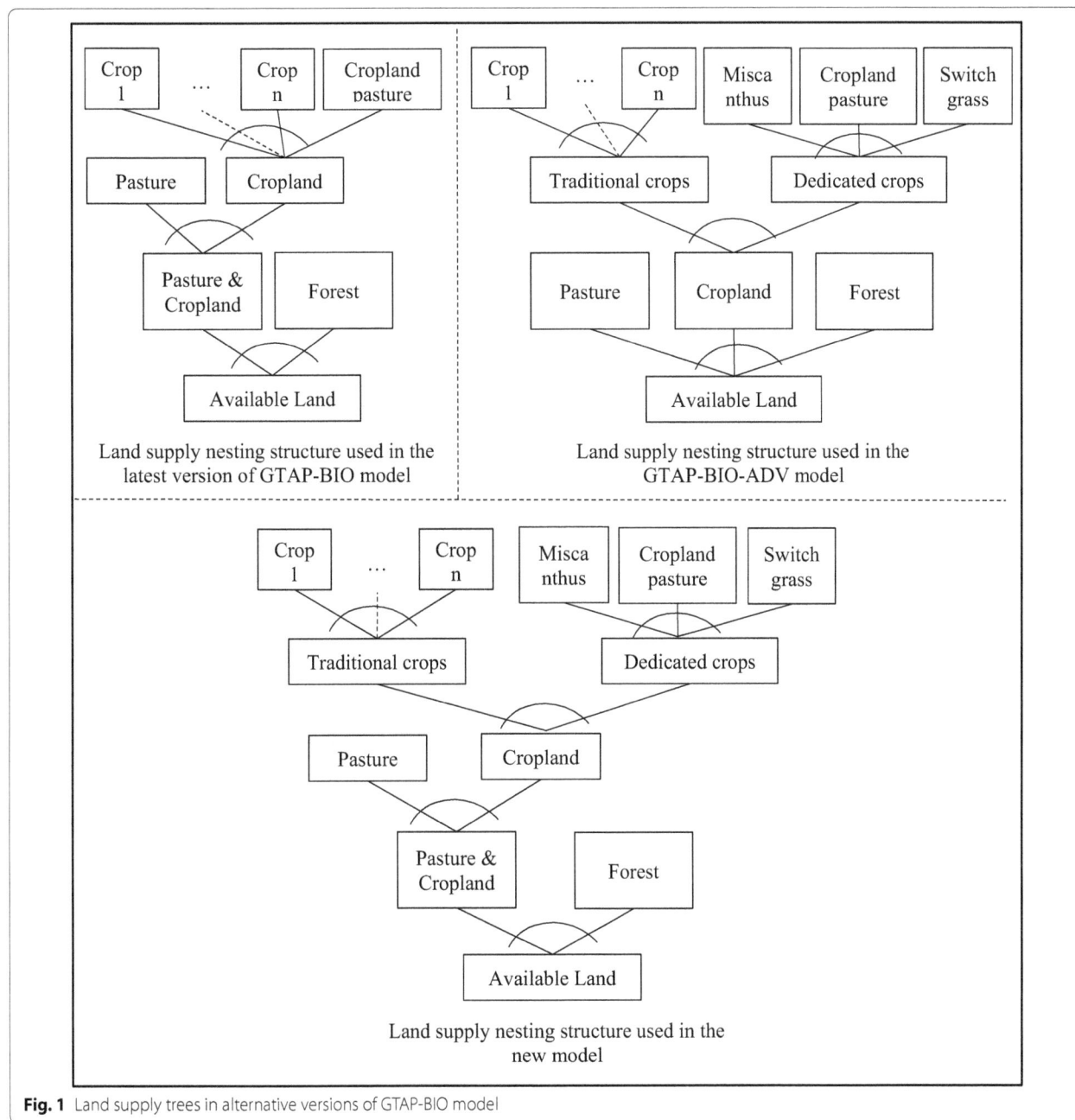

Fig. 1 Land supply trees in alternative versions of GTAP-BIO model

With a one-nest cropland structure used by Taheripour et al. [43], the relationship between changes in harvested area and changes in cropland in the presence of land intensification can be captured by the following equation[2]:

$$h_j = \text{tl} + \theta\left[\text{pl} - \text{ph}_j\right]. \qquad (1)$$

Here, $\text{tl} = l + \text{afs}$, h_j represents changes in the harvested area of crop j, l indicates changes in available cropland due to deforestation (conversion from forest or pasture to cropland and vice versa), afs stands for changes in available land due to intensification (shift factor in land supply), θ shows the land transformation elasticity which governs allocation of land among crops, pl demonstrates changes in the cropland rent, and finally, ph_j denotes changes in the land rent for crop j.

---

[2] This equation only shows the impacts of the shift factor on harvested area. This shift factor appears in several equations of the land supply module. For details, see Taheripour et al. [36].

With a two-nest cropland nesting structure, presented in the bottom panel of Fig. 1, the following four relationships establish the links between changes in cropland and harvested areas in the presence of land intensification:

$$l_1 = \text{tl} + \emptyset\big[\text{pl} - \text{ph}_1\big], \tag{2}$$

$$l_2 = \text{tl} + \emptyset\big[\text{pl} - \text{ph}_2\big], \tag{3}$$

$$h_{1j} = l_1 + \omega_1\big[\text{pl}_1 - \text{ph}_{1j}\big], \tag{4}$$

$$h_{2j} = l_2 + \omega_2\big[\text{pl}_2 - \text{ph}_{2j}\big]. \tag{5}$$

In these equations, tl, afs, and pl carry the same definitions as described above. Other variables are defined as follows:

- l_1 and l_2 represent changes in the first and second branches of cropland.
- ph_1 and ph_2 indicate changes in the rents associated with the first and second branches of cropland.
- h_{1j} and h_{2j} stand for changes in the harvested areas of crops included in the first and second groups of crops.
- ph_{1j} and ph_{2j} show changes in the rents associated with each crop included in the first and second groups of crops.
- $\emptyset$ demonstrates the land transformation elasticity which governs allocation of cropland among the first and second groups of crops.
- ω_1 shows the land transformation elasticity which governs allocation of the first branch of cropland among the first group of crops; and finally.

- ω_2 represents the land transformation elasticity which governs allocation of the second branch of cropland among the second group of crops.

Taheripour et al. [36] used several relationships to introduce land intensification (due to multiple cropping and or conversion of unused land to cropland) and endogenously determine the size of afs by region. Among all modifications, they used to accomplish this task, they introduced a parameter, called intensification factor and denoted by γ_r, which represents the magnitude of intensification by region. This parameter varies between 0 and 1 (i.e. $0 \leq \gamma_r \leq 1$). When $\gamma_r = 1$, there is no land intensification. In this case, any expansion in harvested area leads to an expansion in cropland which comes from conversion of forest and/or pasture. On the other hand, when $\gamma_r = 0$, it shows that an expansion in harvested area will not expand cropland. In this case, the additional harvested area comes from multiple cropping and/or converting unused cropland to crop production. Taheripour et al. [43] determined the regional values for this parameter, according to recent observed trends in land intensification across the world. Figure 2 represents the regional values of this parameter.

As shown in Fig. 2, in China and India, the parameter of land intensification equals 0, indicating that in these two countries, an expansion in harvested area does not lead to an expansion in cropland. On the other hand, in some countries/regions, the parameter of land intensification is close to 1, for instance Japan and East Asia. In these regions, any expansion in harvested area will equal an identical expansion in cropland with no intensification. Finally, in some countries/regions, the land

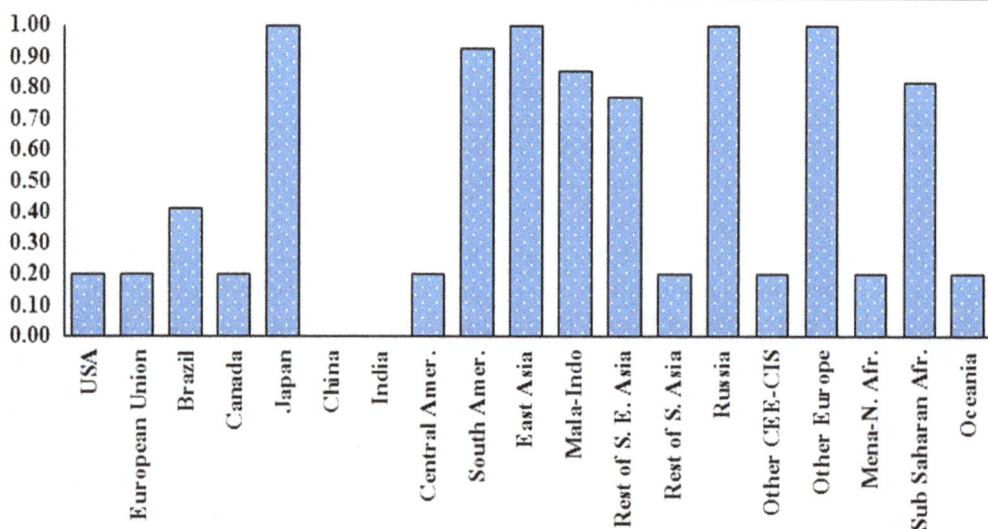

Fig. 2 Tuned regional land intensification parameters (γ_r)

intensification parameter is in between 0 and 1, say in Brazil and sub-Saharan Africa. In these regions, a portion of expansion in harvested area comes from land intensification and a portion from expansion in cropland. We use these values in our new model with one exception. For the case of Malaysia–Indonesia region, while the intensification parameter is less than 1, we assumed no intensification in this region, because it is the main source of palm oil and multiple cropping for palm tree is meaningless.

Following the existing literature [45, 46] which confirms yield improvement due to higher crop prices, Taheripour et al. [43] developed a set of regional elasticities which show yield to price response (known as YDEL) by region. Figure 3 represents these regional yield elasticities. Unlike the earlier version of the GTAP-BIO model which commonly assumed YDEL = 0.25, as shown in Fig. 2, the size of this elasticity varies between 0.175 and 0.325. Several regions including South America, East Asia, and Oceania have the lowest yield response, while Brazil has the highest rate.

Results

We developed several experiments to examine induced land use changes and emissions for the following first- and second-generation biofuel pathways using the GTAP-BIO-ADV11 model:

Experiment 1: Expansion in US corn ethanol by 1.07 BGs (from 13.93 BGs in 2011 to 15 BGs);

Experiment 2: Expansion in US soybean biodiesel by 0.5 BGs;

Experiment 3: Expansion in US miscanthus bio-gasoline by 1 BGs.

The bio-gasoline produced in the third experiment contains 50% more energy compared to corn ethanol. Since producing biofuels from agricultural residue (e.g., corn stover) does not generate noticeable land use changes [44], we did not examine ILUC for these biofuel pathways. We use an improved version of the emissions factor model developed by Plevin et al. [47] to convert the induced land use changes obtained from these simulations to calculate the induced land use emissions for each biofuel pathway. The earlier version of this model was not providing land use emission factors for converting land to dedicated energy crops such as miscanthus and switchgrass. Several papers have shown that producing dedicated energy crops on marginal lands will increase their carbon sequestration capabilities and that helps to sequester more carbon in marginal lands (for example, see [45]). The new emissions factor model provides land use emission factor for converting land to dedicated energy crops and takes into account gains in carbon stocks due to this conversion. The data for calibration of the new component in AEZ-EF were taken from the CCLUB model provided by Argonne National Laboratory [48]. Finally, it is important to note that the emission factor model takes into account carbon fluxes due to conversion of forest, pasture, and cropland pasture to cropland and the reverse.

Land use changes

The induced land use changes obtained from the examined biofuel pathways are presented in Table 1. The

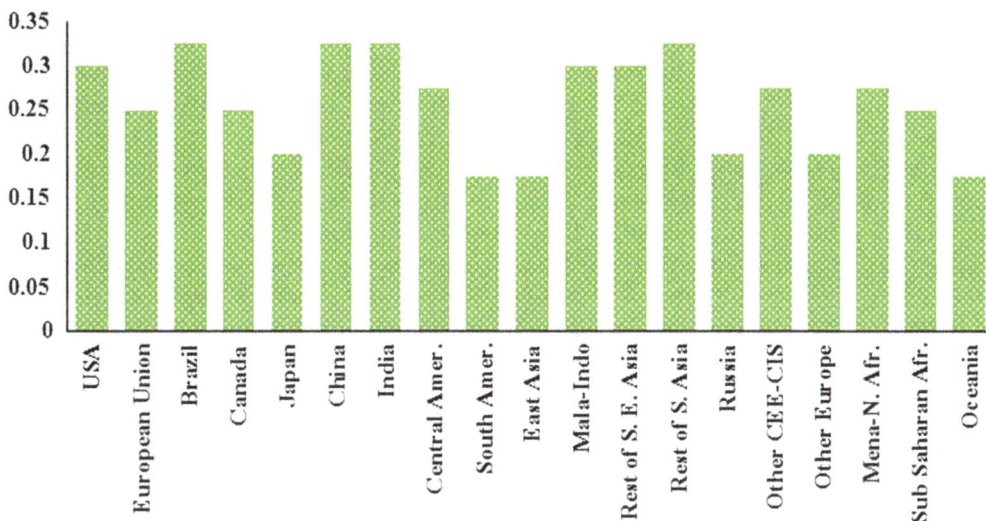

Fig. 3 Tuned regional yield to price elasticities (YDEL$_r$)

Table 1 Induced land use changes for alternative biofuel pathways (thousand hectares)

Description	USA	EU	Brazil	Canada	Sub-Saharan Africa	Others	World
Experiment 1 US corn ethanol							
Forest	3.0	−0.9	−3.8	−0.8	−8.8	−10.6	−22.0
Pasture	−7.3	−2.0	−8.1	−0.4	−22.3	−7.5	−47.5
Cropland	4.3	3.0	11.8	1.2	31.0	18.0	69.4
Harvested area	19.4	15.0	25.8	6.2	40.7	35.6	142.6
Cropland pasture	−74.8	0.0	−39.8	−14.1	0.0	0.0	−128.7
Experiment 2 US soybean biodiesel							
Forest	−0.9	−0.4	−2.0	−0.4	3.6	−3.7	−3.8
Pasture	−1.2	−0.8	−3.4	0.0	−16.4	−10.6	−32.5
Cropland	2.0	1.2	5.4	0.4	12.9	14.3	36.2
Harvested area	9.5	5.8	11.7	2.2	16.6	19.1	64.9
Cropland pasture	−42.6	0.0	−18.1	−5.2	0.0	0.0	−66.0
Experiment 3 US miscanthus bio-gasoline							
Forest	−41.0	−1.4	−5.7	−1.7	−13.7	−20.1	−83.7
Pasture	39.3	−2.2	−8.3	0.2	−23.9	−0.9	4.2
Cropland	1.8	3.7	13.9	1.5	37.6	21.0	79.7
Harvested area	9.1	18.2	30.3	7.7	49.6	41.6	156.4
Cropland pasture	−1017.3	0.0	−45.6	−14.7	0.0	0.0	−1077.6

expansion in US ethanol production from its 2011 to 15 BGs increases the global harvested area of corn by about 621 thousand hectares, after taking into the expansion in DDGS in conjunction with ethanol production. The expansion in demand for corn encourages farmers to switch from other crops (e.g., wheat, soybeans, and several animal feed crops) to corn due to market-mediated responses. That transfers a net of 349 thousand hectares from other crops to corn at the global scale. In addition, the area of cropland pasture (a marginal land used by livestock industry) drops by 129 thousand hectares in the US, Brazil, and Canada. Hence, about 478 (i.e. 349 + 129) thousand hectares of the land requirement for corn production comes from reductions in other crops and cropland pasture. Therefore, at the end, harvested area increases only by 143 (i.e. 621−478) thousand hectares, as shown in Table 1. However, due to intensification, cropland area grows only by 69.4 thousand hectares. This means that about 51% of the need for expansion in harvested area is expected to be covered by multiple cropping and/or using idled cropland. Therefore, the land requirement for 1000 gallons of corn ethanol is about 0.06 hectares in the presence of land intensification. Ignoring intensification, the land requirement increases to 0.13 hectares per 1000 gallons of ethanol.

In addition to changes in land cover, expansion in corn ethanol generates changes in the mix of cropland. In particular, it transfers some cropland pasture to the traditional crops. For the expansion in corn ethanol from 2011 to 15 BGs, about 129 thousand hectares of cropland pasture will be converted to the traditional crops, as shown in the first panel of Table 1. This is about 0.12 hectares per 1000 gallons of ethanol. For the case of corn ethanol, deforestation covers 32% of the land requirement and the rest (68%) is due to conversion of pasture to cropland.

An expansion in soybean biodiesel produced in the US by 0.5 BGs increases the global harvested area by about 64.5 thousand hectares, but only 56% of this expansion transfers to new cropland due to intensification. Therefore, global cropland increases by 36.1 thousand hectares. The index of land requirement for 1000 gallon of soybean biodiesel is about 0.07 hectares. Ignoring the land intensification, this index jumps to 0.13 hectares per 1000 gallons of soybean biodiesel. These indices are similar to their corresponding values for the cases of corn ethanol. For this pathway, the rate of conversion from cropland pasture to traditional crops is about 0.13 hectares per 1000 gallons of biodiesel, very similar to the corresponding rate for corn ethanol.

We now turn to induced land use changes for cellulosic biofuels produced from dedicated energy crops such as miscanthus or switchgrass. The narrative of induced land use changes for these biofuels is entirely different from the description of induced land use changes for the first-generation biofuels producing biofuels (say ethanol) from traditional crops (say corn) generates market-mediated responses such as reduction in consumption of crops in non-biofuel uses, switching among crops, expansion in biofuels by-products (which can be used in livestock feed rations instead of crops), and yield improvement. These market-mediated responses reduce the land use impacts of producing biofuels from traditional crops as described by Hertel et al. [20]. However, producing cellulosic biofuels from energy crops such as miscanthus or switchgrass may not generate these market-mediated responses.

For example, consider producing bio-gasoline from miscanthus, which we examine in this paper. This pathway produces no animal feed by-product. Therefore, an expansion in this biofuel does not lead to a reduction in livestock demand for crops. Miscanthus is not used in other industries. Hence, we cannot divert its current uses to biofuel production. Thus, miscanthus should be produced for every drop of bio-gasoline. For example, if we plan to produce 1 BGs of miscanthus bio-gasoline, then we need about 775 thousand hectares of land (with a conversion rate of 66.1 gallons per metric ton of miscanthus and 19.5 metric tons of miscanthus per hectare as we assumed in developing the GTAP-BIO database). Now, the question is: From where will the required land for miscanthus production come?

It is frequently argued that dedicated energy crops should not compete with the traditional food crops. This means no or little conversion from the traditional food-feed crops to cellulosic energy crops. It is also commonly believed that cellulosic energy crops should be produced on low-quality "*marginal land*". Beside this widespread belief, the definition and availability of "*marginal land*" are subject to debate [49]. If the low-quality marginal land is entirely unused, then producing cellulosic crops on these lands may not significantly affect competition for land. In this case, unused land will be converted to miscanthus as needed to meet the feedstock demand for the stipulated expansion in cellulosic biofuel.

However, if the low-quality marginal land is used by livestock producers as grazing land (e.g., cropland pasture in the US), then producing energy crops on cropland pasture directly and indirectly affects the livestock industry, and that generates some consequences. In this case, the livestock industry demands more feed crops, uses more processed feed, and/or converts natural forest to pasture in response to converting cropland pasture to miscanthus.

Now, consider the induced land use changes for the third experiment which extends production of the US bio-gasoline from miscanthus by 1 BGs. As shown in the bottom panel of Table 1, the anticipated expansion in miscanthus bio-gasoline increases the global harvested area by 156.4 thousand hectares. However, due to intensification, the global cropland area grows only by 79.7 thousand hectares. Therefore, the index of land requirement for 1000 gallons of miscanthus bio-gasoline is about

0.08 hectares in the presence of land intensification. Ignoring intensification, the index of land requirement increases to 0.16 hectares per 1000 gallons of bio-gasoline. These land requirement indices are not very different from the corresponding figures for corn ethanol. However, three is a major difference between corn ethanol and miscanthus bio-gasoline when we compare their impacts on cropland pasture.

As shown in Table 1, an expansion in US miscanthus bio-gasoline by 1 BG converts 1077.6 thousand hectares of cropland pasture to cropland. This is about 1.08 hectares per 1000 gallons of miscanthus bio-gasoline. This figure is approximately 9 times higher than the corresponding figure for corn ethanol. This difference is because producing miscanthus bio-gasoline does not create the market-mediated responses which corn ethanol generates. The change in cropland pasture area (i.e., 1077.6 thousand hectare) is higher than the direct land requirement for producing 1 BG of miscanthus bio-gasoline (i.e., 763 thousand hectares). When the livestock industry gives up cropland pasture at a large scale, it uses more feed crops and/or processed feed items, and that generates some land use changes including more conversion of cropland pasture to traditional crops. Furthermore, a large conversion of cropland pasture to miscanthus increases the rental value of pasture land (a substitute for cropland pasture) significantly, and that generates some incentives for a mild deforestation in the US, as shown in the lowest panel of Table 1. In the third experiment, the price of miscanthus increases by 53% and the livestock price index (excluding non-ruminant) goes up by about 0.5% which is 5 times higher than the corresponding figure for the forestry sector. Pasture rent grows at about 5% across US AEZs, while the corresponding rate for forest is less than 1%. For the case of corn ethanol, which induces mild conversion of cropland pasture forest and pasture rents grow similarly at rates less than 1% across AEZs in the US. Finally, it is important to note that the tuned land transformation elasticity for forest to agricultural land in the US is small, according to recent observations [29]. In conclusion, while producing miscanthus bio-gasoline slightly increases demand for cropland, it induces major shifts in marginal land (say cropland pasture) to miscanthus production.

Land use emissions

First, consider induced land use emissions for the first-generation biofuels including corn ethanol and soybean biodiesel for four alternative modeling and database cases: (1) 2004 database with no intensification; (2) 2004 database with intensification; (3) 2011 with no intensification; and (4) 2011 with intensification. The emission results for the first three cases (i.e., cases 1, 2, 3) are taken from Taheripour et al. [43]. The last case represents the results of the simulations conducted in this paper.

Figure 4 shows the results for corn ethanol. With intensification in cropland, an expansion in US ethanol from its 2011 level to 15 BGs generates 12 g CO_2e/MJ emissions. The corresponding simulation with no intensification generates 23.3 g CO_2e/MJ emissions. This means that the new model which takes into account intensification in cropland and uses tuned regional YDEL parameters generates significantly lower emissions, approximately by half. The corresponding cases obtained from the 2004 databases represent the same pattern, but demonstrate lower emissions rates. An expansion in corn ethanol from its 2004 level to 15 BGs generates 8.7 g CO_2e/MJ emissions with intensification and 13.4 g CO_2e/MJ with no intensification.

These results indicate that the 2011 database generates higher emissions for corn ethanol compared with the 2004 databases, regardless of modeling approach. However, the new model which takes into account intensification in cropland and uses tuned regional YDEL values projects lower emissions, regardless of the implemented database. The 2011 database generates more emissions for corn due to several factors including but not limited to: (1) less availability of cropland pasture in the US in 2011; (2) less flexibility in domestic use of corn in 2011; (3) less flexibility in US corn exports in 2011; (4) smaller US corn yield in 2011; (5) more reductions in US crop exports (in particular soybean and wheat) in 2011; (6) larger DDGS trade share in 2011; (7) smaller capital share in corn ethanol cost structure; and (8) finally, the marginal land use impacts of ethanol in 2011 are much larger than 2004, because the base level of ethanol in 2011 is much larger than 2004.

Figure 5 shows the results for soybean biodiesel. In the presence of intensification in cropland, an expansion

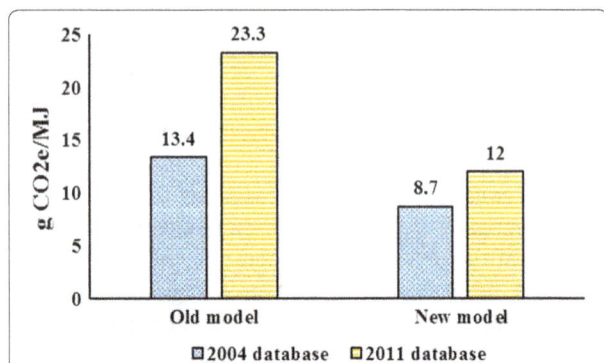

Fig. 5 Induced land use emissions for soybean biodiesel with 2004 and 2011 databases with and without land intensification

in the US soybean biodiesel by 0.5 BGs generates 18 g CO_2e/MJ emissions. The corresponding simulation with no intensification generates 25.5 g CO_2e/MJ emissions. This means that, similar to the cases for corn ethanol, the new model which takes into account intensification in cropland and uses tuned regional YDEL parameters generates significantly lower emissions. The corresponding cases obtained from the 2004 databases represent the same pattern. An expansion in the US soybean biodiesel by 0.5 BGs generates 17 g CO_2e/MJ emissions with intensification and 21.6 g CO_2e/MJ with no intensification. Furthermore, producing soybean biodiesel in the US encourages expansion in vegetable oils produced in some other countries including more production of palm oil in Malaysia and Indonesia on peat land, which entails extremely high emissions. This is one reason why land use change emissions induced by US soybean biodiesel production are generally higher than those induced by US corn ethanol production.

Unlike the case of corn ethanol, these results indicate that the 2011 database generates slightly higher emissions for soybean biodiesel compared with the 2004 databases, regardless of modeling approach. This observation is due to several factors including but not limited to: (1) conversion of a larger portion of US soybean exports to domestic use in 2011 which reduces the size of land conversion in US; (2) Brazil, Canada, and other countries produce more soybeans in 2011; (3) significantly larger oilseed yields across the world (except for US) generates weaker land conversion outside the US; (4) larger availability of oilseed meals in 2011 which contributes to a higher share of pasture in 2011; and larger share of palm oil in total vegetable oils in 2011.

We now turn to induced land use emissions for miscanthus bio-gasoline. Two alternative cases are examined to highlight the role of soil carbon sequestration gained from production of miscanthus on marginal land. First, we assume that producing miscanthus on cropland

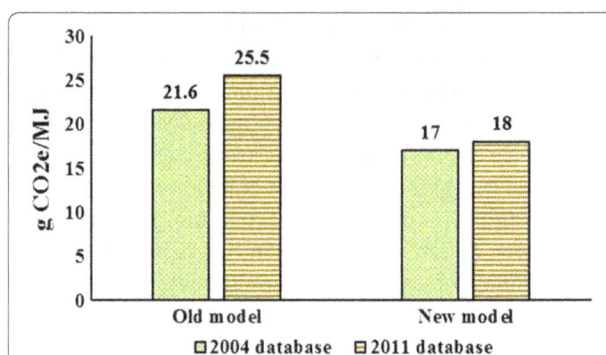

Fig. 4 Induced land use emissions for corn ethanol with 2004 and 2011 databases with and without land intensification

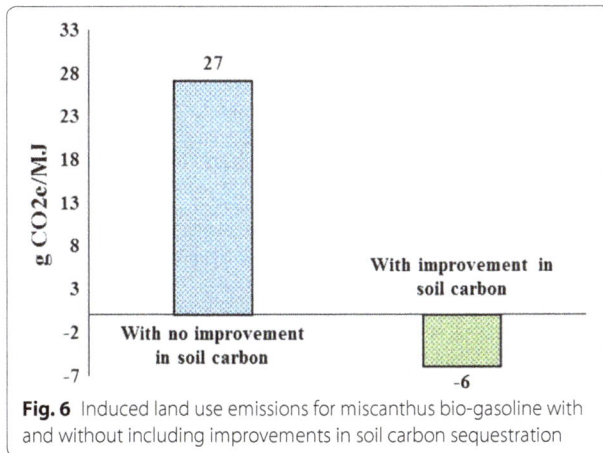

Fig. 6 Induced land use emissions for miscanthus bio-gasoline with and without including improvements in soil carbon sequestration

pasture does not improve soil carbon sequestration. Then, following the literature [48, 49][3], we take into account the fact that producing miscanthus on marginal land improves the soil carbon content. The existing literature confirms that producing miscanthus on marginal land improves its soil carbon content.

For the first case, an expansion in US miscanthus bio-gasoline by 1 BGs generates about 27 g CO_2e/MJ emissions. Compared with corn ethanol and soybean biodiesel, this figure is large. As mentioned before, an expansion in US miscanthus bio-gasoline by 1 BGs transfers about 1117.6 thousand hectares of cropland pasture to miscanthus production and other tradition crops. Only about 70% of this conversion goes to miscanthus. Hence, if we ignore the carbon saving from miscanthus production, then producing bio-gasoline from miscanthus generates more emissions than corn ethanol. For the second case, as shown in Fig. 6, the emissions score for miscanthus to bio-gasoline drops to about −6 g CO_2e/MJ. This figure is in line with the results reported by Wang et al. [50]. These authors used induced land use results obtained from an earlier version of the GTAP model and emissions factors from the CCLUB calculated that producing ethanol from miscanthus generates negative land use emissions by −7 g CO_2e/MJ. On the other hand, Dwivedi et al. [45], who used farm and firm level data in combination with some limited field experiments, reported that converting miscanthus to ethanol generates about −34 to −59 g CO_2e/MJ land use emissions. These results underscore the fact that for the case of cellulosic biofuels, the magnitude of induced land use emissions

varies significantly by the method of calculating land use changes and largely depends on the assigned emission factor to the converted marginal land.

Conclusions

In this paper, we have covered three major modifications to the GTAP-BIO model. First, we reviewed the change from using the 2004 database to 2011. Many changes in the global economy occurred between 2004 and 2011 including the development of the first-generation biofuels in many world regions, changes in crop production area and yields, and vast changes in the levels and mix of GDP in many world regions. All these changes and many others have a profound impact on any simulations that are performed using the 2011 database versus the older 2004 data. Of course, moving forward, we must use the updated data, so it is important to understand the significance of the major changes, particularly as they impact biofuels and land use.

The second major change was a revision of the GTAP-BIO model to better handle intensification. The previous versions of the GTAP model and other similar models assumed that a change in harvested area equals a change in land cover. Examining the FAO data, it was clear that this is not the case, so we used that data to develop and parameterize differences in changes at the intensive and extensive margins for each world region. We also calibrated the yield price elasticity by region, as the FAO data also indicated significant differences in yield response by region.

The third major change was to develop a new version of the model (GTAP-BIO-ADV11) used to evaluate land use changes and emissions for dedicated cellulosic feedstocks such as miscanthus. These dedicated energy crops are not similar to the first-generation feedstocks in the sense that they do not generate the level of market-mediated responses we have seen in the first-generation feedstocks. The major market-mediated responses are reduced consumption, crop switching, changes in trade, changes in intensification, and forest or pasture conversion. There is no current consumption or trade in miscanthus. There are no close crop substitutes. Most of the land needed for miscanthus production comes from cropland pasture. Since that is an input into livestock production, more land is needed to produce the needed livestock inputs (which is a market-mediated response). Thus, miscanthus (and other similar cellulosic feedstocks) will need more land that required to actually grow the feedstock. Then, the emissions for the cellulosic feedstocks depend on what we assume in the emissions factor model regarding soil carbon gained or lost in converting land to miscanthus. Much of the literature suggests miscanthus actually sequesters carbon, when grown on the existing

[3] The authors are grateful to Argonne National Laboratory for providing data on carbon sequestration for cellulosic feedstocks and to Dr. Richard Plevin for his work in revising the CARB Agro-ecological Zone Emission Factor (AEZ-EF) Model to handle cellulosic feedstocks.

cropland or even marginal land. When we take into account this important fact, land use change emissions due to production of bio-gasoline from miscanthus drop to a negative number.

Finally, it is important to note the importance of the new results for the regulatory process. The current CARB carbon scores for corn ethanol and soy biodiesel are 19.8 and 29.1, respectively. The new model and database scores are 12 and 18, respectively, for corn ethanol and soy biodiesel. Thus, the current estimate values are substantially less than the values currently being used for regulatory purposes.

Abbreviations
GTAP: Global Trade Analysis Project; GHG: greenhouse gas; FAO: Food and Agricultural Organization; CARB: California Air Resources Board; ILUC: induced land use change; LCA: life cycle analysis; EIA: Energy Information Administration; FAOSTAT: FAO Statistics Database; gro: coarse grains (in GTAP); osd: oilseeds (in GTAP); vol: vegetable oils and fats (in GTAP); ofd: food (in GTAP); BG: billion gallons; GDP: gross domestic product; EU: European Union; MMT: million metric tons; DDGS: distillers dried grains with solubles; US: United States; TEM: Terrestrial Ecosystem Model.

Authors' contributions
FJ, XZ, and WET contributed to all phases of the research. All authors read and approved the final manuscript.

Acknowledgements
The authors are grateful to Luis Pena Levano for his research assistance on the 2011 database. We are also grateful to Hao (David) Cui for his research assistance on the intensification component of the research.

Competing interests
The authors declare that they have no competing interests.

Funding
This research was partially funded by the US Federal Aviation Administration (FAA) Office of Environment and Energy as a part of ASCENT Project 001 under FAA Award Number 13-C-AJFE-PU. Any opinions, findings, and conclusions or recommendations expressed in this material are those of the authors and do not necessarily reflect the views of the FAA or other ASCENT sponsors.

References
1. Birur D, Hertel T, Tyner WE. Impact of biofuel production on world agricultural markets: a computable general equilibrium analysis. GTAP Working Paper No. 53. Purdue University; 2008.
2. Hertel TW, Tyner WE, Birur DK. The global impacts of biofuel mandates. Energy J. 2010;30:75–100.
3. Taheripour F, Hertel TW, Tyner WE. Implications of biofuels mandates for the global livestock industry: a computable general equilibrium analysis. Agric Econ. 2011;42:325–42.
4. Taheripour F, Hertel TW, Tyner WE, Beckman J, Birur DK. Biofuels and their by-products: global economic and environmental implications. Biomass Bioenerg. 2010;34:278–89.
5. Taheripour F, Zhuang Q, Tyner WE, Lu X. Biofuels, cropland expansion, and the extensive margin. Energy Sustain Soc. 2012;2:25.
6. Tyner W, Taheripour F. Biofuels, policy options, and their implications: analyses using partial and general equilibrium approaches. J Agric Food Ind Organ. 2008;6:9.
7. Tyner WE, Taheripour F. Policy options for integrated energy and agricultural markets. Rev Agric Econ. 2008;30:387–96.
8. Alexandratos N, Bruinsame J. World agriculture towards 2030/2050: the 2012 revision. ESA working paper no. 12–30. Rome, Italy: U.N. Food and Agricultural Organization; 2012.
9. Alston JM, Babcock BA, Pardey PG. The shifting patterns of agricultural production and productivity worldwide. Ames: The Midwest Agribusiness Trade Research and Information Center, Iowa State University; 2010.
10. Babcock BA, Iqbal Z. Using recent land use changes to validate land use change models, staff report 14-ST 109. In: Staff report, Center for Agricultural and Rural Development ISU. Ames, Iowa: Iowa State University; 2014.
11. Borchers A, Truex-Powell E, Wallander S, Nickerson C. Multi-cropping practices: recent trends in double cropping. Economic information bulletin number 125 ed. Washington, DC: Economic Research Service, US Department of Agriculture; 2014.
12. Brady M, Sohngen B: Agricultural productivity, technological change, and deforestation: a global analysis. In: American Agricultural Economics Association Annual Meeting. Orlando, FL; 2008.
13. Cassman K. Ecological intensification of cereal production systems: yield potential, soil quality, and precision agriculture. Proc Natl Acad Sci. 1999;96:5952–9.
14. Foley JA, Ramankutty N, Brauman KA, Cassidy ES, Gerber JS, Johnston M, Mueller ND, O'Connell C, Ray DK, West PC. Solutions for a cultivated planet. Nature. 2011;478:337–42.
15. Ray DK, Foley JA. Increasing global crop harvest frequency: recent trends and future directions. Environ Res Lett. 2013;8:044041 (pp. 044010).
16. Taheripour F, Birur D, Hertel T, Tyner W. Introducing liquid biofuels into the GTAP database. In: GTAP Research Memorandum No. 11. West Lafayette, Purdue University; 2007.
17. Dimaranan BV, editor. Global trade, assistance, and production: the GTAP 6 data base. West Lafayette: Purdue University; 2006.
18. Beckman J, Hertel T, Taheripour F, Tyner WE. Structural change in the biofuels era. Eur Rev Agric Econ. 2012;39:137–56.
19. Birur DK, Hertel TW, Tyner WE. The biofuels boom: implications for world food markets. The Netherlands: Wageningen Academic Publishers; 2009.
20. Hertel T, Golub A, Jones A, O'Hare M, Plevin R, Kammen D. Effects of US maize ethanol on global land use and greenhouse gas emissions: estimating market-mediated responses. Bioscience. 2010;60:223–31.
21. Tyner WE, Taheripour F. Land use changes and consequent co_2 emissions due to US corn ethanol production. In: Encyclopedia of biodiversity, 2nd ed. 2012.
22. California Air Resources Board: California Low Carbon Fuel Standard (LCFS) Documents, Models, and Methods/Instruction. (2009). http://www.arb.ca.gov/regact/2009/lcfs09/lcfs09.htm.
23. Tyner WE, Taheripour F, Zhuang Q, Birur D, Baldos U. Land use changes and consequent co_2 emissions due to US corn ethanol production: a comprehensive analysis. In: Report to Argonne National Laboratory. Purdue University, Department of Agricultural Economics; 2010.
24. Narayanan G, Walmsley T, editors. Global trade, assistance, and production: the GTAP 7 data base. 2008.
25. Taheripour F, Tyner WE. Introducing first and second generation biofuels into GTAP data base version 7. In: GTAP, editor. GTAP Research Memorandum No. 21. Purdue University; 2011.
26. Liu J, Hertel T, Farzad T, Zhu T, Rigal C. International trade buffers the impacts of future irrigation shortfalls. Glob Environ Change. 2014;29:22–31.
27. Taheripour F, Hertel T, Liu J. Role of irrigation in determining the global land use impacts of biofuels. Energy Sustain Soc. 2013;3(4):1–18.
28. Taheripour F, Tyner WE. Global land use changes and consequent co_2 emissions due to US cellulosic biofuel program: a preliminary analysis. Purdue University; 2011.
29. Taheripour F, Tyner WE. Biofuels and land use change: applying recent evidence to model estimates. Appl Sci. 2013;3:14–38.
30. Taheripour F, Tyner WE, Wang MQ. Global land use changes due to the U.S. cellulosic biofuel program simulated with the GTAP Model Argonne National Laboratory and Purdue University; 2011.
31. Liu J, Hertel T, Taheripour F. Analyzing future water scarcity in computable general equilibrium models. Water Econo Policy. 2016;2.
32. California Air Resources Board: Staff report: initial statement of reasons for proposed rulemaking, proposed re-adoption of the low carbon fuel standard. Sacramento, CA; 2014.
33. California Air Resources Board: Staff report: initial statement of reasons for proposed rulemaking, proposed re-adoption of the low carbon fuel

standard, appendix I, Detailed analysis for indirect land use change. Sacramento, CA;2014.

34. Tyner WE, Taheripour F, Golub A. Calculation of indirect land use change (ILUC) values for low carbon fuel standard (LCFS) fuel pathways, interim report to the California Air Resources Board. 2011.

35. Aguiar A, Narayanan BG, McDougall R. Overview of the GTAP 9 data base. J Glob Econ Anal. 2016;1:181–208.

36. Taheripour F, Pena-Levano L, Tyner WE. Introducing first and second generation biofuels into GTAP data base version 9. GTAP Working Paper No. 29. Purdue University; 2016.

37. Pena-Levano L, Taheripour F, Tyner WE. Development of the GTAP land use data base for 2011. In: GTAP Research Memorandum West Lafayette, Purdue University; 2015.

38. ISTA Mielke GmbH. Oil world annual 2014: first global projections for 2014/2015. Germany: Hamburg; 2014.

39. Worth D. Agriculture and Agri-food Canada: Canadian cropland pasture data. Canada AaA; 2016.

40. Horridge M. GTAPAdjust—a program to balance or adjust a GTAP database. Melbourne: Centre of Policy Studies, Monash University; 2011.

41. McDougall R, Golub A. GTAP-E Release 6: a revised energy-environmental version of the GTAP model. In: GTAP technical paper no. 15. West Lafayette, Purdue University; 2007.

42. Golub A, Hertel T. Modeling land use change impacts of biofuels in the GTAP-BIO framework. Clim Change Econ. 2012;3:1250015.

43. Taheripour F, Cui H, Tyner WE. An Exploration of agricultural land use change at the intensive and extensive margins: implications for biofuels

44. Taheripour F, Tyner WE. Induced land use emissions due to first and second generation biofuels and uncertainty in land use emission factors. Econ Res Int. 2013;2013:12. doi:10.1155/2013/315787.

45. Dwivedi P, Wang W, Hudiburg T, Jaiswal D, Parton S, Long S, DeLucia E, Khanna M. Cost of abating greenhouse gas emissions with cellulosic ethanol. Environ Sci Technol. 2015;49:2512–22.

46. Miao R, Khanna M, Huang H. Responsiveness of yield and acreage to climate and prices. Am J Agric Econ. 2016;98:191–211.

47. Plevin RJ, Giggs HK, Duffy J, UYui S, Yeh S. Agro-ecological Zone Emission Factor (AEZ-EF) Model (v47): a model of greenhouse gas emissions from land-use change for use with AEZ-based economic models. In: GTAP technical paper no. 34. West Lafayette, Global Trade Analysis Project; 2014.

48. Argonne National Laboratory: Carbon calculator for land use change from biofuels production (CCLUB), users' manual and technical documentation. 2016. http://www.ipd.anl.gov/anlpubs/2016/09/130387.pdf. Accessed 22 Feb 2017.

49. Emery I, Mueller S, Qin Z, Dunn J. Evaluating the potential of marginal land for cellulosic feedstock production and carbon sequestration in the United States. Environ Sci Technol forthcoming. 2017.

50. Wang M, Han J, Dunn JB, Cai H, Elgowainy A. Well-to-wheels energy use and greenhouse gas emissions of ethanol from corn, sugarcane and cellulosic biomass for US use. Environ Res Lett. 2012;7:045905.

Induced land use change. In: Qin Z, Mishra U, Hastings A, editors. Bioenergy and land use change. American Geophysical Union (Wiley); 2016.

5

A novel transcription factor specifically regulates GH11 xylanase genes in *Trichoderma reesei*

Rui Liu[1,2], Ling Chen[1], Yanping Jiang[1], Gen Zou[1*] and Zhihua Zhou[1*]

Abstract

Background: The filamentous fungus *Trichoderma reesei* is widely utilized in industry for cellulase production, but its xylanase activity must be improved to enhance the accessibility of lignocellulose to cellulases. Several transcription factors play important roles in this progress; however, nearly all the reported transcription factors typically target both cellulase and hemi-cellulase genes. Specific xylanase transcription factor would be useful to regulate xylanase activity directly.

Results: In this study, a novel zinc binuclear cluster transcription factor (jgi|Trire2|123881) was found to repress xylanase activity, but not cellulase activity, and was designated as SxlR (specialized xylanase regulator). Further investigations using real-time PCR and an electrophoretic mobility shift assay demonstrated that SxlR might bind the promoters of GH11 xylanase genes (*xyn1*, *xyn2*, and *xyn5*), but not those of GH10 (*xyn3*) and GH30 (*xyn4*) xylanase genes, and thus regulate their transcription and expression directly. We also identified the binding consensus sequence of SxlR as 5'- CATCSGSWCWMSA-3'. The deletion of SxlR in *T. reesei* RUT-C30 to generate the mutant Δ*sxlr* strain resulted in higher xylanase activity as well as higher hydrolytic efficiency on pretreated rice straw.

Conclusions: Our study characterizes a novel specific transcriptional repressor of GH11 xylanase genes, which adds to our understanding of the regulatory system for the synthesis and secretion of cellulase and hemi-cellulase in *T. reesei*. The deletion of SxlR may also help to improve the hydrolytic efficiency of *T. reesei* for lignocellulose degradation by increasing the xylanase-to-cellulase ratio.

Keywords: Hemi-cellulase, Transcription factor, Xylanase, Glycoside hydrolase, *Trichoderma reesei*, CRISPR/Cas9

Background

Lignocellulosic biomass, consisting mostly of cellulose, hemi-cellulose, and lignin, is the most abundant and renewable energy source on earth [1]. Degradation of lignocellulosic biomass and continuation of the carbon cycle in nature is maintained mainly by microbial action, including different fungal species, such as *Trichoderma*, *Aspergillus*, *and Penicillium*. The biomass-degrading enzymes produced by these organisms also have applications in various fields of industry including food, fodder, paper, and textile industries. *Trichoderma reesei* is a well-known efficient producer of cellulase and hemi-cellulase, and is therefore widely employed by the enzyme industry for production of its own endogenous enzymes as well as production of heterogeneous proteins [2].

Cellulosic biofuel production continues to increase world-wide every year with a consequent rapidly increasing requirement for cellulase production from *T. reesei*. However, the production and optimization of enzyme formulations for lignocellulose degradation are still the major barriers to its extensive application. It is necessary to further enhance the hydrolytic efficiency of *T. reesei* enzyme preparations for lignocellulose and lower their relatively high cost. Due to the complex constitution and structure of native lignocellulose, the enzyme

*Correspondence: zougen@sibs.ac.cn; zhouzhihua@sippe.ac.cn
[1] CAS-Key Laboratory of Synthetic Biology, CAS Center for Excellence in Molecular Plant Sciences, Institute of Plant Physiology and Ecology, Chinese Academy of Science, Fenglin Rd 300, Shanghai 200032, China
Full list of author information is available at the end of the article

preparations produced by *T. reesei* must be supplemented with several types of exogenous enzymes to achieve effective degradation of natural complex lignocellulosic materials. For example, exogenous hemi-cellulase and other auxiliary enzymes are added to commercial cellulase complexes from Danisco or Novozymes [3]. This indicates that increasing the production of the most prominent hemi-cellulase (endo-β-1,4-xylanase), which catalyzes the hydrolysis of 1,4-β-D-xylosidic linkages in xylan to short xylooligosaccharides of varying length in *T. reesei*, is a good way to strengthen its hydrolysis activity [4].

Expression of genes encoding cellulase and hemi-cellulase in *T. reesei* is tightly controlled at the transcriptional level. Therefore, deleting and/or over-expressing transcription factors (TFs) that specifically regulate xylanase gene expression in *T. reesei* is a straightforward way to perform knowledge-based strain design. However, most known TFs regulate expression of both cellulase and xylanase genes in the same way. The most extensively studied TF is the negative TF CRE1 [5], which mediates carbon catabolite repression (CCR). In the cellulase hyperproduction strain *T. reesei* RUT-C30, CRE1 is truncated, which renders this strain carbon catabolite depressed [6]. The global transcriptional activator Xyr1 is obligatory for expression of most cellulase and hemi-cellulase genes [7, 8]. Other recognized TFs are the positively acting factor ACE2 [9, 10] and HAP2/3/5 complex [11], and the negatively acting factor ACE1 [12]. In addition, BglR [13] was identified as a new TF that upregulates expression of specific genes encoding β-glucosidases in *T. reesei*, and a putative methyltransferase, LAE1 [14, 15], was found to be essential for cellulase and hemi-cellulase production in *T. reesei*, although the mechanism is still unclear. ACE3 has been newly characterized to be indispensable for cellulase and xylanase activity in *T. reesei* [16]. All the above TFs have effects on the expression of both cellulase and hemi-cellulase genes.

Recently, a study found that some segmentally aneuploid (SAN) *T. reesei* strains exhibited enhanced growth in xylan-based media and produced higher levels of xylanase [17]. Further analysis confirmed that D segment duplication, not L segment deletion, in the genomes of these SAN strains, was responsible for the growth advantage in xylan-based media, but did not affect the cellulase activity. In fact, the Xpp1, TF, was identified by a pull-down assay based on the *xyn2* promoter and was confirmed as a negative regulator of the *xyn1* and *xyn2* xylanase genes and the *bxl2* β-xylosidase gene [18].

In this study, using bioinformatics analysis and gene deletion with the CRISPR/Cas9 system [19], we identified a gene encoding a protein designated as specialized xylanase regulator (SxlR). Deletion of the *sxlr* gene

resulted in increased xylanase activity while not affecting cellulase activity. This indicated that SxlR might be a novel TF that regulates xylanase expression. According to the results of real-time PCR (qPCR) and electrophoretic mobility shift assay (EMSA) analyses, we demonstrated that SxlR plays a critical role in the inhibition of the expression of xylanases belonging to the glycoside hydrolase 11 family (GH11) in *T. reesei* through binding to the promoter regions of target genes directly. Finally, we identified the binding consensus sequence of SxlR as 5′-CATCSGSWCWMSA-3′.

Results

Screening the putative xylanase-specific TFs

Based on bioinformatics analysis, we chose seven putative TFs as candidates (Additional file 1) for screening. All these genes were located in five different chromosomes [20, 21], and six of candidates were located in the D segment duplication resulting in higher xylanase activities as reported by Chen et al. [17]. We overexpressed these genes in *T. reesei* RUT-C30. After creating monoconidial cultures for genetic stability, we measured the xylanase and cellulase activities of the transformants. The xylanase activity, but not the cellulase activity, of the strain overexpressing the *sxlr* gene (jgi|TrireRUTC30_1|26638, O*sxlr*) decreased significantly (*t* test, *P* < 0.05) (Fig. 1a, b), and its extracellular protein concentration was decreased (Fig. 1c), which might indicate that this gene affects the xylanase activity alone. In contrast, no changes in xylanase activity, cellulase activity, or protein concentration were detected between RUT-C30 and the other six overexpression strains (Fig. 1).

To investigate how the *sxlr* gene is involved in regulation of xylanase activity in *T. reesei*, we deleted *sxlr* gene from the RUT-C30 strain to obtain Δ*sxlr* transformants (Fig. 2a, b). In contrast to the O*sxlr* strain, the xylanase activity of the Δ*sxlr* strains was increased significantly (Fig. 2c). However, the cellulase activity of the Δ*sxlr* strains was almost the same as the parent RUT-C30 strain. We also deleted the homologous *sxlr* gene of *T. reesei* Qm6a (jgi|Trire2|123881, Δ*6a-sxlr*) using the CRISPR/Cas9 system to verify the common regulation of xylanase activity in *T. reesei* strains. As expected, the xylanase activity of Δ*6a-sxlr* transformants was also dramatically increased, similar to the Δ*sxlr* strain (Fig. 2c), while its cellulase activity was not affected. Taken together, these results suggested that *sxlr* encoded a TF that negatively regulates xylanase activity.

The deletion of SxlR in *T. reesei* RUT-C30 results in higher xylanase activity and higher reducing sugar yield

We selected the cellulase hyperproduction strain RUT-C30 to explore the potential regulatory mechanism of

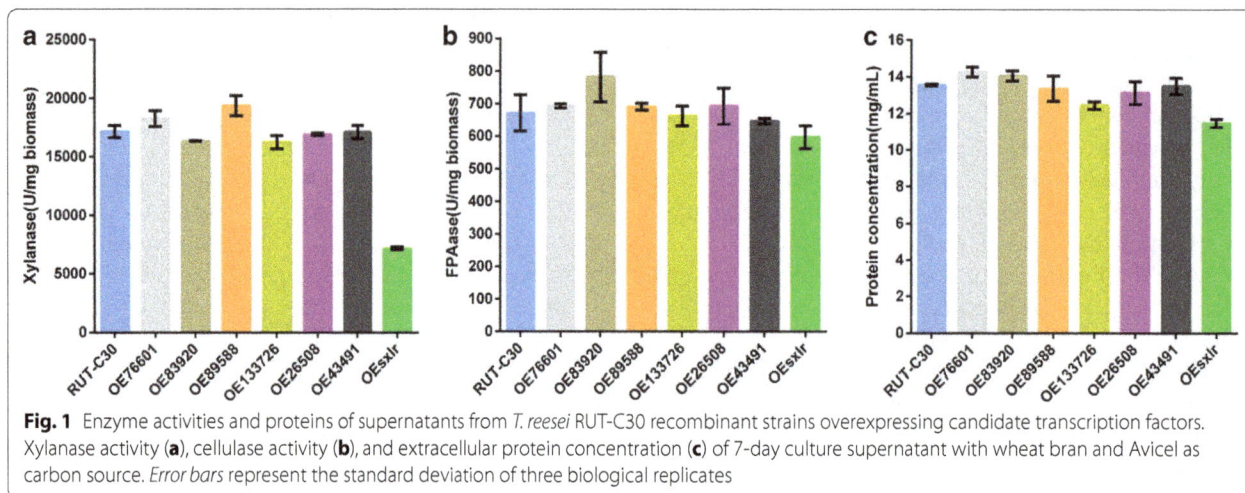

Fig. 1 Enzyme activities and proteins of supernatants from *T. reesei* RUT-C30 recombinant strains overexpressing candidate transcription factors. Xylanase activity (**a**), cellulase activity (**b**), and extracellular protein concentration (**c**) of 7-day culture supernatant with wheat bran and Avicel as carbon source. *Error bars* represent the standard deviation of three biological replicates

SxlR in *T. reesei*. In addition to the *sxlr* deletion strain (Δ*sxlr*, transformant 1) and the *sxlr* overexpression strain (O*sxlr*), we also constructed the in situ re-complementation strain (R*sxlr*) based on the Δ*sxlr* strain (Additional file 2). The xylanase activity of the Δ*sxlr* strain increased relative to that of RUT-C30 by 0.7-fold and 1.4-fold after 3 and 7 days, respectively, of incubation in inducing medium containing wheat bran and Avicel (Fig. 3a). The xylanase activity was also examined when xylan or lactose was used as the inducer (the sole carbon source in the inducing medium). The xylanase activity of Δ*sxlr* was 1.3-, 0.9-, and 0.7-fold higher than RUT-C30 after 1, 2, and 3 days, respectively, of cultivation with xylan as the inducer (Fig. 3b), and 14.2-, 4.7-, and 5.2-fold higher, respectively, with lactose as the inducer (Fig. 3c). The xylanase activity of the *sxlr* re-complementation control strain R*sxlr* reverted to the same level as that of RUT-C30. In contrast, the *sxlr* overexpression strain O*sxlr* demonstrated weaker xylanase activity than RUT-C30 (Fig. 3a–c). However, no significant difference in cellulase activity was detected between the four strains (Fig. 3d).

Degradation of lignocellulose is a growth-associated process. The growth rate affects the secretion of enzymes directly [22]. We observed the growth of RUT-C30 and its derivative strains on potato dextrose agar (PDA) and minimal medium (MM) containing 1% xylose, xylan, glucose, lactose, or Avicel as carbon sources, respectively (Fig. 4a). RUT-C30 and its derivative strains did not show significantly different growths and sporulations on PDA and MM containing 1% glucose, lactose or Avicel (Fig. 4a). However, the Δ*sxlr* strain demonstrated rapid growth and the O*sxlr* strain demonstrated reduced growth on MM containing xylose or xylan when compared to RUT-C30. These results could be explained by differences in xylanase activities.

Enzyme activities are also influenced by the concentration and composition of lignocellulose preparations. The extracellular protein concentrations of the Δ*sxlr*, O*sxlr*, and RUT-C30 strains were 15.275 ± 0.096, 10.943 ± 0.399, and 13.378 ± 0.240 mg/mL, respectively. The Δ*sxlr* strain had the highest extracellular protein concentration, and O*sxlr* had the lowest (Fig. 4b). According to sodium dodecyl sulfate–polyacrylamide gel electrophoresis (SDS-PAGE), the band corresponding to ~20 kDa, which was identified as the xylanase protein XYN2 by MALDI-TOF/TOF, was enhanced dramatically in the Δ*sxlr* strain but decreased in the O*sxlr* strain (Fig. 4c). This indicates that SxlR affects the amount of secreted XYN2 in the lignocellulose preparations produced by *T. reesei*.

To determine whether the deletion of SxlR in *T. reesei* RUT-C30 could improve the strain's efficiency to hydrolyze pretreated lignocellulose, steam-exploded rice straw and steam-exploded rice straw mixed with corn straw were used as saccharification substrates. The crude enzyme complex of the mutant Δ*sxlr* strain produced more reducing sugar than that of RUT-C30 (Fig. 4d). The straw hydrolysis increased to 21% by the supernatant of the Δ*sxlr* strain, while 14.1% by the supernatant of RUT-C30 strain after 3 days of hydrolysis from the pretreated rice and corn straw. Similarly, the straw hydrolysis increased to 11.9% by the supernatant of the Δ*sxlr* strain, while 7.6% by the supernatant of RUT-C30 strain after 3 days of hydrolysis from the pretreated rice straw.

SxlR regulates the GH11 xylanase genes

To further identify the regulation mechanism of SxlR on xylanase activity, we examined the expression levels of five xylanase-encoding genes using qPCR; the xylanases encoded by these genes included three members of GH11

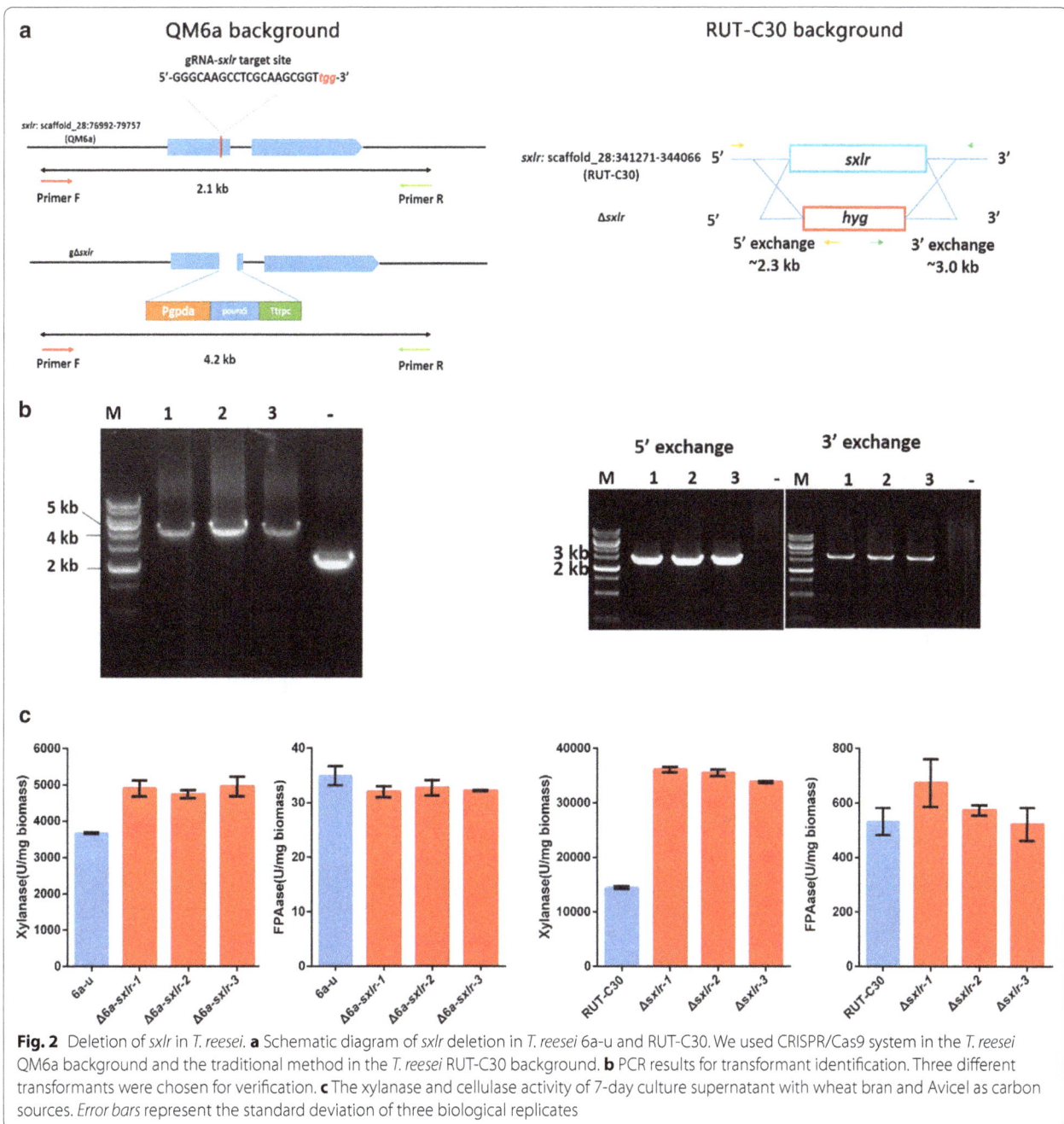

Fig. 2 Deletion of *sxlr* in *T. reesei*. **a** Schematic diagram of *sxlr* deletion in *T. reesei* 6a-u and RUT-C30. We used CRISPR/Cas9 system in the *T. reesei* QM6a background and the traditional method in the *T. reesei* RUT-C30 background. **b** PCR results for transformant identification. Three different transformants were chosen for verification. **c** The xylanase and cellulase activity of 7-day culture supernatant with wheat bran and Avicel as carbon sources. *Error bars* represent the standard deviation of three biological replicates

family (XYN1, XYN2, and newly discovered XYN5), a GH10 family member XYN3 and a GH30 family member XYN4 [23]. The transcription levels of the genes encoding the three GH11 members showed significant differences in the Δ*sxlr* strain (upregulated) and the O*sxlr* strain (downregulated) at all sampling time points using wheat bran and Avicel were used as inducers (Fig. 5a).

The transcription level of the genes encoding XYN3 and XYN4 were also downregulated in the O*sxlr* strain at all sampling time points (Fig. 5a). However, their transcription levels in the Δ*sxlr* strain changed differently (Fig. 5a).

In the Δ*sxlr* strain, the relative expression level of *xyn3* was upregulated compared to the wild-type stain in the first 8 h of incubation in the inducing medium. The relative expression level of *xyn4* was upregulated after 4 h of induction, but downregulated after 8 and 12 h of induction (Fig. 5a). When using xylan as the inducer, the

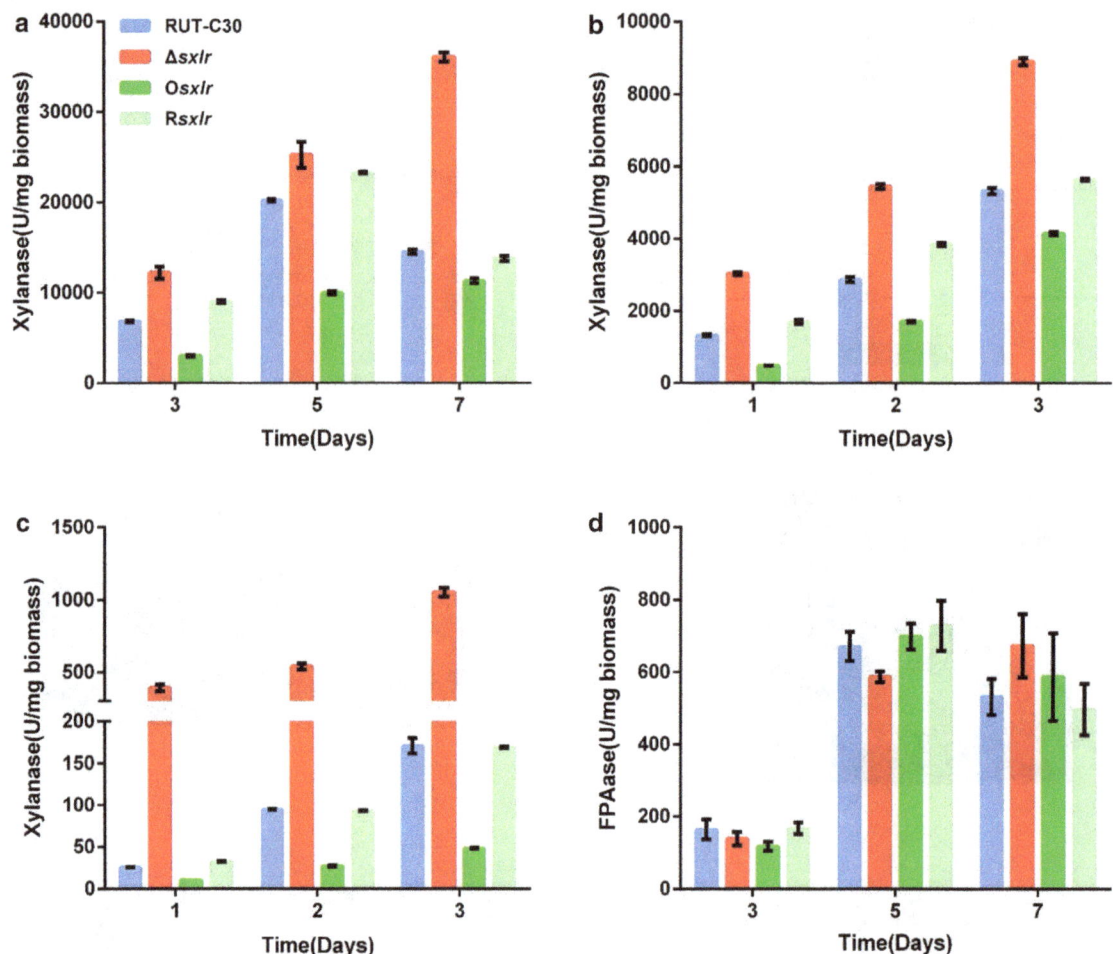

Fig. 3 Enzyme activities of *sxlr* transformants on various carbon sources. The xylanase activity of culture supernatants with wheat bran and Avicel (**a**), xylan (**b**), and lactose (**c**) as carbon sources. **d** The cellulase activity of 3-, 5-, and 7-day culture supernatants with wheat bran and Avicel as carbon source. *Error bars* represent the standard deviation of three biological replicates

transcription level variation tendency of the encoding genes of XYN1, XYN2, XYN5, and XYN3 was similar to that using wheat bran and Avicel as the inducer (Fig. 5b). In contrast, the relative expression level of *xyn4* was upregulated in the O*sxlr* strain after 4 h of induction (Fig. 5b).

To further confirm the qRT-PCR results, qRT-PCR experiments were re-carried out using *rpl6e* (a ribosomal protein encoding gene) [24] and *sar1* (a small GTPase encoding gene) [25] as the reference gene. The transcription level variation tendency of the five xylanase-encoding genes was similar to that using β-actin gene as the reference gene (Additional file 3).

Additionally, we examined the relative expression levels of other genes coding (hemi-) cellulases or their regulators, including three major cellulase genes (*cbh1*, *cbh2*, *egl1)*, the key transcriptional activator *(xyr1)*, and two hemi-cellulase genes (β-mannanase, *man1* and

α-L-arabinofuranosidase, *abf1*). There were no significant differences in the relative expression levels of these genes between RUT-C30 strain and Δ*sxlr* strain (Additional file 3). These results provide the first experimental evidence that SxlR is a negative and specific regulator of GH11 family xylanases.

Xyr1, a major transcription activator in *T. reesei*, regulates expression of most cellulase and hemi-cellulase genes directly, including xylanases [26, 27]. To investigate the relationship between Xyr1 and SxlR, we determined the relative expression levels of *sxlr* in the Δ*xyr1* strain (an *xyr1*-deletion strain derived from RUT-C30) and in the O*xyr1* strain (an *xyr1*-overexpression strain derived from RUT-C30) as well as the relative expression levels of *xyr1* in Δ*sxlr* strain and O*sxlr* strain (Fig. 5c). Comparing with *xyr1*, the transcription level of *sxlr* was quite low in RUT-C30. The *xyr1* transcription level decreased

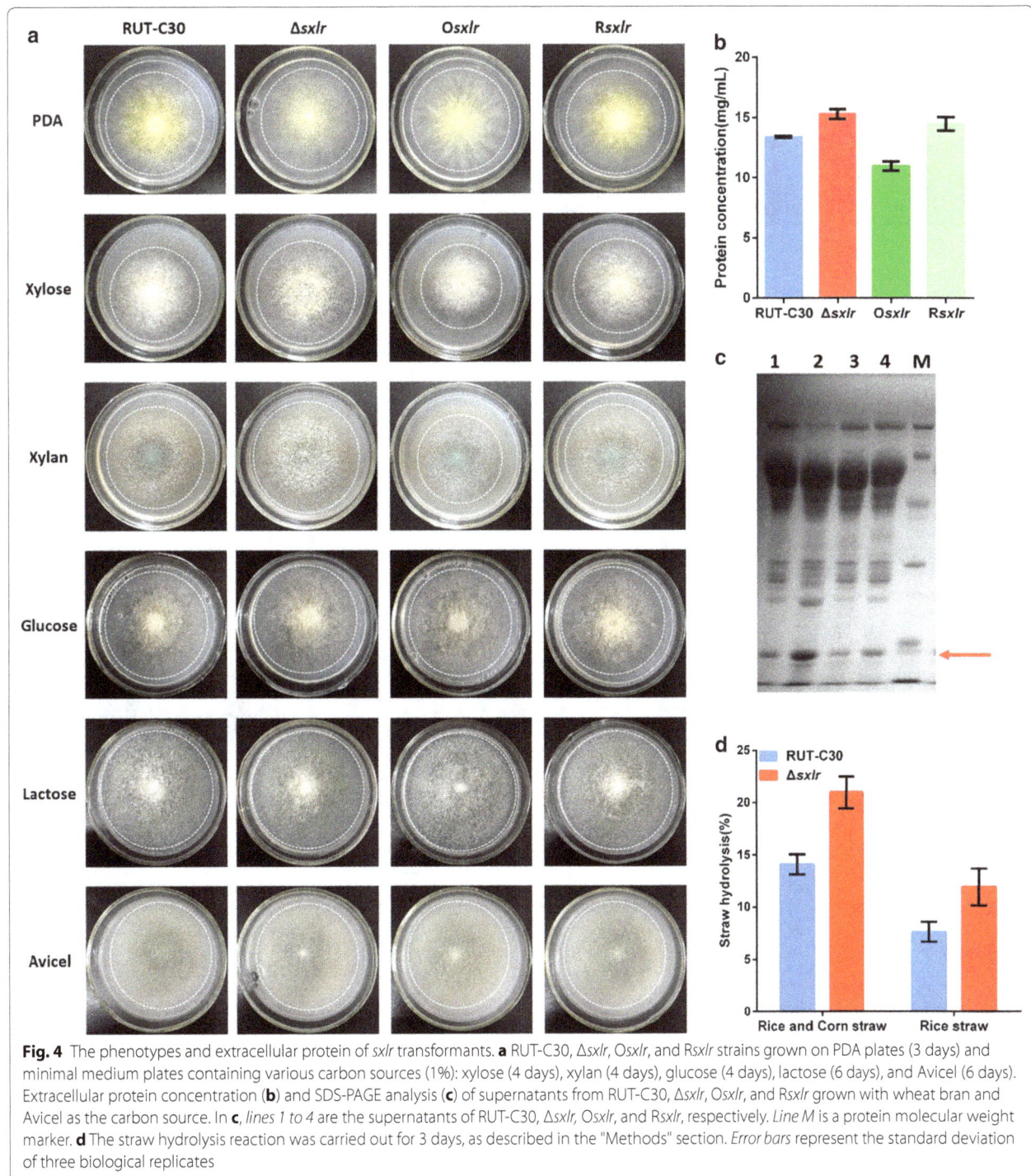

Fig. 4 The phenotypes and extracellular protein of *sxlr* transformants. **a** RUT-C30, Δ*sxlr*, O*sxlr*, and R*sxlr* strains grown on PDA plates (3 days) and minimal medium plates containing various carbon sources (1%): xylose (4 days), xylan (4 days), glucose (4 days), lactose (6 days), and Avicel (6 days). Extracellular protein concentration (**b**) and SDS-PAGE analysis (**c**) of supernatants from RUT-C30, Δ*sxlr*, O*sxlr*, and R*sxlr* grown with wheat bran and Avicel as the carbon source. In **c**, *lines 1 to 4* are the supernatants of RUT-C30, Δ*sxlr*, O*sxlr*, and R*sxlr*, respectively. *Line M* is a protein molecular weight marker. **d** The straw hydrolysis reaction was carried out for 3 days, as described in the "Methods" section. *Error bars* represent the standard deviation of three biological replicates

significantly in the O*sxlr* strain at all the sampling time points. In contrast, its transcription level in the Δ*sxlr* strain significantly increased at 4-h induction, and then decreased from 8-h induction. It seemed that SxlR might repress *xyr1* expression somehow. The transcription level of *sxlr* in the O*xyr1* strain was nearly the same as that in RUT-C30 strain, and its transcription level in the Δ*xyr1* strain was lower or similar to that in RUT-C30 strain. The variation tendency of the *sxlr* transcription level was quite different from that of the downstream genes of Xyr1, of which the transcription were sharply repressed after the deletion of Xyr1 [27].

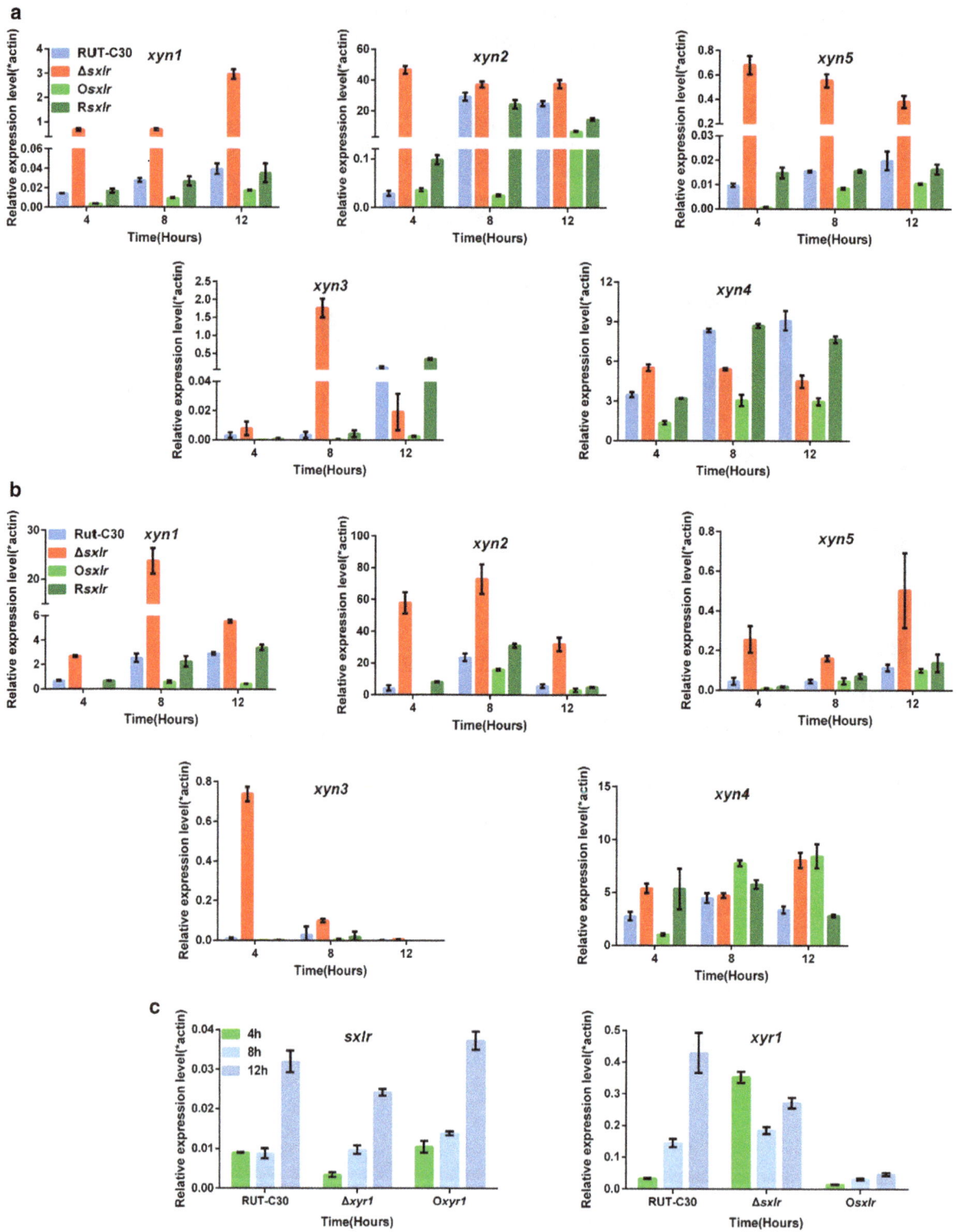

Fig. 5 Quantitative PCR analysis of gene expression levels. The expression levels of *xyn1* (GH11), *xyn2* (GH11), *xyn5* (GH11), *xyn3* (GH10), and *xyn4* (GH30) in RUT-C30, Δ*sxlr*, O*sxlr*, and R*sxlr* when wheat bran and Avicel (**a**) and xylan (**b**) were used as carbon sources, expression levels were normalized to the signal of rpl6e. RNA was extracted after 4, 8, and 12 h. *Error bars* represent the standard deviation of three biological replicates. **c** Expression levels of *sxlr* in RUT-C30, the *xyr1* deletion mutant (Δ*xyr1*), and the *xyr1*-overexpression strain (O*xyr1*); and *xyr1* expression levels in RUT-C30, Δ*sxlr*, and O*sxlr* when wheat bran and Avicel were used as the carbon source. Expression levels were normalized to the signal of β-actin. RNA was extracted after 4, 8, and 12 h. *Error bars* represent the standard deviation of three biological replicates

SxlR binds the promoters of GH11 xylanase genes

According to functional domain analysis, SxlR contains two distinct conserved domains, one GAL4-like Zn_2Cys_6 binuclear cluster DNA binding domain at the N-terminus (cd00067: residues 282–317) and one fungal transcription factor regulatory middle homology region at the C-terminus (cd12148: residues 436–853). To verify the regulation of the GH11 xylanase genes by SxlR, the GAL4-like Zn_2Cys_6 binuclear cluster DNA binding domain was expressed in vitro for EMSAs. The nearly 1500-bp upstream regions (nucleotide position −1500 to −1) assumed to be the promoter regions of the xylanase-encoding genes were divided into six parts (for example, *xyn2*-P1, 268 bp, nucleotide position −268 to −1; *xyn2*-P2, 268 bp, position −516 to −249; *xyn2*-P3, 268 bp, position −764 to −497; *xyn2*-P4, 268 bp, position −1012 to −745; *xyn2*-P5, 268 bp, position −1260 to −993; *xyn2*-P6, 268 bp, position −1508 to −1241). Based on the EMSA gel shifts, SxlR could bind the *xyn1*-P5, *xyn2*-P4, *xyn2*-P5, and *xyn5*-P5 promoter regions of the three GH11 xylanases (Fig. 6a). No specific gel shift was observed for the *xyn3* and *xyn4* promoters, which belonged to the GH10 and GH30 families, respectively (Fig. 5b; Additional file 4). It means that SxlR plays a critical role in the inhibition of GH11 family genes through binding to their promoter regions directly.

To identify the binding motif of SxlR, three promoter fragments (*xyn2*-P4, *xyn1*-P5, and *xyn5*-P5) were each divided into two 144-bp segments (Additional file 5A). SxlR bound to *xyn2*-P4-1, *xyn1*-P5-2, and *xyn5*-P5-2 (Additional file 5B). Based on MEME Suite (http://meme-suite.org/tools/meme) analysis, three candidate consensus motifs were predicted (Additional file 5C, D). However, the deletion of these three motifs did not affect the binding of SxlR to these DNA fragments (Additional file 5E). The 144-bp fragments were further shortened for identification of a consensus motif, and two motifs were predicted (Fig. 7a). Based on the disappearance of the SxlR-DNA complex with the deletion of motif 5 (Fig. 7b), the consensus motif of SxlR was determined as 5′-CATCSGSWCWMSA-3′ (Fig. 7c, d). In this motif, one guanine and one cytosine are present in the core region and the consistent nucleotides are located in the flanks. The potential SxlR binding motif were not detected in the promoter regions of Xyr1 and xylanases XYN3 and XYN4. It suggested that SxlR might not regulate the transcription of Xyr1 and the GH10 family member XYN3 as well as the GH30 family member XYN4 directly.

Discussion

The aim of this study was to gain a comprehensive understanding of how viable SAN progeny produces higher levels of xylanases [17]. We focused on candidate TFs exhibiting differential expression levels between SAN and euploid progeny (six TF-encoding genes: ID in jgi|Trire2: 106677, 65854, 111446, 68930, 111515, 36913), as well as another candidate TF, SxlR. We confirmed that SxlR is a negative regulator of GH11 family xylanases. However, overexpression of the other six tested TFs had no effect on the xylanase activity. It seems that the segmental aneuploidy did not result in a change in *sxlr* gene copy number; however, it is still unknown whether segmental aneuploidy affects the expression of *sxlr* in SAN progeny.

Recently, another negative regulator of hemi-cellulase, xylanase promoter-binding protein 1 (Xpp1) was reported in *T. reesei* [18]. Xpp1 regulates transcription of hemi-cellulase genes only at later stages of cultivation. There was no significant difference in xylanase activity between an *xpp1*-disrupted strain and its parent strain before 72 h [18]. In this study, SxlR regulated expression of GH11 xylanase genes during the entire induction period, and the xylanase activity of the deletion strain was significantly higher than that of the parent strain (Fig. 3a). It has been suggested that SxlR may play an important role in regulating xylanase expression in *T. reesei*. Xpp1 repressed the expression of *xyn1*, *xyn2*, and *bxl2* (encoding a putative β-xylosidase) but like SxlR, did not affect cellulase genes [18]. SxlR bound the *xyn1*, *xyn2*, and *xyn5* promoter regions in vitro (Fig. 6a) and was demonstrated to be a GH11-specific TF. This indicated that the regulation mechanisms of SxlR and Xpp1 are not identical. Additionally, in inducing medium containing xylan, the *xpp1*-disrupted strain showed nearly 2.2-fold increases in transcription levels of *xyn2* at 72 h, while the Δ*sxlr* strain showed 9.9-, 2.1-, and 4.6-fold increases in transcription levels of *xyn2* at 4, 8, and 12 h, respectively. Both the intensity and timing of these two TFs differed.

The binding consensus sequence of SxlR (5′-CATC-SGSWCWMSA-3′) differs from that of the hemi-cellulase regulator Xpp1 (a hexameric palindrome 5′-WCTAGW-3′ together with an inverted AGAA-repeat [18]). In the *xyn1* promoter region, the binding consensus sequence of Xpp1 is located in the *xyn1*-P3 fragment, while SxlR binds to *xyn1*-P5. In the *xyn2* promoter region, the binding consensus sequence of Xpp1 is located in *xyn2*-P1 fragment, while SxlR binds to *xyn2*-P4 and *xyn2*-P5. In the *xyn5* promoter region, only the SxlR binding consensus sequence was found. Meanwhile, in the *bxl2* promoter region, only the Xpp1 binding consensus sequence was found. Therefore, SxlR has a regulon that differs from that of Xpp1.

Most xylanases (endo-β-1, 4-xylanase, EC 3. 2. 1. 8) belong to the GH10 and GH11 families. Compared to GH10, the GH11 family is much more xylan-specific [28]. In *T. reesei*, XYN2 belonging to GH11 family was the dominant extracellular xylanase. The Δ*sxlr* strain grew

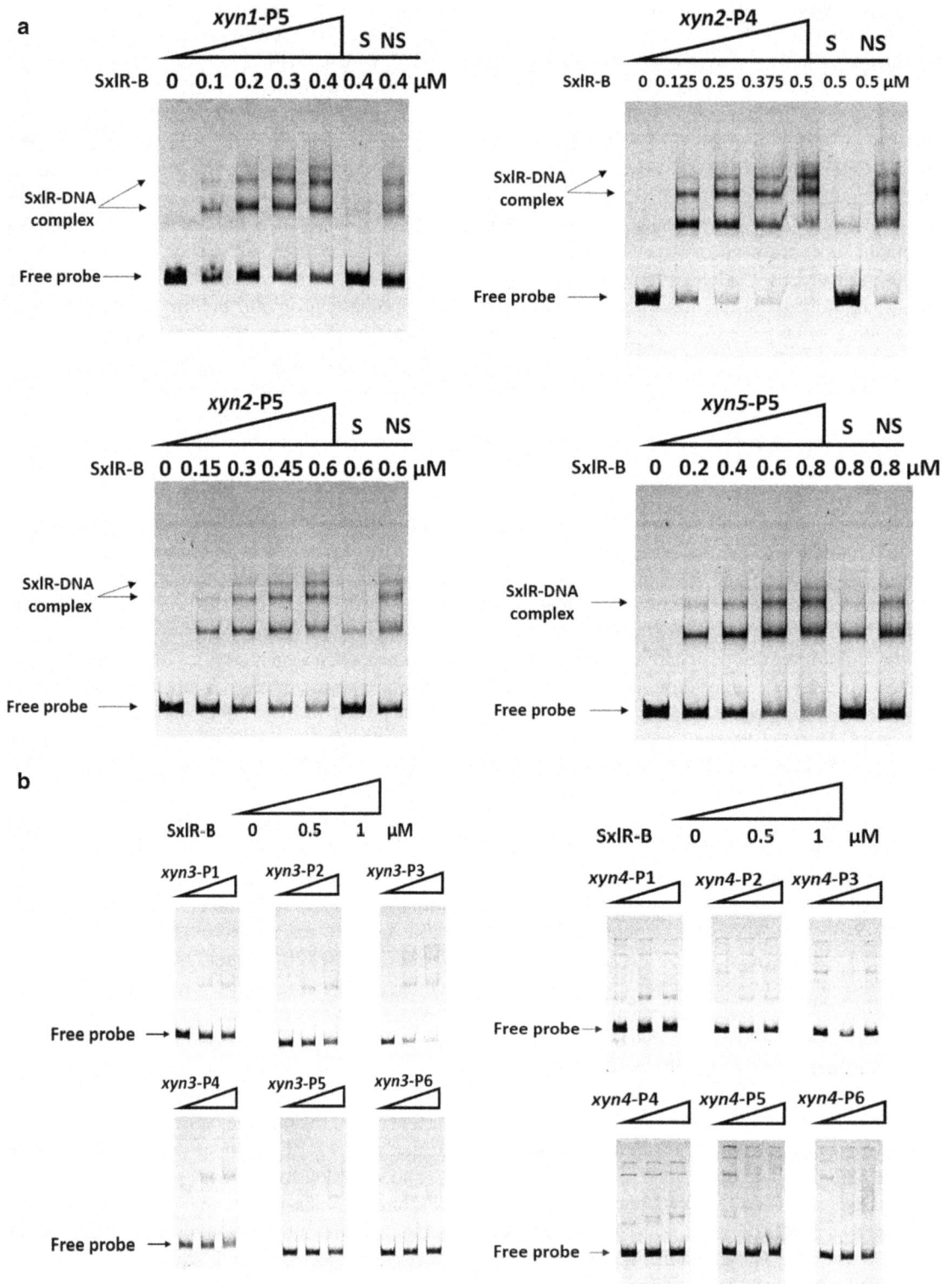

Fig. 6 EMSAs of SxlR binding to the promoter regions of xylanase genes. **a** DNA binding of SxlR to the promoter regions of *xyn1*, *xyn2*, and *xyn5*. The amounts of purified SxlR binding domain (SxlR-B, μM) used were as indicated; ~10 ng of Cy5-labeled probe was added to each reaction. The shifts were verified to be specific by adding 100-fold excess of unlabeled specific (S) and non-specific (NS) competitor DNA. The SxlR-DNA complex is indicated by the *arrow*. **b** DNA binding of SxlR to the *xyn3* and *xyn4* promoter regions. We used three concentrations of SxlR-B: 0, 0.5, and 1 μM; ~10 ng of Cy5-labeled probe was added to each reaction

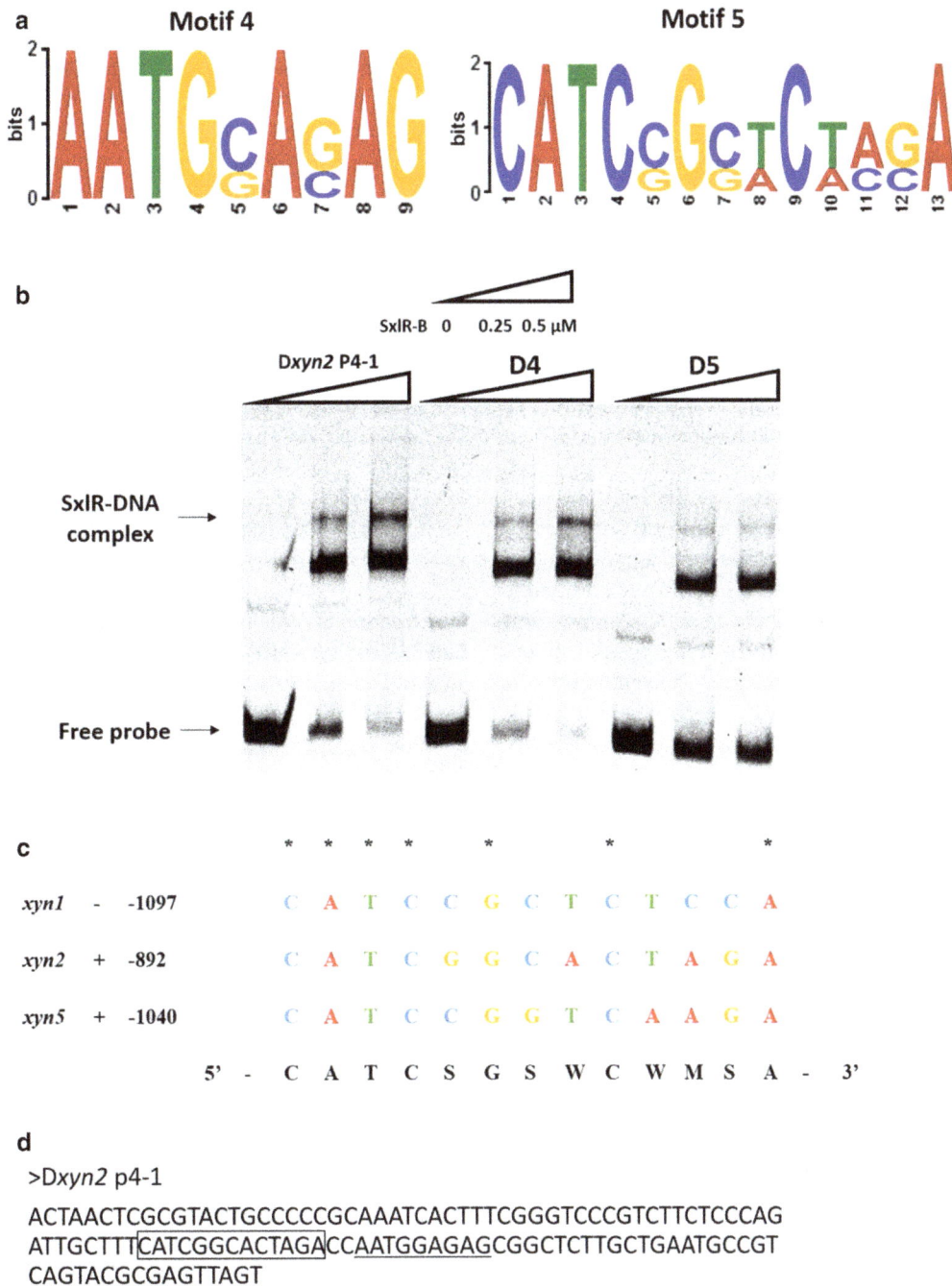

Fig. 7 Identification of the SxlR binding consensus sequence. **a** Sequence motifs of a putative SxlR binding consensus sequence derived by MEME from D*xyn2*-P4-1, D*xyn1*-P5-2 and D*xyn5*-P5-2. Two putative SxlR binding consensus sequences were obtained. **b** EMSA results of SxlR binding to D*xyn2* P4-1, D4 (Motif 4 deletion), and D5 (Motif 5 deletion); the SxlR-DNA complex is indicated by the *arrow*. The amounts of purified SxlR binding domain (SxlR-B, µM) used were as indicated; ~10 ng of Cy5-labeled probe was added to each reaction. **c** Alignment of SxlR binding consensus sequence on sense (+) or antisense (−) strand in the upstream regions of *xyn1*, *xyn2*, and *xyn5*. The *numbers following the gene name* indicated the point of the 5′ starting nucleotide relative to the translation start point and the same nucleotide was indicated by *asterisk*. **d** The location of motif 4 and motif 5 in D*xyn2* P4-1 probe. Motif 4 was indicated by *underline* while motif 5 was indicated by *pane*

more vigorously than the wild-type on MM with xylan or xylose as the sole carbon source. This indicates that the deletion of *sxlr* results in the high ability to utilize xylan or xylose. According to our hypothesis, the role of SxlR is the downregulation of the main xylanases to save energy when cellulose or glucose is present in the habitat of

T. reesei. Hexose is, after all, the preferred carbon resource of these organisms [29]. The novel TF SxlR is indicative of the sophisticated regulatory network evolved by *T. reesei* to adapt to complex environments.

Unlike the previously reported global regulators of carbohydrate-active enzymes (CAZymes), SxlR, as a specific regulator of GH11 family xylanases, might provide new leads for strain engineering to enhance hemi-cellulase activity in *T. reesei*. In the straw hydrolysis assay, more reducing sugar was produced by the Δ*sxlr* culture supernatant than by the parent strain. Compared with steam-exploded rice straw, higher straw hydrolysis was achieved by steam-exploded rice and corn straw mixture (Fig. 4d). This was probably because the hemi-cellulose content of corn straw was much higher than that of rice straw (41% vs 22%) [30, 31]. It has been suggested that the utilization of hemi-cellulose-rich substrate was significantly improved by the modified enzyme preparation. Because most TFs regulate cellulase and hemi-cellulase genes concurrently [32], the ratio of cellulase and hemi-cellulase is difficult to optimize by genetic manipulation of TFs. Our results shown that the appropriate proportion of the enzyme preparation can be conveniently optimized via engineering of *sxlr*. The derived strain would have a higher xylanase activity, while the cellulase activity would not be affected.

Through phylogenetic analysis of SxlR, we found its homologous genes in other cellulase producer including *Trichoderma virens* (XP_013961599.1), *T. atroviride* (XP_013948752.1), *Neurospora crassa* (XP_960943.2), *Penicillium oxalicum* (EPS34484.1), *Aspergillus nidulans* (XP_681446.1), *A. oryzae* (KOC12472.1), and *A. niger* (CAK40371.1), which indicates that its function may be conserved (Additional file 6).

Conclusions

SxlR appears to be a major xylanase regulator in *T. reesei* that represses the expression of xylanases belonging to the GH11 family and does not affect cellulase activity. Deletion of *sxlr* dramatically increases xylanase activity. SxlR is a good target candidate for *T. reesei* strain modification, especially for hydrolyzing substrates rich in hemi-cellulose.

Methods

Strains and culture conditions

Trichoderma reesei strains including QM6a (ATCC 13631) and RUT-C30 (ATCC 56765) were maintained on potato dextrose agar plate (PDA) at 28 °C for 7 days for spore collection. *Escherichia coli* DH5α used as cloning host was culture at 37 °C in Luria–Bertani (LB) medium. *Agrobacterium tumefaciens* AGL1 was used to transform the gene to *T. reesei* strains. To induce

enzyme production, the conidial suspension (0.5 mL, 1×10^7 conidia/mL) was inoculated into a 50-mL Erlenmeyer flask containing 10 mL of Sabouraud Dextrose Broth (SDB) and incubated for 40 h on an orbital shaker at 200 rpm at 28 °C. The culture was then transferred into a flask containing 10 mL of inducing fermentation medium at 10% inoculum ratio (v/v). The flasks were incubated on an orbital shaker at 200 rpm at 28 °C for 1 week. The wheat bran and Avicel medium were prepared as follows: 0.4% KH_2PO_4, 0.28% $(NH_4)_2SO_4$, 0.06% $MgSO_4 \cdot 7H_2O$, 0.05% $CaCl_2$, 0.06% urea, 0.3% tryptone, 0.1% Tween-80, 0.5% $CaCO_3$, 0.001% $FeSO_4 \cdot 7H_2O$, 0.00032% $MnSO_4 \cdot H_2O$, 0.00028% $ZnSO_4 \cdot 7H_2O$, 0.0004% $CoCl_2$, 2% wheat bran, 3% microcrystalline cellulose. We also used minimal medium (MM) containing 0.5% $(NH_4)_2SO_4$, 1.5% KH_2PO_4, 0.06% $MgSO_4$, 0.06% $CaCl_2$, 0.0005% $FeSO_4 \cdot 7H_2O$, 0.00016% $MnSO_4 \cdot H_2O$, 0.00014% $ZnSO_4 \cdot 7H_2O$, and 0.0002% $CoCl_2$ with 1% lactose or 1% xylan used as carbon source for 3 days of fermentation.

Construction of Δ*sxlr* in *T. reesei* QM6a strain using the CRISPR/Cas9 system

T. reesei 6a-u (a uridine-dependent strain derived from 6a-pc) [19] was used as host. The guide RNA (gRNA) cassette including the synthetic gRNA sequence and target DNA of the *sxlr* gene (5′-GGGCAAGCCT CGCAAGCGGT*tgg*-3′, PAM is shown in italics) was driven by T7 promoter and transcribed into RNA in vitro with the MEGAscript T7 Kit (Ambion, Austin, TX, USA). Donor DNA-*sxlr* (dDNA-*sxlr*) containing the 5′- and 3′ flanking sequences of *sxlr* (jgi|Trire2|123881) and the selectable marker cassette (the *ura5* gene from *P. oxalicum* controlled by the Pgpda promoter and Ttrpc terminator, Pgpda-*poura5*-Ttrpc) was ligated into the pMD-18T vector (Takara, Dalian, China). The gRNA and dDNA-*sxlr* were co-transformed using a modified polyethylene glycol-mediated protoplast transformation procedure [33]. The transformants were selected using MM plates with 1% glucose as the carbon source.

Deletion, overexpression, and re-complementation of *sxlr* in *T. reesei* RUT-C30

To delete the *sxlr* gene (jgi|TrireRUTC30_1|26638) of *T. reesei*, the 2.8-kb *sxlr* coding region was replaced by the *hph* (hygromycin phosphotransferase) gene. This was performed by amplifying 1.5 kb from upstream and 1.7 kb from downstream of *sxlr* from genomic DNA of *T. reesei* using the primer pairs listed in Additional file 7. Then, the two resulting PCR fragments were ligated into the *Hind*III (upstream) and *Xho*I (downstream) sites of the linearized pXBthg vector [30] using the ClonExpress II One Step Cloning Kit (Vazyme, Nanjing, China).

For overexpression of *sxlr* under a strong constitutive promoter in *T. reesei*, we fused the *T. reesei* translation-elongation factor 1α (*tef1*) promoter, the *sxlr* coding region and the Ttrpc terminator from *A. nidulans* and inserted this fragment into the *Hind*III and *Xba*I sites of pXBthg using the ClonExpress II One Step Cloning Kit. Besides, we overexpressed 76601, 83920, 89588, 133726, 26508, and 43491 (gene ID in jgi|TrireRUTC30_1) with the same method.

For re-complementation of *sxlr in* the Δ*sxlr* strain, the promoter, coding region, and terminator of *sxlr* were inserted into the *Hind*III and *Xba*I sites of the pXBt vector [the sequence is similar to pXBthg, except the *hph* marker gene was replaced by the bleomycin resistance (*ble*) gene] and the sequence downstream of *sxlr* was inserted into the *Xho*I site by the same method mentioned above.

All vectors constructed were verified by sequencing. *Agrobacterium*-mediated transformation was conducted as described previously by Ma et al. [34].

Biochemical assays

FPAase activities were determined using cellulose filter paper as described previously [35]. Xylanase activities were tested using 2% xylan from beechwood (Sigma, St. Louis, USA) as described previously [36]. Protein concentrations in the culture supernatant were determined by RC-DC protein assay (Bio-Rad, Hercules, CA, USA) (1976). Biomass was tested by the diphenylamine colorimetric method [37].

For straw hydrolysis, different pretreated biomasses such as rice straw and a corn stover plus rice straw mixture (3:7 ratio) [38] were used as substrates. The standard hydrolysis assay was carried out in a volume of 1 mL of 50 mM sodium acetate, pH 4.8, and 5% (w/v) pretreated straw as substrate in a 2-mL FastPrep tube (MP Biomedicals, Santa Ana, CA, USA). The enzyme dosage was 20 U of cellulase activity/g substrate. Incubation was typically at 50 °C with shaking at 200 rpm for 3 days, followed by centrifugation at 12,000 rpm for 10 min to obtain the supernatant to determine the reducing sugar with DNS. Each sample was examined in triplicate.

Growth of *sxlr* transformants on plates

Spores of *T. reesei* RUT-C30, Δ*sxlr*, O*sxlr*, and R*sxlr* were first prepared by growth on PDA plates at 28 °C and harvesting spores after 7 days. Spores were counted using a hemocytometer and 10^5 spores of each strain were inoculated onto PDA plates and MM plates containing 1% xylose, xylan, glucose, lactose, or Avicel. Double-layer Avicel plates were prepared by first casting an MM agarose bottom layer containing no carbon source, followed by an MM agarose top layer containing 1% Avicel; each

strain was grown in triplicates. These plates were incubated at 28 °C before phenotypic examination.

Protein identification with MALDI-TOF/TOF analysis

After 7 days' culture with wheat bran and Avicel medium, the supernatant of RUT-C30, Δ*sxlr*, O*sxlr*, and R*sxlr* strain was collected. The samples were loaded into 12.5% PAGE gel, 70 min at 120 V were preferred for proteins separation. Coomassie brilliant blue R250 (Sangon, Shanghai, China) was used to color the gel for 1 h. Destainer (distilled water: ethanol: acetic acid, 5:4:1, v/v) was added for another 2–4 h. Finally, the protein band was excised from the gel and send to Shanghai Applied Protein Technology Co. Ltd, protein analysis was conducted with a MALDI-TOF/TOF mass spectrometer 5800 Proteomics Analyzer (Applied Biosystems, Framingham, MA, USA), NCBI *Trichoderma* (59929 protein sequence) was selected as database.

RNA extraction and real-time PCR

Mycelia were harvested after induction by wheat bran and Avicel for 4, 8, and 12 h, then transferred into TRIzol reagent (Invitrogen, USA) with standard protocol. Three biological replicates were prepared in the process. Real-time PCR was performed with primers listed in Additional file 7 with standard method using β-actin gene, *rpl6e* (a ribosomal protein encoding gene) and *sar1* (a gene encoding a small GTPase) as the reference genes, respectively.

Expression and purification of the DNA binding domain of SxlR

The DNA binding domain (residues 252–347) of SxlR (SxlR-B) was expressed by the pGEX system according to the manufacturer's guidelines. First-strand cDNA was used as a template to amplify the fragment encompassing the DNA binding domain of SxlR using the primers indicated in Additional file 7. The fragment was then ligated into plasmid pGEX-4T-1 via *Bam*HI and *Xho*I double digestion to produce pGEX-4T-SxlR-B, then subsequently introduced into *E. coli* BL21 (DE3) for protein production. The recombinant protein was purified using a GST-Bind resin column (Merck, Darmstadt, Germany) according to the supplier's recommendations. SDS-PAGE was used to verify the purified protein.

Electrophoretic mobility shift assays

For EMSA, the labeling of probes containing the promoter regions of specific genes was performed via PCR using the primers listed in Additional file 7. The nearly 1500-bp upstream regions (nucleotide position −1500 to −1) assumed to be the promoter regions of the xylanase-encoding genes from *T. reesei* RUT-C30 were divided into six parts and each part was about 268 bp (Additional file 4A). The genomic DNA of *T. reesei* RUT-C30 was used as the template and the PCR products were purified by gel electrophoresis and quantified using a Bio-Photometer plus (Eppendorf, Hamburg, Germany). The experiment was then performed as described previously by Chen et al. [39]. Motif 1 (5′-AMTGSAGAG-3′) was located in −1034 to −1026 position of *xyn1* promoter, −890 to −882 position of *xyn2* promoter and −1091 to −1083 position of *xyn5* promoter. Motif 2 (5′-TGAW-GAG-3′) was located in −1022 to −1016 position of *xyn1* promoter, −957 to −951 position of *xyn2* promoter and −1021 to −1015 position of *xyn5* promoter. Motif 3 (5′-WTATAT-3′) was located in −998 to −993 position of *xyn2* promoter and −1117 to −1112 position of *xyn5* promoter. Motif 4 (5′-AATGSASAG-3′) was located in −1119 to −1111 position of *xyn1* promoter, −890 to −882 position of *xyn2* promoter and −1091 to −1083 position of *xyn5* promoter. Motif 5 (5′-CATCSGSW-CWMSA-3′) was located in −1110 to −1098 position of *xyn1* promoter, −905 to −893 position of *xyn2* promoter and −1053 to −1041 position of *xyn5* promoter.

Additional files

Additional file 1. Putative specific xylanase transcription factors in *T. reesei*.

Additional file 2. *sxlr* transformants verification. (A) PCR result for R*sxlr* transformants identification. Three different transformants were chosen to verify. (B) Verification of random insertion in R*sxlr* transformants. During the *Agrobacterium*-mediated transformation, the T-DNA will randomly insert into genome with LB and RB, so we can verify it with PCR. The expected PCR product length of 5′ and 3′ verification was 6.1 kb and 2.6 kb. The *sxlr* re-complementation vector was used as positive control. Three different transformants were chosen to verify. The *sxlr* transcription levels in *sxlr* transformants were normalized to the signal of β-actin (C) or rpl6e (D), a gene encoding a ribosomal protein. RNA was extracted after 12 h induction by wheat bran and Avicel. Error bars represent the standard deviation of three biological replicates.

Additional file 3. Quantitative PCR analysis of cellulase and hemi-cellulase gene expression levels. The expression level of five xylanase genes in RUT-C30, Δ*sxlr*, O*sxlr* and R*sxlr* when wheat bran and Avicel were used as carbon sources. Expression levels were normalized to the signal of rpl6e, (A) or sar1 (B), a gene encoding a small GTPase. (C) The expression levels of cbh1, cbh2, egl1, xyr1, man1 and abf1 in RUT-C30 and Δ*sxlr*. Expression levels were normalized to the signal of β-actin. RNA was extracted after 24, 48, and 72 h after induction by wheat bran and Avicel. Error bars represent the standard deviation of three biological replicates.

Additional file 4. The description of xylanase genes promoter and EMSAs of SxlR binding to *xyn3* and *xyn4*. (A) The nearly 1500-bp upstream regions (nucleotide position –1500 to –1) assumed to be the promoter regions of the xylanase-encoding genes from *T. reesei* RUT-C30 were divided into six parts and each part was about 268 bp. (B) DNA binding of SxlR to the *xyn3* and *xyn4* promoter regions. We used three concentrations of SxlR-B: 0, 0.5, and 1 μM; ~ 10 ng of Cy5-labeled probe was added to each reaction. For specific (S) and non-specific (NS) control experiment, 100-fold excess of unlabeled S and NS competitor DNA were added.

Additional file 5. Truncation prmoter sequence of GH11 xylanase genes. (A) The description of DNA sequence truncation. Each sequence was divided into two parts. (B) EMSAs of SxlR binding to *xyn2*-P4, *xyn1*-P5 and *xyn5*-P5 truncation sequence, respectively. (C) Three putative SxlR binding consensus sequences derived by MEME. (D)The location of three putative SxlR binding consensus sequences in *xyn2*-P4-1, *xyn1*-P5-2 and *xyn5*-P5-2. Motif 1, 2 and 3 was labeled as red, blue and green, respectively. D*xyn2*-P4-1, D*xyn1*-P5-2 and D*xyn5*-P5-2 was the sequence after truncation. (E) EMSAs of SxlR binding to D*xyn2*-P4-1, D*xyn1*-P5-2 and D*xyn5*-P5-2, the SxlR-DNA complex was indicated by *arrow*. The amounts of purified SxlR binding domain (SxlR-B, μM) used were as indicated; ~ 10 ng of Cy5-labeled probe was added to each reaction.

Additional file 6. Phylogenetic relationship between SxlR and putative orthologs. The phylogenetic tree was inferred using the neighbor-joining method. Evolutionary analyses were conducted in MEGA 5. The gene ID in NCBI is ETR97987.1 (*T. reesei*), XP_013961599.1 (*T. virens*), XP_013948752.1 (*T. atroviride*), XP_960943.2 (*Neurospora crassa*), EPS34484.1 (*Penicillium oxalicum*), XP_681446.1 (*Aspergillus nidulans*), CAK40371.1 (*A. niger*) and KOC12472.1 (*A. oryzae*).

Additional file 7. Oligonucleotides used in this study.

Abbreviations

SxlR: specialized xylanase regulator; EMSA: electrophoretic mobility shift assay; TFs: transcriptional factors; CCR: carbon catabolite repression; SAN: segmentally aneuploid; Xpp1: xylanase promoter-binding protein 1; CRISPR: clustered regularly interspaced short palindromic repeats; SDS-PAGE: sodium dodecyl sulfate–polyacrylamide gel electrophoresis; GH11: glycoside hydrolase family 11; CAZymes: carbohydrate-active enzymes.

Authors' contributions

GZ and ZHZ designed the study. RL, LC, and YPJ performed the experiments. RL and GZ analyzed the data and wrote the manuscript. GZ and ZHZ revised the manuscript. All authors read and approved the final manuscript.

Author details

[1] CAS-Key Laboratory of Synthetic Biology, CAS Center for Excellence in Molecular Plant Sciences, Institute of Plant Physiology and Ecology, Chinese Academy of Science, Fenglin Rd 300, Shanghai 200032, China. [2] University of Chinese Academy of Sciences, Beijing 100049, China.

Acknowledgements

We thank Professor TingFang Wang (Institute of Molecular Biology, Academia Sinica) for research discussion and providing gene information of 106677, 65854, 111446, 68930, 111515, 36913 for test.

Competing interests

The authors declare that they have no competing interests.

Funding

This work was financially supported by High-tech Research and Development Program of China (863:2013AA102806) and the National Natural Science Foundation of China (31470201, 31300073).

References

1. Rubin EM. Genomics of cellulosic biofuels. Nature. 2008;454:841–5.
2. Singh A, Taylor LE, Vander Wall TA, Linger J, Himmel ME, Podkaminer K, Adney WS, Decker SR. Heterologous protein expression in *Hypocrea jecorina*: a historical perspective and new developments. Biotechnol Adv. 2015;33:142–54.
3. Harris PV, Welner D, McFarland KC, Re E, Navarro Poulsen JC, Brown K, Salbo R, Ding H, Vlasenko E, Merino S, et al. Stimulation of lignocellulosic biomass hydrolysis by proteins of glycoside hydrolase family 61: structure and function of a large, enigmatic family. Biochemistry. 2010;49:3305–16.

4. Hu JG, Arantes V, Saddler JN. The enhancement of enzymatic hydrolysis of lignocellulosic substrates by the addition of accessory enzymes such as xylanase: is it an additive or synergistic effect? Biotechnol Biofuels. 2011;4:36–48.

5. Portnoy T, Margeot A, Linke R, Atanasova L, Fekete E, Sandor E, Hartl L, Karaffa L, Druzhinina IS, Seiboth B, et al. The CRE1 carbon catabolite repressor of the fungus Trichoderma reesei: a master regulator of carbon assimilation. Bmc Genom. 2011;12:269–80.

6. Ilmén M, Thrane C, Penttilä M. The glucose repressor gene cre1 of Trichoderma: isolation and expression of a full-length and a truncated mutant form. Mol Gen Genet. 1996;251:451–60.

7. Stricker AR, Grosstessner-Hain K, Wurleitner E, Mach RL. Xyr1 (xylanase regulator 1) regulates both the hydrolytic enzyme system and D-xylose metabolism in Hypocrea jecorina. Eukaryot Cell. 2006;5:2128–37.

8. Derntl C, Gudynaite-Savitch L, Calixte S, White T, Mach RL, Mach-Aigner AR. Mutation of the Xylanase regulator 1 causes a glucose blind hydrolase expressing phenotype in industrially used Trichoderma strains. Biotechnol Biofuels. 2013;6:62–72.

9. Aro N, Saloheimo A, Ilmén M, Penttilä M. ACEII, a novel transcriptional activator involved in regulation of cellulase and xylanase genes of Trichoderma reesei. J Biol Chem. 2001;276:24309–14.

10. Stricker AR, Trefflinger P, Aro N, Penttilä M, Mach RL. Role of Ace2 (activator of cellulases 2) within the xyn2 transcriptosome of Hypocrea jecorina. Fungal Genet Biol. 2008;45:436–45.

11. Zeilinger S, Ebner A, Marosits T, Mach R, Kubicek CP. The Hypocrea jecorina HAP 2/3/5 protein complex binds to the inverted CCAAT-box (ATTGG) within the cbh2 (cellobiohydrolase II-gene) activating element. Mol Genet Genom. 2001;266:56–63.

12. Aro N, Ilmén M, Saloheimo A, Penttilä M. ACEI of Trichoderma reesei is a repressor of cellulase and xylanase expression. Appl Environ Microb. 2003;69:56–65.

13. Nitta M, Furukawa T, Shida Y, Mori K, Kuhara S, Morikawa Y, Ogasawara W. A new Zn(II)$_2$Cys$_6$-type transcription factor BglR regulates beta-glucosidase expression in Trichoderma reesei. Fungal Genet Biol. 2012;49:388–97.

14. Seiboth B, Karimi RA, Phatale PA, Linke R, Hartl L, Sauer DG, Smith KM, Baker SE, Freitag M, Kubicek CP. The putative protein methyltransferase LAE1 controls cellulase gene expression in Trichoderma reesei. Mol Microbiol. 2012;84:1150–64.

15. Fekete E, Karaffa L, Karimi Aghcheh R, Németh Z, Fekete É, Orosz A, Paholcsek M, Stágel A, Kubicek CP. The transcriptome of lae1 mutants of Trichoderma reesei cultivated at constant growth rates reveals new targets of LAE1 function. Bmc Genom. 2014;15:447–56.

16. Häkkinen M, Valkonen MJ, Westerholm-Parvinen A, Aro N, Arvas M, Vitikainen M, Penttilä M, Saloheimo M, Pakula TM. Screening of candidate regulators for cellulase and hemicellulase production in Trichoderma reesei and identification of a factor essential for cellulase production. Biotechnol Biofuels. 2014;7:14–34.

17. Chuang YC, Li WC, Chen CL, Hsu PW, Tung SY, Kuo HC, Schmoll M, Wang TF. Trichoderma reesei meiosis generates segmentally aneuploid progeny with higher xylanase-producing capability. Biotechnol Biofuels. 2015;8:30–44.

18. Derntl C, Rassinger A, Srebotnik E, Mach RL, Mach-Aigner AR. Xpp1 regulates the expression of xylanases, but not of cellulases in Trichoderma reesei. Biotechnol Biofuels. 2015;8:12–22.

19. Liu R, Chen L, Jiang Y, Zhou Z, Zou G. Efficient genome editing in filamentous fungus Trichoderma reesei using the CRISPR/Cas9 system. Cell Discov. 2015;1:15007–17.

20. Marie-Nelly H, Marbouty M, Cournac A, Flot JF, Liti G, Parodi DP, Syan S, Guillen N, Margeot A, Zimmer C, Koszul R. High-quality genome (re) assembly using chromosomal contact data. Nat Commun. 2014;5:5695–704.

21. Druzhinina IS, Kopchinskiy AG, Kubicek EM, Kubicek CP. A complete annotation of the chromosomes of the cellulase producer Trichoderma reesei provides insights in gene clusters, their expression and reveals genes required for fitness. Biotechnol Biofuels. 2016;9:75–90.

22. Arvas M, Pakula T, Smit B, Rautio J, Koivistoinen H, Jouhten P, Lindfors E, Wiebe M, Penttilä M, Saloheimo M. Correlation of gene expression and protein production rate—a system wide study. Bmc Genom. 2011;12:616–40.

23. Herold S, Bischof R, Metz B, Seiboth B, Kubicek CP. Xylanase gene transcription in Trichoderma reesei is triggered by different inducers representing different hemicellulosic pentose polymers. Eukaryot Cell. 2013;12:390–8.

24. Tisch D, Kubicek CP, Schmoll M. New insights into the mechanism of light modulated signaling by heterotrimeric G-proteins: ENVOY acts on gna1 and gna3 and adjusts cAMP levels in Trichoderma reesei (Hypocrea jecorina). Fungal Genet Biol. 2011;48:631–40.

25. Steiger MG, Mach RL, Mach-Aigner AR. An accurate normalization strategy for RT-qPCR in Hypocrea jecorina (Trichoderma reesei). J Biotechnol. 2010;145:30–7.

26. Portnoy T, Margeot A, Seidl-Seiboth V, Le Crom S, Ben Chaabane F, Linke R, Seiboth B, Kubicek CP. Differential regulation of the cellulase transcription factors XYR1, ACE2, and ACE1 in Trichoderma reesei strains producing high and low levels of cellulase. Eukaryot Cell. 2011;10:262–71.

27. Castro LD, de Paula RG, Antonieto ACC, Persinoti GF, Silva-Rocha R, Silva RN. Understanding the role of the master regulator XYR1 in Trichoderma reesei by global transcriptional analysis. Front Microbiol. 2016;7:175–90.

28. Paës G, Berrin JG, Beaugrand J. GH11 xylanases: structure/function/properties relationships and applications. Biotechnol Adv. 2012;30:564–92.

29. Aristidou A, Penttilä M. Metabolic engineering applications to renewable resource utilization. Curr Opin Biotechnol. 2000;11:187–98.

30. Shawky BT, Mahmoud MG, Ghazy EA, Asker MMS, Ibrahim GS. Enzymatic hydrolysis of rice straw and corn stalks for monosugars production. J Genet Eng Biotechnol. 2011;9:59–63.

31. Sumphanwanich J, Leepipatpiboon N, Srinorakutara T, Akaracharanya A. Evaluation of dilute-acid pretreated bagasse, corn cob and rice straw for ethanol fermentation by Saccharomyces cerevisiae. Ann Microbiol. 2008;58:219–25.

32. Wang SW, Liu G, Wang J, Yu JT, Huang BQ, Xing M. Enhancing cellulase production in Trichoderma reesei RUT C30 through combined manipulation of activating and repressing genes. J Ind Microbiol Biotechnol. 2013;40:633–41.

33. Liu T, Wang T, Li X, Liu X. Improved heterologous gene expression in Trichoderma reesei by cellobiohydrolase I gene (cbh1) promoter optimization. Acta Biochim Biophys Sin (Shanghai). 2008;40:158–65.

34. Ma L, Zhang J, Zou G, Wang CS, Zhou ZH. Improvement of cellulase activity in Trichoderma reesei by heterologous expression of a beta-glucosidase gene from Penicillium decumbens. Enzyme Microb Technol. 2011;49:366–71.

35. Xiao Z, Storms R, Tsang A. Microplate-based filter paper assay to measure total cellulase activity. Biotechnol Bioeng. 2004;88:832–7.

36. Bailey MJ, Biely P, Poutanen K. Interlaboratory testing of methods for assay of xylanase activity. J Biotechnol. 1992;23:257–70.

37. Zhao Y, Xiang S, Dai X, Yang K. A simplified diphenylamine colorimetric method for growth quantification. Appl Microbiol Biotechnol. 2013;97:5069–77.

38. Jin SY, Chen HZ. Fractionation of fibrous fraction from steam-exploded rice straw. Process Biochem. 2007;42:188–92.

39. Chen L, Zou G, Zhang L, de Vries RP, Yan X, Zhang J, Liu R, Wang C, Qu Y, Zhou Z. The distinctive regulatory roles of PrtT in the cell metabolism of Penicillium oxalicum. Fungal Genet Biol. 2014;63:42–54.

6

Improving ethanol productivity through self-cycling fermentation of yeast: a proof of concept

Jie Wang[1], Michael Chae[1], Dominic Sauvageau[2]* and David C. Bressler[1]*

Abstract

Background: The cellulosic ethanol industry has developed efficient strategies for converting sugars obtained from various cellulosic feedstocks to bioethanol. However, any further major improvements in ethanol productivity will require development of novel and innovative fermentation strategies that enhance incumbent technologies in a cost-effective manner. The present study investigates the feasibility of applying self-cycling fermentation (SCF) to cellulosic ethanol production to elevate productivity. SCF is a semi-continuous cycling process that employs the following strategy: once the onset of stationary phase is detected, half of the broth volume is automatically harvested and replaced with fresh medium to initiate the next cycle. SCF has been shown to increase product yield and/or productivity in many types of microbial cultivation. To test whether this cycling process could increase productivity during ethanol fermentations, we mimicked the process by manually cycling the fermentation for five cycles in shake flasks, and then compared the results to batch operation.

Results: Mimicking SCF for five cycles resulted in regular patterns with regards to glucose consumption, ethanol titer, pH, and biomass production. Compared to batch fermentation, our cycling strategy displayed improved ethanol volumetric productivity (the titer of ethanol produced in a given cycle per corresponding cycle time) and specific productivity (the amount of ethanol produced per cellular biomass) by 43.1 ± 11.6 and $42.7 \pm 9.8\%$, respectively. Five successive cycles contributed to an improvement of overall productivity (the aggregate amount of ethanol produced at the end of a given cycle per total processing time) and the estimated annual ethanol productivity (the amount of ethanol produced per year) by 64.4 ± 3.3 and $33.1 \pm 7.2\%$, respectively.

Conclusions: This study provides proof of concept that applying SCF to ethanol production could significantly increase productivities, which will help strengthen the cellulosic ethanol industry.

Keywords: Cellulosic ethanol, Batch, Self-cycling fermentation, Manual cycling fermentation, Ethanol volumetric productivity, Specific productivity, Overall productivity, Annual ethanol productivity, Production cost, Capital cost

Background

The global interest in cellulosic ethanol has surged due to the abundance of feedstock [1], increasing concerns for environmental sustainability and security of energy supplies [2], and the reduction of greenhouse gas emissions compared to first-generation ethanol [3]. Production of cellulosic ethanol requires a pretreatment to open the complex structure of lignocellulosic materials, enzymatic hydrolysis to digest polymers into monomer sugars, microbial propagation to generate inoculum, fermentation of monomer sugars to produce ethanol, and distillation to acquire ethanol. However, according to Chen et al. [4], the cellulosic ethanol industry, as compared to mature first-generation ethanol, is still faced with economic challenges such as high production costs. Therefore, technologies for the production of cellulosic ethanol

*Correspondence: dominic.sauvageau@ualberta.ca;
david.bressler@ualberta.ca
[1] Department of Agricultural, Food and Nutritional Science, University of Alberta, Edmonton T6G 2P5, Canada
[2] Department of Chemical and Materials Engineering, University of Alberta, Edmonton T6G 1H9, Canada

still need extensive development. Various approaches have been attempted to offset costs, which have been primarily focused on development of effective pretreatment methods to facilitate hydrolysis and fermentation (i.e., efficient sugar digestion and inhibitors reduction, respectively) [5], reduction of enzyme costs/usage [6], and modification/improvement of strains that are efficient in co-fermentation of pentose and hexose sugars under inhibition conditions [7]. Researchers are also working on processing configurations, which are mainly focused on the relationship between hydrolysis and batch fermentation, such as separate hydrolysis and fermentation (SHF), simultaneous saccharification and fermentation (SSF), hybrid hydrolysis and fermentation (HHF), and consolidated bioprocessing (CBP); with SHF and HHF currently being more applicable [8]. Yet, much less effort has been spent on the development of bioprocessing strategies that increase productivity through fermentation methodologies.

Batch operation is a widely used and preferred method for ethanol fermentation [9, 10]. However, batch fermentation incorporates lag and stationary phases, during which ethanol is not being produced at substantial levels. Furthermore, significant downtime is necessary after each fermentation to clean up the reactor and prepare for the next campaign. Thus, one approach to improving productivity of batch fermentation would be to reduce fermentation time and downtime. In addition, to achieve the desired levels of ethanol production, industrial ethanol facilities require a number of large batch bioreactors that operate intermittently to ensure a continuous supply of fermentation product for distillation [11]. Correspondingly, microbial propagation, a lengthy and multi-stage scale-up process that provides fermenters with seed culture, is needed for every batch fermentation cycle [10]. Therefore, batch fermentation and its associated seed cultivation contribute to high capital and operating costs. Altogether, capital and operating costs account for 34 and 33% of the total production costs of cellulosic biofuel, respectively [12]. One approach to address these cost issues is to develop a novel fermentation strategy that will improve productivity.

Self-cycling fermentation (SCF) was developed in the 1990s to facilitate the synchronization of cells. SCF is a semi-continuous cycling process where an online monitoring parameter is used to identify the onset of stationary phase. This identification automatically triggers the removal of half of the fermenter contents, which is immediately replaced with fresh, sterile medium to start a subsequent cycle of growth [13]. Through the operation of SCF, cells are synchronized, which means that all, or almost all the cells are divided at the same time. The actual growth rate of cells will vary depending on

growth conditions, which will impact the time required to reach stationary phase, linked to the depletion of a limiting nutrient. Nevertheless, regardless of the time it takes for cells to enter stationary phase, an indicative real-time parameter can be used to trigger the removal and replacement of fermentation broth. Therefore, compared to batch operation, SCF (starting from cycle 2) avoids lag and stationary phases, which means that cells are always in exponential growth, and cycle time equals to generation time [13]. Dissolved oxygen, redox potential, and carbon dioxide evolution rate are commonly monitored parameters in batch reactors, and have all been used as real-time parameters to indicate cell growth and trigger the automation process of SCF [14–16]. Theoretically, SCF can continue indefinitely, with a successful demonstration by Wentworth et al. of more than 100 consecutive cycles for the production of citric acid [14]. Compared to batch fermentation, SCF has also demonstrated increased product yield and/or productivity for many microbial production systems, such as citric acid [14], bioemulsifier [17], shikimic acid [18], and recombinant protein β-galactosidase [19]. Despite these achievements, SCF has not yet been successfully employed for ethanol production. Therefore, considering the relatively low productivity and high production costs associated with batch fermentation of ethanol, we aim to apply SCF to automate the fermentation process and improve productivity for cellulosic ethanol production, thus offsetting high production costs and helping strengthen the cellulosic ethanol industry. The present work provides proof of concept that applying SCF to ethanol fermentation can improve productivity.

Methods

Yeast, medium, and inoculum

Superstart™ active distillers dry yeast, *Saccharomyces cerevisiae*, was purchased from Lallemand Ethanol Technology (Milwaukee, WI, USA). The yeast powder was hydrated, and after dilution, cell suspensions were spread on yeast extract peptone dextrose (YPD) agar plates [10 g/L yeast extract (Sigma-Aldrich, St. Louis, MO, USA); 20 g/L peptone (Sigma-Aldrich, St. Louis, MO, USA); 20 g/L D-glucose (Sigma-Aldrich, St. Louis, MO, USA); 14 g/L agar (Thermo Fisher Scientific, Waltham, MA, USA)] and cultivated for 2 days at 30 °C. Individual colonies were transferred to YPD liquid medium (no agar) in glass tubes for overnight cultivation at 30.0 °C and 230 rpm. Some of the overnight culture was transferred to fresh YPD medium to obtain an optical density at 600 nm (OD_{600}) of roughly 0.3, and was then allowed to grow under the same conditions until the OD_{600} reached 1.0. The broth was then mixed with 50% (v/v) glycerol at a ratio of 1:1, and stored in vials at −80 °C to produce glycerol stock strains. When required, the stock strain was streaked on a YPD agar plate and allowed to cultivate for 2 days at 30.0 °C, and then stored in a 4 °C fridge. Colonies were transferred to a fresh YPD agar plate monthly.

For all seed cultures and fermentations performed in this work, chemically defined medium was used: 50 g/L D-glucose (Sigma-Aldrich, Oakville, ON, Canada), 6.7 g/L yeast nitrogen base with amino acids (YNB, Sigma-Aldrich, St. Louis, MO, USA), and 0.1 M sodium phosphate buffer ($NaH_2PO_4 \cdot 2H_2O/Na_2HPO_4 \cdot 2H_2O$, Thermo Fisher Scientific, Waltham, MA, USA) at pH 6.0. The medium was filter sterilized (Sartolab™ P20 Plus Filter Systems: 0.2 μm, Thermo Fisher Scientific, Waltham, MA, USA) prior to being used.

To prepare the inoculum, isolated colonies on YPD plates were transferred to 10 mL of chemically defined medium in glass tubes and incubated overnight at 30.0 °C with shaking at 200–250 rpm. A portion of this starter culture was transferred to a 1-L shake flask containing 180 or 600 mL fresh medium to obtain an OD_{600} of ≈0.2 for further incubation under the same condition. When an OD_{600} of ~0.5 was achieved in the shake flask, the culture was used to inoculate the fermentation experiments described below. The inoculum volume for all experiments in the report was ~8% (v/v) of the fermentation medium.

Dynamic study of yeast fermentation

To baseline the dynamic changes that occur during batch ethanol fermentation using our fermentation system, a total of 24 shake flasks (500 mL) with 270 mL of chemically defined medium (described above) were inoculated with yeast (8%, v/v). The shake flasks were incubated at 30.0 °C with shaking at 200 rpm. Each flask was attached to an S-lock filled with distilled water to minimize air from flowing in the flask and to release gas out of the flask. At eight specific time points, three flasks were taken out of the incubator and sacrificed for analysis, allowing for the analyses to be carried out in triplicate.

Cycling fermentation

This experiment, in which cycling was performed manually, was designed to mimic the process of SCF, and test whether our fermentation system could result in a stable process of reproducible cycles. The initial cycle had a working volume of 280 mL in a 500-mL shake flask. Additional shake flasks were incubated in parallel to allow for analysis of glucose levels at a given time point. Within fermentation cycles, the additional flasks were taken out from the incubator and monitored to determine the time at which glucose was virtually depleted (<1 g/L; analytical method described below). At this point, experimental flasks were taken out of the

incubator and half of the broth volume (140 mL) was manually removed, and immediately replaced with an equal volume of sterile medium to start the next cycle. Immediately after the sterile medium was added, the flask was gently mixed and a 10-mL sample was removed for analysis. This process was repeated until the end of 5th cycle. It should be noted that for each successive cycle, a smaller amount of broth was removed/replaced due to the drop in fermentation broth volume resulting from withdrawal of 10 mL of samples taken for analysis. For example, at the end of cycle 2, there was a total working volume of 270 mL, and thus, only 135 mL were removed/replaced. All shake flasks were capped with an S-lock filled with distilled water. This experiment was performed in triplicate.

Analytical methods

Optical density (OD_{600}) was measured using a spectrophotometer (Ultrospec 4300 pro, Biochrom, England, UK). High OD_{600} values of fermentation broth were diluted with medium to fall within the range of 0.2–0.9, and cell concentration was calculated according to the appropriate dilution factor. The broth pH was measured using a pH meter (Accumet® AB 15, Fisher Scientific, Thermo Fisher Scientific, Waltham, MA, USA). The concentrations of glucose and ethanol were quantified according to Parashar et al. [20]. Briefly, ethanol content was determined by gas chromatography with a flame ionization detector, using 1-butanol as the internal standard. Glucose content was quantified through high-performance liquid chromatography using an HPX-87H column coupled with a refractive index detector. For samples with glucose content lower than 1 g/L or for quick confirmation of glucose depletion during manual cycling fermentation experiments, a Megazyme D-Glucose (glucose oxidase/peroxidase; GOPOD) assay kit (Bray, Ireland) was used, and the whole procedure took no more than 20 min. In this method, glucose concentration was determined through an absorbance reading at 510 nm, which quantified the amount of a quinoneimine dye derived through enzymatic processing of glucose. Samples were filtered (0.22 μm), mixed with GOPOD reagent, incubated at 50.0 °C, and the absorbance reading was compared against both blank and standard samples. Fermentation efficiency was calculated by using the following equation:

Fermentation efficiency

$$= \left(\frac{\text{Amount of ethanol produced}}{\text{Amount of glucose consumed}} \div 0.511 \right) \times 100.$$

Theoretically, 0.511 g of ethanol is produced per gram of glucose. Fermentation samples were examined under a microscope to confirm the lack of bacterial contamination.

Statistical analysis

Analysis of variance (ANOVA) was conducted with Tukey test set at 95% confidence level by GraphPad Prism 5.04 software (La Jolla, CA, USA). Suspected outliers were evaluated by Q test (95% confidence) within triplicate results. The OD_{600} measurement for one of the three flasks examined at the end of cycle 4 of the manual cycling experiment was confirmed as an outlier, and was therefore excluded from our data analyses; the other parameters assessed passed the test and were kept.

Results

Dynamic study of batch fermentation

The primary goal of the present work was to explore whether a self-cycling fermentation strategy can be incorporated into ethanol production to improve productivity and/or increase product yield. As a baseline comparison for our system, we first performed batch fermentation, monitoring several parameters (OD_{600}, pH, glucose, and ethanol concentrations) at various time points (Fig. 1). During fermentation, glucose, the main carbon source, was consumed by yeast for growth and ethanol production. Glucose was depleted at ~20.5 h, when cell concentration (measured by OD_{600}) and ethanol yield reached maximum values. The pH of the fermentation broth dropped while the yeast was growing and stabilized before the onset of stationary phase. The fermentation efficiency at 20.5 h was 86.1 ± 0.4%. Based on the growth curve generated through OD_{600} readings, we estimated the generation time to be approximately 6 h. Inspection of cells under a microscope confirmed that there was no bacterial contamination.

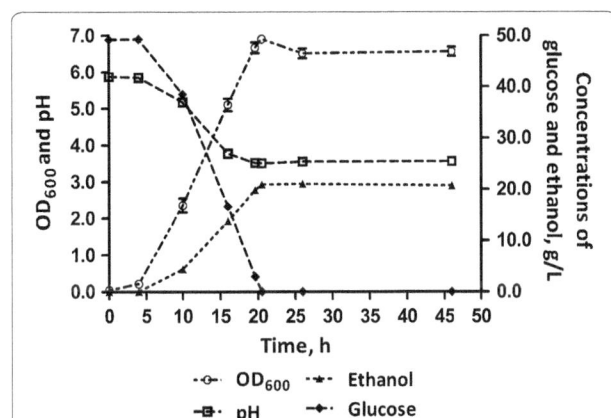

Fig. 1 Dynamic study of batch fermentation. Optical density (OD_{600}), pH, glucose consumption, and ethanol production were monitored over a 46-h period. *Error bars* represent standard deviation of triplicate experiments

Cycling study

Following the batch fermentation experiments, we performed cycling fermentations to determine the impact of incorporating this methodology into an ethanol production system. In this work, cycle time is defined as the time used only for fermentation, excluding the harvest and replacement steps. Once the depletion of glucose had been confirmed, half the fermentation broth was removed and replaced with an equal amount of fresh growth medium, initiating the next fermentation cycle. This was repeated for a total of 5 fermentation cycles. Based on this strategy, in cycle 1, the input content of glucose, as well as the produced amount of ethanol, was roughly twice as much as corresponding values from cycles 2 to 5 (Fig. 2; Table 1). As shown in Fig. 2c, glucose was completely consumed at the end of all cycles. For cycles 2–5, although a smaller amount of ethanol was generated in each cycle compared to cycle 1 (Table 1), the final concentration of ethanol (g/L) was statistically equal in all cycles (Fig. 2d). Additionally, when compared on a per glucose input basis, the yield of ethanol produced was statistically similar in all cycles (Table 1). Thus, the key significance of these experiments is the dramatic decrease in fermentation time required to produce ethanol when the SCF approach was employed. For example, cycle 1 produced 5.6 ± 0.0 g of ethanol in 21.9 ± 0.1 h, while cycles 2, 3, and 4 together produced 7.2 ± 0.2 g of ethanol in 18.8 ± 0.0 h. Thus, cycles 2, 3, and 4 produced 129.2 ± 2.3% of the ethanol generated in cycle 1, but in only 86.0 ± 0.5% of the time.

Figure 2b shows that the first cycle started with a pH of 5.9 ± 0.0 and dropped to 3.5 ± 0.1 by the end of the cycle. After the first manual removal and broth replacement, the buffer capacity of the added medium was not strong enough to bring the pH back to the original value, stabilizing at around pH 5.0. For successive cycles, the pH fluctuated roughly from pH 5.0 at the beginning of the cycles to pH 3.5 at their ends. In terms of yeast growth (Fig. 2a), cultures from all cycles ended with a statistically similar optical density, except for cycle 5, which generated a statistically higher value than the other cycles. However, the starting OD_{600} values of cycles generally increased as a function of cycle number, with cycle 5 being the highest among all cycles. Furthermore, when the change in OD_{600} was compared among cycles 2–5, there was no significant difference.

Ethanol volumetric productivity (Fig. 3a) represents the ethanol produced (g/L) at each cycle per corresponding cycle time. Compared to the 1st cycle, which is essentially a normal batch fermentation, successive manual cycling significantly improved ethanol volumetric productivity (Fig. 3a). For example, cycle 2 displayed an ethanol volumetric productivity increase of 60.4 ± 12.1

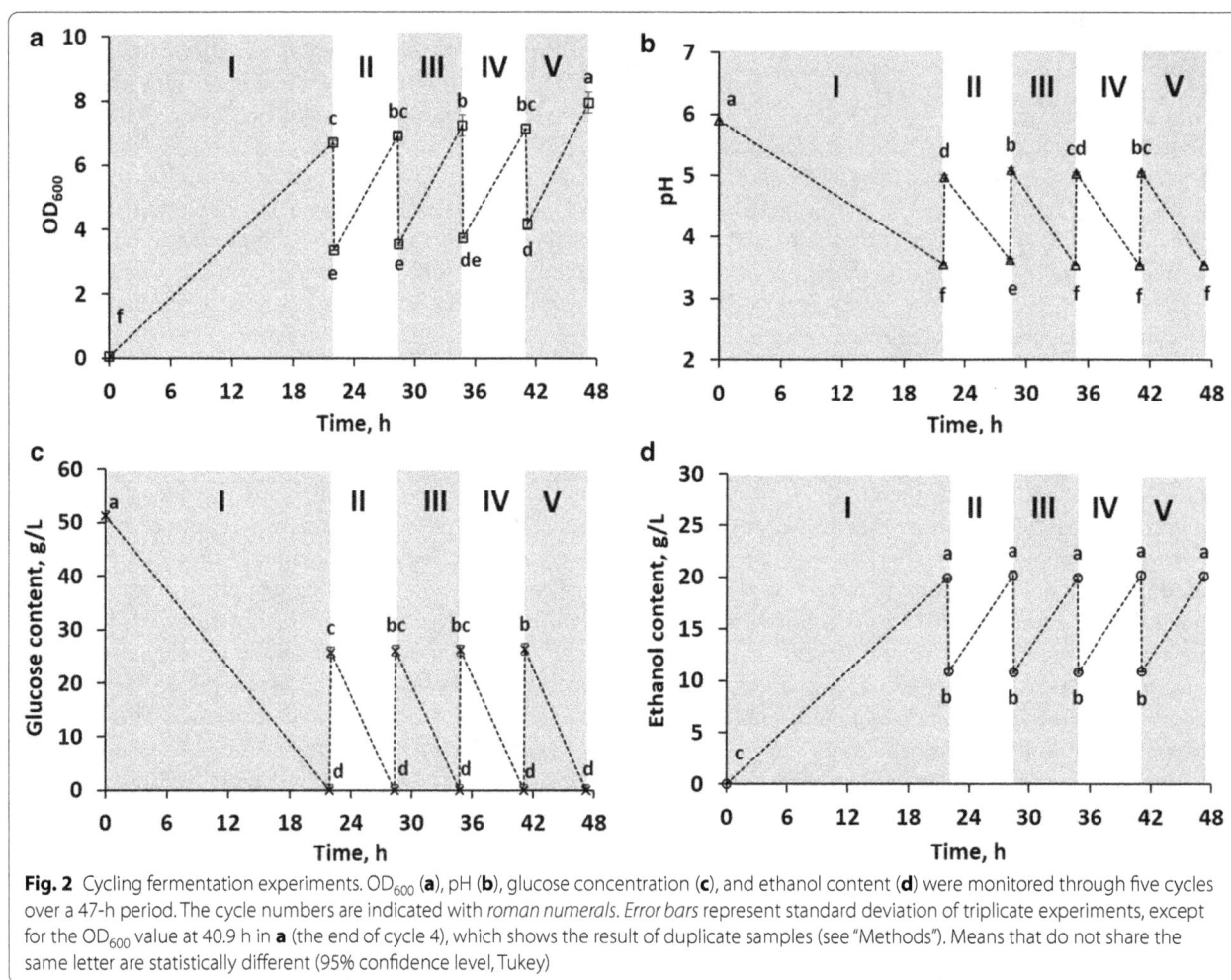

Fig. 2 Cycling fermentation experiments. OD_{600} (**a**), pH (**b**), glucose concentration (**c**), and ethanol content (**d**) were monitored through five cycles over a 47-h period. The cycle numbers are indicated with *roman numerals*. *Error bars* represent standard deviation of triplicate experiments, except for the OD_{600} value at 40.9 h in **a** (the end of cycle 4), which shows the result of duplicate samples (see "Methods"). Means that do not share the same letter are statistically different (95% confidence level, Tukey)

Table 1 Cycling fermentations and overall productivity improvement

Cycle number	Cycle time (h)	Glucose available at onset of cycle (g)	Amount of ethanol produced in a given cycle (g)	Yield of ethanol produced in a given cycle per glucose fed (g/g)	Overall productivity improvement compared to batch (%)
1	21.9 ± 0.1^a	14.3 ± 0.0^a	5.6 ± 0.0^a	0.4 ± 0.0^a	-9.7 ± 0.6^a
2	6.4 ± 0.0^b	6.9 ± 0.1^b	2.5 ± 0.2^b	0.4 ± 0.0^a	15.6 ± 3.4^b
3	6.3 ± 0.0^{bc}	6.8 ± 0.0^c	2.4 ± 0.1^{bc}	0.3 ± 0.0^a	34.9 ± 2.0^c
4	6.2 ± 0.0^{cd}	6.5 ± 0.0^d	2.3 ± 0.1^{bc}	0.4 ± 0.0^a	51.1 ± 2.7^d
5	6.1 ± 0.0^d	6.3 ± 0.1^e	2.2 ± 0.1^c	0.3 ± 0.0^a	64.4 ± 3.3^e

Numbers indicate the mean ± standard deviation of triplicate experiments. Within the same column, values with different superscript letters are statistically different

and $43.1 \pm 11.6\%$, compared to cycle 1 and batch fermentation, respectively (Fig. 1). Specific productivity (Fig. 3b)—representing the ethanol volumetric productivity per biomass content (based on OD_{600} readings)—was 55.1 ± 9.7 and $42.7 \pm 9.8\%$ greater in cycle 2 than in the first cycle and batch fermentation, respectively. These values did not differ significantly in cycles 2–5. To obtain

an approximation of the influence of self-cycling fermentation on overall production efficiency, we calculated the overall productivity based on the laboratory conditions used (Fig. 3c). Overall productivity for a cycle considers the ethanol (g/L) accumulated at the end of the cycle per total process time—which includes medium preparation, the cumulative fermentation cycle time, as well as the

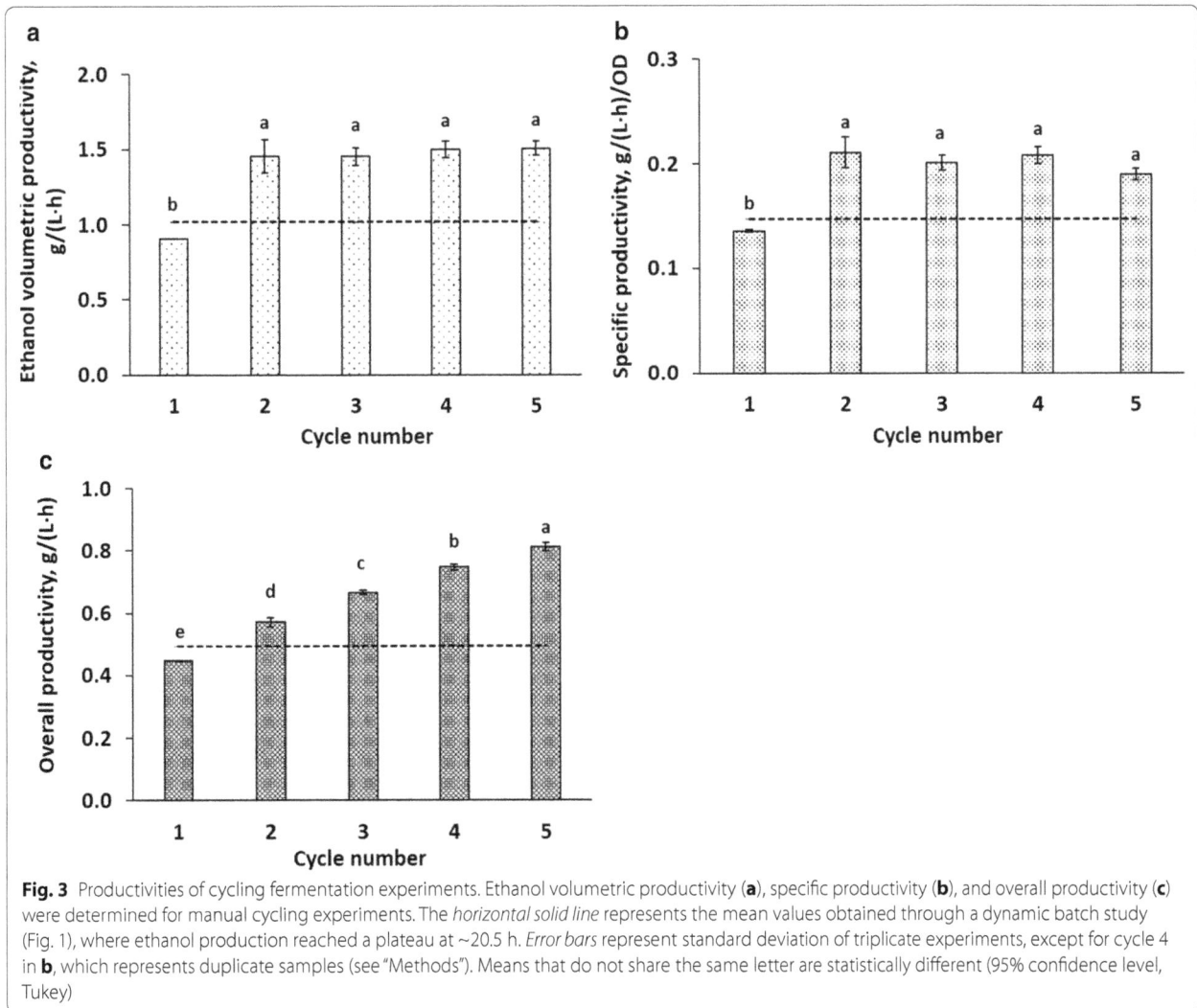

Fig. 3 Productivities of cycling fermentation experiments. Ethanol volumetric productivity (**a**), specific productivity (**b**), and overall productivity (**c**) were determined for manual cycling experiments. The *horizontal solid line* represents the mean values obtained through a dynamic batch study (Fig. 1), where ethanol production reached a plateau at ~20.5 h. *Error bars* represent standard deviation of triplicate experiments, except for cycle 4 in **b**, which represents duplicate samples (see "Methods"). Means that do not share the same letter are statistically different (95% confidence level, Tukey)

time required for the harvesting and refilling steps (3 min each in lab conditions). For a single-batch fermentation, medium preparation, sterilization of media and equipment, and seed cultivation took a total of 21.8 h, while slightly longer was spent for manual cycling fermentation runs (22.7 h); this increase was due to longer time necessary for filter sterilization of a larger volume of medium and a longer period of seed culture cultivation. The length of batch fermentation (20.5 h) was adapted from our dynamic study as similar laboratory procedures, reagents, and glassware were used for both. For the total process, 42.2 ± 0.0 and 69.9 ± 0.1 h were spent for batch and manual cycling fermentation (5 cycles), respectively, which means that manual cycling for 5 fermentation cycles took $65.4 \pm 0.2\%$ longer than batch. Compared to batch, an increase of cycle number in manual cycling

fermentation significantly improved overall productivity (Fig. 3c), and a $64.4 \pm 3.3\%$ improvement was observed when 5 cycles were involved (Table 1).

Annual ethanol productivity for potential scale-up

To appreciate how implementation of a self-cycling strategy could potentially impact annual ethanol production goals at large scale, we compared SCF with batch in terms of annual ethanol productivity, which represents the ethanol produced per year (P, ton/year). Feng et al. determined the annual fermentation operation time (t_{annual}) for an ethanol plant to be 7920 h (330 days) [21]. For a reactor with a working volume (V) of 10^5 L, downtime between cycles was estimated at 6.0 h ($t_{d-batch}$) and 0.25 h for batch and SCF methodology, respectively [21]. Residence time ($t_{f-batch}$, t_{f-SCF}) and ethanol produced (C_{batch},

C_{SCF}) per campaign for batch and SCF were adapted from our dynamic batch and manual cycling studies (cycles 1 and 2), respectively. For demonstration of the calculations below, mean values of triplicate experimental results were used.

For batch fermentation, the number of campaigns (N_{batch}) possible in a year would be:

$$N_{batch} = \frac{t_{annual}}{t_{f\text{-batch}} + t_{d\text{-batch}}}$$
$$= \frac{7920 \text{ h/year}}{20.5 \text{ h/campaign} + 6.0 \text{ h/campaign}}$$
$$= 299 \text{ campaign/year.}$$

Thus, the annual ethanol productivity for batch fermentation (P_{batch}) would be:

$$P_{batch} = N_{batch} \times C_{batch} \times V$$
$$= 299 \text{ campaign/year} \times 20.9 \text{ g/L/campaign} \times 10^5 \text{ L}$$
$$= 6.25 \times 10^8 \text{ g/year}$$
$$= 625 \text{ ton/year.}$$

We assume that a plant can continuously run SCF for x ($x \geq 1$) cycle numbers (with 0.25 h downtime between cycles) each campaign, after which point, the reactor will need to be cleaned and set up for a new campaign (6 h downtime following the last SCF cycle; assumed to be the same as batch).

For a single SCF campaign, the total of residence time ($t_{f\text{-SCF}}$) would be the sum of all x cycles:

$$t_{f\text{-SCF}} = t_{f\text{-cycle 1}} + (t_{f\text{-subsequent cycles}})(x-1) \text{ cycles}$$
$$= 21.9 \text{ h} + (6.4 \text{ h/cycle})(x-1) \text{ cycles}$$
$$= 15.5 \text{ h} + 6.4x \text{ h.}$$

Similarly, the total downtime ($t_{d\text{-SCF}}$) for a single SCF campaign can be summarized as follows:

$$t_{d\text{-SCF}} = (t_{d\text{-}(x-1)\text{ cycles}})(x-1) \text{ cycles} + t_{d\text{-cycle } x}$$
$$= (0.25 \text{ h/cycle})(x-1) \text{ cycles} + 6.0 \text{ h}$$
$$= 5.75 \text{ h} + 0.25x \text{ h.}$$

Therefore, the total number of SCF campaigns that can be run each year (N_{SCF}) can then be determined as follows:

$$N_{SCF} = \frac{t_{annual}}{t_{f\text{-SCF}} + t_{d\text{-SCF}}}$$
$$= \frac{7920 \text{ h/year}}{[(15.5 \text{ h} + 6.4x \text{ h}) + (5.75 \text{ h} + 0.25x \text{ h})]/\text{campaign}}$$
$$= \frac{7920 \text{ h/year}}{[21.25 \text{ h} + 6.65x \text{ h}]/\text{campaign}}.$$

Using the SCF strategy, the total amount of ethanol produced per campaign with x cycles (E_{SCF}) can be calculated as shown below:

$$E_{SCF} = (C_{f\text{-cycle 1}})(V) + (C_{f\text{-subsequent cycles}})(V)(x-1) \text{ cycles}$$
$$= V \times [(C_{f\text{-cycle 1}}) + (C_{f\text{-subsequent cycles}})(x-1) \text{ cycles}]/\text{campaign}$$
$$= 10^5 \text{ L} \times [19.9 \text{ g/L} + (9.3 \text{ g/L/cycle})(x-1) \text{ cycles}]/\text{campaign}$$
$$= 10^5 \text{ L} \times [10.6 \text{ g/L} + 9.3x \text{ g/L}]/\text{campaign.}$$

Therefore, the annual ethanol productivity (P_{SCF}) would be:

$$P_{SCF} = N_{SCF} \times E_{SCF} = \frac{7920 \text{ h/year}}{[21.25 + 6.65x \text{ h}]/\text{campaign}} \times 10^5 \text{ L}$$
$$\times [10.6 \text{ g/L} + 9.3x \text{ g/L}]/\text{campaign}$$
$$= \frac{792 \times (10.6 + 9.3x)}{21.25 + 6.65x} \text{ ton/year.}$$

If cycling fermentation is operated for 5 consecutive cycles ($x = 5$), as was the case in our manual cycling study, P_{SCF} would be 830 ± 41 ton/year, representing a $33.1 \pm 7.2\%$ improvement in annual ethanol productivity compared to batch (P_{batch}, 624 ± 3 ton/year). As implied by Fig. 4, as the number of consecutive cycles (x) in each SCF campaign increases, the annual ethanol productivity (P_{SCF}) initially increases sharply before the increase becomes almost negligible (as the fraction of downtime to production time becomes negligible). Moreover, annual ethanol productivity in SCF (P_{SCF}) is expected to be significantly greater than that of batch fermentation (P_{batch}), even when only 2 cycles ($x \geq 2$) are operated for each SCF campaign.

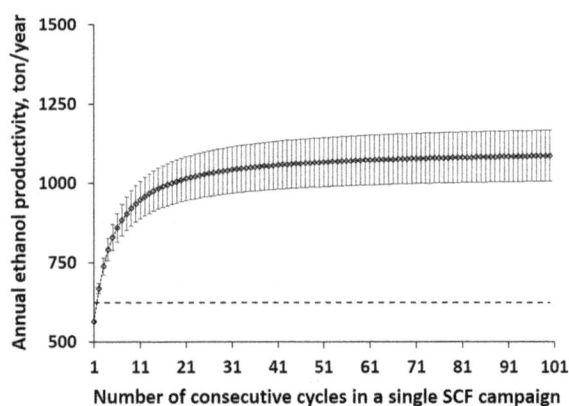

Fig. 4 Annual ethanol productivity. Annual ethanol productivity was calculated assuming the number of consecutive cycles operated for each SCF campaign could range from 1 to 100. The *horizontal solid line* represents the mean values obtained through a dynamic batch study (Fig. 1), where ethanol production reached a plateau at ~20.5 h. *Error bars* were calculated from the errors in ethanol yield and cycle time of SCF cycles and represent standard deviation of triplicate experiments

Examined from a different perspective, the goal of SCF application may be to achieve the same annual ethanol productivity as batch fermentation (based on 625 ton/year), but in a shorter amount of time (i.e., fewer campaigns). Using the equation above, if SCF operation is based on 5 consecutive cycles ($x = 5$), each SCF campaign will produce 5.7 tons of ethanol (E_{SCF}). Therefore, the number of SCF campaigns required to produce 625 tons of ethanol through SCF is roughly 110 per year (P_{batch}/E_{SCF}). Given that each SCF campaign would require a total of 54.5 h ($t_{f\text{-}SCF} + t_{d\text{-}SCF}$), the total time required for 110 campaigns is roughly 6000 h. This is approximately 1900 h (~80 days) shorter than the annual fermentation time required for batch fermentation to produce the same amount of ethanol.

Discussion

Dynamic study

Ethanol production is tightly associated with cell growth. As such, when the limiting nutrient is depleted under anaerobic conditions, yeast stops growing and producing ethanol, entering into stationary phase. In the present study, the ethanol titer was lower than what has been achieved in industry [10]; this is because a defined medium (6.7 g/L YNB, 0.1 mol/L phosphate buffer, and 50 g/L glucose), where glucose was the main carbon source, was used at a low concentration, instead of directly using a typical hydrolysate of lignocellulosic material that contains a mixture of pentose and hexose sugars, corn steep liquor, and inhibitors [10]. This was done in order to simplify the implementation of SCF operation for the study at hand.

As is typical during ethanol fermentation, the pH of our batch system decreased, likely because the uptake of buffering materials such as amino nitrogen compounds, the excretion of organic acids [22], the utilization of ammonium—which releases hydrogen ions outside of the cell [23]—and the production of carbonic acid due to the reaction of carbon dioxide (released by yeast) with water. Nevertheless, given the data on glucose consumption and ethanol production in batch fermentation (Fig. 1), the medium system was adequately buffered and was able to achieve relatively high fermentation efficiency and biomass yield. It should be noted that the fermentation efficiency observed in the batch fermentation is lower than those typically observed in fermentations using wheat grain as feedstock (roughly 90–94%) [20, 24]. One explanation for this may be the presence of oxygen in the headspace of shake flasks, which would enable yeast to momentarily grow aerobically to produce biomass, rather than ethanol. Furthermore, the medium used was not optimized for ethanol production, as is the case with ethanol fermentations using grains. Despite the

suboptimal conditions, the fermentation efficiency still reached 86.1 ± 0.4%. Furthermore, we are exploring the use of other media that could be employed in SCF operation to further improve fermentation efficiency.

Cycling study

Since Fig. 1 revealed that the onset of stationary phase was tightly linked to the depletion of glucose, identification of the specific time point where glucose is depleted is important for SCF systems. Doing so may allow for harvest and fresh medium addition right as cells would enter stationary phase, where ethanol production ends and cell metabolism begins to change. As suggested in the manual cycling study, sugar was depleted by the end of each cycle (Fig. 2c), which would help bioethanol producers avoid unnecessary sugar losses and improve process economics. This also gives SCF an advantage over chemostat operation, where some of the nutrients are washed out throughout the process.

It should be noted that, for cycles 2–5, there was a slight gradual increase in starting cell concentration (Fig. 2a), yet no significant difference in OD_{600} change was found among the four cycles. This is possibly due to the settling of cells during manual broth removal, which made the cell concentration of removed sample slightly lower than that of the broth left inside of shake flasks. The settling might be the reason why one of the three samples at the end of cycle 4 was rejected as an outlier (Q test) of its parallel samples in cell concentration (as measured by OD_{600}). Whereas the measurements of the other parameters (pH, glucose, and ethanol concentrations) used techniques that are not related to cell concentration and the values were retained by Q test, we still incorporated that sample for the results of parameters not based on OD_{600}. All in all, this settling phenomenon will likely not be an issue in scale-up due to continual stirring during broth removal.

It should also be noted that due to the sampling required for analysis, the fermentation broth volume decreased by 10 mL in each cycle, which led to a reduction in total glucose input (g) and also the total amount of ethanol produced (g) from cycles 1 to 5 (Table 1). While such sampling may have slightly decreased total ethanol production in our shake flask studies, this would not be significant in bioreactor operation, as sampling volumes are negligible in larger vessels. Nevertheless, our data clearly provide proof of concept that our SCF approach to ethanol production can retain ethanol yield and increase ethanol yield per fermentation time.

As shown in the cycling experiments, which mimicked SCF, at the end of each cycle, half of the cell population was harvested with the other half serving as the "inoculum" (50% (v/v) of the working volume) for the next cycle.

SCF can contribute several benefits to the ethanol production process.

Firstly, in current cellulosic ethanol plants, a few steps are typically required to gradually scale-up a seed culture for inoculation, which is a common practice for batch fermentation [10]. Whereas for SCF operation, once inoculated for the first cycle, the yeast propagation process is no longer required for subsequent cycles, and is only necessary when a new SCF campaign is initiated. Thus, the more cycles incorporated into an SCF campaign, the fewer microbial propagation steps would be required. This will save nutrients, energy, and work hours spent on the propagation stage. Furthermore, the cycling strategy of SCF is easily compatible with current processing infrastructure, since the removed volume of broth can be fed continuously into a distillation column, and fresh medium could be pumped from the hydrolysis section (SHF) of an integrated process.

Secondly, as shown in the cycling experiments (Table 1), compared to batch operation, fermentation time is dramatically reduced in SCF, without compromising the ethanol yield. This is likely because the lag and stationary phases are removed from SCF operation [13], and therefore, cells are always in exponential growth. It should be noted that only two data points are shown for each cycle (Fig. 2), so there is a possibility that the substrate was depleted earlier than reported, which could make the fermentation cycle times shorter and productivity higher. Also, cycle times varied among cycles 2–5 (Table 1). These cycle times are based on the confirmation of glucose disappearance (using the GOPOD method) from additional shake flasks incubated in parallel to minimize volume change of the main experimental flasks and avoid exposure to air during fermentation. Thus, this analytical procedure may have introduced a slight delay, and the cycling times reported may not be absolutely reflective of what happened in the main experimental flasks. Furthermore, although some of the cycling times were statistically different, they were only different by a few minutes. In the implementation of a fully automated SCF system, the overestimation of cycle time is unlikely, since the fermentation will be monitored by a real-time parameter, which will automatically trigger the cycling process once cells enter stationary phase.

Finally, for cellulosic ethanol production, the pretreatment of feedstocks can form or release inhibitors, such as furfural, phenolic compounds, and weak acids, that can inhibit cell growth and ethanol production [8]. It has been reported that inhibition can be biochemically mitigated through exposing microbe seed cultures to inhibitors during propagation [25]. Therefore, for SCF, it may be worthwhile in the future to test whether the "inoculum" (i.e., half of the fermentation broth from the previous cycle), which has been grown in the presence of any potential inhibitors, will help the following cycle achieve better inhibitor tolerance and therefore better ethanol production.

According to Table 1, the yield of ethanol produced per glucose fed was statistically similar for all cycles. Thus, the 43.1 ± 11.6% improvement in ethanol volumetric productivity (g/(L h)) observed in the cycling fermentation study was due to the reduced fermentation time. This result, even though performed in shake flasks—which may lead to higher variability than controlled bioreactors, is still consistent with those of reported SCF systems—where bioreactors were used and improvements in productivity were achieved primarily due to shorter fermentation time than batch [17, 19]. It should be noted that cell synchrony was not assessed in this study as synchrony has been shown to be established after 5–10 SCF cycles. Therefore, the reduction of fermentation time in the present study is unlikely to be due to cell synchrony. Whichever is the case, significant improvements in productivities are observed and optimization of cell synchronies could possibly further enhance these results. This supports the argument that application of SCF in industrial ethanol production may reduce the fermentation time necessary to reach current production goals, without changing existing infrastructures. The reduction of fermentation time leads to lower operation costs, which currently make up 33% of total production costs [12]. Alternatively, this improvement also suggests that current production levels could be met by employing smaller bioreactors in an SCF strategy. In this way, new cellulosic ethanol plants may be able to reduce their capital costs, which typically account for 34% of the total production cost [12]. Given the similar biomass yields of all cycles (Fig. 2a), the specific productivity of all cycles is clearly most impacted by the ethanol volumetric productivity. Again, this strongly suggests that the improvement in specific productivity is mainly due to the reduced cycle time. Overall productivity in a laboratory setting indicates how the cycling strategy could impact the overall process. Examining the cycling process in a single shake flask, results support that when more cycles are incorporated in a campaign, higher overall productivity is achieved (Fig. 3c). Note that there was a 9.7 ± 0.6% reduction in overall productivity compared to batch when cycle 1 (essentially a batch cycle) was performed. This probably results from the extra time required to confirm the disappearance of glucose in parallel flasks prior to performing the manual cycling, which results in a slight overestimation of cycling times in the manual cycling study. In dynamic batch study, flasks were directly taken out from the incubator and sacrificed for dynamic analysis throughout the fermentation process.

In addition, this difference could result from batch to batch variations as batch operation is known to be variable [17].

Annual ethanol productivity

Currently, cost reductions of cellulosic ethanol production primarily come from improvements in pretreatment [5], hydrolysis [6], and strain improvement [7]. However, much less effort has been spent on improving productivity and reducing costs by directly changing processing strategies of fermentation. To get an idea whether applying cycling strategies to ethanol fermentation could increase the total amount of ethanol that could be produced per year (annual ethanol productivity) at large scale, we assumed that, with the exception of the length of downtime, SCF would operate under the same conditions as batch. According to Feng et al. downtime between cycles is approximately 6 h for batch fermentation [21], which includes the time used to harvest broth, clean, sterilize, and refill the reactor. However, only 0.25 h will be needed to exchange volumes between SCF cycles [21], since only half the volume of the broth will be harvested, and no cleaning or sterilization steps are necessary between cycles. Furthermore, the time required to add fresh medium to the reactor will actually be part of the cycle time because cells continue to grow as soon as the nutrients are added to the reactor. These calculations indicate that, compared to batch operation, if a five-cycle SCF strategy were implemented for each campaign performed at a plant, either the amount of ethanol produced annually would be greatly increased or annual fermentation time would be dramatically reduced for the same production level. These improvements, which would help reduce production cost, are mainly attributable to the reduced fermentation time, as well as the reduced fraction of downtime.

While theoretically SCF can run indefinitely, there is a concern that in long-term continuous operation of SCF, a non-beneficial mutation or severe bacterial contamination may occur and affect ethanol productivity. This can be averted by implementing SCF operation with number of cycles (x) that is low enough to minimize the probability of mutations or contamination affecting productivity, but large enough to significantly increase annual ethanol productivity. Figure 4 provides a basis for the determination of an optimal number of cycles. Based on these results, we found that, with regard to annual ethanol productivity, operation of SCF for approximately 20 sequential cycles (essentially 20 generations) would provide a good balance between improved productivity and reduced risks of mutation/contamination.

To the best of our knowledge, the present study is the first to provide proof of concept that SCF could

be employed for ethanol production towards elevated productivities. Feng et al. attempted to implement SCF operation by using redox potential as a feedback control parameter for ethanol fermentation [21]. Air was purged in the reactor when the redox potential of the broth fell below a certain level, so that redox potential could generate a transient response. However, this switch between anaerobic and aerobic conditions during fermentation likely disrupted cell metabolism, and thus affected ethanol production. Therefore, this artificial manipulation of redox potential during SCF for ethanol production led to longer fermentation time and reduced ethanol volumetric productivity compared to batch operation.

Conclusions

By mimicking the SCF process in manual cycling experiments at the shake flask scale, the required fermentation time was greatly reduced, while maintaining statistically equivalent glucose to ethanol conversion. With respect to batch operation, our cycling strategy improved ethanol volumetric productivity by $43.1 \pm 11.6\%$, overall productivity by $64.4 \pm 3.3\%$, and estimated annual ethanol productivity by $33.1 \pm 7.2\%$. These elevated productivities may lead to reduced capital costs (i.e., the number and/or size of fermenters required) or operation costs (i.e., the fermentation time required), increased amounts of ethanol production per year, and could eventually lower production costs, relative to batch fermentation. This work, even though performed under suboptimal conditions, has successfully provided proof of concept that adoption of an SCF strategy for cellulosic ethanol could increase productivities, thereby opening up a great possibility for applying novel cycling fermentation strategies to strengthen the cellulosic ethanol industry.

Abbreviations

SCF: self-cycling fermentation; SHF: separate hydrolysis and fermentation; SSF: simultaneous saccharification and fermentation; HHF: hybrid hydrolysis and fermentation; CBP: consolidated bioprocessing; GOPOD: glucose oxidase/peroxidase; OD_{600}: optical density at 600 nm; YNB: yeast nitrogen base with amino acids; YPD: yeast extract peptone dextrose; P_{batch}, P_{SCF}: annual ethanol productivity (ton/year) for batch and SCF, respectively; t_{annual}: annual fermentation operation time, h; $t_{f\text{-}batch}$, $t_{d\text{-}batch}$: for a single batch campaign, the residence time (h) and downtime (h), respectively; C_{batch}, C_{SCF}: ethanol titer produced per cycle (g/L) in batch and SCF, respectively; N_{batch}, N_{SCF}: the number of campaigns possible in a year for batch and SCF, respectively; V: working volume, L; x: for a single SCF campaign, assumed numbers of cycles that a plant can continuously run; $t_{f\text{-}cycle\,1}$, $C_{f\text{-}cycle\,1}$: for the first cycle of an SCF campaign, the residence time (h) and the titer of ethanol produced (g/L), respectively; $t_{f\text{-}subsequent\,cycles}$, $C_{f\text{-}subsequent\,cycles}$: for cycles following the first one in an SCF campaign, the residence time of the cycle (h) and the titer of ethanol produced (g/L), respectively; $t_{d\text{-}(x-1)\,cycles}$: down time (h) for each cycle in a single SCF campaign, except cycle x; $t_{d\text{-}cycle\,x}$: down time for cycle x (the last cycle) in a single SCF campaign; $t_{f\text{-}SCF}$, $t_{d\text{-}SCF}$, E_{SCF}: for a single SCF campaign with x cycles, the total residence time (h), the total downtime (h), and the total amount of ethanol produced (g), respectively.

Authors' contributions
JW designed, performed, and analyzed the experiments, and wrote the manuscript. MC was involved in experiment design and discussion, and helped write the manuscript. DCB and DS discussed and designed experiments, and proofread the manuscript. In addition, DCB and DS are co-supervisors for JW. All authors read and approved the final manuscript.

Acknowledgements
Not applicable.

Competing interests
The authors declare that they have no competing interests.

Funding
The project is funded by BioFuelNet Canada and the Natural Sciences and Engineering Research Council of Canada (NSERC).

References
1. Jarvis M. Chemistry: cellulose stacks up. Nature. 2003;426:611–2.
2. Goldemberg J. Ethanol for a sustainable energy future. Science. 2007;315:808–10.
3. Kumar R, Tabatabaei M, Karimi K, Horváth IS. Recent updates on lignocellulosic biomass derived ethanol—a review. Biofuel Res J. 2016;3:347–56.
4. Chen M, Smith PM, Wolcott MP. US biofuels industry: a critical review of opportunities and challenges. Bioprod Bus. 2016;1:42–59.
5. Chen X, Shekiro J, Pschorn T, Sabourin M, Tao L, Elander R, et al. A highly efficient dilute alkali deacetylation and mechanical (disc) refining process for the conversion of renewable biomass to lower cost sugars. Biotechnol Biofuels. 2014;7:98.
6. Teter SA. DECREASE final technical report: development of a commercial ready enzyme application system for ethanol. Davis, CA. Prepared for the U.S. Department of. Energy. 2012. doi:10.2172/1039767.
7. Mohagheghi A, Linger JG, Yang S, Smith H, Dowe N, Zhang M, et al. Improving a recombinant *Zymomonas mobilis* strain 8b through continuous adaptation on dilute acid pretreated corn stover hydrolysate. Biotechnol Biofuels. 2015;8:55.
8. dos Santos LV, de Grassi MCB, Gallardo JCM, Pirolla RAS, Calderón LL, de Carvalho-Neto OV, et al. Second-generation ethanol: the need is becoming a reality. Ind Biotechnol. 2016;12:40–57.
9. Godoy A, Amorim HV, Lopes ML, Oliveira AJ. Continuous and batch fermentation processes: advantages and disadvantages of these processes in the Brazilian ethanol production. Int Sugar J. 2008;110:175–82.
10. Humbird D, Davis R, Tao L, Kinchin C, Hsu D, Aden A, Schoen P, Lukas J, Olthof B, Worley M, Sexton D, Dudgeon D. Process design and economics for biochemical conversion of lignocellulosic biomass to ethanol: dilute-acid pretreatment and enzymatic hydrolysis of corn stover. Golden, CO: National Renewable Energy Laboratory Technical Report (NREL/TP-5100-47764); 2011.
11. de Vasconcelos JN. Ethanol Fermentation. In: Santos F, Borém A, Caldas C, editors. Sugarcane: agricultural production, bioenergy and ethanol. London: Academic Press; 2015. p. 311–40.
12. Jin M, da Costa SL, Schwartz C, He Y, Sarks C, Gunawan C, et al. Toward lower cost cellulosic biofuel production using ammonia based pretreatment technologies. Green Chem. 2016;18:957–66.
13. Brown WA. The self-cycling fermentor: development, applications, and future opportunities. Recent Res Dev Biotechnol Bioeng. 2001;4:61–90.
14. Wentworth SD, Cooper DG. Self-cycling fermentation of a citric acid producing strain of Candida lipolytica. J Ferment Bioeng. 1996;81:400–5.
15. Brown WA, Cooper DG, Liss SN. Adapting the self-cycling fermentor to anoxic conditions. Environ Sci Technol. 1999;33:1458–63.
16. Sauvageau D, Storms Z, Cooper DG. Synchronized populations of *Escherichia coli* using simplified self-cycling fermentation. J Biotechnol. 2010;149:67–73.
17. Brown WA, Cooper DG. Self-cycling fermentation applied to *Acinetobacter calcoaceticus* RAG-1. Appl Environ Microbiol. 1991;57:2901–6.
18. Agustin RV. The impact of self-cycling fermentation on the production of shikimic acid in populations of engineered Saccharomyces cerevisiae. Master thesis, University of Alberta, 2015.
19. Storms ZJ, Brown T, Sauvageau D, Cooper DG. Self-cycling operation increases productivity of recombinant protein in *Escherichia coli*. Biotechnol Bioeng. 2012;109:2262–70.
20. Parashar A, Jin Y, Mason B, Chae M, Bressler DC. Incorporation of whey permeate, a dairy effluent, in ethanol fermentation to provide a zero waste solution for the dairy industry. J Dairy Sci. 2015;99:1859–67.
21. Feng S, Srinivasan S, Lin YH. Redox potential-driven repeated batch ethanol fermentation under very-high-gravity conditions. Process Biochem. 2012;47:523–7.
22. Bamforth CW. pH in brewing: an overview. Tech Q Master Brew Assoc Am. 2001;38:1–9.
23. Shuler M, Filkret K. Bioprocess engineering: basic concepts. 2nd ed. Upper Saddle River: Prentice Hall; 2002.
24. Bai FW, Anderson WA, Moo-Young M. Ethanol fermentation technologies from sugar and starch feedstocks. Biotechnol Adv. 2008;26:89–105.
25. Tomás-pejó E, Olsson L. Influence of the propagation strategy for obtaining robust Saccharomyces cerevisiae cells that efficiently co-ferment xylose and glucose in lignocellulosic hydrolysates. Microb Biotechnol. 2015;8:999–1005.

7

Time-resolved transcriptome analysis and lipid pathway reconstruction of the oleaginous green microalga *Monoraphidium neglectum* reveal a model for triacylglycerol and lipid hyperaccumulation

Daniel Jaeger[1], Anika Winkler[2], Jan H. Mussgnug[1], Jörn Kalinowski[2], Alexander Goesmann[3] and Olaf Kruse[1,4*]

Abstract

Background: Oleaginous microalgae are promising production hosts for the sustainable generation of lipid-based bioproducts and as bioenergy carriers such as biodiesel. Transcriptomics of the lipid accumulation phase, triggered efficiently by nitrogen starvation, is a valuable approach for the identification of gene targets for metabolic engineering.

Results: An explorative analysis of the detailed transcriptional response to different stages of nitrogen availability was performed in the oleaginous green alga *Monoraphidium neglectum*. Transcript data were correlated with metabolic data for cellular contents of starch and of different lipid fractions. A pronounced transcriptional down-regulation of photosynthesis became apparent in response to nitrogen starvation, whereas glucose catabolism was found to be up-regulated. An in-depth reconstruction and analysis of the pathways for glycerolipid, central carbon, and starch metabolism revealed that distinct transcriptional changes were generally found only for specific steps within a metabolic pathway. In addition to pathway analyses, the transcript data were also used to refine the current genome annotation. The transcriptome data were integrated into a database and complemented with data for other microalgae which were also subjected to nitrogen starvation. It is available at https://tdbmn.cebitec.uni-bielefeld.de.

Conclusions: Based on the transcriptional responses to different stages of nitrogen availability, a model for triacylglycerol and lipid hyperaccumulation is proposed, which involves transcriptional induction of thioesterases, differential regulation of lipases, and a re-routing of the central carbon metabolism. Over-expression of distinct thioesterases was identified to be a potential strategy to increase the oleaginous phenotype of *M. neglectum*, and furthermore specific lipases were identified as potential targets for future metabolic engineering approaches.

Keywords: *Monoraphidium neglectum*, mRNA-seq, Biodiesel, Lipid metabolism, Nitrogen starvation, TAG accumulation, Pathway analysis, Fatty acid, Lipase, Central carbon metabolism

Background

The production of bulk bio-commodities in a sustainable way is a key target of many biotechnological processes.

Owing to their phototrophic growth characteristics, microalgae have been considered to be promising candidates for the production of biofuels such as biodiesel, bioethanol, biogas, or biohydrogen (H_2), as well as of high-value products such as terpenoids, polyunsaturated fatty acids (FAs), or recombinant proteins [1–3]. In the context of biodiesel production however, microalgal lipid productivity needs to be improved for overall economic

*Correspondence: olaf.kruse@uni-bielefeld.de
[4] Present Address: Algae Biotechnology and Bioenergy, Faculty of Biology, Center for Biotechnology (CeBiTec), Bielefeld University, Universitaetsstrasse 27, 33615 Bielefeld, Germany
Full list of author information is available at the end of the article

feasibility [3–5]. A valuable strategy to reach this goal is metabolic engineering [6, 7]. However, the current understanding of the algal lipid metabolism is still incomplete, although it has been progressively investigated in the model green alga *Chlamydomonas reinhardtii* [8]. For the design of rational metabolic engineering strategies, a valuable approach is to follow the cell's endogenous regulation of carbon partitioning under conditions of high lipid and especially triacylglycerol (TAG) productivity. TAG accumulation in microalgae is efficiently induced by nitrogen starvation (−N) [9] and transcriptome studies yield an initial overview of pathway regulation, which is scalable to the single-gene level. Therefore, the investigation of the transcriptome profiles during nitrogen limitation is an appropriate strategy for the identification of gene targets.

Previous transcriptome studies with the aim to investigate the molecular mechanisms of TAG accumulation under −N conditions were performed for the green algae *C. reinhardtii* [10–15], *Chlorella sorokiniana* [16], *Neochloris oleoabundans* [17], and *Neodesmus* sp. [18], as well as the diatom *Phaeodactylum tricornutum* [19] and the eustigmatophyceae *Nannochloropsis oceanica* [20] and *Nannochloropsis gaditana* [21]. These studies have shown a transcriptional induction of nitrogen assimilation [10, 11, 17, 19], of pyruvate kinase [13, 17, 19, 20, 22] indicating a redirection of carbon flow towards pyruvate generation, of the tricarboxylic acid cycle [17, 19, 20], and of a subset of diacylglycerol acyltransferases which are key enzymes for TAG synthesis [10, 12, 19, 20]. These changes were accompanied by a general transcriptional repression of the cellular processes photosynthesis [10, 11, 17, 19, 20, 23], translation (ribosomes) [11, 17, 20, 23], and gluconeogenesis [10, 19, 20].

Today, the most detailed time course experiments were performed for the model green alga *C. reinhardtii* [12–14, 22], where the nutrient starvation phase was investigated in great detail at several time points and in different mutants, such as the starchless strain *sta6* [14]. However, in those works, the reverse phase of nutrient resupply after the starvation phase, triggering degradation of TAGs [24] and other storage compounds such as starch [25], has not been investigated so far. Furthermore, *C. reinhardtii* is not considered as the optimal choice for large-scale biofuel production [26]. At the same time, dynamic transcript changes in other chlorophyceae have not been investigated in great detail. In this regard, only single time point analyses are available for the oleaginous chlorophyceae *N. oleoabundans* (11 days of −N) [17] and *Neodesmus* sp. (a single pool of samples from 3 and 6 h of −N) [18], as well as for the squalene-rich chlorophyceae *Botryosphaerella sudetica* (3 days of −N) [23]. Although more extensive transcriptome data were acquired for *C.*

sorokiniana, such that nitrogen limitation was investigated in both heterotrophic and autotrophic conditions, only one time point from each condition was sampled and biological replicates were not performed [16].

The chlorophyceae *Monoraphidium neglectum* was recently identified as a promising strain for the sustainable production of lipid-derived bioproducts [27]. The species was demonstrated to exhibit robust growth characteristics and a neutral lipid productivity of $52 \pm 6 \, \text{mg L}^{-1} \, \text{day}^{-1}$ under autotrophic conditions, which is four times the productivity of the model chlorophyceae *C. reinhardtii* [27]. When exposed to −N treatment, neutral lipids accumulated to ca. 33% of the total dry biomass [28], with fatty acid profiles being well suited for biodiesel production [27]. Furthermore, the genome has recently been sequenced [27], and it was shown that genetic transformation and stable recombinant protein expression are possible [29]. We therefore chose this promising species as a target for a time-resolved investigation of its transcriptome profiles under nutrient replete and nutrient starvation conditions. In contrast to transcriptome studies with *C. reinhardtii* [10–14, 22], we applied fully autotrophic conditions with excess CO_2, therefore more closely representing the conditions for sustainable biofuel production. Our goals for the transcriptome analysis were to (a) elucidate the physiological pathways important for cellular lipid turnover processes and (b) identify potential bottlenecks for metabolic pathway engineering to further improve the capacity for neutral lipid production with this microalga. Towards this end, we performed a time course experiment consisting of a −N phase and a subsequent +N resupply step. The −N phase was subdivided into two stages, an early −N stage (e−N), characterized by increased starch production, and late −N stage (l−N), characterized by increased lipid production. The third stage investigated in this work constituted the N resupply treatment (r+N). During each of the stages, multiple samples were taken and analyzed by mRNA sequencing (mRNA-seq). With this experimental setup, lipid accumulation (−N) and lipid degradation (+N) were both analyzed at the transcriptional level for the first time by mRNA-seq in the context of microalgal lipid accumulation. As an additional benefit of the transcriptome sequencing, we used the extensive data to improve the currently available genome annotation with the aim to facilitate future genetic engineering approaches.

Results
Overview of experiments

One major goal of this study was the identification of gene targets that could be promising for subsequent genetic engineering approaches with the aim to increase the microalgal triacylglycerol (TAG) accumulation.

Towards this end, we applied transcriptomics of *M. neglectum* under alternating phases of nitrogen (N) availability, as microalgal TAG accumulation is efficiently induced by −N treatment [9].

As a preparatory step, a long-term −N experiment of 17 days of autotrophic −N conditions was performed, in which the dynamics of starch, TAG, and total lipid accumulation in *M. neglectum* were identified (Fig. 1, exp1). As a result, the −N phase could be subdivided into an early stage of starch accumulation (e−N stage), and a subsequent, late stage where TAG and total lipid levels increased (l−N stage).

For transcriptome analysis, we decided to further extend the scope by not only investigating the conditions of TAG accumulation, but also including conditions of TAG degradation, in order to obtain a profound understanding of the transcriptional regulation of the TAG metabolism in *M. neglectum*. TAG accumulation can be reversed by resupplementation of N [24, 30, 31]. Therefore, the experiment one was repeated, and an additional

N resupply treatment (r+N stage) after 48 h of −N conditions was included (Fig. 1, exp2).

Characterization of the cellular response of *M. neglectum* to nitrogen starvation to define the timing of starch, TAG, and total lipid accumulation

Starch and TAGs are the major storage compounds in chlorophyceae [32, 33]. For the model microalga *C. reinhardtii*, a biphasic pattern of starch and TAG accumulation under −N conditions has been described, with starch accumulation preceding TAG accumulation [34]. In order to investigate how far this pattern also applies to *M. neglectum*, the dynamics of starch and lipid accumulation were determined during a long-term −N trial (Fig. 1).

Removal of N resulted in cessation of cell doubling after approximately 2 days of starvation, while biomass concentrations continuously increased from initially 0.236 ± 0.024 g L^{-1} SE (standard error, $n = 3$) to 1.583 ± 0.06 g L^{-1} SE after 11 days of −N conditions. Increasing biomass concentrations and cessation of cell

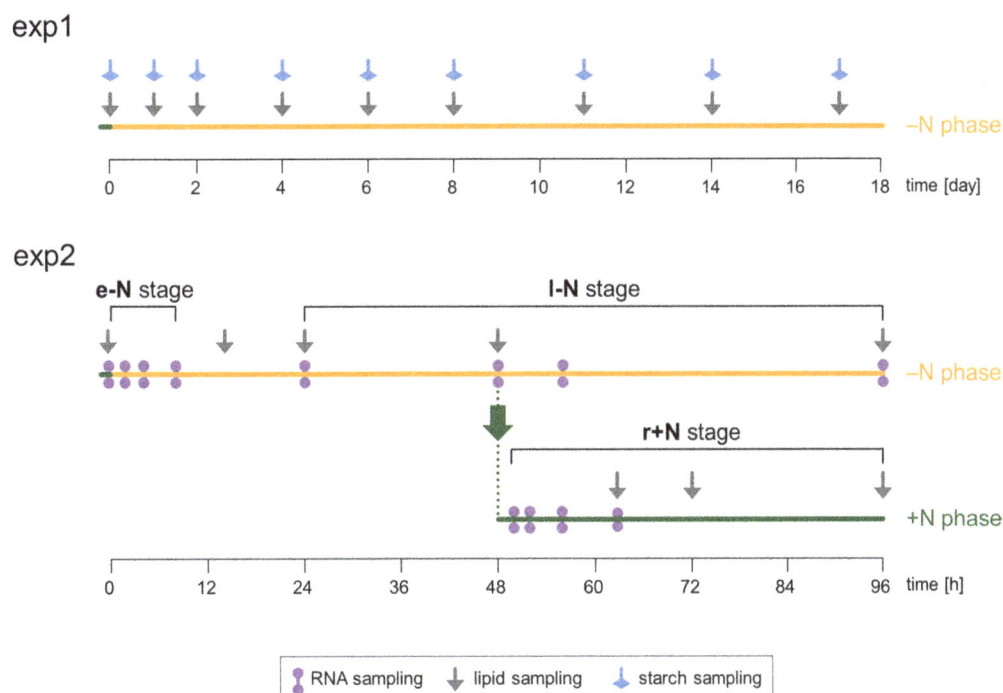

Fig. 1 Experimental design to elucidate the transcriptional mechanism of microalgal TAG accumulation, triggered by nitrogen starvation, in *M. neglectum*. In a first experiment, the dynamic of cellular starch and lipid accumulation in response to −N treatment by *M. neglectum* was investigated in a long-term −N experiment. The individual sampling time points for starch and lipid determination are depicted; the corresponding metabolic data are shown in Fig. 2. In order to elucidate the transcriptional mechanisms correlating with starch and lipid accumulation, a second −N experiment was performed. In this experiment, an additional N resupply treatment was included, in order to induce the end of cellular quiescence and to consequently trigger the reversal of storage compound accumulation. By this combined treatment of alternating phases of N availability, the transcriptional program for both accumulation and degradation of TAGs was monitored, which facilitated the identification of central transcriptional responses underlying microalgal TAG accumulation. As starch accumulation precedes TAG accumulation under −N conditions, the −N phase was subdivided into two stages, i.e., e−N = starch accumulation stage and l−N = TAG and lipid hyperaccumulation stage. Upon N resupply, storage compounds were degraded, i.e., r+N = TAG degradation stage. The individual sampling time points for RNA and lipid isolation are depicted; the corresponding metabolic data are shown in Additional file 1: Figure S1

doubling indicate storage compound accumulation and translated into an increase in cell weight (Fig. 2a).

The cellular starch level was found to peak at day 1 and slowly declined afterwards (Fig. 2b). Such a pattern was also described for *Chlorella zofingiensis* [35], but is in contrast to *C. reinhardtii*, for which a continuously increasing starch level was reported, however with a decreasing slope after 2 days of −N conditions [34, 36]. The neutral lipid content of *M. neglectum* increased almost linearly until day 8, after which it remained approximately constant (Fig. 2c).

During the first 2 days of −N cultivation, no net increase of the cellular total lipid content was observed (Fig. 2c, black line). Interestingly, a clear accumulation of the neutral lipid fraction was observed (Fig. 2c, dark

gray line), whereas the fraction of polar lipids decreased (Fig. 2c, light gray line). These opposing tendencies indicate that during this period TAG accumulation was connected to acyl chain recycling from membrane lipids into the TAG pool. At the same time, we noticed that FA synthesis remained active, because the volumetric total lipid content increased from 51 ± 7 mg L^{-1} SE (day 0) to 155 ± 5 mg L^{-1} SE (day 2) (data not shown). This did, however, not translate into an increase of the cellular total lipid content, because the cell concentration concomitantly increased from $3.7 \pm 0.2 \times 10^6$ cells mL^{-1} SE (day 0) to $12.9 \pm 0.9 \times 10^6$ cells mL^{-1} SE (day 2) (data not shown). In the following days, both the total lipid levels as well as the neutral lipid levels strongly increased, while the polar lipid amount remained approximately constant

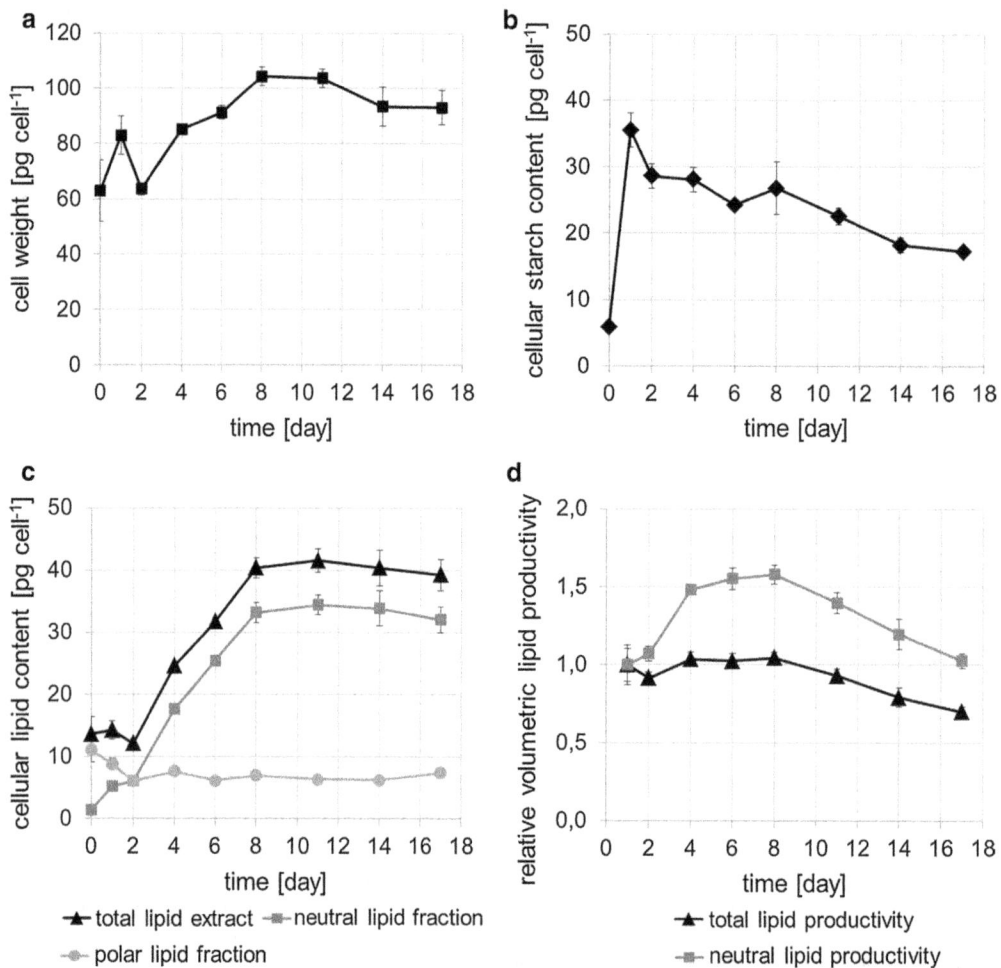

Fig. 2 Cell weight, starch, and lipid profiles of *M. neglectum* during 17 days of nitrogen starvation (exp1). **a** Cell weight during the time course of −N treatment. **b** Starch content per cell, determined by enzymatic reaction of solubilized starch with amyloglucosidase, hexokinase, and glucose-6-phosphate dehydrogenase measuring NADPH production. **c** Gravimetrically determined total (*black, triangles*), neutral (*dark gray, squares*), and polar (*light gray, circles*) lipid content per cell. **d** Relative volumetric total (*black, triangles*) and neutral (*gray, squares*) lipid productivities. In **a–d**, mean values and standard errors (*n* = 3) are shown

(Fig. 2c). This indicates that during this period TAG accumulation was directly fueled by de novo synthesis of cellular FAs. Additionally, a metabolic state allowing for total lipid hyperaccumulation had been established. Importantly, starch degradation apparently was not a major contributor to the total lipid accumulation in *M. neglectum*, because cellular starch levels declined by only ~4 pg from day 2 to day 8, while the total lipid content increased by ~28 pg in the same period of time (Fig. 2b, c).

Interestingly, in this period the cell weight even increased by ~40 pg (Fig. 2a), therefore by more than the sum of the lipid and starch levels. Accordingly, the observed increase of the cell weight cannot be attributed solely to lipid accumulation, indicating that additional carbon sinks must exist, possibly associated with cell wall components.

The optimal harvesting time for maximum volumetric neutral lipid production was found to be between day 4 and day 8, while the volumetric productivity of total lipids remained constant until day 8 (Fig. 2d). Note that low cell concentrations of approximately 4 million cells mL^{-1} were used for inoculation, in order to ensure optimal light penetration and consequently rapid neutral lipid accumulation [27, 37]. Low cell concentrations, however, might not be biotechnologically most relevant. Accordingly, we did not calculate absolute volumetric neutral lipid productivity values, as those would certainly be misleading by underestimating the amounts expected from cultures with higher cell concentrations.

In summary, the long-term −N experiment revealed a similar biphasic pattern of starch accumulation preceding TAG accumulation in *M. neglectum*, as has been reported for other chlorophyceae [34–36, 38].

Design of the transcriptome experiment and data evaluation

For the transcriptome study, we extended the analysis from −N conditions to include N resupply conditions, in order to analyze both the accumulation and the degradation phase of the TAGs. Towards this end, the −N experiment was repeated and after 48 h an aliquot of N-starved cells was resuspended in N-containing media, while the remaining cells were kept under continuous −N conditions (Fig. 1, exp2). As expected, the N resupply treatment resulted in decreased levels of both the neutral lipid and the total lipid contents (Additional file 1: Figure S1), indicating TAG and FA degradation, respectively.

RNA samples for deep mRNA sequencing were taken from all three stages of N availability (e−N, l−N, and r+N stages), constituting a total of twelve time points (Fig. 1, exp2, purple circles). The first sample was taken immediately prior to the −N treatment and is referred

to as the reference time point N_0 representing the transcriptome from exponential growth conditions. For the e−N stage, three samples were taken, after 2, 4, and 8 h of −N conditions, referred to as N_2, N_4, and N_8, respectively. These time points were chosen to represent the transcriptional basis of starch accumulation and of acyl chain recycling from membrane lipids for early TAG accumulation. For the l−N stage, four samples were taken, after 24, 48, 56, and 96 h of −N conditions, referred to as N_24, N_48, N_56, and N_96, respectively. This was to elucidate the transcriptional basis for TAG and total lipid hyperaccumulation. The relatively large time interval was chosen to cover putatively different phases of TAG accumulation, as well as to differentiate between transient and stable transcriptional responses. For the r+N stage, another four samples were taken, after 2, 4, 8, and 14 h of N resupply conditions, referred to as R_2, R_4, R_8, and R_14, respectively. It was accordingly possible to identify different timings of transcript changes triggering the end of cellular quiescence and to comprehensively investigate the reversal of storage compound accumulation.

As the transcriptome of *M. neglectum* was sequenced for the first time, we aimed to acquire a great sequencing depth, in order to also obtain read support for genes with low expression, allowing accurate reconstruction of transcript models for the majority of genes. To reach this goal, we limited the number of samples being sequenced, so that more reads per individual sample could be obtained. Since we expected that transcript abundance changes were rather accurately monitored by mRNA-seq because mRNA-seq was often reported to exhibit low technical variability [39, 40] and to correlate well with RT-qPCR data [15, 19, 20, 41–44], sequencing replicates were not performed. Sequencing replicates have also been omitted in other studies investigating transcriptome changes of microalgae subjected to −N conditions [15, 21, 23]. To mitigate biological variance, a pool of equal amounts of total RNA from two biological replicates was sequenced for each individual time point. The approach of sequencing a pool of RNA from biological replicates was also conducted in other studies [41, 45, 46]. We note that using this approach we could not quantify the biological variance. However, the setup comprised two directly opposing culture conditions, each including several harvesting time points (Fig. 1, −N vs +N resupply). Therefore, despite this limitation, we chose this approach to obtain both a structural genome annotation refinement (Additional file 1: Results) and a first approximation of the transcriptome response of *M. neglectum* to alternating phases of N availability.

A total of 796 million 100 nt paired-end reads were obtained, translating into approximately 33 million fragments for each time point, and thus a coverage of >200-fold assuming a 32 mbp transcriptome. For data processing, the Tuxedo protocol was followed [47]. Accordingly, a genome- and reference-guided approach was conducted, based on read mapping by TopHat2 [48] and transcript assembly and quantification by the Cufflinks suite [49]. Both software tools used the genome assembly from [27] and the improved genome annotation obtained in this study by BRAKER1 [50] as additional input (Additional file 1: Results). Of the Cufflinks tools, we used cuffquant and cuffnorm, instead of cuffdiff due to the absence of sequencing replicates. As a result, normalized transcript abundance values, expressed as fragments per kilobase of exon per million fragments mapped (FPKM), were obtained.

Changes in transcript abundances relative to the reference time point N_0 were expressed as log2-transformed fold change (FC) values. We selected an absolute log2-FC > 1 to define a gene as responsive to the treatment, whereas an absolute log2-FC between 0 and 1 classified a gene as not-responsive, as has been done in previous studies [15, 46, 51, 52]. Interestingly, we observed in MA plot from previous studies that most absolute log2-FC values greater than ~1 were also statistically significant [53–56]. However, we stress that FC is not indicative of any statistical significance, and that furthermore less abundant transcripts might require higher absolute FC values to detect statistical significance than more abundant transcripts [56]. The FC threshold used in our study was supported by our own data for the expression of housekeeping genes. For those, the absolute log2-FC of at most one of the eight time points of −N conditions was >1, whereas the remaining FC values of −N conditions were ≤1. Although most housekeeping genes showed a transient transcriptional response in the r+N stage, their expression usually relaxed to the range of absolute log2-FC ≤ 1 after 8 h of N resupply (Additional file 1: Figure S2a).

The data were integrated into a database, which is available at https://tdbmn.cebitec.uni-bielefeld.de. For *M. neglectum*, the structural and functional annotation of the queried transcript locus is displayed, as well as the transcript abundance profile during the three different stages of N availability. In addition, published transcriptome datasets from other microalgal species that were also subjected to −N treatment were integrated into this database, enabling inter-species comparisons of transcript changes to −N conditions (Additional file 1: Results).

Transcriptome reconstruction and quantification
20,751 transcript loci were assembled by Cufflinks [47], which contained a total of 35,146 isoforms. The higher number of isoforms was predominantly due to the presence and absence of untranslated regions (UTRs). While UTRs were predicted by Cufflinks, they were not included in the reference annotation. Therefore, 85% of all loci had either a single or two isoforms attributed, i.e., the provided "UTR-free CDS isoform" and the fully annotated version including UTRs. For the remaining 15%, it was checked whether evidence for alternative splicing and dominant isoform switching [49] could be detected. This was found for at least four genes (Additional file 1: Results, Figures S3–S6); however, a more detailed analysis is required to further investigate the extent and effect of alternative splicing and dominant isoform switching in *M. neglectum*.

The top 100 most abundant transcripts under exponential growth conditions were annotated as proteins involved in the cellular processes translation (52%) and photosynthesis (28%) (Additional file 2). Two examples are a putative 60 S ribosomal protein L13a and a putative chlorophyll a-b binding protein (XLOC_013860, FPKM at N_0 = 4715 and XLOC_000814, FPKM at N_0 = 10,978, respectively). This has also been noted for *B. sudeticus* [23]. Interestingly, although the majority of transcripts assigned to photosynthesis were strongly decreased in the l−N stage (Additional file 2), a few exhibited both a high abundance and a stable expression pattern under −N conditions. A notable example is one isoform of RuBisCo small subunit (*rbcS2*, XLOC_007679, median FPKM = 10,028). Further examples for transcripts with highest abundances and stable expression patterns under −N conditions are a putative component of the cytosolic large 60 S ribosomal subunit (*rpl7ae*, XLOC_000987, median FPKM = 3838) and a putative elongation factor (putative fragment pair XLOC_005939 and XLOC_012699, median FPKM = 2413 and 2393, respectively) (Additional file 1: Figure S2b). Importantly, these genes are ideal candidates for cloning of the promoter regions to efficiently drive transgene expression in subsequent genetic engineering studies.

The range of FPKM values covered more than four orders of magnitude, and this range distribution was approximately similar in the different stages of N availability (Fig. 3a). The log2-FC values (relative to N_0) of most transcripts were between −1 and 1 at most time points (gray box in Fig. 3b). As noted above, this interval was accordingly defined as the "not-responsive range" (Additional file 1: Figures S2 and S10a). No alteration of the expression of the majority of genes by the experimental conditions is in accordance with the null hypothesis for the determination of significant differentially expressed genes [55]. The upper values of the 1.5-fold interquartile ranges of log2-FC values were for most time points less than four (dotted blue line in Fig. 3b). Therefore, transcripts with log2-FC values of more than four were summarized as highly regulated, and this was visually highlighted by a darker color (see Fig. 3c for the extent of up-regulation of MLDP in the l−N stage, which is indicated in Fig. 4 by a dark red color for the MLDP transcript at the time points from the l−N stage). In contrast, the range between absolute log2-FC values larger than one and less than four was linearly color-coded (see figure keys in Figs. 4, 5, 6).

Effect of N resupply on gene expression of selected examples

Nitrogen resupply treatment was applied to reverse the phenotypic effects of nutrient starvation and the efficiency of the selected procedure can be exemplified by the profile of the transcript encoding for the major lipid droplet protein (MLDP, Fig. 3c). TAGs, which are accumulated in response to −N treatment, are stored as lipid droplets inside the cell [28, 57], of which MLDP is a major structural protein [30, 58]. Therefore, as expected, a strong increase of the MLDP transcript levels was detected during the two stages of N starvation (e−N and l−N), while relaxation of gene expression occurred immediately upon N resupply in the r+N stage (Fig. 3c).

The N resupply treatment facilitated the assignment of potential functions to genes whose gene products putatively catalyze similar enzymatic reactions. For instance, two transcripts were annotated as ferredoxin-NADP$^+$-reductase, a central enzyme of photosynthesis. While the first transcript was strongly up-regulated in the l−N stage (putative fragment pair XLOC_015550 and XLOC_016383), the second transcript was strongly down-regulated in this stage (XLOC_001499) (Additional file 1: Figure S2c). Interestingly, this pattern was reversed when nitrogen was resupplied and expression reverted to pre-starvation levels of the reference time point (Additional file 1: Figure S2c). This indicates that during nutrient starvation the former enzyme might be involved in maintaining photosynthetic electron flow at reduced availability of NADP$^+$ levels or in photoprotective release of excitation pressure when demand for NADPH is low, for instance by redistribution of electrons to various redox reactions, which has been described for one of two leaf ferredoxin-NADP$^+$-reductase genes of *A. thaliana* [59]. The latter ferredoxin-NADP$^+$-reductase of *M. neglectum* could be important under environmental conditions supporting fast cell growth when demand for photosynthetically provided NADPH is high. The opposing expression patterns might also be indicative of a modulation of photosynthetic electron flow towards increased cyclic electron flow under −N conditions. Cyclic electron flow generates ATP [60] and was found

to be important for neutral lipid accumulation and to be increased under both autotrophic and mixotrophic −N conditions in *C. reinhardtii* [61]. A second example is the transcriptional regulation of two isoamylase genes. Both genes were expressed under −N conditions in a stable pattern, except for the first time point of −N conditions, at which a gentle transcriptional regulation (~threefold) in opposite directions was observed (Additional file 1: Figure S2d). In the r+N stage however, both genes were contrastingly up- or down-regulated at three of four time points, respectively (Additional file 1: Figure S2d). As the r+N stage was characteristic of storage compound degradation, it is tempting to speculate that the up-regulated isoamylase candidate is implicated in starch degradation (XLOC_001619), whereas the down-regulated one is implicated in starch synthesis (putative fragments XLOC_004804 and XLOC_012040).

It should be mentioned that the N resupply approach furthermore allowed better interpretation of some of the −N transcriptome data. As an example, the transcript abundances of the subunits of the plastidial pyruvate dehydrogenase complex (cpPDHC) were approximately constant in both the e−N and l−N stages (Additional file 1: Figure S2e). However, in the r+N stage, a transient strong down-regulation was observed (Additional file 1: Figure S2e). This shows that cpPDHC was subjected to transcriptional regulation, which in our setup would not have been detected without the r+N treatment. Therefore, the differential transcriptional response of cpPDHC under −N and N resupply conditions indicates that under −N conditions cpPDHC expression is actively maintained, which is reasonable, because cpPDHC converts pyruvate to acetyl-CoA (CoA, coenzyme A), the carbon precursor for FA synthesis (see "Discussion").

Global transcriptional responses in the three different stages of N availability

The differentiation into three distinct stages of N availability was the backbone for the interpretation of transcriptional patterns in this study. However, this differentiation was initially solely based on metabolic data, i.e., on the timing of starch and TAG accumulation, as well as of TAG degradation (Fig. 2b, c; Additional file 1: Figure S1). We therefore investigated whether the transcriptomes from the twelve time points would accordingly cluster into the three stages. Towards this end, we applied hierarchical clustering based on Jensen−Shannon distance [49, 62]. From visual evaluation of the resulting dendrogram, four clusters could be distinguished (Fig. 3d). The first cluster consisted of the transcriptomes from the time points of the e−N stage, whereas the second cluster consisted of those from the l−N stage (Fig. 3d). The third and the fourth clusters each contained two of the four transcriptomes from the time points of the r+N stage (Fig. 3d). The reason for the unexpected division of the r+N data into two separate clusters was likely of technical nature, most likely due to the higher magnitude of transcript abundance values at the R_2 and R_4 time points compared to R_8 and R_14 (larger box size and whiskers in Fig. 3a), because the distance metric relies on the extent of change in relative expression [49]. Therefore, the three different stages of nitrogen availability also manifested on the level of the transcriptome data sets.

In order to dissect the transcriptional responses into those being shared between the e−N and l−N stages, and those restricted to either the e−N or l−N stage, as well as those unique to the r+N stage, an analysis of shared responsive genes was performed (Fig. 3e; Additional file 1: Figure S7). As expected, more responsive genes were shared between the e−N and l−N stages, compared to the r+N stage, which was true for both up- and down-regulated genes (Fig. 3e). To subsequently identify the cellular processes in which the respective genes were involved, a gene ontology (GO) term enrichment analysis was performed (Additional file 1: Table ST1). GO terms represent unified vocabulary to annotate genes and gene products [63]. GO terms are organized in a hierarchical structure, with broader vocabulary at the higher level (e.g., "*signal transduction*") and more specific vocabulary at the lower levels (e.g., "*cAMP biosynthesis*") [63]. Three different categories ("roots") of GO terms are defined, which are biological process, molecular function, and cellular component. An example for the category biological process is the GO term "*translation*," which is a child of the GO term "*gene expression*," and a parent to the GO terms "*translational initiation*," "*translation elongation*," and "*translation termination*" [63]. We restricted our analysis to the category biological process, because we were interested in the cellular processes that were subjected to transcriptional regulation in the three stages.

As a result, the enriched GO terms of the genes whose up-regulation was restricted to the e−N stage were indicative of an induction of cell division (Additional file 1: Table ST1). Furthermore, the GO term "*microtubule-based processes*" was enriched (Additional file 1: Table ST1). GO terms enriched among genes down-regulated specifically in the e−N stage included "*chlorophyll biosynthesis process*," "*aromatic amino acid family biosynthetic process*," and "*translation*" (Additional file 1: Table ST1).

Enriched GO terms of genes whose up-regulation was restricted to the l−N stage were "*protein phosphorylation*," "*fatty acid biosynthetic process*," "*chlorophyll catabolic process*," and "*lipid catabolism process*" (Additional file 1: Table ST1). As the l−N stage was characterized by

TAG and total lipid hyperaccumulation and thus central to this study, the top 100 genes showing the highest degree of up-regulation in this stage were additionally determined (Additional file 2). This was to identify the processes that were subjected to the highest transcriptional induction. These genes encoded almost the complete set of N assimilation proteins (Additional file 1: Results), including genes annotated as urea carboxylase. This indicates that *M. neglectum* is able to use urea as an external nitrogen source, which was also confirmed phenotypically in a separate experiment (Additional file 1: Figure S8). Interestingly, an MYB-domain containing transcription factor (XLOC_013389) was also found in this set, possibly implicated in core metabolic regulation (Additional file 1: Results). GO terms enriched among genes down-regulated specifically in the l−N stage included "*photosystem II assembly*" and "*proline cis−trans isomerization*" (Additional file 1: Table ST1).

The GO term enrichment profile of genes up-regulated in both the e−N and the l−N stages suggested an induction of the tricarboxylic acid cycle, glycolysis, and arginine biosynthesis (Additional file 1: Table ST1). Interestingly, the GO term "*ATP hydrolysis coupled proton transport*" was also enriched (Additional file 1: Table ST1), which was assigned to eleven genes that encoded putative vacuolar-type (V-type) ATPases. The GO term enrichment profile of genes down-regulated in both stages consisted mostly of photosynthesis-associated processes, such as light-harvesting and the non-oxidative pentose-phosphate shunt (Additional file 1: Table ST1). Interestingly, the GO term "*cysteine biosynthetic process*" was also found to be enriched in this set.

The resupplementation with N restored the unstressed cellular state allowing for exponential growth (Additional file 1: Figure S1). GO terms enriched among genes up-regulated specifically in the r+N stage were indicative of the respective processes, such as "*ribosome biogenesis*" and "*photosynthesis*" (Additional file 1: Table ST1). Enriched GO terms of genes whose down-regulation was restricted to the r+N stage included "*tricarboxylic acid cycle*," "*fructose 6-phosphate metabolic process*," and "*ATP hydrolysis coupled proton transport*" (Additional file 1: Table ST1). Since they were previously enriched in the set of genes up-regulated in both the e−N and l−N stages, it is tempting to speculate that these cellular processes are central to cope with −N conditions.

Reconstruction of pathway maps and the glycerolipid metabolism of *M. neglectum*

For reconstruction of metabolic pathway maps, tBLASTx comparison of known enzymes from *C. reinhardtii* and *A. thaliana* was performed with the transcriptome of *M.*

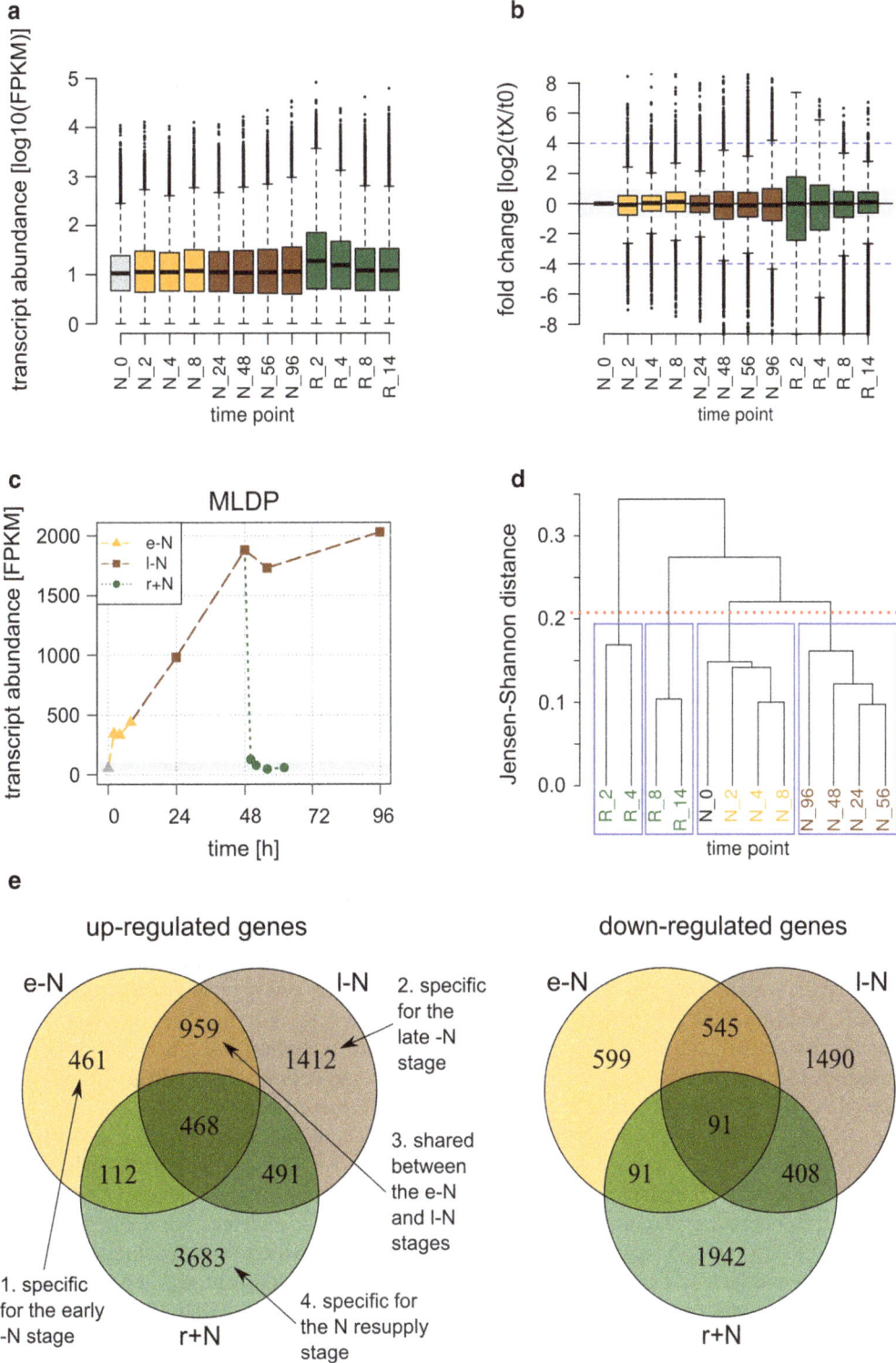

a transcript abundance [log10(FPKM)] vs time point (N_0, N_2, N_4, N_8, N_24, N_48, N_56, N_96, R_2, R_4, R_8, R_14)

b fold change [log2(tX/t0)] vs time point (N_0, N_2, N_4, N_8, N_24, N_48, N_56, N_96, R_2, R_4, R_8, R_14)

c MLDP — transcript abundance [FPKM] vs time [h]
- e-N
- l-N
- r+N

d Jensen-Shannon distance vs time point (R_2, R_4, R_8, R_14, N_0, N_2, N_4, N_8, N_96, N_48, N_24, N_56)

e

up-regulated genes

e-N, l-N, r+N Venn diagram:
461; 959; 1412; 468; 112; 491; 3683

1. specific for the early -N stage
2. specific for the late -N stage
3. shared between the e-N and l-N stages
4. specific for the N resupply stage

down-regulated genes

e-N, l-N, r+N Venn diagram:
599; 545; 1490; 91; 91; 408; 1942

(See figure on previous page.)
Fig. 3 Overview of the transcriptome dataset analysis. **a** Distribution of absolute transcript abundances as FPKM values from all time points on a half-log scale (log10). In the *box-whisker plot* representation, the *thick line* represents the median value, the *colored box* represents the interval between the first and third quartiles, the *two whiskers* indicate the respective 1.5× interquartile ranges, and the *black dots* mark the outliers. **b** Distribution of relative transcript abundance changes normalized to the reference time point N_0, expressed as fold change (FC) on a half-log scale (log2); *box-whisker plots* are as in (**a**). The *black vertical line* highlights the zero line (no regulation), while the *light gray box* indicates the threshold range where absolute log2-FC < 1 indicating no response to −N or N resupply. The *dashed blue line* indicates the threshold above which a transcript was considered as transcriptionally highly regulated, and above which the log2-FC was restricted to 4 and appropriately highlighted by *darker color* in Figs. 4, 5, 6 and Additional file 1: Figure S10. **c** Expression profile of the MLDP gene (XLOC_008097). *Different colors* represent the three stages of nitrogen availability: early −N (e−N, *orange*), late −N (l−N, *brown*), and N resupply (r+N, *green*). The reference time point N_0 is shown in gray. The *light gray box* highlights the threshold range where the absolute log2-FC relative to N_0 was less than one. **d** Dendrogram of a hierarchical clustering of time points using the CummeRbund package [154] with default settings. The *red dotted line* indicates the applied threshold yielding the four clusters (*blue frames*). **e** Venn diagrams of shared genes between the three stages of N availability. The sets on the left consist of genes classified as up-regulated by mean FC during the e−N stage (*orange*), during the l−N stage (*brown*), and during the r+N stage (*green*). On the right, the same was done for down-regulated genes

neglectum. After tBLASTx application, redundancy in the list of candidate genes was minimized by the identification of fragmented genes, which were the result of the assembly status of the genome of *M. neglectum* as 6739 scaffolds [27]. For instance, a protein AB might be encoded by two individual genes on two different scaffolds, such that part A is encoded as a first individual gene at the end of a first scaffold, while part B is encoded as a second individual gene at the beginning of a second scaffold. Importantly, however, as both genes are fragments of the same protein AB, their transcriptional profiles are identical. Approximately half of all gene models were located at a scaffold margin, and accordingly tagged as putatively truncated. For these genes, the transcript data acquired in this study allowed the assignment of gene fragments to fragment pairs. As an example, two candidate genes for the α-carboxyltransferase subunit of the acetyl-CoA carboxylase (ACCase) complex, a central enzyme of FA synthesis, were tagged as likely fragmented and had very similar expression patterns (XLOC_015237 and XLOC_12365, Additional file 1: Figure S9). In addition, their domain structures matched, because both had a predicted crotonase-like superfamily domain, appropriately truncated at the C- and N-terminus, respectively (Additional file 1: Figure S9). Accordingly, we considered these two sequences as a fragment pair. To minimize redundancy, only the fragment containing the N-terminus of the respective protein was retained for further analysis (XLOC_015237), because it optionally encodes the targeting peptide, hence allowing for localization prediction [32]. An obvious consequence of gene fragmentation was an over-estimation of the gene content of *M. neglectum* for a specific enzymatic step. For instance, two transcript loci were identified by tBLASTx search for phosphoribulokinase, which, however, were identified as a fragment pair, due to almost identical expression patterns (Additional file 3, XLOC_002711 and XLOC_006065, respectively) and matching domain

structures (C- and N-terminal truncated Udk superfamily hit, respectively) [64]. Accordingly, *M. neglectum* most likely contains only one phosphoribulokinase enzyme, similar to *C. reinhardtii* [65], *N. oceanica* [66], and *P. tricornutum* [67]. A further consequence was over-estimation of the FPKM values of the individual parts of a fragment pair, because transcript length is taken into account for FPKM calculation [47]. This, however, is a systematic bias and cancels out during FC calculation, as long as transcript length remains unchanged during the time course experiment. Therefore, the over-estimation of FPKM values of fragment pairs did not affect the evaluation of relative patterns (intra-gene comparisons), which was the central element for subsequent pathway analysis.

Following this approach, the glycerolipid metabolism of *M. neglectum* was reconstructed (see "Discussion"). Briefly, FA synthesis in the chloroplast generates acyl chains, which are esterified to glycerol-3-phosphate in the Kennedy pathway to yield various membrane lipids or TAGs (Fig. 4). The direct precursor for most glycerolipids, including TAG, is diacylglycerol (DAG). DAG accordingly represents a central intermediate of the glycerolipid metabolism [8]. The acylation of DAG yields TAG, and an important group of enzymes catalyzing this reaction are the diacylglycerol acyltransferases, which use DAG and acyl-CoA as substrates [8]. Diacylglycerol acyltransferase enzymes are divided into several classes, and the first two are responsible for the bulk of TAG synthesis in plants [68], abbreviated as DGAT and DGTT in *C. reinhardtii*, respectively [10]. This route of TAG formation is also referred to as the acyl-CoA-dependent route. The alternative, the acyl-CoA independent route, refers to the transacylation of DAG and an acyl donor glycerolipid molecule (phospholipid, galactolipid or DAG [69]). This reaction produces a TAG and a lyso-lipid molecule, and is catalyzed by phospholipid:diacylglycerol acyltransferase (PDAT) [8].

Transcriptional regulation of the glycerolipid metabolism of *M. neglectum* during the two stages of nitrogen starvation and the stage of nitrogen resupply

We developed a modified heat map visualization (Additional file 1: Results), which contains both the extent of differential transcriptional regulation at the individual time points of the three stages of N availability as well as the absolute transcript abundance at the reference time point (N_0), represented by one of five abundance categories (category I–V). As additional information, the putative protein localization as predicted by PredAlgo software [70] is indicated (Fig. 4). Using this modified heat map visualization, the transcriptional regulation of the glycerolipid metabolism of *M. neglectum* during the three stages of N availability was analyzed. In the e−N stage, 3- to 8-fold increased transcript abundances were observed for MLDP, one stearoyl-ACP (ACP, acyl carrier protein) desaturase (SAD) candidate and two long-chain acyl-CoA synthetase (LACS) candidates (Fig. 4, MLDP, SAD, and LACS, respectively). Genes down-regulated to a high extent (≥eightfold) were not found in the e−N stage, except for a hypothetical subunit of the ACCase complex (Fig. 4, additional β-CT). However, its transcript abundance under logarithmic growth conditions (N_0) was only 20% of a second candidate (Additional file 3, XLOC_016656 and XLOC_015237, respectively). Therefore, the overall contribution of the second gene's product to the flux of FA synthesis might be negligible. Seven genes were subjected to more subtle decreases in transcript abundances in the e−N stage; those were implicated in FA synthesis, thylakoid membrane assembly, and FA degradation (Fig. 4, KASIII, KAR, ENR, ACP, and MGDGS as well as ECH, respectively). Regarding the latter, however, down-regulation of the respective gene was also observed in the r+N stage (Fig. 4, ECH), rendering interpretation difficult.

In the l−N stage, the highest increase in transcript abundance was noted for MLDP, correlating with the increasing number and diameter of lipid droplets under −N conditions [28]. Additionally strongly up-regulated (10- to >16-fold) were acyl-ACP thioesterase (FAT), SAD, and the putative lipase PGD1 (Fig. 4). More gentle up-regulation was noted for CTP:phosphocholine cytidylyltransferase (~fourfold) putatively involved in phosphatidylcholine (PtdCho) synthesis, as well as for three of four central subunits of the ACCase complex (~twofold) (Fig. 4, CCT and ACCase, respectively). Contrasting the transcriptional induction of PtdCho synthesis, reduced transcript abundances were noticed for the genes implicated in the synthesis of phosphatidylserine and phosphatidylglycerol in the l−N stage (Fig. 4, PSS and PGPS,

respectively). Thylakoid membrane lipid synthesis was also differentially regulated, with monogalactosyldiacylglycerol (MGDG) synthesis being transcriptionally repressed and digalactosyldiacylglycerol (DGDG) synthesis induced (Fig. 4, MGDGS and DGDGS, respectively). This might be indicative of an alteration of the thylakoid membrane architecture, because MGDG has a conical shape, does not form bilayers, and its accumulation results in a negative membrane curvature, while DGDG has a cylindrical shape and forms bilayers [71]. Three steps of FA synthesis were down-regulated in the l−N stage (Fig. 4, MCMT, KASIII, and KAR). The remaining down-regulated genes had 5–40% lower transcript abundance values than other genes associated with the same function at the reference time point N_0 (Fig. 4, additional β-CT, BCCP, and SAD); therefore, the effect of their down-regulation in the l−N stage was considered to be negligible.

In the r+N stage, the synthesis of MGDG was transcriptionally induced, while DGDG and diacylglycerol-N,N,N-trimethylhomoserine (DGTS) synthesis was transcriptionally repressed (Fig. 4, MGDGS, DGDGS, and BTA, respectively). FA degradation was transiently sharply up-regulated (Fig. 4, ACX, ECH, and KAT), which is in accordance with the decreasing total lipid content upon N resupply (Additional file 1: Figure S1). Contrastingly, FA synthesis was transcriptionally repressed at almost all individual enzymatic steps during the first 4 to 8 h of N resupply (Fig. 4, ACCase, MCMT, KASII, KAR, HAD, ENR, SAD, FAT, and ACP). Two putative LACS transcripts were transiently induced in the r+N stage, and two of three phosphatidic acid phosphatase candidate genes were transiently down-regulated (Fig. 4, LACS and PAP, respectively).

Transcriptional regulation of the committed step of TAG synthesis

The only committed reaction to TAG synthesis is the addition of a third acyl chain to DAG [8]. In respect of the acyl-CoA-dependent pathway, four of nine DGTT candidate genes showed a transcriptional induction under −N conditions, while approximately constant abundances were noted for the single putative DGAT transcript (Fig. 4). In the r+N stage, a transient down-regulation was observed for three of the aforementioned DGTT transcripts, as well as for two additional putative DGTT transcripts, and for the DGAT transcript (Fig. 4). In respect of the acyl-CoA independent pathway, the transcript level of the PDAT candidate gene was approximately constant in the three stages of N availability, except for a transient down-regulation in the r+N stage after 2 h of N resupply (Fig. 4). The transacylation reaction catalyzed by PDAT generates a lyso-lipid. Two

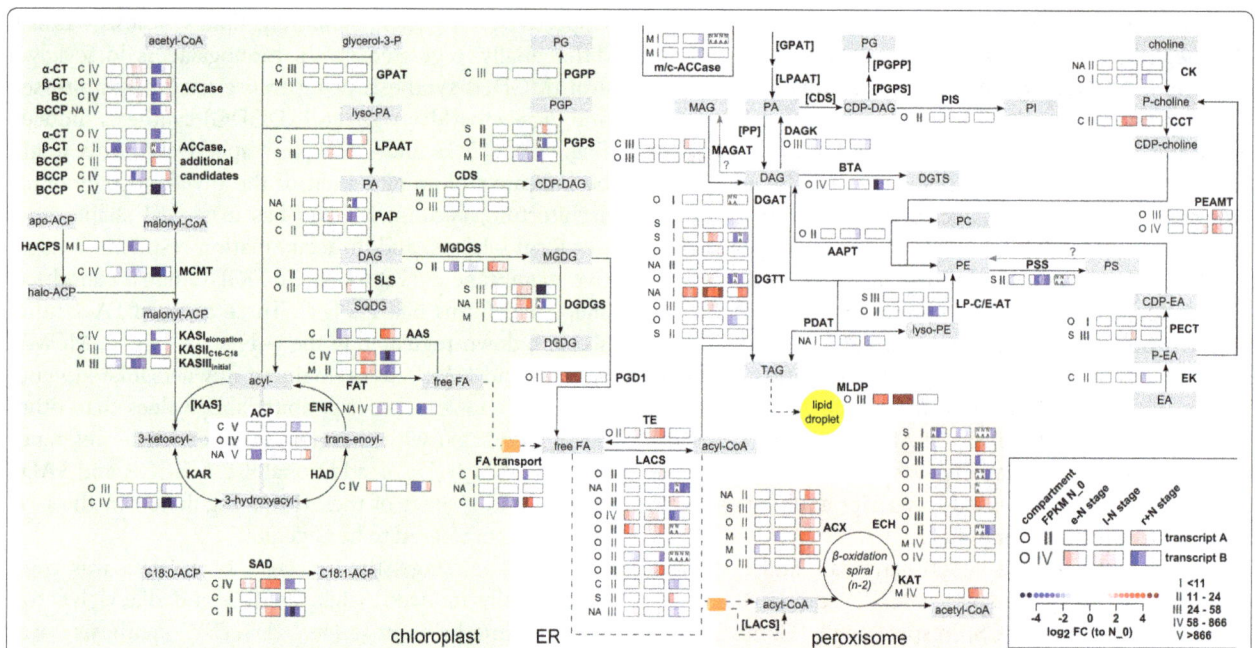

Fig. 4 Schematic representation of the putative enzymatic steps of the glycerolipid metabolism in *M. neglectum*, including the transcriptional responses to N starvation (stages e−N and l−N) or N resupply (stage r+N). Enzymatic steps are represented by *solid arrows* and transport processes by *dashed lines*. For simplicity, PDAT is drawn utilizing PE, but has been shown to also use other lipid substrates [69]. Fatty acid desaturation steps are not shown, except for the generation of oleic acid (C18:1). The localization is drawn according to [8] and for additional reactions according to [87]. Each step has at least one transcript associated, and the putative localization is indicated on the *left* (C chloroplast, M mitochondrion, O other, S secretory pathway, NA localization prediction not possible due to truncation). The section with the *gray Roman numerals* next to the predicted localization shows the binned transcript abundance at the reference time point N_0. Five abundance categories are defined: I = below 50% percentile abundance, II = 50–75% percentile abundance, III = 75–90% percentile abundance, IV = 90–99% percentile abundance, V = >99% percentile abundance; see also legend on the *bottom right* and Fig. 3a for the distribution of FPKM values at N_0. *Bold Roman numerals* indicate that the respective gene is likely not fragmented, whereas normal font style indicates that only the transcript abundance of the putative fragment containing the 5′ end is shown. The transcription profile of each enzyme is represented by *three color boxes*, representing the three different cultivation stages investigated in this work (e−N, l−N, r+N). In each of the *boxes*, the transcriptional regulation at the individual harvesting time points relative to time point zero (N_0) is indicated by *color-coded bars* (red up-regulation, *blue* down-regulation compared to N_0). *White bars* are shown if the change in relative transcript abundance was between 50 and 200% (absolute log2-FC < 1). The tag "NA" (not available) is added if the absolute transcript abundance (as FPKM) at that time point was less than 1.0, which was set as the minimum threshold for reliable transcript abundance estimation. The full annotations of the corresponding genes are given in Additional file 4. *ACCase* acetyl-CoA carboxylase, *ACP* acyl carrier protein, *ACX* acyl-CoA oxidase, *AAPT* aminoalcoholphosphotransferase (putatively dual substrate specificity producing PC and PE), *BTA* betaine lipid synthase, *CCT* CTP:phosphorylcholine cytidylyltransferase, *CDS* CDP-DAG synthase, *CK* choline kinase, *DGAT* diacylglycerol acyltransferase type 1, *DGDGS* digalactosyldiacylglycerol synthase, *DGTT* diacylglycerol acyltransferase type 2, *ECH* multifunctional protein containing a 2E-enoyl-CoA hydratase and a 3S-hydroxyacyl-CoA dehydrogenase, *EK* ethanolamine kinase, *ENR* enoyl-ACP reductase, *FAT* acyl-ACP thioesterase, *GPAT* glycerol-3-phosphate acyltransferase, *HAD* hydroxyacyl-ACP dehydrase, *KAR* ketoacyl-ACP reductase, *KAS* ketoacyl-ACP synthase, *KAT* 3-ketoacyl-CoA thiolase, *LACS* long-chain acyl-CoA synthetase, *LPAAT* lysophosphatidic acid acyltransferase, *LP-C/E-AT* lysophosphatidylcholine/ethanolamine acyltransferases, *m/c-ACCase* mitochondrial or cytosolic ACCase, *MCMT* malonyl-CoA:acyl carrier protein malonyltransferase, *MGDGS* monogalactosyldiacylglycerol synthase, *MLDP* major lipid droplet protein, *PAP* phosphatidic acid phosphatase, *PDAT* phospholipid:diacylglycerol acyltransferase, *PEAMT* phosphoethanolamine N-methyltransferase, *PECT* CTP:phosphorylethanolamine cytidylyltransferase, *PGD1* plastid galactoglycerolipid degradation lipase, *PGPP* phosphatidylglycerol phosphate, *PGPS* phosphatidylglycerophosphate synthase, *PIS* phosphatidylinositol synthase, *PSD* phosphatidylserine decarboxylase, *PSS* phosphatidylserine synthase, SAD, Δ^9 stearoyl-ACP desaturase, *SLS* sulfolipid synthase, *TE* acyl-CoA thioesterase, *CDP* cytidine diphosphate, *CoA* coenzyme A, *DAG* diacylglycerol, *DGDG* digalactosyldiacylglycerol, *DGTS* diacylglycerol-N,N,N-trimethylhomoserine, *EA* ethanolamine, *ER* endoplasmic reticulum, *FA* fatty acid, *LPE* lysophosphatidylethanolamine, *MGDG* monogalactosyldiacylglycerol, *PA* phosphatidic acid, *PC* phosphatidylcholine, *PE* phosphatidylethanolamine, *PGP* phosphatidylglycerolphosphate, *PG* phosphatidylglycerol, *PI* phosphatidylinositol, *SQDG* sulfoquinovosyldiacylglycerol, *TAG* triacylglycerol

enzymes putatively catalyzing the re-acylation of this lyso-lipid were identified in *M. neglectum*. The first candidate was transcriptionally induced (~twofold) in the e+N stage, whereas the second candidate was transiently repressed in the r+N stage (Fig. 4, LP-C/E-AT).

Clustering of transcripts annotated as lipases reveals candidates likely involved in TAG accumulation and TAG degradation

Lipases hydrolyze the ester bond between the glycerol backbone and the acyl chain of TAG and other glycerolipid molecules, yielding a free FA and the corresponding lyso-lipid [8]. The released acyl chain can be subsequently incorporated into other glycerolipids after activation by CoA. By this process, acyl chains can be shuttled between the membrane lipid and TAG pool (another route is transacylation). Acyl chain recycling apparently contributed significantly to early TAG accumulation in *M. neglectum*, because the polar lipid content decreased during the first 2 days of −N conditions (Fig. 2c; see above).

We reasoned that lipases involved in the process of de novo TAG accumulation can be identified according to a transcriptional induction under −N conditions. In contrast, opposing TAG lipases that degrade storage lipids are expected to be down-regulated under −N conditions, but up-regulated upon N resupply. Accordingly in this work, the transcriptional pattern of a lipase candidate gene was used as an indicator for its putative function in storage lipid metabolism.

In *C. reinhardtii*, a correlation between transcriptional regulation and metabolic function was observed for the lipases CrLIP1 and PGD1 (plastid galactoglycerolipid degradation 1) [72, 73]. Whereas CrLIP1 was down-regulated under −N conditions and indirectly implicated in TAG turnover [72], PGD1 was up-regulated under −N conditions and implicated in de novo TAG accumulation [73]. For both genes, a putative homologue was identified in the transcriptome of *M. neglectum*. As expected, the CrLIP1 transcript exhibited a transient sharp increase in the r+N stage (~sixfold at R_2, Fig. 5, XLOC_016073), in agreement with a putative role in storage lipid degradation. Likewise, the PGD1 transcript in *M. neglectum* exhibited a strong induction in the l−N stage (>16-fold, Fig. 5, XLOC_012515), corroborating a putative role in de novo TAG accumulation. Although the expression patterns are clear indicators that these proteins indeed represent homologous enzymes of both microalgal species, biochemical characterization would be required to prove this model and to further determine substrate specificity in *M. neglectum*.

Additional putative lipases were identified according to the gene annotation or their GO term description. This revealed a total of 68 putative lipase transcripts in *M. neglectum*. In order to group these transcripts based on their transcriptional profiles, hierarchical clustering was performed [74], and the resulting dendrogram divided into four clusters, supported by values for the average silhouette width as cluster quality parameter [75, 76] (data not shown).

Most putative lipase genes of the first cluster were characterized by a constant or slightly up-regulated gene expression in the e−N stage, were most strongly up-regulated in the l−N stage, and induction relaxed when nitrogen was supplemented in the r+N stage (Fig. 5, c1). This profile is similar to the PGD1 gene expression pattern which was also part of this cluster (Fig. 5, c1, XLOC_012515), therefore indicating a putative involvement in the process of de novo TAG accumulation under −N conditions. The second cluster contained several lipase candidates which were strongly up-regulated in the r+N stage, including the aforementioned CrLIP1 transcript of *M. neglectum* (Fig. 5, c2, XLOC_016073). Since up-regulation was concomitant with the induction of putative FA degradation genes (beta-oxidation, Fig. 4), it seems likely that these lipase candidates are involved in the process of TAG degradation. The third cluster contained transcripts with approximately stable expression in the e−N and l−N stages, but decreased expression in the r+N stage (Fig. 5, c3). This down-regulation might indicate that the corresponding enzymes are not required for the process of thylakoid membrane reassembly occurring in the r+N stage (Additional file 1: Table ST1); alternatively, they might impair the process of thylakoid membrane reassembly, therefore necessitating down-regulation in the r+N stage. The fourth cluster consisted of only two genes, which were strongly down-regulated in the e−N and l−N stages (Fig. 5, c4). This suggests that the corresponding enzymes might act as suppressors of TAG accumulation under exponential growth conditions.

Reconstruction and prediction of compartmentalization of the central carbon metabolism of *M. neglectum*

Central to the development of metabolic engineering strategies for improved TAG accumulation is not only the glycerolipid metabolism, but also the central carbon metabolism, as it determines the availability of acetyl-CoA for FA synthesis. This has been demonstrated for *C. reinhardtii*, for which carbon precursor supply was reported to be a key metabolic factor controlling oil biosynthesis under mixotrophic −N conditions [77]. Therefore, we next reconstructed the central carbon metabolism of *M. neglectum* (Additional file 1: Results). Supported by localization prediction, we propose a compartmentalization similar to *C. reinhardtii* [32]. Accordingly, the oxidative pentose-phosphate pathway (OPPP) is entirely plastidial, whereas glycolysis is highly compartmentalized, such that the initial steps of glycolysis take place in the chloroplast, while the later steps from 3-phosphoglycerate to pyruvate are located in the cytosol [32] (Fig. 6). However, *M. neglectum* might be able to perform the initial steps of glycolysis from glucose to the triose phosphates additionally in the cytosol, because *M. neglectum* can utilize glucose as a sole carbon source (Additional file 1: Figure S11), whereas *C. reinhardtii* cannot [78]; this model, however, requires further localization studies.

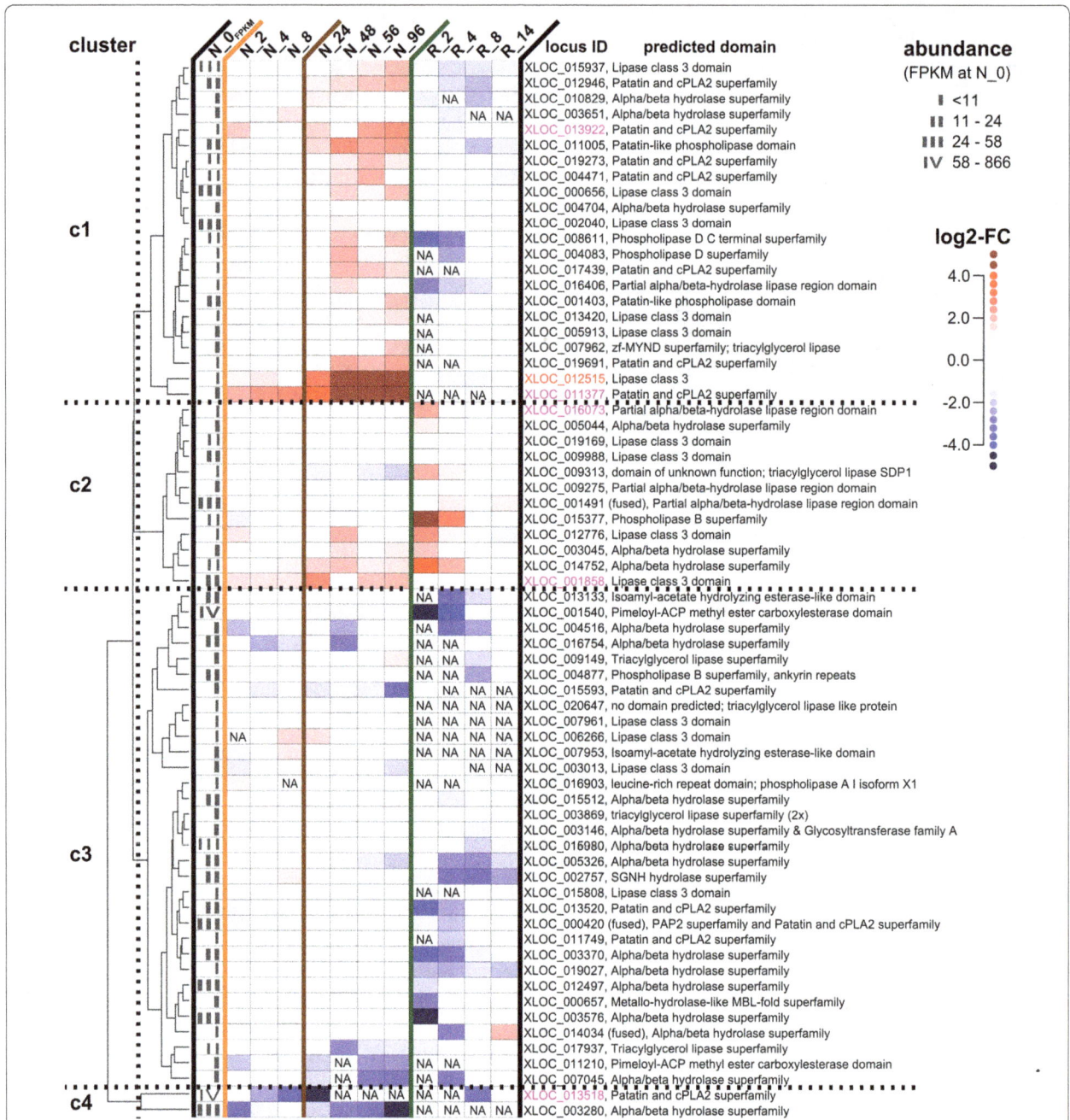

Fig. 5 Identification of lipase candidates possibly implicated in lipid accumulation. The transcription profiles of putative lipases of *M. neglectum* were subjected to hierarchical clustering with complete linkage based on log2-FC values with distance metric defined as Euclidean distance. The resulting dendrogram is shown on the *left*. The dendrogram was divided into four clusters as indicated by the *dotted lines*. The *first column* of the heatmap (N_0$_{FPKM}$) indicates the binned transcript abundance value (FPKM) at the reference time point N_0, i.e., the abundance category. The category "I" represents the range of FPKM values of $1 \leq FPKM \leq 11$ (below median expression), category "II" the range $11 \leq FPKM \leq 24$ (between 50 and 75% percentile expression), category "III" the range $24 \leq FPKM \leq 58$ (between 75 and 90% percentile expression), and category "IV" the range $58 \leq FPKM \leq 866$ (between 90 and 99% percentile expression); see Fig. 3a for the corresponding *box plot*. The remaining columns show the transcriptional regulation at the individual harvesting time points relative to time point zero (N_0), given as *color-coded boxes*. *Red color* represents higher transcript abundance and *blue* lower transcript abundance compared to time point N_0. The tag "NA" (not available) is used when the absolute transcript abundance (as FPKM) at the respective time point was less than 1.0. The three stages of N availability are separated by *vertical lines*, where *orange* represents the e−N stage, *brown* the l−N stage, and *green* the r+N stage, respectively. The locus ID and the predicted domain of each lipase transcript are given on the *right*. As an example for transcriptional regulation of lipase candidates possibly implicated in TAG accumulation rather than TAG degradation, the PGD1 candidate of *M. neglectum* is highlighted in *red color*. Other transcripts mentioned in the text are shown in *magenta*

Differentially regulated genes of the central carbon metabolism during the three stages of N availability

Most genes of the central carbon metabolism that were responsive in the e−N stage exhibited the same direction of transcriptional regulation also in the l−N stage (see below). Transcriptional induction that was restricted to the e−N stage was observed for the putative mitochondrial pyruvate dehydrogenase complex (Fig. 6, PDHC). A dependence of starch and lipid accumulation on mitochondrial respiration was shown for *C. reinhardtii*, because mitochondrial mutants exhibit impaired starch and lipid accumulation under −N and −S conditions, respectively [79, 80]. Down-regulation of genes specifically in the e−N stage was observed for a putative phosphoenolpyruvate (PEP) transporter, and for one of six acetyl-CoA synthetase candidates (Fig. 6, PPT and ACS, respectively). Up-regulation restricted to the l−N stage was observed for one of two glucose-6-phosphate isomerase candidates, for one of four malate dehydrogenase candidates, and for a second of six acetyl-CoA synthetase candidates (Fig. 6, PGI, MDH, and ACS, respectively). Transcripts with exclusively decreased abundances in the l−N stage were not found. For the two acetyl-CoA synthetase candidates with differential expression under −N conditions, enzymatic activity should be confirmed in future studies. This is because *M. neglectum* can use acetate only to a limited extent for growth (Additional file 1: Figure S11). However, a likely functional glyoxylate cycle is encoded, since both key enzymes, malate synthase and isocitrate lyase, had transcript support and responded to the N resupply treatment by a transient up-regulation (Additional file 3, XLOC_013435 and XLOC_002446, respectively).

Of the transcriptional responses shared between the e−N and the l−N stages, the most pronounced up-regulation (>16-fold) was observed for the OPPP (Fig. 6, G6PDH and 6PGDH). A similarly strong up-regulation was noted for fermentative reactions, which have the production of acetyl-CoA from pyruvate in common (Fig. 6, PFL, PFOR, PDC, and bifunctional ADH). Furthermore, the conversion of 2-phosphoglycerate to PEP, catalyzed by enolase, was transcriptionally strongly induced (4- to 16-fold) (Fig. 6, ENO). 4- to 8-fold up-regulation was detected for one of three glycerol-3-phosphate dehydrogenase candidate genes (Fig. 6, GPDH); the respective enzyme generates glycerol-3-phosphate, which is a substrate for the Kennedy pathway and the backbone of TAG. A less pronounced induction (2- to 5-fold) was noted for PEP carboxylase (Fig. 6, PEPC). This enzyme is characteristic for C4 plants and catalyzes the fixation of CO_2 by the generation of oxaloacetate from PEP [81].

For glycolysis and gluconeogenesis, an opposing transcriptional regulation was observed under −N conditions. While two committed steps of glycolysis were both up-regulated, one committed step of gluconeogenesis was down-regulated (Fig. 6, PFK, PK, and FBP, respectively). Another transcript implicated in gluconeogenesis, a putative PEP carboxykinase, was also decreased in abundance in the l−N stage, although absolute transcript abundance values were already low at the reference time point N_0 (Fig. 6, PEPCK).

Interestingly, opposite transcriptional patterns within the same enzymatic step under −N conditions were also observed. This applied to both glyceraldehyde-3-phosphate dehydrogenase candidates, as well as to all four carbonic anhydrase candidates (Fig. 6, GAPDH and CAH, respectively). Each putative carbonic anhydrase was specifically responsive in either the e−N or the l−N stage (Fig. 6, CAH), suggesting a tight transcriptional regulation of the carbon concentration mechanism under −N conditions in *M. neglectum*.

Upon N resupply in the r+N stage, only a few enzymatic steps of the central carbon metabolism were transcriptionally induced, while most others were repressed. Interestingly, the aforementioned transcriptional induction of the OPPP under −N conditions was maintained during the first 4 h of N resupply (Fig. 6, G6PDH and PGL). Furthermore, one putative PEP transporter, one of two putative small subunits of RuBisCo, two of four putative malate dehydrogenase enzymes, as well as two of six putative malic enzyme proteins were transcriptionally up-regulated in the r+N stage (Fig. 6, PPT, rbcS2, MDH, and MME, respectively). The transcriptional repression of both putative phosphoglycerate kinase enzymes that had been noticed for the −N phase was continued for the first 4 h in the r+N stage, after which transcript levels normalized to pre-starvation levels (Fig. 6, PGK). A transient sharp down-regulation in the r+N stage was observed for cpPDHC, as well as for four of 15 putative triose phosphate transporters (Fig. 6, PDHC and TPT, respectively). Glycolysis was down-regulated upon N resupply at several enzymatic steps, contrasting the transcriptional induction under −N conditions (Fig. 6, HK, PFK, FBA, PGM, and PK).

Transcriptional regulation of starch metabolism

Finally, the starch metabolism of *M. neglectum* was reconstructed (Additional file 1: Results). Starch is the major carbon storage molecule in green microalgae and plants, and can amount up to 50% of the dry biomass in *C. reinhardtii* under −N conditions [82]. In *M. neglectum*, the cellular starch content after 1 day of −N conditions amounted to ~35 pg cell^{-1}, which was similar to the cellular neutral lipid content at day 8 with ~33 pg cell^{-1} (Fig. 2b, c). We analyzed the transcriptional regulation of both pathways in order to understand the interplay of both storage compound production processes. In

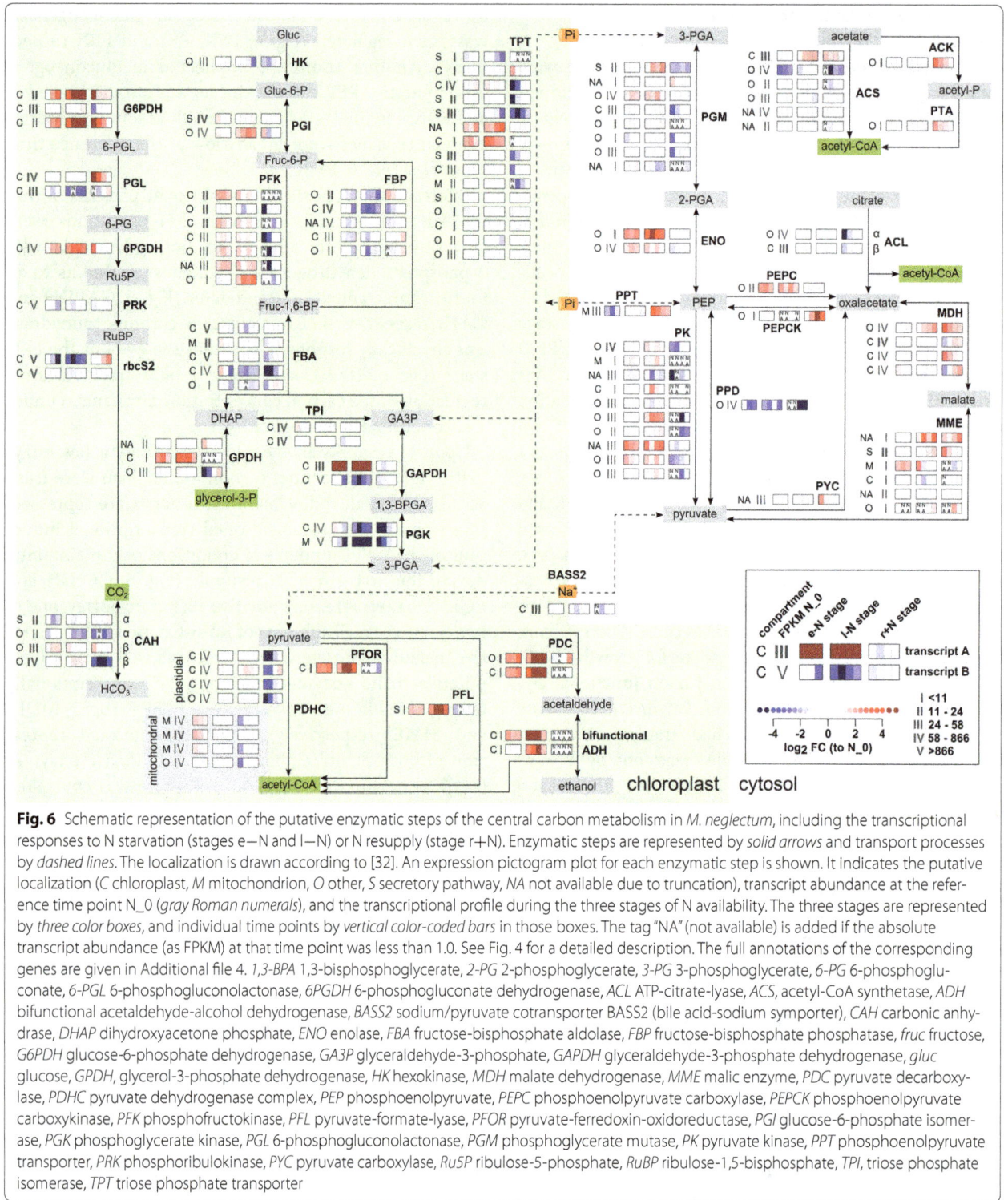

Fig. 6 Schematic representation of the putative enzymatic steps of the central carbon metabolism in *M. neglectum*, including the transcriptional responses to N starvation (stages e−N and l−N) or N resupply (stage r+N). Enzymatic steps are represented by *solid arrows* and transport processes by *dashed lines*. The localization is drawn according to [32]. An expression pictogram plot for each enzymatic step is shown. It indicates the putative localization (*C* chloroplast, *M* mitochondrion, *O* other, *S* secretory pathway, *NA* not available due to truncation), transcript abundance at the reference time point N_0 (*gray Roman numerals*), and the transcriptional profile during the three stages of N availability. The three stages are represented by *three color boxes*, and individual time points by *vertical color-coded bars* in those boxes. The tag "NA" (not available) is added if the absolute transcript abundance (as FPKM) at that time point was less than 1.0. See Fig. 4 for a detailed description. The full annotations of the corresponding genes are given in Additional file 4. *1,3-BPA* 1,3-bisphosphoglycerate, *2-PG* 2-phosphoglycerate, *3-PG* 3-phosphoglycerate, *6-PG* 6-phosphogluconate, *6-PGL* 6-phosphogluconolactonase, *6PGDH* 6-phosphogluconate dehydrogenase, *ACL* ATP-citrate-lyase, *ACS*, acetyl-CoA synthetase, *ADH* bifunctional acetaldehyde-alcohol dehydrogenase, *BASS2* sodium/pyruvate cotransporter BASS2 (bile acid-sodium symporter), *CAH* carbonic anhydrase, *DHAP* dihydroxyacetone phosphate, *ENO* enolase, *FBA* fructose-bisphosphate aldolase, *FBP* fructose-bisphosphate phosphatase, *fruc* fructose, *G6PDH* glucose-6-phosphate dehydrogenase, *GA3P* glyceraldehyde-3-phosphate, *GAPDH* glyceraldehyde-3-phosphate dehydrogenase, *gluc* glucose, *GPDH*, glycerol-3-phosphate dehydrogenase, *HK* hexokinase, *MDH* malate dehydrogenase, *MME* malic enzyme, *PDC* pyruvate decarboxylase, *PDHC* pyruvate dehydrogenase complex, *PEP* phosphoenolpyruvate, *PEPC* phosphoenolpyruvate carboxylase, *PEPCK* phosphoenolpyruvate carboxykinase, *PFK* phosphofructokinase, *PFL* pyruvate-formate-lyase, *PFOR* pyruvate-ferredoxin-oxidoreductase, *PGI* glucose-6-phosphate isomerase, *PGK* phosphoglycerate kinase, *PGL* 6-phosphogluconolactonase, *PGM* phosphoglycerate mutase, *PK* pyruvate kinase, *PPT* phosphoenolpyruvate transporter, *PRK* phosphoribulokinase, *PYC* pyruvate carboxylase, *Ru5P* ribulose-5-phosphate, *RuBP* ribulose-1,5-bisphosphate, *TPI*, triose phosphate isomerase, *TPT* triose phosphate transporter

summary, most genes assigned to the starch metabolism were subjected to transcriptional regulation (Additional file 1: Figure S10b). However, the changes in transcript abundances did not yield a completely conclusive picture. For instance, genes indicative of starch synthesis and starch degradation were consistently up-regulated during the −N treatment (Additional file 1: Figure S10b), yet cellular net starch levels decreased slightly (Fig. 2b). This discrepancy might be attributed to currently unknown post-transcriptional or post-translational regulation steps of the corresponding catabolic enzymes. Interestingly, the most strongly induced transcripts in the l−N stage were annotated as putative starch phosphorylases, and these transcripts were transiently sharply repressed in the r+N stage (Additional file 1: Figure S10b). This indicates that starch phosphorylase might play a key role in starch metabolism in *M. neglectum*.

Discussion

In this study, the transcriptional changes in different pathways in the context of microalgal lipid accumulation were investigated in *M. neglectum* under three different stages of N availability. The first two stages represented cellular acclimation processes from nitrogen replete to nitrogen-free conditions. Removal of nitrogen resulted in a cessation of cell doubling (Additional file 1: Figure S1), while biomass concentrations continuously increased in total 6.7-fold until day 11 of the −N treatment. This is close to the reported 7.8-fold change of *Acutodesmus obliquus* (UTEX 393) after 13–14 days of −N conditions with 5% CO_2 [38], highlighting the strong biomass productivity of *M. neglectum* [27]. The two stages were the early, starch accumulation stage (e−N stage) and the later, TAG and total lipid hyperaccumulation stage (l−N stage) (Figs. 1 and 2). The third stage was selected to allow investigation of the reverse reactions, when nitrogen-limited cells (48 h starvation period) were resupplied with the essential nutrient (r+N stage; Fig. 1). This alternating treatment allowed the analysis of both storage compound accumulation and subsequent degradation, respectively, which significantly increased the reliability of the interpretation of gene expression changes (see above and Additional file 1: Figure S2c–f). This is not only the first time that the transcriptome of *M. neglectum* has been sequenced, but also, to our knowledge, the first time that a microalgal transcriptome has been analyzed by mRNA-seq under both −N and N resupply conditions in a time course experiment to elucidate the molecular mechanisms of TAG accumulation. Although a previous study with the dinoflagellate *Karenia brevis* analyzed the effect of N re-addition using microarrays, this study had a different focus, because the authors were interested in determining if the transcriptome of *K. brevis* is responsive to nitrogen and phosphorus concentrations due to the prevalence of post-transcriptional regulation in dinoflagellates [83]. Accordingly, the effects of nutrient re-addition on individual pathways, such as TAG and lipid metabolism, were not investigated [83]. Furthermore, the transcriptome changes under −N conditions were not investigated in great detail [83].

We used the standard criterion of absolute log2-FC > 1 to define a gene as responsive to our treatment. This was supported by the expression profiles of housekeeping genes, which were classified as not-responsive under −N conditions according to this definition (Additional file 1: Fig. 2a). However, it should be noted that the importance of a gene with regard to a phenotypic effect is not necessarily directly correlated with the factor of up- or down-regulation, but can strongly depend on potential post-transcriptional regulation and on the function of the corresponding protein. In this context, protein kinases were specifically up-regulated in the l−N stage in *M. neglectum* (GO term "*protein phosphorylation*" in Additional file 1: Table ST1), which has also been reported for *C. reinhardtii* after 48 h of mixotrophic −N conditions [10], indicating that protein phosphorylation might be a factor for post-transcriptional regulation under −N conditions. With respect to protein function, a pronounced transcriptional regulation can be expected for structural proteins such as MLDP (Fig. 3c), because these are abundantly required in the cell to exert a specific function, in this case to determine the size of lipid droplets [58]. In contrast, enzymes involved, e.g., in the distribution of carbon flow can be expected to react more moderately, but still to an extent that allows sufficient protein synthesis to direct metabolic flux. Regulatory proteins such as transcription factors, however, might only show a low degree of transcriptional regulation, which could still be metabolically extremely important. Therefore, the FC-based threshold has the limitation that the classification of individual genes might not in all cases represent physiological importance. However, this classification was chosen because this study primarily focused on the interpretation of transcriptional regulation of enzymatic reactions to identify gene targets for subsequent genetic engineering approaches.

Global characterization of the three stages of N availability on the level of the transcriptome by GO term enrichment analysis

To first characterize the individual stages of N availability on the transcriptome level, a GO term enrichment analysis of shared responsive genes was conducted (Fig. 3e, Additional file 1: Table ST1). As a result, biological processes were identified, which were subjected to transcriptional regulation in the three stages. The GO term enrichment profile of genes up-regulated continuously during the e−N and l−N stages is indicative of a transcriptional induction of the tricarboxylic acid cycle and glycolysis pathways. This has also been reported for other microalgae as a response to −N conditions, such as *N. oleoabundans* [17], *N. oceanica* [20], and *P. tricornutum* [19]. The induction of the tricarboxylic acid cycle under −N conditions might be bolstered by the consistent up-regulation of PEP carboxylase generating oxaloacetate from PEP (Fig. 6, PEPC), which might help regulate the flux through the tricarboxylic acid cycle by replenishing oxaloacetate. Alternatively, the up-regulation of PEP carboxylase could indicate that this route of CO_2 fixation became increasingly important under −N conditions. The opposing transcriptional regulation of glycolysis (up-regulation) and gluconeogenesis (down-regulation) indicates that *M. neglectum* switched from a primarily gluconeogenetic to glycolytic state under −N conditions (Fig. 6), which has also been described for *C. reinhardtii*, *P. tricornutum*, and *N. oceanica* [10, 19, 20]. The enrichment profile of down-regulated genes in both the e−N and l−N stages consisted mostly of photosynthesis-associated processes, which has also been observed in *C. reinhardtii*, *N. oleoabundans*, *B. sudeticus*, *P. tricornutum*, and *N. oceanica* under −N conditions [11, 17, 19, 20, 23].

According to the GO term analysis, a biphasic response of the tetrapyrrole pathway for *M. neglectum* becomes apparent, because the transcriptional repression of chlorophyll biosynthesis preceded the transcriptional induction of chlorophyll catabolism in the e−N and the l−N stages, respectively (Additional file 1: Table ST1). A biphasic response of the tetrapyrrole pathway was also described for *C. reinhardtii* under mixotrophic −N conditions [11]. The enrichment profile of the e−N stage is indicative of a transcriptional induction of cell division (Additional file 1: Table ST1). This might simply be seen in accordance with the increase in cell concentration during the first 2 days of −N treatment (Additional file 1: Figure S1). Alternatively, in combination with the aforementioned down-regulation of chlorophyll biosynthesis in the e−N stage, this might reveal a transcriptional program that aims at diluting the cellular chlorophyll pool. This scenario has been described for *C. reinhardtii*, in which the decrease of cellular chlorophyll contents under mixotrophic −N conditions was not only due to cessation in chlorophyll synthesis, but also due to dilution by cellular growth [84].

The GO term *"microtubule-based processes"* was enriched among genes specifically up-regulated in the e−N stage (Additional file 1: Table ST1), which was attributed to seven genes in this set. Six of these encoded structural components of microtubules (tubulin), while the seventh encoded the microtubule motor protein dynein. The up-regulation of these seven genes might either be put into the aforementioned context of cell division. Alternatively, it might be interpreted as a hint towards a putative remodeling of the cytoskeleton as a preparatory step for subsequent lipid droplet formation for TAG storage. Microtubules play a role in directing MLDP to lipid droplets in *C. reinhardtii* [30]. Furthermore, cytoskeleton remodeling was required to elicit the obesity phenotype of high lipid transformants of the oleaginous yeast *Yarrowia lipolytica* [85].

The −N treatment aimed at the cessation of cell doubling as a result of inhibited protein biosynthesis. Therefore, we expected a transcriptional repression of cellular processes involved in protein biosynthesis, e.g., synthesis of ribosomal proteins. In accordance with this, we found that the GO term *"translation"* was enriched among genes down-regulated in the e−N stage (Additional file 1: Table ST1). Transcriptional repression of ribosomal proteins in response to −N treatment has also been noted in *C. reinhardtii* [11], *B. sudeticus* [23], *N. oleoabundans* [17], *P. tricornutum* [19], *N. oceanica* [20], and *N. gaditana* [21]. Interestingly, despite the general transcriptional repression of ribosomal proteins, the GO term analysis revealed that arginine biosynthesis was up-regulated in both the e−N and the l−N stages, whereas cysteine biosynthesis was down-regulated in both stages; in contrast, aromatic amino acid biosynthesis was specifically down-regulated in the e−N stage (Additional file 1: Table ST1). However, this might not necessarily translate into a metabolic effect, because no increased flux through arginine biosynthesis was detected under −N conditions in *C. reinhardtii*, despite a transcriptional up-regulation [22].

Putative glycerolipid metabolism of *M. neglectum*

FA synthesis generates the acyl chains that can be used for glycerolipid synthesis (Fig. 4). The synthesis of C16 and C18 FAs is presumed to be exclusively plastidial in *C. reinhardtii* [8], which most likely also applies to *M. neglectum*. This was supported by localization prediction (Fig. 4), and by a comparison of overall transcript abundance values. Transcript levels were 13- to 46-fold higher for the heteromeric ACCase, compared to the homomeric ACCase (Additional files 3 and 4), which are characteristic for plastidial and cytosolic/mitochondrial FA synthesis, respectively. ACCase catalyzes the first committed step of FA synthesis and is considered a key rate-limiting enzyme [86]. The product of FA synthesis is acyl-ACP, which can directly be used for the synthesis of chloroplast membrane glycerolipids by sequential acylation of glycerol-3-phosphate [8]. Alternatively, acyl-ACP can be separated into the ACP moiety and a free FA by FAT [8]. It is postulated that the free FA is transported across the chloroplast membrane into the cytosol, where it is activated by CoA to acyl-CoA, catalyzed by LACS [8]. The activated acyl-CoA moiety can then be used for the synthesis of, for instance, ER membrane lipids [8]. This pathway could be completely reconstructed for *M. neglectum* (Fig. 4).

In higher plants, TAG synthesis is restricted to the ER [87], while in microalgae TAG synthesis does likely not only take place in the ER, but also in the chloroplast [8].

However, none of the putative DGAT, DGTT, and PGAT enzymes of *M. neglectum* were predicted to be chloroplast localized (Fig. 4). Therefore, localization studies are required to confirm whether TAG synthesis in *M. neglectum* is both chloroplast- and ER-localized, analogous to other microalgae.

For higher plants, PtdCho is a key intermediate for TAG synthesis [87]. For developing soybean embryos, it has been shown that 60% of newly synthesized FAs are incorporated directly into the *sn*-2 position of lyso-PtdCho yielding PtdCho, rather than being used for sequential acylation of glycerol-3-phosphate in the Kennedy pathway [88]. This reaction is catalyzed by lysophosphatidylcholine acyltransferase. An acyl chain can subsequently be released from PtdCho by different lipases, which can then be used for sequential acylation of glycerol-3-phosphate and TAG synthesis [88]. This pathway does not apply for the model green alga *C. reinhardtii*, because it lacks PtdCho [8]. However, it might apply to other microalgae to some extent, because genomic evidence for PtdCho synthesis was reported for *Chlorella pyrenoidosa* [42] and PtdCho has been detected in *N. oceanica* [89]. Transcript evidence for PtdCho was also detected for *M. neglectum*, because one CTP:phosphocholine cytidylyltransferase and two phosphatidylethanolamine N-methyltransferase homologues could be identified (Fig. 4, CCT and PEAMT, respectively), which are both absent from the *C. reinhardtii* genome [65]. CTP:phosphocholine cytidylyltransferase is considered as a key rate-limiting enzyme for de novo synthesis of PthCho, and phosphatidylethanolamine N-methyltransferase catalyzes an alternative route of PthCho synthesis, which is by methylation of phosphatidylethanolamine [90]. In *C. reinhardtii*, PthCho is replaced by DGTS [8]. Interestingly, the enzyme required for DGTS synthesis was also identified in the transcriptome of *M. neglectum* (Fig. 4, BTA). Accordingly, similarly to what has been recently suggested for *C. pyrenoidosa* [42], *M. neglectum* might be capable of both PthCho and DGTS syntheses, a hypothesis that has to be confirmed biochemically.

The transcriptional regulation of the glycerolipid metabolism reveals several endogenous factors for TAG and total lipid hyperaccumulation

A first factor underlying TAG and total lipid hyperaccumulation in the l−N stage becomes evident from the transcriptional regulation of FA synthesis. Its regulation was not uniform in the l−N stage, in contrast to the r+N stage, in which almost all enzymatic steps of FA synthesis were uniformly transiently down-regulated (Fig. 4, ACCase, MCMT, KASI, KASII, KAR, HAD, ENR, ACP, and FAT). In the l−N stage, a strong transcriptional induction of FAT and a gentle induction of ACCase were

observed (Fig. 4), while several intermediate steps of FA synthesis were repressed, although the extent was moderate compared to the r+N stage (Fig. 4, MCMT, KASIII, and KAR). At the same time, the transcript levels of ACP were maintained in the l−N stage, which were among the most abundant transcripts (Additional file 2). ACP prevents the growing acyl chain from degradation [91]. High expression of ACP is in accordance with data for *E. coli* showing that ACP belongs to the most abundant proteins [91]. As an abundant protein, however, a strong transcriptional down-regulation of ACP would be expected, if FA synthesis was significantly inhibited under −N conditions. Since this was not observed, but ACP transcript levels were maintained (Fig. 4), it is tempting to speculate that the down-regulation of intermediate enzymatic steps of FA synthesis (e.g., KASIII) results in a modulation of FA synthesis towards increased production of oleic acid (C18:1). Oleic acid is the most abundant FA in the neutral lipid fraction of *M. neglectum* [27, 28]. This is supported by studies for *E. coli* that reported increased and decreased levels of C18 species upon deletion and over-expression of 3-ketoacyl synthase III (KASIII, *fabH*), respectively, at the expense of C16 FAs [92, 93]. In accordance with this, the gentle transcriptional induction of ACCase might aim at replenishing the pool of malonyl-CoA required for FA elongation. The transcriptional repression of the immediate downstream step might seem contradictory (Fig. 4, MCMT); however, both genes had approximately equal expression levels after 56 h of −N conditions (Additional file 3, MCMT, XLOC_016566, 175 FPKM and β-CT, XLOC_015237, 177 FPKM, respectively). Accordingly, the gentle transcriptional repression of MCMT in the l−N stage might aim at balancing both enzyme levels.

A second factor can be deduced from the transcriptional regulation of FAT and SAD. FAT cleaves ACP from acyl-ACP yielding a free FA, and SAD catalyzes the desaturation of stearic acid (C18:0) to oleic acid (C18:1). The transcriptional up-regulation of both genes might result in a deregulation of FA synthesis. This is because in plants and bacteria acyl-ACP has an inhibitory effect on ACCase [91, 94]. Accordingly, up-regulation of FAT in *M. neglectum* might decrease the pool of inhibitory acyl-ACP, thereby relieving feedback inhibition and in turn keeping FA synthesis in an active state. SAD was postulated to be a metabolic lipid regulator in mammals and shown to co-limit lipid production in oleaginous yeast [85]. In *C. reinhardtii*, SAD over-expression was reported to increase the total lipid content by approximately 28% [95].

A third factor becomes evident from the transcriptional regulation of the putative PGD1 lipase in *M. neglectum* (Fig. 4). For *C. reinhardtii*, it was shown that de novo TAG accumulation involves a significant flux through the membrane lipid pool [73]. In this scenario, rather than direct channeling of de novo synthesized FAs into TAG synthesis, at least some are first incorporated into membrane lipids, from which free FAs are released by the action of PGD1, which are in turn incorporated into TAGs [36, 73]. This indirect channeling of FAs into TAGs through an intermediate membrane lipid step likely also applies to *M. neglectum*, because PGD1 and other lipases were strongly up-regulated (Fig. 5), and the GO term *"lipid catabolic process"* was enriched among genes up-regulated specifically in the l−N stage (Additional file 1: Table ST1). Interestingly, the pronounced transcriptional up-regulation of the PGD1 lipase was reserved for the l−N stage (Fig. 4). This suggests that PGD1 has a specific role in de novo TAG accumulation, because acyl chain recycling from membrane lipid degradation was finished after two days (Fig. 2c). The transcriptional regulation of PGD1 furthermore indicates that some acyl-shuttling processes, for instance those involving PGD1, are less relevant in N-containing media, because PGD1 induction was immediately relaxed in the r+N stage (Fig. 4). In accordance with a possibly diminished pool of lyso-glycerolipids during the first hours of N resupply, one lysophosphatidylcholine acyltransferase candidate was transiently down-regulated in the r+N stage (XLOC_009408, LP-C/E-AT in Fig. 4).

The strong transcriptional regulation of PGD1 led us to explore the transcriptional profiles of further lipase candidate genes (Fig. 5). Notable examples of lipases that were also up-regulated in the l−N stage are XLOC_011377, XLOC_001858, and XLOC_013922 (Table 1; Fig. 5). Interestingly, the homologues of XLOC_001858 and XLOC_013922 were also up-regulated in *C. reinhardtii* under −N conditions, indicating potentially conserved functions (Cre09. g388763, 6e−65; Cre03.g195500, 1e−36; transcript data in [11]). A highly interesting lipase candidate is XLOC_013518, a putative patatin-like phospholipase (Table 1). Under exponential growth conditions, its transcript levels were the highest among all lipases (Fig. 5, XLOC_013518, category IV). Under −N conditions in contrast, this gene was down-regulated to an extent that expression was hardly detectable (Fig. 5, XLOC_013518, tag "NA" at time points N_48, N_56, and N_96). In the r+N stage, initial transcript abundance values were restored after 14 h of N resupply (Fig. 5, XLOC_013518, R_14). This relatively late relaxation of repression in the r+N stage indicates that this lipase is unlikely a TAG lipase, which otherwise should be transcriptionally induced early in the r+N stage. Therefore, this protein could be involved in preserving membrane lipid integrity under exponential growth

conditions, although postulation of its function is complicated, because homologues could be identified neither in *C. reinhardtii*, nor in *A. thaliana*.

Interestingly, the putative CTP:phosphocholine cytidylyltransferase enzyme characteristic for PthCho synthesis was up-regulated in the l−N stage in *M. neglectum* (Fig. 4, CCT). Considering that in developing soybean embryos DAG moieties used for TAG synthesis were mostly derived from PthCho [88], this transcriptional induction might be indicative of a similar flux pattern in *M. neglectum*. Consequently, a considerable amount of TAGs would be obtained via the PthCho route, rather than via sequential acylation of glycerol-3-phosphate. However, flux analyses are required to investigate how far this model applies to *M. neglectum*.

The strong transcriptional induction of FAT, PGD1, and other lipases under −N conditions likely yields an elevated pool of free FAs, which need to be activated by CoA to acyl-CoA, before they can be incorporated into glycerolipids, such as TAGs. This activation is catalyzed by LACS enzymes, and a central role of LACS in TAG accumulation became recently evident for *C. reinhardtii*, because loss of LACS2 decreased TAG content by 50% [96]. In *M. neglectum*, eleven LACS transcripts were identified, and those LACS enzymes with increased transcript abundances under −N conditions might play a role in FA activation for subsequent incorporation into TAGs (e.g., XLOC_003478), whereas those with increased abundances upon N resupply might activate FAs for subsequent β-oxidation (e.g., XLOC_002550) (Fig. 4).

Transcriptional regulation of the acylation of DAG, the committed step of TAG synthesis

The committed step in the acyl-CoA-dependent pathway of TAG formation is catalyzed by DGAT and DGTT. Although transcriptional induction of DGTT transcripts was observed under −N conditions in *M. neglectum*, induction was moderate compared to *C. reinhardtii* (>100-fold induction for DGTT1 [10]) and *N. oceanica* (5.7-fold induction for DGAT-2B [20]). The only putative DGTT gene that was strongly up-regulated in −N conditions surprisingly was also strongly induced under N resupply, a profile that is not expected for an enzyme involved in TAG accumulation (Fig. 4, sixth DGTT transcript from top). A gentle transcriptional induction of DGAT/DGTT genes under −N conditions was reported for *N. oleoabundans* and *P. tricornutum* [17, 19]. Accordingly, other regulation levels than transcriptional control might be more relevant for DGAT/DGTT-mediated TAG accumulation in *M. neglectum*. This might also be an explanation of the limited success of DGTT overexpression to increase TAG accumulation in *C. reinhardtii* [97, 98]. Interestingly, transcript levels of PDAT, characteristic for the acyl-CoA independent route of TAG formation, were approximately constant under −N conditions (Fig. 4). A gentle induction of ~twofold has been reported for *C. reinhardtii* [12] and *P. tricornutum* [19] under −N conditions, and a marginal up-regulation of ~0.5-fold for *N. oceanica* [20]. Accordingly, the extent of transcriptional regulation of PDAT was similar for *M. neglectum*, indicating that PDAT may contribute to a similar extent to TAG formation in *M. neglectum* as has been reported for *C. reinhardtii* [12]. In summary, it seems likely that the predominant way of TAG formation in *M. neglectum* is by the acyl-CoA-dependent pathway, rather than by the acyl-CoA independent pathway, as indicated by the strong transcriptional induction of putative FAT, PGD1, and lipase enzymes, as well as by the gentle induction of putative LACS and DGTT enzymes under −N conditions, while PDAT was approximately constantly expressed (Fig. 4).

Sources of acetyl-CoA for FA synthesis in the I−N stage

Acetyl-CoA is the carbon precursor for FA synthesis. There are several routes for acetyl-CoA formation, and two are considered the most important in chloroplasts of higher plants [87]. Those are the oxidative decarboxylation of pyruvate catalyzed by the multisubunit cpPDHC (plastidial pyruvate dehydrogenase complex) and, to a lesser extent, the activation of free acetate to acetyl-CoA by acetyl-CoA synthetase [87]. Acetate can be obtained by fermentative decarboxylation of pyruvate catalyzed by pyruvate decarboxylase. Alternative fermentative reactions directly yield acetyl-CoA from pyruvate, such as those catalyzed by pyruvate-formate-lyase and pyruvate-ferredoxin-oxidoreductase. In microalgae, the preferred route of acetyl-CoA generation for FA synthesis under −N conditions has not yet been determined, but is presumed to be via cpPDHC [99], analogous to higher plants [87].

The transcriptional regulation of the putative cpPDHC subunits in *M. neglectum* was outlined in the Results section (Additional file 1: Figure S2e) and suggests that cpPDHC activity might be approximately maintained under −N conditions (Fig. 6). High transcript abundances for cpPDHC were noted (Fig. 6, category IV), indicating that cpPDHC might contribute significantly to the plastidial acetyl-CoA pool. A strong transcriptional induction in the I−N stage was observed for fermentative reactions, which was immediately relaxed upon N resupply (Fig. 6, PFOR, PFL, PDC, and ADH). This might have two effects. First, the corresponding enzymes might increase acetyl-CoA supply in the I−N stage, yet likely to a lesser extent than cpPDHC, because transcript levels in the I−N stage of the former were lower than those of the latter (Additional file 3). Second, the transcriptional induction might

aim to optimize the flux of pyruvate reaching cpPDHC, in order to maintain optimal acetyl-CoA production rates by cpPDHC to ensure sufficient precursor supply for FA synthesis, as has been postulated for *Chlorella desiccata* [100, 101].

Transcriptional regulation of two enzymatic steps in the I−N stage of the central carbon metabolism might result in a re-routing that fuels into fatty acid synthesis

The most abundant FA of the neutral lipid fraction in *M. neglectum* is oleic acid (18:1) [27, 28], and its synthesis requires 19 NADPH molecules (18 for the C18 acyl chain, and an additional molecule for the subsequent desaturation of stearic acid to oleic acid). Accordingly, FA synthesis requires a high supply of NADPH. A first pathway for NADPH regeneration exists with the linear electron flow in photosynthesis [102]. The OPPP provides two NADPH molecules per glucose 6-phosphate molecule and was predicted to be localized to the chloroplast of *M. neglectum*, as has been described for *C. reinhardtii* [32]. Our data, obtained on the transcriptional level, are feasible to postulate that under −N conditions the transcriptional induction of the OPPP (Fig. 6) offers the opportunity to provide an additional pathway for the delivery of NADPH. Transcriptional induction of the OPPP under −N conditions has also been reported for *C. reinhardtii* and *N. oleoabundans*, which correlated with increased protein abundances in *C. reinhardtii* [103], and has been interpreted similarly [13, 14, 17, 103]. In this context, by analyzing the response of the starchless mutant *sta6* of *C. reinhardtii* under mixotrophic −N conditions, it was proposed that increasing the reductant pool is a promising strategy to increase lipid productivity [14]. This would be contradictory to the overflow hypothesis, according to which TAGs are "overflow products," i.e., to serve as a sink for excess photosynthetic energy [104]. However, it was recently elaborated that this hypothesis is insufficient for carbon compound accumulation [36, 84].

Phosphoglycerate kinase catalyzes the interconversion between 3-phosphoglycerate and 1,3-bisphosphoglycerate. Both phosphoglycerate kinase candidate genes of *M. neglectum* were strongly down-regulated in the I−N stage (Fig. 6, PGK). This was also observed in *C. reinhardtii* [11], which correlated with decreased protein abundances in [103] but not in [11]. Assuming that the transcriptional repression of phosphoglycerate kinase under −N conditions in *M. neglectum* manifests on the metabolic level, a redirection of the carbon flow would be obtained. Accordingly, carbon flow might be shifted away from the formation of 1,3-bisphosphoglycerate for replenishment of the Calvin cycle, towards the production of 2-phosphoglycerate, such that carbon flow is directed to pyruvate generation by subsequent glycolytic reactions. The down-regulation of phosphoglycerate kinase on the transcript level might thus represent an important determinant to redirect fixed carbon towards pyruvate synthesis, and was also observed in other microalgae under −N conditions (Additional file 1: Results). However, although the expression patterns clearly indicate this alternative metabolic route, labeling studies would be required to investigate how far they reflect the metabolic flux in *M. neglectum* under −N conditions.

Transcriptional regulation of glycerol-3-phosphate supply for TAG biosynthesis

Glycerol is the backbone of TAG and other glycerolipids. Glycerol-3-phosphate is obtained from the reduction of dihydroxyacetone phosphate, catalyzed by glycerol-3-phosphate dehydrogenase (Fig. 6). Three glycerol-3-phosphate dehydrogenase candidate transcripts were identified in the transcriptome of *M. neglectum* (Additional file 4). One of these had predicted chloroplast localization and was highly up-regulated in response to N starvation (Fig. 6, GPDH). A strong transcriptional induction of a subset of glycerol-3-phosphate dehydrogenase enzymes has also been noted for *C. reinhardtii* [13]. In rape (*Brassica napus*), glycerol-3-phosphate supply was proposed to co-limit oil accumulation, because over-expression of a yeast glycerol-3-phosphate dehydrogenase increased seed oil content [105]. In the diatom *P. tricornutum*, over-expression of glycerol-3-phosphate dehydrogenase increased the neutral lipid content by 60% [106]. Therefore, the respective up-regulated enzyme of *M. neglectum* (XLOC_014979) might have a central role in TAG accumulation and represents a promising target for genetic engineering approaches (Table 1). However, over-expression of glycerol kinase to increase glycerol-3-phosphate supply in the diatom *Fistulifera solaris* was of mixed success, because the total lipid content was both increased and decreased in a first and a second transformant, respectively, under autotrophic as well as mixotrophic conditions (external glycerol supply) [107].

Transcriptional regulation of malic enzyme in *M. neglectum* suggests that this reaction likely does not have a central role in lipid hyperaccumulation in the l−N stage

A central component of lipid accumulation in oleaginous fungi is the induction of ATP-citrate lyase and of malic enzyme. ATP-citrate-lyase converts citrate into acetyl-CoA and oxaloacetate. Oxalacetate can be converted to malate by malate dehydrogenase. Malic enzyme decarboxylates malate to the central intermediate pyruvate, and this reaction additionally provides NADPH [108–110]. We note that there is a significant difference between oleaginous fungi and microalgae, because the former are obligate heterotrophs, thus lacking chloroplasts, while most microalgae are capable of growing photoautotrophically. A key difference therefore is the localization of FA synthesis, which generally takes place in the cytosol in non-photosynthetic species and in the chloroplast in eukaryotic photosynthetic species.

Interestingly, citrate was shown to accumulate under −N conditions in *C. reinhardtii*, and its exogenous supply increased lipid accumulation [111]. In *M. neglectum* however, transcript levels of ATP-citrate lyase were approximately constant under −N conditions (Fig. 6, ACL), suggesting that this reaction is not a preferred route for acetyl-CoA generation in *M. neglectum*.

In the diatom *P. tricornutum*, one of the most strongly up-regulated genes under −N conditions was malic enzyme [19], and its subsequent over-expression in both *P. tricornutum* (homologous expression) and the chlorophyceae *C. pyrenoidosa* (heterologous expression) was claimed to increase total lipid contents by 230–250 and 240–322% in the stationary phase, respectively [112, 113]. However, transcriptional regulation and absolute abundances of malic enzyme in *M. neglectum* in the l−N stage were modest (Fig. 6, MME), suggesting that malic enzyme does not play a major role during lipid accumulation in this stage. This is in accordance with a metabolome study of *Chlorella protothecoides* concluding that malic enzyme had little to no activity under both +N and −N conditions despite robust lipid accumulation under −N conditions [114].

Enolase and C3 carbon transporters are subjected to differential transcriptional regulation in response to alternating nitrogen availability

Enolase catalyzes the reversible conversion of 2-phosphoglycerate to PEP and was recently postulated to represent an important determinant for carbon flux in chlorophyceae [115]. Two putative enolase transcripts were identified in the transcriptome of *M. neglectum*, which had both increased abundances under −N conditions (Fig. 6, ENO). Interestingly, a gentle up-regulation (~twofold) of enolase could also be found in the oleaginous chlorophyceae *N. oleoabundans* after 11 days of −N conditions (transcript data in [17]), whereas transcript levels in the non-oleaginous chlorophyceae *C. reinhardtii* and *B. sudeticus* remained approximately constant during 48 and 72 h of −N conditions, respectively (transcript data in [11, 23]). Accordingly, the transcriptional induction of enolase in *M. neglectum* could potentially be of importance for its oleaginous phenotype, which would be interesting to investigate in future studies.

Besides, two of 15 putative triose phosphate transporters were up-regulated under −N conditions, of which one was predicted to localize to the chloroplast (Fig. 6, TPT).

In contrast, five different triose phosphate transporters were transiently down-regulated in the r+N stage, of which one with high expression under initial exponential growth conditions was predicted to be chloroplast localized (Fig. 6, TPT). Furthermore, a putative PEP transporter was strongly down-regulated in the e −N stage, while this down-regulation relaxed in the l −N stage, and a strong up-regulation was subsequently noted for the r+N stage (Fig. 6, PPT). This indicates a differential transporter composition of, for instance, the chloroplast membrane under both N regimes and suggests a possibly important role of carbon transport in the response to alternating N availability.

Thylakoid membrane lipid synthesis is likely stimulated on two different levels during the N resupply treatment

A central characteristic of the r+N stage was the reestablishment of photosynthesis, including thylakoid membrane reassembly (Additional file 1: Table ST1, Figure S1). The major thylakoid membrane lipid of *C. reinhardtii* is MGDG [8]. MGDG is derived from DAG by galactosylation, catalyzed by MGDG synthase [8]. MGDG synthesis in the r+N stage in *M. neglectum* might be stimulated on two different levels: first on the transcript level, because the respective gene was up-regulated in the r+N stage (Fig. 4, MGDGS); second, the transient transcriptional repression of two of three phosphatidic acid phosphatase candidate genes in the r+N stage might reduce the turnover of phosphatidic acid to DAG (Fig. 4, PAP), which would result in a transient accumulation of phosphatidic acid in this stage. Accumulation of phosphatidic acid was postulated to hyperstimulate MGDG synthase activity in *C. reinhardtii* [116]. Therefore, the second level of stimulation of MGDG synthesis in the r+N stage in *M. neglectum* might be on the metabolic level by the putative transient accumulation of phosphatidic acid. Although reduced turnover would also decrease the availability of de novo DAG for MGDG synthesis, it seems unlikely that this limits MGDG synthesis, since the overall DAG pool in the r+N stage should be saturated due to TAG hydrolysis yielding DAG moieties.

The restoration of photosynthetic efficiency upon N resupply involves several steps with different timings

The immediate transcriptional responses of *M. neglectum* in the r+N stage were the up-regulation of TAG hydrolysis, membrane lipid synthesis, and FA degradation, as well as the concomitant down-regulation of FA synthesis, which were most pronounced after 2 h of N resupply (Fig. 4, ACX, MGDGS, and ACCase, respectively, Fig. 5). Next, the transcriptional repression of light-harvesting protein genes that was detected during the complete −N phase relaxed after 4 h of N resupply (Additional file 3,

XLOC_000814, XLOC_006556, XLOC_016372, and XLOC_003245). Finally, the transcriptional repression of the Calvin cycle under −N conditions relaxed after 8 to 14 h of N resupply (Fig. 6, rbcS2 and PGK). Taken together, this indicates that during N resupplementation the reestablishment of full photosynthetic capacity takes place in a coordinated manner and in temporally different patterns.

Metabolic engineering strategies to optimize TAG production in *M. neglectum*

A major goal of this differential transcriptome study was to understand the transcriptional basis for TAG accumulation and TAG degradation in *M. neglectum*, in order to identify potential gene targets for future metabolic engineering approaches. The most promising candidates are summarized in Table 1. One option is to relieve feedback inhibition of FA synthesis by the over-expression of chloroplast-targeted FAT (acyl-ACP thioesterase) genes, a strategy that has been successfully applied before in *P. tricornutum* [117], *C. reinhardtii* [118], and other photosynthetic [119] and heterotrophic species [91, 120]. Since in this work a clear correlation between transcriptional up-regulation of both putative FAT enzymes and lipid hyperaccumulation was observed (Figs. 2c and 4), over-expression of these genes represents a prime target. Heterologous expression of FAT from a closely related species, such as *C. reinhardtii* (Additional file 1: Figure S12), might also be of interest, in order to circumvent possible species-specific regulations of the endogenous enzymes.

Furthermore, lipases represent promising targets, because they are implicated in different processes of the glycerolipid metabolism, such as membrane lipid turnover and TAG degradation under −N and N resupply conditions, respectively [72, 73]. In accordance with a central role of lipases in the glycerolipid metabolism, the lipid content of the diatom *Thalassiosira pseudonana* could be increased 2.4- to 4.1-fold upon *knock-down* of a multifunctional lipase/phospholipase/acyltransferase enzyme under both nutrient replete and nutrient starvation conditions [121]. Interestingly, to our knowledge, overexpression approaches of lipases were not yet explored for microalgae. Therefore, both *knock-down* and overexpression approaches seem highly interesting for *M. neglectum*. Several putative lipase genes were transcriptionally regulated in opposite directions under −N and N resupply conditions in *M. neglectum* (Fig. 5), characterizing them as highly promising targets.

An alternative approach for increased TAG accumulation would be the repression of competitive catabolic pathways such as FA degradation, i.e., β-oxidation. In *C. reinhardtii*, disruption of one acyl-CoA oxidase gene

Table 1 List of potential gene targets for genetic engineering approaches to increase lipid production in *M. neglectum*

Annotation	Locus ID	Putative enzyme function	Transcript profile	Approach	Postulated effect	References in other microalgae
Acyl-ACP thioesterase (FAT, Fig. 4)	XLOC_007529 or XLOC_011123	Cleaves acyl-ACP into a free FA and ACP, thus relieving feedback inhibition of FA synthesis; requires functional interaction with ACP [148]	Up-regulated in the l−N stage, down-regulated in the r+N stage	Over-expression	Deregulation of FA synthesis allowing for total lipid hyperaccumulation under +N conditions	[91, 117–120]
Lipases, such as PGD1 (Fig. 5)	XLOC_12515 (PGD1) or XLOC_011377 or XLOC_013518	XLOC_12515: lipase that may act on membrane lipids such as de novo MGDG XLOC_011377: unknown function and no domain predicted; XLOC_013518: unknown function with a predicted patatin_cPLA2 superfamily domain	XLOC_12515 and XLOC_011377: both are strongly up-regulated in the l−N stage; expression of XLOC_011377 was highly correlated to MLDP (>0.99) XLOC_013518: abundant transcript; strongly repressed in the l−N stage, and repression was continued during the first 8 h of N resupply	Over-expression (XLOC_12515 and XLOC_011377) Knock-down (XLOC_013518)	XLOC_12515 and XLOC_011377: membrane lipid turnover putatively triggering TAG accumulation under +N conditions XLOC_01351: might be involved in maintaining membrane lipid integrity; down-regulation could lead to an increased TAG content under +N conditions	[73, 121]
Acyl-CoA oxidase (ACX2, Fig. 4)	XLOC_015398	Implicated in FA degradation by oxidizing acyl-CoA	Up-regulated in the r+N stage	Knock-down, deletion	Increased TAG content by reducing the rate of FA degradation (repression of catabolic pathways)	[31]
Phosphoglycerate kinase (PGK, Fig. 6)	XLOC_018937	Converts 3-phosphoglycerate to 1,3-bisphosphoglycerate and vice versa	Strongly repressed in the l−N stage, which was continued for the first 4 h in the r+N stage	Knock-down	Redirecting carbon flow towards pyruvate generation, away from glyceraldehyde-3-phosphate for replenishment of the Calvin cycle	NA
PEP carboxylase (PEPC, Fig. 6)	XLOC_004101	Carboxylation of PEP to oxalacetate	Up-regulated under −N conditions; approximately unaltered transcript levels in the r+N stage	Knock-down	PEP can be increasingly used for pyruvate generation, rather than for replenishment of the tricarboxylic acid cycle	[149–151]
Glycerol-3-phosphate dehydrogenase (GPDH, Fig. 6)	XLOC_014979	Converts dihydroxyacetone phosphate into glycerol-3-phosphate	Up-regulated under −N conditions	Over-expression	Increased TAG accumulation due to increased supply of glycerol-3-phosphate under −N conditions	[106]
Enolase (ENO, Fig. 6)	XLOC_012275	Converts 2-phosphoglycerate to PEP	Up-regulated under −N conditions	Over-expression	Possibly altered carbon partitioning towards increased PEP generation under +N conditions	[115]
Transcription factor (Additional file 5)	XLOC_013389 or XLOC_005581	XLOC_013389: transcription factor cf MYB family XLOC_005581: transcription factor cf GATA family	XLOC_013389: strongly up-regulated under −N conditions; XLOC_005581: strongly repressed under −N conditions	Over-expression (XLOC_013389), knock-down (XLOC_005581)	Mimicking parts of the transcriptional regulation from −N conditions under +N conditions	[125, 126]

Table 1 continued

Annotation	Locus ID	Putative enzyme function	Transcript profile	Approach	Postulated effect	References in other microalgae
Small subunit of RuBisCo (rbcS2) or elongation factor or ribosomal protein (RPL7aE) (Additional file 1: Figure S2b)	XLOC_007679 or XLOC_005939 or XLOC_000987	XLOC_007679: rbcS2, central enzyme of photosynthesis catalyzing carbon fixation XLOC_005939 and XLOC_000987: implicated in protein biosynthesis	XLOC_007679: the third highest rate of expression under +N conditions, and only moderately affected by the −N treatment; the first intron could putatively contain an enhancer motif as in [152] XLOC_005939 and XLOC_000987: very high and stable expression under −N conditions	Cloning	Strong constitutive promoter for both +N and −N conditions	[152, 153]

The table is sorted according to pathways, starting with the glycerolipid metabolism, followed by the central carbon metabolism, finishing with other targets

greatly impaired the rate of β-oxidation, resulting in a 20% increased TAG content under −N conditions [31]. A putative homologue of the respective gene was identified in the transcriptome of *M. neglectum* (XLOC_015398). It exhibited a sharp transient up-regulation in the r+N stage (Fig. 4, ACX), suggesting that it is actively involved in FA degradation, therefore representing another prime target for gene *knock-down/knock-out* strategies. Alternatively, complete loss of peroxisomes by deletion of a peroxisome biogenesis factor was demonstrated to exhibit a cooperative effect on lipid hyperaccumulation in the heterotrophic host *Y. lipolytica* [122].

Investigations of the starchless *C. reinhardtii* strain *sta6* [123] have indicated that carbon precursor supply is not a limiting factor that prevents TAG accumulation under +N conditions. Compared to the parental strain, 18- and 27-fold increased intracellular malonyl-CoA levels were reported for the *sta6* strain under photoautotrophic +N conditions with low and high light intensities, respectively [124]. Cellular lipid levels, however, were similar to the parental strain [124]. Therefore, a currently unknown "−N signal" (or multiple "−N signals") seems to be decisive for TAG accumulation, while carbon precursor supply augments TAG accumulation, once the "−N signal" is established. Accordingly, transcription factors are additional gene targets. The effectiveness of transcription factor over-expression was recently demonstrated with PSR1 (phosphorus starvation response 1) in *C. reinhardtii*, resulting in a "liporotund phenotype," which exhibited lipid levels under +N conditions similar to the parental strain under −N conditions [125]. A *knock-down* strategy targeting a transcription factor was similarly effective, which doubled lipid content and volumetric lipid productivity under autotrophic +N conditions in *N. gaditana* [126]. Several transcription factors of *M. neglectum* showed clear transcriptional responses to alternating N availability, defining them as promising gene targets (Additional file 1: Results, Additional file 5). Although their efficient over-expression might be challenging because of the large size of transcription factor genes (e.g., 5191 bp for XLOC_013389), a strategy has been recently described that allows high expression of large transgenes [127]. It includes an even distribution of regulatory introns into the transgene [127], and this approach might be also effective for *M. neglectum*, because *M. neglectum* has a similarly intron-rich CDS composition as *C. reinhardtii* (Additional file 1: Results).

An interesting further gene target would be a chloroplast-targeted phosphoketolase [128]. This enzyme is absent from *M. neglectum* and other microalgae, except for *P. tricornutum* [129]. The xylulose-5-phosphate type of phosphoketolase catalyzes the cleavage of xylulose-5-phosphate into acetyl-phosphate and glyceraldehyde-3-phosphate [130]. Xylulose-5-phosphate is an intermediate of the Calvin cycle, which has a high flux under autotrophic conditions, and was reported to accumulate under mixotrophic −N conditions in *C. reinhardtii* [11]. Of the two products, acetyl-phosphate is the substrate for phosphotransacetylase, which generates acetyl-CoA that can be used for FA synthesis. Compared to the classical route of pyruvate-to-acetyl-CoA conversion, the phosphoketolase–phosphotransacetylase route does not involve a decarboxylation step, which increases the efficiency of carbon use. The over-expression of phosphoketolase and phosphotransacetylase in an engineered strain of the oleaginous yeast *Y. lipolytica* did not only increase lipid content to >60% of dry biomass, but importantly also uncoupled lipid accumulation from nutrient deprivation [131]. Although one candidate for phosphotransacetylase was identified in the transcriptome of *M. neglectum* (putative fragments XLOC_019320, XLOC_019178, and XLOC_011051), which was transcriptionally coherently regulated with β-oxidation in the r+N stage (Fig. 6, PTA), its transcript levels were low (FPKM = 4–14). Therefore, a two-gene approach, i.e., combined over-expression of both phosphoketolase and phosphotransacetylase, is likely necessary to establish an efficient metabolic shortcut for acetyl-CoA formation from xylulose-5-phosphate in *M. neglectum*.

Conclusions

This study for the first time provides a comprehensive overview of the transcriptome responses to three different stages of N availability in the oleaginous chlorophyceae *M. neglectum*. The three stages early −N, late −N, and N resupply (e−N, l−N, and r+N stage; Fig. 1) were characterized by net starch accumulation, net lipid and TAG accumulation, and subsequent storage compound degradation, respectively (Fig. 2b, c; Additional file 1: Figure S1). The explorative analysis of transcriptional profiles was focused on the lipid and central carbon metabolism, and the starch metabolism was included as an associated pathway (Figs. 4 and 6; Additional file 1: Figure S10b). The transcript data were first used to refine the structural genome annotation of *M. neglectum* (Additional file 1: Results). Next, the quantitative evaluation of transcript data revealed that distinct putative lipases exhibited contrastingly different expression patterns (Fig. 5). Accordingly, the transcript profiles of lipases made it possible to distinguish candidates putatively involved in TAG accumulation from those likely involved in TAG degradation, which were up-regulated under −N and N resupply conditions, respectively (Fig. 5). The GO term enrichment analysis revealed a transcriptional induction of the tricarboxylic acid cycle and of glycolysis under $-N$ conditions in the e$-$N and l$-$N stages (Additional file 1: Table ST1). Three main factors for lipid hyperaccumulation in the l$-$N stage could be deduced in *M. neglectum*. Those were (a) increasing carbon precursor supply, (b) relieving feedback inhibition of FA synthesis, and (c) remodeling of membrane lipid homeostasis, suggested by transcriptional induction of heteromeric ACCase, acyl-ACP thioesterase, and lipases such as PGD1, respectively (Fig. 4). A major goal of this work was the identification of target genes for future metabolic engineering approaches, and the detailed transcriptome analysis of *M. neglectum* cells during lipid accumulation and subsequent lipid remobilization stages allowed identification of several potential targets, which are summarized in Table 1. Reverse genetic approaches can now be performed to test the concept that altered expression of these target genes indeed leads to improved lipid accumulation phenotypes in *M. neglectum*.

Methods

Cultivation conditions

To define the timing of starch and lipid accumulation in *Monoraphidium neglectum* (SAG 48.87) and accordingly subdivide the $-N$ phase in the e$-$N and l$-$N stages for subsequent transcriptome analysis, a long-term $-N$ experiment was performed (Fig. 1, exp1). For this purpose, 900 mL ProF medium was inoculated with 10 mL of a mixotrophic growing pre-culture and grown at room temperature (24 °C) under autotrophic conditions with 3% CO_2 bubbling with gentle stirring under 350–400 μmol m^{-2} s^{-1} constant illumination with white light from both the front and back sides until early mid-logarithmic phase was reached ($\sim$10 $\times$ 10^6 cells mL^{-1}). Cells were washed twice with N-free ProF and adjusted to $\sim$3 $\times$ 10^6 cells mL^{-1} with N-free ProF medium in a total volume of 2.8 in 3-L vertical Schott bottles. Cultivation was performed for 17 days under autotrophic conditions as above. After 1, 2, 4, 6, 8, 11, 14, and 17 days, 200 mL was removed for sample analysis. Three biological replicates were performed. Cell concentration was determined with a Z-series Coulter Counter cell and particle counter (Beckmann Coulter) and dry biomass according to [28].

For the transcriptome experiment (Fig. 1, exp2), *M. neglectum* was inoculated from agar plates and grown in 2.5 L Provasoli-based minimal media (ProF) [132] in 3-L vertical Schott bottles (Schott, USA) in two biological replicates. Cultivation was performed at room temperature (24 °C) under autotrophic conditions with 3% CO_2 bubbling with gentle stirring under 350–400 μmol m^{-2} s^{-1} constant illumination with white light from only the front side until late exponential phase (40 $\times$ 10^6 cells mL^{-1}) was reached. Both cultures were

diluted to $\sim$4 $\times$ 10^6 cells mL^{-1} with fresh ProF medium in a total volume of 2.5 L. Cultivation was continued for 2 days at 350–400 μmol m^{-2} s^{-1} constant illumination with white light from both the front and back sides, until a cell concentration of $\sim$10 $\times$ 10^6 cells mL^{-1} was reached.

From these cultures, samples for the time point zero (N_0) were taken to represent the untreated exponential growth phase. For N starvation treatment ("N"), the cells were centrifuged and washed twice with N-free ProF medium, resuspended in this medium to a cell concentration of $\sim$4 $\times$ 10^6 cells mL^{-1} in a total volume of 2.5 L, and cultivated under the same conditions for 4 days. Samples for RNA isolation were taken after 2, 4, 8, 24, 48, 56, and 96 h of $-N$ conditions constituting the transcriptome sampling time points "N_0, N_2, N_4, N_8, N_24, N_48, N_56, and N_96," respectively. Samples of 50–150 mL for lipid isolation were taken after 14, 24, 48, and 96 h of $-N$ treatment.

For N resupply treatment ("R"), an aliquot of the starved cultures was removed after 48 h of $-N$ treatment, washed with ProF, and adjusted to $\sim$4 $\times$ 10^6 cells mL^{-1} with fresh ProF medium (containing N), and cultivation was continued for an additional 2 days. Samples for RNA isolation were taken after 2, 4, 8, and 14 h, yielding the transcriptome sampling time points "R_2, R_4, R_8, and R_14," respectively. Additional samples for lipid isolation were taken after 14, 24, and 48 h of N resupply. Cell concentration was determined microscopically with a hemocytometer and cell dry biomass according to [28].

Cell weight was obtained by dividing the biomass concentration by cell concentration.

Lipid extractions and chromatography

Lipid extractions and chromatography were performed from 30 to 50 mg lyophilized biomass according to [28]. For calculation of the net lipid production rate, the volumetric lipid content at day 0 was subtracted from the volumetric lipid content at day X and this value was divided by the cultivation time, yielding the volumetric lipid productivity. These values were normalized to day 1 to retrieve the relative volumetric lipid productivity.

Isolation of total RNA and DNase digest

For RNA isolation, the biomass was resuspended in 2 mL RNA extraction buffer (1:1 mix of aqua-phenol and buffer L [0.5% SDS, 10 mM EDTA, 0.2 M sodium acetate (pH 5), and 1:100 β-mercaptoethanol)]. Samples were frozen in liquid nitrogen and subsequently lysed with a Ribolyzer (PRECELLYS 24, Precellys, France) with three cycles of 45 s at 6500 rpm with 15-s break and 0.3 mL silica beads (diameter 0.1 mm). The lysate was incubated on ice for 10 min and 0.5 volumes of chloroform added. For phase separation, samples were centrifuged at 3000$\times g$

for $\geq$30 min and 16 °C. The upper phase was transferred to a new tube and treated with DNase I to remove contaminating DNA according to the manufacturer's instructions (Promega, USA). The solution was extracted with equal volumes of aqua-phenol and chloroform for a second time, and a third time with an equal volume of chloroform only. RNA in the upper phase was precipitated by the addition of 0.1 volumes of 3 M sodium acetate and 1 volume of isopropanol. After overnight incubation at −20 °C, RNA was pelleted by centrifugation at 16,000×g for 30 min at 16 °C. The RNA washed once with 70% ethanol, resuspended with DEPC-water, and stored at −80 °C until use.

Starch determination

50 mL culture was centrifuged at 500×g for 5 min, vortexed with 5 mL methanol, and centrifuged again at 3000×g for 10 min. The pellet was resuspended with methanol to a final concentration of approximately 20×10^6 cells mL^{-1}, amounting to 0.4–0.7 g L^{-1} starch. 500 µL of this solution was lysed with a Ribolyzer as above. Samples were cooled on ice and the methanol was evaporated under N$_2$ flow. Next, 500 µL of 50 mM sodium acetate (pH 5.2) was added, the mixture was transferred to a new 2-mL tube, and the ribolyze tube was washed with 500 µL sodium acetate (this introduced a 1:2 dilution). The tube was wrapped in an aluminum foil and autoclaved at 121 °C for 15 min to solubilize starch. The tubes were shaken vigorously and centrifuged at 20,000×g for 2 min. Starch was enzymatically determined with a starch assay kit (Roche, Germany) using a down-scaled protocol: 10 µL supernatant was mixed with 20 µL solution 1 and incubated at 60 °C for 15 min. 200 µL of 1:2 diluted solution 2 was added and incubated at room temperature for 3 min. Absorbance was measured at 340 nm with a Tecan plate reader (infinite M200, TECAN) yielding the values for $E0$. Absorbance was again measured after incubation for 15 min at room temperature upon the addition of 2 µL solution 3, yielding the values for $E1$. The difference in absorbance ($E1 - E0$) was indicative of the starch content (NADPH synthesis). The starch concentration was calculated according to a simultaneously performed calibration curve with known amounts of starch treated similarly as the cell samples (starting with the ribolyze step).

Preparation of cDNA libraries and Illumina sequencing

For each time point, 1 µg total RNA from each of the two biological replicates (thus 2 µg in total) was mixed in equal amounts to mitigate biological variance. The resulting 12 pools were converted into cDNA libraries by poly-A fishing utilizing a TruSeq Stranded mRNA Library Prep Kit (Illumina, USA) according to the manufacturer's instructions. RNA quality and RNA concentrations were determined using an Agilent RNA Nano 6000 kit on an Agilent 2100 Bioanalyzer (Agilent Technologies, Germany) and Trinean Xpose (Gentbrugge, Belgium), respectively. The library pool was sequenced 2× 100 nt paired-end on Illumina's HiSeq 1500 (Illumina, USA) using rapid run mode. For each time point, approximately 33 million 2× 100 nt fragments were obtained (Additional file 1: Table ST2).

Read mapping and transcript quantification

The quality of the reads was inspected with FastQC [133] (Additional file 1: Figure S13). Trimming was performed with Trimmomatic (version 0.32) requesting a phred33 score of 30 for the leading and trailing bases, and only trimmed reads with at least 75 bp were retained. This length was chosen because TopHat2, which was used for mapping, has been developed with 75-bp reads [48]. For read mapping, transcriptome assembly, and transcript quantification, the TopHat2-Cufflinks-protocol published in [47] was followed. As reference annotation, the refined annotation (BRAKER1-annotation, see Additional file 1: Results) was used. TopHat2 ([48], version 2.1.0) was used with default setting except for min-intron-length = 5, max-intron-length = 1418, mate-inter-dist = 53, mate-std-dev = 124, and library-type = fr-firststrand. The mapping rate was consistently approximately 90% for each time point. Cufflinks ([49, 134], version 2.2.1) was executed with default settings except for library-type = fr-firststrand, intron-overhand-tolerance = 5, min-intron-length = 5, 3-overhang-tolerance = 100, overlap-radius = 5, and max-intron-length = 1418. Parameters adjusted for Cuffquant were library-type = fr-firststrand, multi-read-correct = True, and frag-bias-correct = refined annotation file. The parameter adjusted for Cuffnorm was library-type = fr-firststrand. For quantitative analysis, evaluation was performed on the level of loci, thus not on the level of isoforms, CDS, or transcription starts sites, i.e., the FPKM values for each locus were used.

The full table of locus FPKM values, log2 fold changes relative to the pre-starvation reference time point N_0, and the respective annotations are given in Additional file 3.

Functional annotation of the transcriptome

Annotation for the longest isoform of each locus was obtained by Blast2GO ([135], version 4.0.2) searching the "nr" database at NCBI using the algorithm "blastx-fast" with an expectation threshold of 10^{-10} reporting the 20 best blast hits. For all other settings, default values were used. For each locus, the obtained one-sentence annotation (defline), associated GO terms, and EC numbers were used for subsequent analyses. Furthermore, the best BLASTx hit for the longest isoform of each locus in the proteome of *C. reinhardtii* (Phytozome 11 [136], Creinhardtii_281_v5.5.protein.fasta) and *A. thaliana* (TAIR [137], TAIR10_pep_20101214.fasta) was obtained.

For subcellular localization prediction, the software PredAlgo [70] was used, because it was developed specifically for microalgae. As it only accepts amino acid sequences, transcripts were first translated into proteins, which was performed for only the isoform starting with a start codon ("ATG"); if multiple isoforms of a locus had a start codon as their first base triplet, only the longest of those isoforms was translated; if no isoform contained a start codon, the respective locus was excluded from analysis and the tag "NA" added. Three compartments were predicted: chloroplast, mitochondrion, and secretory pathway; if the resulting localization score was below the threshold defined by PredAlgo, the value "other" was used.

Determination of gene sets and their GO term enrichment analysis

To dissect the transcriptional responses into those restricted to one of the three stages of alternating N availability and those shared between individual stages, an analysis of shared genes was performed. Towards this end, the mean transcript abundance in a specific stage, A_{stage}, was calculated as the mean of FPKM values of the time points attributed to the respective stage. A_{stage} was related to the reference time point N_0, yielding the relative mean transcript abundance, R_{stage}. Its log2-transformed variant was used to classify responsiveness of a gene in the respective stage, which was given if absolute log2-R_{stage} > 1. Consequently, for each stage, a first set with up-regulated genes (log2-R_{stage} > 1) was obtained, and a second one with down-regulated genes (log2-R_{stage} < 1), thus in total six sets.

GO terms were retrieved by Blast2GO (see above). For enrichment analysis, the GNU R (version 3.3.1) [138] package "topGO" (version2.26.0) [139] was used. To initialize the object, the function "GOdata" < −new("topGOdata,"...) was called with default settings, except for ontology = BP. To derive enriched biological processes, the function "resultWeigthFis < −runTest(GOdata,...)" was called with default settings, except for algorithm = weight01 and statistic = fisher. Correction for multiple testing was not performed, as suggested in [139]. Significantly enriched GO terms (p < 10^{-4}) were obtained by the function "GenTable(resultWeigthFis,...)" with default settings.

Pathway reconstruction

The starch, lipid, and central carbon metabolism of *M. neglectum* was reconstructed based on pathways described for *A. thaliana* [87, 140, 141] and *C. reinhardtii* [8, 25, 32, 142, 143]. Several filters were applied to ensure a high-quality pathway reconstruction for subsequent analysis of transcriptional regulation.

For each enzymatic step, a tBLASTx search against the *M. neglectum* transcriptome was performed, yielding a list of candidate transcript loci $L\_transcripts_{enzymatic\_step}$. If multiple isoforms were attributed to a transcript locus, only the longest of those isoforms was considered for subsequent validation. For validation, a reverse tBLASTx search was performed in the two model organisms *A. thaliana* and *C. reinhardtii* for each transcript in $L\_transcripts_{enzymatic\_step}$, and only if the best tBLASTx hits matched the respective enzymatic step, the transcript was retained. Additionally, the predicted domain structure had to match the respective enzymatic step; domain prediction was performed by the NCBI conserved domain search web interface [64]. For instance, if the template sequence encoded an UDK (uridine kinase) superfamily domain (phosphoribulokinase, Cre12.g554800), the candidate transcripts of *M. neglectum* accordingly had to encode the UDK superfamily domain. Next, transcripts likely constituting fragment pairs were identified in a three-step approach. Towards this end, $L\_transcripts_{enzymatic\_step}$ was first searched for transcripts with very similar regulation patterns (Pearson correlation >0.9). Second, only those highly correlated transcripts were retained, whose corresponding genes were classified as putatively fragmented. A gene was defined as fragmented, if it was located within 500 nt distance to the scaffold margin. The value of 500 nt was chosen, because this was the upper size of most introns of *M. neglectum* (Additional file 1: Figure S14c). Third, the predicted domain structures were inspected visually to determine whether both matched each other. This was given, for instance, when the same truncated domain was predicted, such that two transcripts had appropriately C- and N-terminal truncated domains, respectively (Additional file 1: Figure S9). If these requirements were met, the respective transcripts were considered as a fragment pair. Of those, only the leftmost fragment, i.e., containing the start codon, was retained in $L\_transcripts_{enzymatic\_step}$, because this fragment encodes the putative targeting sequence for chloroplast, mitochondrion, or secretory pathway [70].

Fold changes were only calculated if the FPKM values from the reference time point N_0 and the time point of interest were both at least 1.0. Values below this threshold were highlighted by "NA" in Figs. 4, 5, and 6 and Additional file 1: Figure S10 indicating too low read support to allow for reliable transcript quantification.

Identification of putative lipases and clustering of their transcriptional profiles

Lipase candidates were identified by searching the annotation of transcripts for either containing the keyword "lipase" or the keywords "hydrolase" and "beta" in their name, or if attributed GO terms included the term "GO:0016298" (lipase activity) or the terms "GO:0016787" and "GO:0016042" (hydrolase activity and lipid catabolic process, respectively). Hierarchical clustering with Euclidean distance and complete linkage of the log2-transformed fold change values relative to the pre-starvation time point N_0 were performed with R ([138], version 3.3.1; function "hcluster" in package "amap" [144], version 0.8.14). Euclidean distance has been recommended for log ratio data [145], and complete linkage was shown to outperform average linkage in gene expression studies [146]. The average silhouette width as a cluster quality control parameter [75] was determined for $k = 3-7$ clusters. The heatmap visualization of the dendrogram was plotted by the function "heatmap.2" of the R package "gplots" ([147], version 3.0.1), saved as SVG file, and the colors replaced according to the color scheme used in this publication.

Additional files

> **Additional file 1.** Additional Methods, Results, Tables ST1–ST3 and Figures S1–S17.
>
> **Additional file 2.** The 100 most induced genes in the I-N stage, sorted by their log2 mean-FC (R_{I-N}) values in the I-N stage in descending order; The 100 most repressed genes in the I-N stage, sorted according to their log2 mean-FC (R_{I-N}) values in ascending order; The top 100 most expressed genes at the time point N_0 that were classified as not fragmented.
>
> **Additional file 3.** The annotation and additional information for each locus; FPKM values of each locus; Log2-FC values in respect to N_0 of each locus.
>
> **Additional file 4.** All transcripts considered for the pathway analysis of starch metabolism (Additional file 1: Figure S10b); All transcripts considered for the pathway analysis of glycerolipid metabolism (Figure 4); All transcripts considered for the pathway analysis of central carbon metabolism (Figure 6).
>
> **Additional file 5.** The annotation and additional information of all putative transcription factors, sorted by their log2 mean-FC (R_{I-N}) values in the I-N stage in ascending order.

Abbreviations

ACCase: heteromeric acetyl-CoA carboxylase complex; ACP: acyl carrier protein; CDS: coding DNA sequence; CoA: coenzyme A; DAG: diacylglycerol; DGAT: diacylglycerol acyltransferase type 1; DGDG: digalactosyldiacylglycerol; DGTS: diacylglycerol-N,N,N-trimethylhomoserine; DGTT: diacylglycerol acyltransferase type 2; ER: endoplasmic reticulum; FA: fatty acid; FAT: acyl-ACP thioesterase; FC: fold change; FPKM: fragments per kilobase of exon per million fragments mapped; GO: gene ontology; LACS: long-chain acyl-CoA synthetase; MGDG: monogalactosyldiacylglycerol; MLDP: major lipid droplet protein; NA: not available; OPPP: oxidative pentose-phosphate pathway; PDAT: phospholipid:diacylglycerol acyltransferase; cpPDHC: plastidial pyruvate dehydrogenase complex; PEP: phosphoenolpyruvate; PGD1: plastid galactoglycerolipid degradation 1; PS: photosystem; PthCho: phosphatidylcholine; SAD: stearoyl-ACP desaturase; SE: standard error of the mean; TAG: triacylglycerol; UTR: untranslated region.

Authors' contributions

DJ performed the experiments and analyzed the data. AW prepared the cDNA libraries and performed the deep mRNA sequencing. DJ, JHM, OK, AG, and JK wrote the manuscript. JHM, AG, and OK supervised the work and assessed the data. All authors reviewed the manuscript. All authors read and approved the final manuscript.

Author details

[1] Algae Biotechnology and Bioenergy, Faculty of Biology, Center for Biotechnology (CeBiTec), Bielefeld University, 33615 Bielefeld, Germany. [2] Microbial Genomics and Biotechnology, Center for Biotechnology (CeBiTec), Bielefeld University, 33615 Bielefeld, Germany. [3] Bioinformatics and Systems Biology, Justus-Liebig-Universität, 35392 Gießen, Germany. [4] Present Address: Algae Biotechnology and Bioenergy, Faculty of Biology, Center for Biotechnology (CeBiTec), Bielefeld University, Universitaetsstrasse 27, 33615 Bielefeld, Germany.

Acknowledgements

The authors thank Christian Bogen for his excellent assistance during the cultivation of *M. neglectum* which was used for transcriptome analysis. Additional thanks are expressed to Oliver Rupp for fruitful discussions and critical review, and to Kyle Lauersen for critical discussions.

Competing interests

The authors declare that they have no competing interests.

Funding

DJ acknowledges the Cluster Industrial Biotechnology Graduate Cluster (CLIB-GC) (Federal Ministry of Science & Technology North Rhine-Westphalia), AW, JHM, JK, AG, and OK acknowledge the German state North Rhine-Westphalia and Hesse for funding. The authors acknowledge the support for the Article Processing Charge by the Deutsche Forschungsgemeinschaft and the Open Access Publication Fund of Bielefeld University.

References

1. Pulz O, Gross W. Valuable products from biotechnology of microalgae. Appl Microbiol Biotechnol. 2004;65:635–48.
2. Jones CS, Mayfield SP. Algae biofuels: versatility for the future of bioenergy. Curr Opin Biotechnol. 2012;23:346–51.
3. Georgianna DR, Mayfield SP. Exploiting diversity and synthetic biology for the production of algal biofuels. Nature. 2012;188:329–35.
4. Hannon M, Gimpel J, Tran M, Rasala B, Mayfield S. Biofuels from algae: challenges and potential. Biofuels. 2010;1:763–84.
5. Pienkos PT, Darzins A. The promise and challenges of microalgal-derived biofuels. Biofuels, Bioprod Biorefin. 2009;3:431–40.
6. Gimpel JA, Specht EA, Georgianna DR, Mayfield SP. Advances in microalgae engineering and synthetic biology applications for biofuel production. Curr Opin Chem Biol. 2013;17:489–95.
7. Mussgnug JH. Genetic tools and techniques for *Chlamydomonas reinhardtii*. Appl Microbiol Biotechnol. 2015;99:5407–18.
8. Li-Beisson Y, Beisson F, Riekhof W. Metabolism of acyl-lipids in *Chlamydomonas reinhardtii*. Plant J. 2015;82:504–22.
9. Rodolfi L, Zittelli GC, Bassi N, Padovani G, Biondi N, Bonini G, Tredici MR. Microalgae for oil: Strain selection, induction of lipid synthesis and outdoor mass cultivation in a low-cost photobioreactor. Biotechnol Bioeng. 2009;102:100–12.
10. Miller R, Wu G, Deshpande RR, Vieler A, Gärtner K, Li X, Moellering ER, Zäuner S, Cornish AJ, Liu B. Changes in transcript abundance in *Chlamydomonas reinhardtii* following nitrogen deprivation predict diversion of metabolism. Plant Physiol. 2010;154:1737–52.
11. Schmollinger S, Mühlhaus T, Boyle NR, Blaby IK, Casero D, Mettler T, Moseley JL, Kropat J, Sommer F, Strenkert D. Nitrogen-sparing mechanisms in Chlamydomonas affect the transcriptome, the proteome, and photosynthetic metabolism. Plant Cell Online. 2014;26:1410–35.
12. Boyle NR, Page MD, Liu B, Blaby IK, Casero D, Kropat J, Cokus SJ, Hong-Hermesdorf A, Shaw J, Karpowicz SJ. Three acyltransferases and nitrogen-responsive regulator are implicated in nitrogen starvation-induced triacylglycerol accumulation in Chlamydomonas. J Biol Chem. 2012;287:15811–25.

13. Goodenough U, Blaby I, Casero D, Gallaher SD, Goodson C, Johnson S, Lee J-H, Merchant SS, Pellegrini M, Roth R, et al. The path to triacylglyceride obesity in the sta6 strain of *Chlamydomonas reinhardtii*. Eukaryot Cell. 2014;13:591–613.

14. Blaby IK, Glaesener AG, Mettler T, Fitz-Gibbon ST, Gallaher SD, Liu B, Boyle NR, Kropat J, Stitt M, Johnson S, et al. Systems-level analysis of nitrogen starvation-induced modifications of carbon metabolism in a *Chlamydomonas reinhardtii* starchless mutant. Plant Cell. 2013;25:4305–23.

15. Lv H, Qu G, Qi X, Lu L, Tian C, Ma Y. Transcriptome analysis of <i> Chlamydomonas reinhardtii </i > during the process of lipid accumulation. Genomics. 2013;101:229–37.

16. Li L, Zhang G, Wang Q. De novo transcriptomic analysis of *Chlorella sorokiniana* reveals differential genes expression in photosynthetic carbon fixation and lipid production. BMC Microbiol. 2016;16:223.

17. Rismani-Yazdi H, Haznedaroglu BZ, Hsin C, Peccia J. Transcriptomic analysis of the oleaginous microalga *Neochloris oleoabundans* reveals metabolic insights into triacylglyceride accumulation. Biotechnol Biofuels. 2012;5:74.

18. Chang WC, Zheng HQ, Chen cNN. Comparative transcriptome analysis reveals a potential photosynthate partitioning mechanism between lipid and starch biosynthetic pathways in green microalgae. Algal Res. 2016;16:54–62.

19. Yang ZK, Niu YF, Ma YH, Xue J, Zhang MH, Yang WD, Liu JS, Lu SH, Guan Y, Li HY. Molecular and cellular mechanisms of neutral lipid accumulation in diatom following nitrogen deprivation. Biotechnol Biofuels. 2013;6:67.

20. Li J, Han D, Wang D, Ning K, Jia J, Wei L, Jing X, Huang S, Chen J, Li Y. Choreography of transcriptomes and lipidomes of nannochloropsis reveals the mechanisms of oil synthesis in microalgae. Plant Cell Online. 2014;26:1645–65.

21. Radakovits R, Jinkerson RE, Fuerstenberg SI, Tae H, Settlage RE, Boore JL, Posewitz MC. Draft genome sequence and genetic transformation of the oleaginous alga *Nannochloropsis gaditana*. Nat Commun. 2012;3:686.

22. Park JJ, Wang H, Gargouri M, Deshpande RR, Skepper JN, Holguin FO, Juergens MT, Shachar-Hill Y, Hicks LM, Gang DR. The response of *Chlamydomonas reinhardtii* to nitrogen deprivation: a systems biology analysis. Plant J. 2015;81:611–24.

23. Sun D, Zhu J, Fang L, Zhang X, Chow Y, Liu J. De novo transcriptome profiling uncovers a drastic downregulation of photosynthesis upon nitrogen deprivation in the nonmodel green alga *Botryosphaerella sudeticus*. BMC Genom. 2013;14:715.

24. Tsai C-H, Warakanont J, Takeuchi T, Sears BB, Moellering ER, Benning C. The protein compromised hydrolysis of triacylglycerols 7 (CHT7) acts as a repressor of cellular quiescence in Chlamydomonas. Proc Natl Acad Sci. 2014;111:15833–8.

25. Tunçay H, Findinier J, Duchêne T, Cogez V, Cousin C, Peltier G, Ball SG, Dauvillée D. A forward genetic approach in *Chlamydomonas reinhardtii* as a strategy for exploring starch catabolism. PLoS ONE. 2013;8:e74763.

26. Scranton MA, Ostrand JT, Fields FJ, Mayfield SP. Chlamydomonas as a model for biofuels and bio-products production. Plant J. 2015;82:523–31.

27. Bogen C, Al-Dilaimi A, Albersmeier A, Wichmann J, Grundmann M, Rupp O, Lauersen KJ, Blifernez-Klassen O, Kalinowski J, Goesmann A. Reconstruction of the lipid metabolism for the microalga *Monoraphidium neglectum* from its genome sequence reveals characteristics suitable for biofuel production. BMC Genom. 2013;14:926.

28. Jaeger D, Pilger C, Hachmeister H, Oberländer E, Wördenweber R, Wichmann J, Mussgnug JH, Huser T, Kruse O. Label-free in vivo analysis of intracellular lipid droplets in the oleaginous microalga *Monoraphidium neglectum* by coherent Raman scattering microscopy. Sci Rep. 2016;6:35340.

29. Jaeger D, Hübner W, Huser T, Mussgnug JH, Kruse O. Nuclear transformation and functional gene expression in the oleaginous microalga *Monoraphidium neglectum*. J Biotechnol. 2017;249:10–5.

30. Tsai CH, Zienkiewicz K, Amstutz CL, Brink BG, Warakanont J, Roston R, Benning C. Dynamics of protein and polar lipid recruitment during lipid droplet assembly in *Chlamydomonas reinhardtii*. Plant J. 2015;83:650–60.

31. Kong F, Liang Y, Légeret B, Beyly-Adriano A, Blangy S, Haslam RP, Napier JA, Beisson F, Peltier G, Li-Beisson Y. Chlamydomonas carries out fatty

acid β-oxidation in ancestral peroxisomes using a bona fide acyl-CoA oxidase. Plant J.2017

32. Johnson X, Alric J. Central carbon metabolism and electron transport in *Chlamydomonas reinhardtii*: metabolic constraints for carbon partitioning between oil and starch. Eukaryot Cell. 2013;12:776–93.

33. Hildebrand M, Abbriano RM, Polle JEW, Traller JC, Trentacoste EM, Smith SR, Davis AK. Metabolic and cellular organization in evolutionarily diverse microalgae as related to biofuels production. Curr Opin Chem Biol. 2013;17:506–14.

34. Siaut M, Cuiné S, Cagnon C, Fessler B, Nguyen M, Carrier P, Beyly A, Beisson F, Triantaphylidès C, Li-Beisson Y, Peltier G. Oil accumulation in the model green alga *Chlamydomonas reinhardtii*: characterization, variability between common laboratory strains and relationship with starch reserves. BMC Biotechnol. 2011;11:1–15.

35. Zhu S, Huang W, Xu J, Wang Z, Xu J, Yuan Z. Metabolic changes of starch and lipid triggered by nitrogen starvation in the microalga *Chlorella zofingiensis*. Biores Technol. 2014;152:292–8.

36. Juergens MT, Disbrow B, Shachar-Hill Y. The relationship of triacylglycerol and starch accumulation to carbon and energy flows during nutrient deprivation in *Chlamydomonas reinhardtii*. Plant Physiol. 2016;171:2445–57.

37. Merchant SS, Kropat J, Liu B, Shaw J, Warakanont J. TAG, you're it! Chlamydomonas as a reference organism for understanding algal triacylglycerol accumulation. Curr Opin Biotechnol. 2012;23:352–63.

38. Breuer G, Lamers PP, Martens DE, Draaisma RB, Wijffels RH. The impact of nitrogen starvation on the dynamics of triacylglycerol accumulation in nine microalgae strains. Bioresour Technol. 2012;124:217–26.

39. Marioni JC, Mason CE, Mane SM, Stephens M, Gilad Y. RNA-seq: an assessment of technical reproducibility and comparison with gene expression arrays. Genome Res. 2008;18:1509–17.

40. Wang L, Feng Z, Wang X, Wang X, Zhang X. DEGseq: an R package for identifying differentially expressed genes from RNA-seq data. Bioinformatics. 2010;26:136–8.

41. Muller C, Cacaci M, Sauvageot N, Sanguinetti M, Rattei T, Eder T, Giard JC, Kalinowski J, Hain T, Hartke A. The intraperitoneal transcriptome of the opportunistic pathogen *Enterococcus faecalis* in mice. PLoS ONE. 2015;10:e0126143.

42. Fan J, Ning K, Zeng X, Luo Y, Wang D, Hu J, Li J, Xu H, Huang J, Wan M, et al. Genomic foundation of starch-to-lipid switch in Oleaginous Chlorella spp. Plant Physiol. 2015;169:2444–61.

43. Toepel J, Illmer-Kephalides M, Jaenicke S, Straube J, May P, Goesmann A, Kruse O. New insights into *Chlamydomonas reinhardtii* hydrogen production processes by combined microarray/RNA-seq transcriptomics. Plant Biotechnol J. 2013;11:717–33.

44. Peng H, Wei D, Chen G, Chen F. Transcriptome analysis reveals global regulation in response to CO2 supplementation in oleaginous microalga Coccomyxa subellipsoidea C-169. Biotechnol Biofuels. 2016;9:151.

45. Irla M, Neshat A, Brautaset T, Ruckert C, Kalinowski J, Wendisch VF. Transcriptome analysis of thermophilic methylotrophic *Bacillus methanolicus* MGA3 using RNA-sequencing provides detailed insights into its previously uncharted transcriptional landscape. BMC Genom. 2015;16:73.

46. Krober M, Verwaaijen B, Wibberg D, Winkler A, Puhler A, Schluter A. Comparative transcriptome analysis of the biocontrol strain *Bacillus amyloliquefaciens* FZB42 as response to biofilm formation analyzed by RNA sequencing. J Biotechnol. 2016;231:212–23.

47. Trapnell C, Roberts A, Goff L, Pertea G, Kim D, Kelley DR, Pimentel H, Salzberg SL, Rinn JL, Pachter L. Differential gene and transcript expression analysis of RNA-seq experiments with TopHat and Cufflinks. Nat Protoc. 2012;7:562–78.

48. Kim D, Pertea G, Trapnell C, Pimentel H, Kelley R, Salzberg SL. TopHat2: accurate alignment of transcriptomes in the presence of insertions, deletions and gene fusions. Genome Biol. 2013;14:R36.

49. Trapnell C, Williams BA, Pertea G, Mortazavi A, Kwan G, Van Baren MJ, Salzberg SL, Wold BJ, Pachter L. Transcript assembly and quantification by RNA-Seq reveals unannotated transcripts and isoform switching during cell differentiation. Nat Biotechnol. 2010;28:511–5.

50. Hoff KJ, Lange S, Lomsadze A, Borodovsky M, Stanke M. BRAKER1: unsupervised RNA-Seq-based genome annotation with GeneMark-ET and AUGUSTUS. Bioinformatics. 2015;32:767–9.

51. DeRisi J, Penland L, Brown PO, Bittner ML, Meltzer PS, Ray M, Chen Y, Su YA, Trent JM. Use of a cDNA microarray to analyse gene expression patterns in human cancer. Nat Genet. 1996;14:457–60.
52. Schena M, Shalon D, Heller R, Chai A, Brown PO, Davis RW. Parallel human genome analysis: microarray-based expression monitoring of 1000 genes. Proc Natl Acad Sci USA. 1996;93:10614–9.
53. Kerkhoven EJ, Pomraning KR, Baker SE, Nielsen J. Regulation of amino-acid metabolism controls flux to lipid accumulation in Yarrowia lipolytica. npj Syst Biol Appl. 2016;2:16005.
54. Haas BJ, Papanicolaou A, Yassour M, Grabherr M, Blood PD, Bowden J, Couger MB, Eccles D, Li B, Lieber M, et al. De novo transcript sequence reconstruction from RNA-seq using the Trinity platform for reference generation and analysis. Nat Protocols. 2013;8:1494–512.
55. Love MI, Huber W, Anders S. Moderated estimation of fold change and dispersion for RNA-seq data with DESeq2. Genome Biol. 2014;15:550.
56. Anders S, Huber W. Differential expression analysis for sequence count data. Genome Biol. 2010;11:R106.
57. Goodson C, Roth R, Wang ZT, Goodenough U. Structural correlates of cytoplasmic and chloroplast lipid body synthesis in Chlamydomonas reinhardtii and stimulation of lipid body production with acetate boost. Eukaryot Cell. 2011;10:1592–606.
58. Moellering ER, Benning C. RNA interference silencing of a major lipid droplet protein affects lipid droplet size in Chlamydomonas reinhardtii. Eukaryot Cell. 2010;9:97–106.
59. Lintala M, Allahverdiyeva Y, Kangasjarvi S, Lehtimaki N, Keranen M, Rintamaki E, Aro EM, Mulo P. Comparative analysis of leaf-type ferredoxin-NADP oxidoreductase isoforms in Arabidopsis thaliana. Plant J. 2009;57:1103–15.
60. Munekage Y, Hashimoto M, Miyake C, Tomizawa K-I, Endo T, Tasaka M, Shikanai T. Cyclic electron flow around photosystem I is essential for photosynthesis. Nature. 2004;429:579–82.
61. Chen H, Hu J, Qiao Y, Chen W, Rong J, Zhang Y, He C, Wang Q. Ca2+ -regulated cyclic electron flow supplies ATP for nitrogen starvation-induced lipid biosynthesis in green alga. Sci Rep. 2015;5:15117.
62. Fuglede B, Topsoe F. Jensen-Shannon divergence and Hilbert space embedding. In International Symposium on Information Theory, 2004 ISIT 2004 Proceedings; 27 June–2 July 2004; 2004. p. 31.
63. Ashburner M, Ball CA, Blake JA, Botstein D, Butler H, Cherry JM, Davis AP, Dolinski K, Dwight SS, Eppig JT, et al. Gene Ontology: tool for the unification of biology. Nat Genet. 2000;25:25–9.
64. Marchler-Bauer A, Derbyshire MK, Gonzales NR, Lu S, Chitsaz F, Geer LY, Geer RC, He J, Gwadz M, Hurwitz DI, et al. CDD: NCBI's conserved domain database. Nucleic Acids Res. 2015;43:D222–6.
65. Merchant S, Prochnik S, Vallon O, Harris E, Karpowicz S, Witman G, Terry A, Salamov A, Fritz-Laylin L, Marechal-Drouard L, et al. The chlamydomonas genome reveals the evolution of key animal and plant functions. Science. 2007;318:245–50.
66. Vieler A, Wu G, Tsai C-H, Bullard B, Cornish AJ, Harvey C, Reca I-B, Thornburg C, Achawanantakun R, Buehl CJ. Genome, functional gene annotation, and nuclear transformation of the heterokont oleaginous alga Nannochloropsis oceanica CCMP1779. PLoS Genet. 2012;8:e1003064.
67. Bowler C, Allen AE, Badger JH, Grimwood J, Jabbari K, Kuo A, Maheswari U, Martens C, Maumus F, Otillar RP. The Phaeodactylum genome reveals the evolutionary history of diatom genomes. Nature. 2008;456:239–44.
68. Durrett TP, McClosky DD, Tumaney AW, Elzinga DA, Ohlrogge J, Pollard M. A distinct DGAT with sn-3 acetyltransferase activity that synthesizes unusual, reduced-viscosity oils in Euonymus and transgenic seeds. Proc Natl Acad Sci USA. 2010;107:9464–9.
69. Yoon K, Han D, Li Y, Sommerfeld M, Hu Q. Phospholipid:diacylglycerol acyltransferase is a multifunctional enzyme involved in membrane lipid turnover and degradation while synthesizing triacylglycerol in the unicellular green microalga Chlamydomonas reinhardtii. Plant Cell. 2012;24:3708–24.
70. Tardif M, Atteia A, Specht M, Cogne G, Rolland N, Brugiere S, Hippler M, Ferro M, Bruley C, Peltier G, et al. PredAlgo: a new subcellular localization prediction tool dedicated to green algae. Mol Biol Evol. 2012;29:3625–39.
71. Dörmann P. Galactolipids in plant membranes. New York: Wiley; 2001.
72. Li X, Benning C, Kuo M-H. Rapid triacylglycerol turnover in Chlamydomonas reinhardtii requires a lipase with broad substrate specificity. Eukaryot Cell. 2012;11:1451–62.
73. Li X, Moellering ER, Liu B, Johnny C, Fedewa M, Sears BB, Kuo MH, Benning C. A galactoglycerolipid lipase is required for triacylglycerol accumulation and survival following nitrogen deprivation in Chlamydomonas reinhardtii. Plant Cell. 2012;24:4670–86.
74. Murtagh F, Contreras P. Algorithms for hierarchical clustering: an overview. Wiley Interdiscip Rev: Data Mining Knowl Discov. 2012;2:86–97.
75. Rousseeuw PJ. Silhouettes: a graphical aid to the interpretation and validation of cluster analysis. J Comput Appl Math. 1987;20:53–65.
76. Handl J, Knowles J, Kell DB. Computational cluster validation in post-genomic data analysis. Bioinformatics. 2005;21:3201–12.
77. Fan J, Yan C, Andre C, Shanklin J, Schwender J, Xu C. Oil accumulation is controlled by carbon precursor supply for fatty acid synthesis in Chlamydomonas reinhardtii. Plant Cell Physiol. 2012;53:1380–90.
78. Doebbe A, Rupprecht J, Beckmann J, Mussgnug JH, Hallmann A, Hankamer B, Kruse O. Functional integration of the HUP1 hexose symporter gene into the genome of C. reinhardtii: impacts on biological H2 production. J Biotechnol. 2007;131:27–33.
79. BultÉ L, Wollman F-A. Evidence for a selective destabilization of an integral membrane protein, the cytochrome b6/f complex, during gametogenesis in Chlamydomonas reinhardtii. Eur J Biochem. 1992;204:327–36.
80. Lecler R, Godaux D, Vigeolas H, Hiligsmann S, Thonart P, Franck F, Cardol P, Remacle C. Functional analysis of hydrogen photoproduction in respiratory-deficient mutants of Chlamydomonas reinhardtii. Int J Hydrogen Energy. 2011;36:9562–70.
81. Hatch MD. C4 photosynthesis: discovery and resolution. Photosynth Res. 2002;73:251.
82. Chisti Y. Biodiesel from microalgae. Biotechnol Adv. 2007;25:294–306.
83. Morey JS, Monroe EA, Kinney AL, Beal M, Johnson JG, Hitchcock GL, Van Dolah FM. Transcriptomic response of the red tide dinoflagellate, Karenia brevis, to nitrogen and phosphorus depletion and addition. BMC Genom. 2011;12:346.
84. Juergens MT, Deshpande RR, Lucker BF, Park J-J, Wang H, Gargouri M, Holguin FO, Disbrow B, Schaub T, Skepper JN, et al. The regulation of photosynthetic structure and function during nitrogen deprivation in Chlamydomonas reinhardtii. Plant Physiol. 2015;167:558–73.
85. Qiao K, Imam Abidi SH, Liu H, Zhang H, Chakraborty S, Watson N, Kumaran Ajikumar P, Stephanopoulos G. Engineering lipid overproduction in the oleaginous yeast Yarrowia lipolytica. Metab Eng. 2015;29:56–65.
86. Goncalves EC, Wilkie AC, Kirst M, Rathinasabapathi B. Metabolic regulation of triacylglycerol accumulation in the green algae: identification of potential targets for engineering to improve oil yield. Plant Biotechnol J. 2016;14:1649–60.
87. Li-Beisson Y, Shorrosh B, Beisson F, Andersson MX, Arondel V, Bates PD, Baud S, Bird D, Debono A, Durrett TP, et al. Acyl-lipid metabolism. Arabidopsis Book. 2010;8:e0133.
88. Bates PD, Durrett TP, Ohlrogge JB, Pollard M. Analysis of acyl fluxes through multiple pathways of triacylglycerol synthesis in developing soybean embryos. Plant Physiol. 2009;150:55–72.
89. Jia J, Han D, Gerken HG, Li Y, Sommerfeld M, Hu Q, Xu J. Molecular mechanisms for photosynthetic carbon partitioning into storage neutral lipids in Nannochloropsis oceanica under nitrogen-depletion conditions. Algal Res. 2015;7:66–77.
90. Gibellini F, Smith TK. The Kennedy pathway—De novo synthesis of phosphatidylethanolamine and phosphatidylcholine. IUBMB Life. 2010;62:414–28.
91. Janßen HJ, Steinbüchel A. Fatty acid synthesis in Escherichia coli and its applications towards the production of fatty acid based biofuels. Biotechnol Biofuels. 2014;7:1–26.
92. Yao Z, Davis RM, Kishony R, Kahne D, Ruiz N. Regulation of cell size in response to nutrient availability by fatty acid biosynthesis in Escherichia coli. Proc Natl Acad Sci USA. 2012;109:E2561–8.
93. Tsay JT, Oh W, Larson TJ, Jackowski S, Rock CO. Isolation and characterization of the beta-ketoacyl-acyl carrier protein synthase III gene (fabH) from Escherichia coli K-12. J Biol Chem. 1992;267:6807–14.
94. Shintani DK, Ohlrogge JB. Feedback inhibition of fatty acid synthesis in tobacco suspension cells. Plant J. 1995;7:577–87.
95. Hwangbo K, Ahn J-W, Lim J-M, Park Y-I, Liu JR, Jeong W-J. Overexpression of stearoyl-ACP desaturase enhances accumulations of oleic acid

in the green alga *Chlamydomonas reinhardtii*. Plant Biotechnol Rep. 2014;8:135–42.

96. Li X, Zhang R, Patena W, Gang SS, Blum SR, Ivanova N, Yue R, Robertson JM, Lefebvre PA, Fitz-Gibbon ST, et al. An indexed, mapped mutant library enables reverse genetics studies of biological processes in *Chlamydomonas reinhardtii*. Plant Cell. 2016;28:367–87.

97. La Russa M, Bogen C, Uhmeyer A, Doebbe A, Filippone E, Kruse O, Mussgnug JH. Functional analysis of three type-2 DGAT homologue genes for triacylglycerol production in the green microalga *Chlamydomonas reinhardtii*. J Biotechnol. 2012;162:13–20.

98. Iwai M, Ikeda K, Shimojima M, Ohta H. Enhancement of extraplastidic oil synthesis in *Chlamydomonas reinhardtii* using a type-2 diacylglycerol acyltransferase with a phosphorus starvation–inducible promoter. Plant Biotechnol J. 2014;12:808–19.

99. Shtaida N, Khozin-Goldberg I, Boussiba S. The role of pyruvate hub enzymes in supplying carbon precursors for fatty acid synthesis in photosynthetic microalgae. Photosynth Res. 2015;125:407–22.

100. Avidan O, Brandis A, Rogachev I, Pick U. Enhanced acetyl-CoA production is associated with increased triglyceride accumulation in the green alga *Chlorella desiccata*. J Exp Bot. 2015;66:3725–35.

101. Avidan O, Pick U. Acetyl-CoA synthetase is activated as part of the PDH-bypass in the oleaginous green alga *Chlorella desiccata*. J Exp Bot. 2015;66:7287–98.

102. Nelson N, Ben-Shem A. The complex architecture of oxygenic photosynthesis. Nat Rev Mol Cell Biol. 2004;5:971–82.

103. Valledor L, Furuhashi T, Recuenco-Muñoz L, Wienkoop S, Weckwerth W. System-level network analysis of nitrogen starvation and recovery in *Chlamydomonas reinhardtii* reveals potential new targets for increased lipid accumulation. Biotechnol Biofuels. 2014;7:171.

104. Roessler PG. Environmental control of glycerolipid metabolism in microalgae: commercial implications and future research directions. J Phycol. 1990;26:393–9.

105. Vigeolas H, Waldeck P, Zank T, Geigenberger P. Increasing seed oil content in oil-seed rape (*Brassica napus* L.) by over-expression of a yeast glycerol-3-phosphate dehydrogenase under the control of a seed-specific promoter. Plant Biotechnol J. 2007;5:431–41.

106. Yao Y, Lu Y, Peng K-T, Huang T, Niu Y-F, Xie W-H, Yang W-D, Liu J-S, Li H-Y. Glycerol and neutral lipid production in the oleaginous marine diatom *Phaeodactylum tricornutum* promoted by overexpression of glycerol-3-phosphate dehydrogenase. Biotechnol Biofuels. 2014;7:110.

107. Muto M, Tanaka M, Liang Y, Yoshino T, Matsumoto M, Tanaka T. Enhancement of glycerol metabolism in the oleaginous marine diatom *Fistulifera solaris* JPCC DA0580 to improve triacylglycerol productivity. Biotechnol Biofuels. 2015;8:4.

108. Wynn JP, Ratledge C. Malic enzyme is a major source of NADPH for lipid accumulation by *Aspergillus nidulans*. Microbiology. 1997;143:253–7.

109. Evans CT, Ratledge C. Possible regulatory roles of ATP: citrate lyase, malic enzyme, and AMP deaminase in lipid accumulation by *Rhodosporidium toruloides* CBS 14. Can J Microbiol. 1985;31:1000–5.

110. Zhang Y, Adams IP, Ratledge C. Malic enzyme: the controlling activity for lipid production? Overexpression of malic enzyme in *Mucor circinelloides* leads to a 2.5-fold increase in lipid accumulation. Microbiology. 2007;153:2013–25.

111. Wase N, Black PN, Stanley BA, DiRusso CC. Integrated quantitative analysis of nitrogen stress response in *Chlamydomonas reinhardtii* using metabolite and protein profiling. J Proteome Res. 2014;13:1373–96.

112. Xue J, Niu Y-F, Huang T, Yang W-D, Liu J-S, Li H-Y. Genetic improvement of the microalga *Phaeodactylum tricornutum* for boosting neutral lipid accumulation. Metab Eng. 2015;27:1–9.

113. Xue J, Wang L, Zhang L, Balamurugan S, Li D-W, Zeng H, Yang W-D, Liu J-S, Li H-Y. The pivotal role of malic enzyme in enhancing oil accumulation in green microalga *Chlorella pyrenoidosa*. Microb Cell Fact. 2016;15:120.

114. Xiong W, Liu L, Wu C, Yang C, Wu Q. 13C-tracer and gas chromatography-mass spectrometry analyses reveal metabolic flux distribution in the oleaginous microalga *Chlorella protothecoides*. Plant Physiol. 2010;154:1001–11.

115. Polle J, Neofotis P, Huang A, Chang W, Sury K, Wiech E. Carbon partitioning in green algae (chlorophyta) and the enolase enzyme. Metabolites. 2014;4:612.

116. Warakanont J, Tsai C-H, Michel EJS, Murphy GR, Hsueh PY, Roston RL,

Sears BB, Benning C. Chloroplast lipid transfer processes in Chlamydomonas reinhardtii involving a TRIGALACTOSYLDIACYLGLYCEROL 2 (TGD2) orthologue. Plant J. 2015;84:1005–20.

117. Gong Y, Guo X, Wan X, Liang Z, Jiang M. Characterization of a novel thioesterase (PtTE) from *Phaeodactylum tricornutum*. J Basic Microbiol. 2011;51:666–72.

118. Tan KWM, Lee YK. Expression of the heterologous *Dunaliella tertiolecta* fatty acyl-ACP thioesterase leads to increased lipid production in *Chlamydomonas reinhardtii*. J Biotechnol. 2017;247:60–7.

119. Liu X, Sheng J, Curtiss R 3rd. Fatty acid production in genetically modified cyanobacteria. Proc Natl Acad Sci USA. 2011;108:6899–904.

120. Tang X, Feng H, Chen WN. Metabolic engineering for enhanced fatty acids synthesis in *Saccharomyces cerevisiae*. Metab Eng. 2013;16:95–102.

121. Trentacoste EM, Shrestha RP, Smith SR, Glé C, Hartmann AC, Hildebrand M, Gerwick WH. Metabolic engineering of lipid catabolism increases microalgal lipid accumulation without compromising growth. Proc Natl Acad Sci. 2013;110:19748–53.

122. Blazeck J, Hill A, Liu L, Knight R, Miller J, Pan A, Otoupal P, Alper HS. Harnessing *Yarrowia lipolytica* lipogenesis to create a platform for lipid and biofuel production. Nat Commun. 2014;5:3131.

123. Zabawinski C, Van Den Koornhuyse N, D'Hulst C, Schlichting R, Giersch C, Delrue B, Lacroix J-M, Preiss J, Ball S. Starchless mutants of *Chlamydomonas reinhardtii* lack the small subunit of a heterotetrameric ADP-glucose pyrophosphorylase. J Bacteriol. 2001;183:1069–77.

124. Krishnan A, Kumaraswamy GK, Vinyard DJ, Gu H, Ananyev G, Posewitz MC, Dismukes GC. Metabolic and photosynthetic consequences of blocking starch biosynthesis in the green alga *Chlamydomonas reinhardtii* sta6 mutant. Plant J. 2015;81:947–60.

125. Ngan CY, Wong C-H, Choi C, Yoshinaga Y, Louie K, Jia J, Chen C, Bowen B, Cheng H, Leonelli L, et al. Lineage-specific chromatin signatures reveal a regulator of lipid metabolism in microalgae. Nat Plants. 2015;1:15107.

126. Ajjawi I, Verruto J, Aqui M, Soriaga LB, Coppersmith J, Kwok K, Peach L, Orchard E, Kalb R, Xu W, et al. Lipid production in *Nannochloropsis gaditana* is doubled by decreasing expression of a single transcriptional regulator. Nat Biotech. 2017.

127. Lauersen KJ, Baier T, Wichmann J, Wordenweber R, Mussgnug JH, Hubner W, Huser T, Kruse O. Efficient phototrophic production of a high-value sesquiterpenoid from the eukaryotic microalga *Chlamydomonas reinhardtii*. Metab Eng. 2016;38:331–43.

128. Xiong W, Lee T-C, Rommelfanger S, Gjersing E, Cano M, Maness P-C, Ghirardi M, Yu J. Phosphoketolase pathway contributes to carbon metabolism in cyanobacteria. Nat Plants. 2015;2:15187.

129. Fabris M, Matthijs M, Rombauts S, Vyverman W, Goossens A, Baart GJE. The metabolic blueprint of *Phaeodactylum tricornutum* reveals a eukaryotic Entner-Doudoroff glycolytic pathway. Plant J. 2012;70:1004–14.

130. Sanchez B, Zuniga M, Gonzalez-Candelas F, de los Reyes-Gavilan CG, Margolles a. Bacterial and eukaryotic phosphoketolases: phylogeny, distribution and evolution. J Mol Microbiol Biotechnol. 2010;18:37–51.

131. Xu P, Qiao K, Ahn WS, Stephanopoulos G. Engineering *Yarrowia lipolytica* as a platform for synthesis of drop-in transportation fuels and oleochemicals. Proc Natl Acad Sci. 2016;113:10848–53.

132. Provasoli L, McLaughlin J, Droop M. The development of artificial media for marine algae. Archiv für Mikrobiologie. 1957;25:392–428.

133. Andrews S. FastQC: a quality control tool for high throughput sequence data; 2010. https://www.bioinformatics.babraham.ac.uk/projects/fastqc/.

134. Roberts A, Trapnell C, Donaghey J, Rinn JL, Pachter L. Improving RNA-Seq expression estimates by correcting for fragment bias. Genome Biol. 2011;12:R22.

135. Conesa A, Gotz S, Garcia-Gomez JM, Terol J, Talon M, Robles M. Blast2GO: a universal tool for annotation, visualization and analysis in functional genomics research. Bioinformatics. 2005;21:3674–6.

136. Goodstein DM, Shu S, Howson R, Neupane R, Hayes RD, Fazo J, Mitros T, Dirks W, Hellsten U, Putnam N, Rokhsar DS. Phytozome: a comparative platform for green plant genomics. Nucleic Acids Res. 2012;40:D1178–86.

137. Berardini TZ, Reiser L, Li D, Mezheritsky Y, Muller R, Strait E, Huala E. The

Arabidopsis information resource: making and mining the "gold standard" annotated reference plant genome. Genesis. 2015;53:474–85.

138. R Core Team. R A language and environment for statistical computing. Vienna: R Core Team; 2016.

139. Alexa A, Rahnenfuhrer J. topGO: enrichment analysis for gene ontology. R package version 2260; 2016.

140. Streb S, Zeeman SC. Starch metabolism in Arabidopsis. Arabidopsis Book/American Society of Plant Biologists. 2012;10:e0160.

141. Smith AM, Zeeman SC, Smith SM. Starch degradation. Annu Rev Plant Biol. 2005;56:73–98.

142. Ball SG, Deschamps P. Chapter 1-starch metabolism A2-Harris, Elizabeth H. In: Stern DB, Witman GB, editors. The chlamydomonas sourcebook (Second Edition). London: Academic Press; 2009. p. 1–40.

143. Radakovits R, Jinkerson RE, Darzins A, Posewitz MC. Genetic engineering of algae for enhanced biofuel production. Eukaryot Cell. 2010;9:486–501.

144. Lucas A. amap: another multidimensional analysis package. 2014.

145. D'Haeseleer P. How does gene expression clustering work? Nat Biotech. 2005;23:1499–501.

146. Gibbons FD, Roth FP. Judging the quality of gene expression-based clustering methods using gene annotation. Genome Res. 2002;12:1574–81.

147. Warnes GR, Bolker B, Bonebakker L, Gentleman R, Liaw WHA, Lumley T, Maechler M, Magnusson A, Moeller S, Schwartz M, Venables B. gplots: Various R Programming Tools for Plotting Data; 2016.

148. Blatti JL, Beld J, Behnke CA, Mendez M, Mayfield SP, Burkart MD. Manipulating fatty acid biosynthesis in microalgae for biofuel through protein-protein interactions. PLoS ONE. 2012;7:e42949.

149. Kao PH, Ng IS. CRISPRi mediated phosphoenolpyruvate carboxylase regulation to enhance the production of lipid in *Chlamydomonas reinhardtii*. Bioresour Technol. 2017.

150. Wang C, Chen X, Li H, Wang J, Hu Z. Artificial miRNA inhibition of phosphoenolpyruvate carboxylase increases fatty acid production in a green microalga *Chlamydomonas reinhardtii*. Biotechnol Biofuels. 2017;10:91.

151. Deng X, Cai J, Li Y, Fei X. Expression and knockdown of the PEPC1 gene affect carbon flux in the biosynthesis of triacylglycerols by the green alga *Chlamydomonas reinhardtii*. Biotechnol Lett. 2014;36:2199–208.

152. Lumbreras V, Stevens DR, Purton S. Efficient foreign gene expression in Chlamydomonas reinhardtii mediated by an endogenous intron. Plant J. 1998;14:441–7.

153. Stevens DR, Purton S, Rochaix J-D. The bacterial phleomycin resistance geneble as a dominant selectable marker in Chlamydomonas. Mol General Genet MGG. 1996;251:23–30.

154. Goff L, Trapnell C, Kelley D. cummeRbund: analysis, exploration, manipulation, and visualization of Cufflinks high-throughput sequencing data. R package version 2160; 2013.

8

Propionic acid production from corn stover hydrolysate by *Propionibacterium acidipropionici*

Xiaoqing Wang[1], Davinia Salvachúa[1], Violeta Sànchez i Nogué[1], William E. Michener[1], Adam D. Bratis[1], John R. Dorgan[2] and Gregg T. Beckham[1*] (iD)

Abstract

Background: The production of value-added chemicals alongside biofuels from lignocellulosic hydrolysates is critical for developing economically viable biorefineries. Here, the production of propionic acid (PA), a potential building block for C3-based chemicals, from corn stover hydrolysate is investigated using the native PA-producing bacterium *Propionibacterium acidipropionici*.

Results: A wide range of culture conditions and process parameters were examined and experimentally optimized to maximize titer, rate, and yield of PA. The effect of gas sparging during fermentation was first examined, and N_2 was found to exhibit improved performance over CO_2. Subsequently, the effects of different hydrolysate concentrations, nitrogen sources, and neutralization agents were investigated. One of the best combinations found during batch experiments used yeast extract (YE) as the primary nitrogen source and NH_4OH for pH control. This combination enabled PA titers of 30.8 g/L with a productivity of 0.40 g/L h from 76.8 g/L biomass sugars, while successfully minimizing lactic acid production. Due to the economic significance of downstream separations, increasing titers using fed-batch fermentation was examined by changing both feeding media and strategy. Continuous feeding of hydrolysate was found to be superior to pulsed feeding and combined with high YE concentrations increased PA titers to 62.7 g/L and improved the simultaneous utilization of different biomass sugars. Additionally, applying high YE supplementation maintains the lactic acid concentration below 4 g/L for the duration of the fermentation. Finally, with the aim of increasing productivity, high cell density fed-batch fermentations were conducted. PA titers increased to 64.7 g/L with a productivity of 2.35 g/L h for the batch stage and 0.77 g/L h for the overall process.

Conclusion: These results highlight the importance of media and fermentation strategy to improve PA production. Overall, this work demonstrates the feasibility of producing PA from corn stover hydrolysate.

Keywords: Lignocellulosic biomass, Fermentation, Biochemicals, Biorefinery, Organic acids

Background

Lignocellulosic biomass is a promising feedstock for the production of sustainable biofuels [1] and building block chemicals [2]. Indeed, manufacturing high-value biobased chemicals alongside biofuels can both reduce reliance on petroleum-based feedstocks and simultaneously de-risk the financial viability of the integrated biorefinery [3–7]. Propionic acid (PA), an aliphatic C3 carboxylic acid, is a potentially promising value-added chemical that can be produced from lignocellulosic sugars. PA is a preservative and chemical intermediate in the food, pharmaceutical, and herbicide industries, and can serve as a potential building block for the production of various C3-based chemicals. Presently, PA is predominantly produced from fossil-based sources at industrial scales. However, recent efforts for PA production have focused on biological production using bacteria in the

*Correspondence: Gregg.Beckham@nrel.gov
[1] National Bioenergy Center, National Renewable Energy Laboratory, Golden, CO 80401, USA
Full list of author information is available at the end of the article

Propionibacterium genus, including *P. acidipropionici*, *P. freudenreichii*, and *P. shermanii*, which are able to metabolize a broad diversity of carbon sources and produce PA anaerobically. To date, multiple feedstocks have been explored in PA fermentation, including whey permeate [8–10], corn steep liquor (CSL) [11, 12], hydrolyzed corn meal [12, 13], glycerol [14–16], and Jerusalem artichoke tuber hydrolysate [17]. For PA production from lignocellulosic biomass, sugarcane bagasse hydrolysate [18, 19], enzymatically hydrolyzed aspen [20], and corncob molasses [21] have been used as substrates. In North America and elsewhere, corn stover is abundant and its utilization as a feedstock for PA production warrants investigation.

It is well known that PA is produced through the dicarboxylic acid pathway in *Propionibacteria* under anaerobic conditions [22, 23]. Unfortunately, this group of microorganisms simultaneously converts sugars to other carboxylic acids leading to a decreased PA yield. Succinic acid (SA) and acetic acid (AA) are the major byproducts [12–14, 16, 17, 24–26]. AA production from glucose in *Propionibacteria* is associated with redox balance, energy generation, and cell growth [22, 27]. Therefore, AA production is closely tied to substrate utilization rather than the process conditions. Lactic acid (LA) has also been previously demonstrated as an intermediate during PA fermentation [23, 28]. LA can be then converted to PA under glucose-limiting conditions by extending the fermentation time after glucose depletion. In addition, Stowers et al. demonstrated that 150 kPa of headspace pressure in the fermentor can maintain the LA titer under 3 g/L in batch fermentation [28].

Due to the strong product inhibition of acids, PA fermentations traditionally result in low volumetric productivity and titer [22]. Thus, some efforts have focused on in situ PA removal from the culture broth through extractive fermentation [11, 24, 29, 30], or development of high cell density (HCD) fermentation by means of cell immobilization [10, 12–14, 17, 19, 25] or through cell recycling via an external ultrafiltration system [9, 18]. Although these processes have been demonstrated as efficient approaches to alleviate inhibition and enhance PA productivity, the process complexity of controlling these bioreactors may ultimately limit their large-scale application [26]. On the other hand, densifying the concentration of the cell inoculum to conduct HCD fermentation is a simple and efficient way to reach high product titer and productivity. Recently, by increasing cell density, Stowers et al. reported a PA productivity to 2 g/L h from glucose [28]. Additionally, a PA productivity of 1.42 g/L h was reached from 50 g/L glycerol in HCD sequential batches [15]. Wang et al. also demonstrated a PA titer over 55 g/L with a productivity of 2.23 g/L h from pure glucose by conducting fed-batch HCD fermentation [26].

The aim of this work was to develop an efficient bioprocess for PA production using hydrolysate from lignocellulosic biomass—specifically from corn stover. The hydrolysate is produced via deacetylation, dilute acid pretreatment, and enzymatic hydrolysis (DDAPH) of corn stover, and includes sugars from both cellulose and hemicellulose fractions alongside various microbial inhibitors [e.g., acetate, furfural, and 5-hydroxymethylfurfural (HMF)]. PA production was first evaluated in mixed pure sugar streams containing inhibitors while N_2 or CO_2 was used to maintain an anaerobic environment. The feasibility of producing PA in DDAPH was subsequently examined in batch cultures, and the influence of different nitrogen sources and pH control reagents as well as byproduct accumulation were investigated. To increase PA titer, fed-batch fermentations using concentrated DDAPH in feed media were conducted and LA accumulation was also examined at different levels of nutrient supplementation. Finally, a fed-batch HCD fermentation mode was applied to further improve fermentation performance. This study demonstrates that an efficient, high-productivity, and high-titer process for PA production via fermentation of lignocellulosic hydrolysate is feasible.

Results and discussion

Evaluation of *P. acidipropionici* in mock DDAPH substrate and effects of CO_2 and N_2 on PA production

Dilute acid and hydrothermal pretreatment processes, such as the one utilized in the current study, often result in the formation of toxic compounds such as HMF, furfural, and acetic acid, and these compounds can inhibit microbial growth [31]. To evaluate the bacterial tolerance to such compounds, PA fermentation was first evaluated in mock DDAPH, which mimics the DDAPH composition, at an initial sugar concentration of ~60 g/L. Component concentrations in the mock DDAPH are provided in Table 1. Additionally, as *Propionibacterium* produces PA anaerobically, the effect of sparging N_2 or CO_2 on PA production during the fermentation was also analyzed. *P. acidipropionici* has been reported to be able to fix CO_2 in the conversion step of pyruvate to oxaloacetate through pyruvate carboxylate [32, 33], which may allow for CO_2 uptake and a corresponding PA yield enhancement during fermentation.

Figure 1 shows the fermentation performance in mock DDAPH with both CO_2 and N_2-sparging. Sparging CO_2 or N_2 does not significantly affect cell growth (Figs. 1a, 2a), and sparging CO_2 decreased sugar utilization rates (Fig. 1b). Figure 1b also shows that glucose, xylose, and arabinose can be used simultaneously by *P. acidipropionici*, with a higher glucose utilization rate than for the other sugars. As shown in Fig. 1c (and Fig. 2b), 26.4 g/L

Table 1 Composition of DDAPH at the different dilution folds and mock DDAPH utilized in the current study

Compounds (g/L)	100% DDAPH	84% DDAPH	70% DDAPH	50% DDAPH	40% DDAPH	30% DDAPH	Mock DDAPH
Total sugars	153.6	129	107.6	76.8	61.4	46.3	58.9
Glucose	89.1	74.8	62.4	44.5	35.6	27.0	37.0
Xylose	56.7	47.6	39.7	28.3	22.7	17.0	19.0
Arabinose	7.8	6.6	5.5	3.9	3.1	2.3	2.9
Inhibitors							
Acetic acid	1.8	1.5	1.3	0.9	0.7	0.5	0.81
Furfural	0.9	0.8	0.6	0.4	0.4	0.3	0.34
HMF	0.14	0.12	0.1	0.07	0.06	0.04	0.05

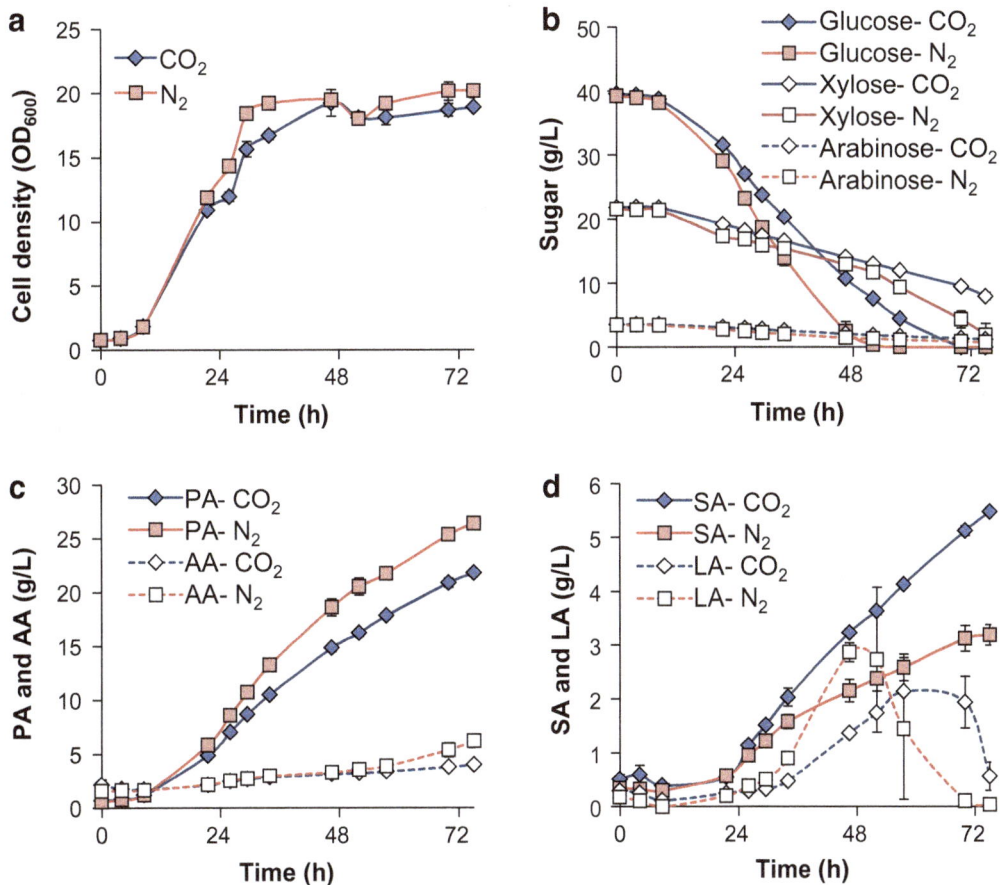

Fig. 1 Batch fermentation kinetics of *P. acidipropionici* under CO_2-sparging and N_2-sparging conditions in Mock DDAPH. Scatter plots show **a** cell growth (OD_{600}), **b** sugar utilization, **c** propionic and acetic acid production, and **d** succinic and lactic acid production. PA, propionic acid; AA, acetic acid; SA, succinic acid; LA, lactic acid

PA and 6.2 g/L AA were achieved at 75 h with N_2, compared to the 21.8 g/L PA and 5.5 g/L AA obtained under CO_2 sparging. The overall PA volumetric productivity (0.35 g/L h) and PA yields from sugars (0.45 g/g) under N_2 were 20.7 and 12.5% higher than those obtained under CO_2, respectively (Fig. 2c, d). Although the ability of *P.*

acidipropionici ATCC 4875 to fix CO_2 has been reported [32], the effect of CO_2 assimilation on PA production was demonstrated to be highly dependent on the carbon source used in the fermentation [33]. Namely, Zhang et al. reported that CO_2 exhibited minimal effect on PA production from glucose, but significantly enhanced cell

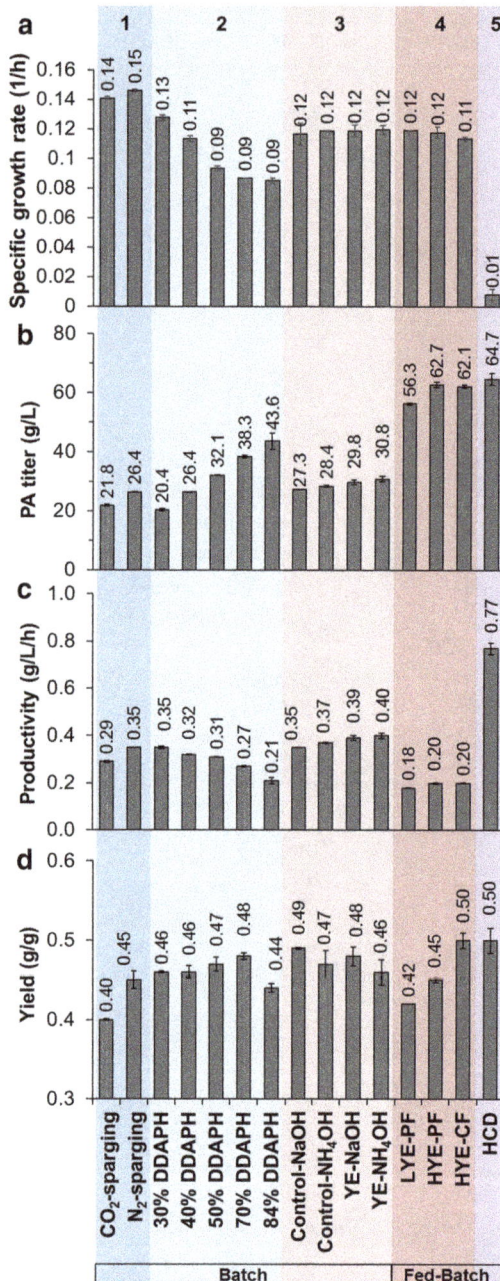

Fig. 2 Kinetics of propionic acid fermentations by *P. acidipropionici* from the different experiments conducted in the current study. **a** Specific growth rate (1/h), **b** PA maximum titers (g/L), **c** overall productivities (g/L h), and **d** yields (g PA/g sugars) from all of the fermentations conducted in this work. Color blocks highlight different experiments: (1) effect of CO_2 and N_2 on PA production in mock DDAPH substrates, (2) effect of different initial DDAPH concentrations on PA production, (3) effect of different nitrogen sources and pH control reagents on PA production, (4) evaluation of PA production in different fed-batch modes and feeding media, and (5) fed-batch fermentation with high cell density. YE, yeast extract; Control, YE + TSB; LYE-PF, 13.9 g/L YE in feeding medium and pulsed feeding; HYE-PF, 86.8 g/L YE in feeding medium and pulsed feeding; HYE-CF, 86.8 g/L YE in feeding medium and continuous feeding; HCD, high cell density

growth and PA and SA production when using glycerol as a sole carbon source [33]. In the current study, we observed reduced sugar utilization rates, lower PA titers, and lower PA productivity and yield from mixed sugars under CO_2-sparging than with N_2.

Propionibacteria natively produce other products, such AA, SA, and LA (vide supra). Since the AA production pathway can generate energy in the form of ATP and, together with PA production, achieve redox balance [22], it is expected and observed that AA is produced along with PA (Fig. 1c). In parallel, LA accumulation was observed during glucose utilization (Fig. 1d). However, when glucose was depleted, the LA concentration decreases to zero. Under N_2-sparging, the maximum LA concentration was 2.9 g/L (at 46.5 h) when the glucose concentration was 2.4 g/L. Consistent with the slower glucose consumption rate under CO_2-sparging, the maximum LA concentration (2.1 g/L) peaked later (57.5 h). Unlike the other acids, SA accumulation exhibited the opposite trend under the two conditions. After 26 h, SA production rates were enhanced with CO_2, and the final SA concentration was 71% higher with CO_2 than N_2. This type of increase in SA production when utilizing CO_2 or CO_2 donors has been previously demonstrated in other capnophilic and SA-producing organisms [34]. Overall, these results demonstrate that (i) *P. acidipropionici* tolerates some of the known primary inhibitors in corn stover hydrolysate and that (ii) using N_2 to maintain anaerobic conditions is advantageous for selective PA production. Thus, N_2 sparging was applied in the subsequent experiments.

Effect of different initial DDAPH concentrations on *P. acidipropionici*

PA production was subsequently evaluated utilizing different DDAPH concentrations from 30 to 84% (Table 1). The resulting fermentation data are shown in Fig. 3 and fermentation parameters are summarized in Fig. 2. The effect of minor lignocellulosic inhibitors (apart from furfural, HMF, and acetic acid) on the growth of *P. acidipropionici* can be observed by comparing the specific growth rates when cultivated on mock hydrolysate (Table 1), and diluted DDAPH (in this case 40%, to have the same initial concentration) (Fig. 2a). A reduction of 26.7% (from 0.15 1/h to 0.11 1/h) was observed, demonstrating the contribution of additional compounds, such as lignin-derived aromatics [4, 31], to the inhibitory level of lignocellulosic hydrolysate. While an expected decrease of specific growth rate occurred when increasing DDAPH content (Figs. 2a, 3a) due to both increasing inhibitors and sugar concentrations, a similar growth rate (0.09 1/h) was obtained from 50 to 84% DDAPH (Fig. 2a), highlighting the tolerance of *P. acidipropionici* towards hydrolysate.

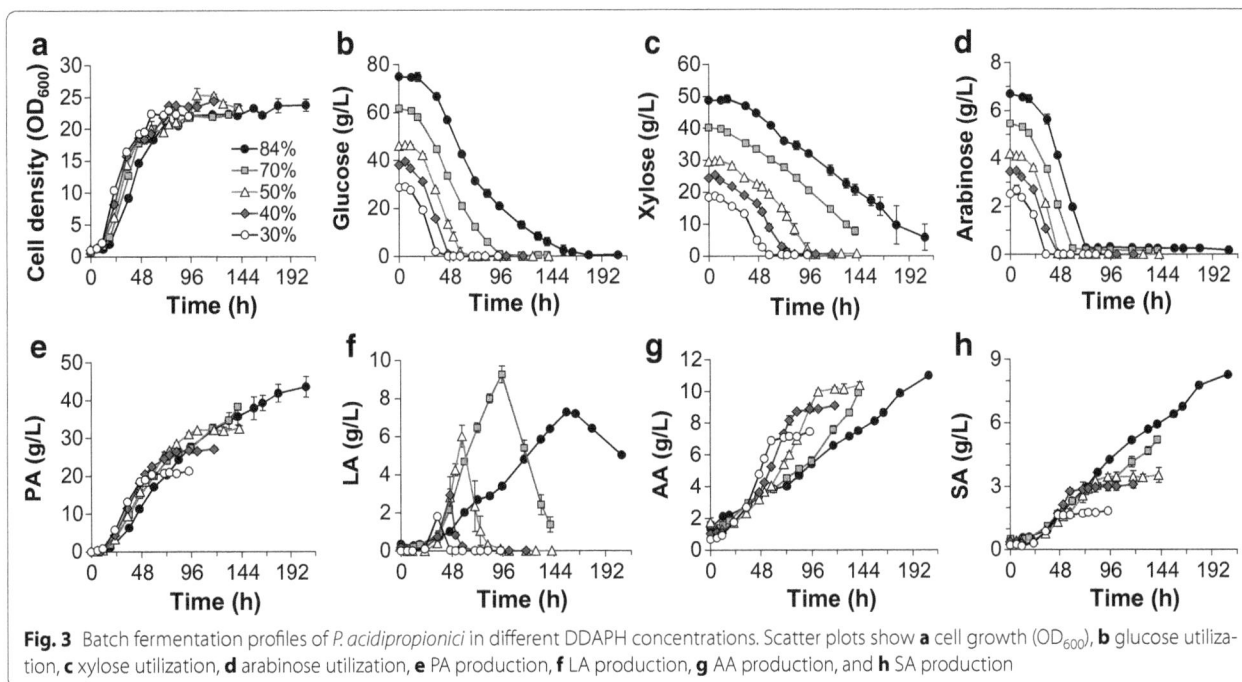

Fig. 3 Batch fermentation profiles of *P. acidipropionici* in different DDAPH concentrations. Scatter plots show **a** cell growth (OD$_{600}$), **b** glucose utilization, **c** xylose utilization, **d** arabinose utilization, **e** PA production, **f** LA production, **g** AA production, and **h** SA production

Rather than initial DDAPH concentration, a more severe inhibition on cell growth was observed from PA accumulation. At ~70 h, when the PA concentration reached >20 g/L, a concentration previously reported as inhibitory [35], cell growth ceased and entered a stationary phase (Fig. 3a). As shown in Fig. 3b–d, the consumption rates of each sugar were similar from 30 to 50% DDAPH, whereas decreased glucose and xylose utilization rates were observed at 70 and 84% DDAPH dilutions. Moreover, utilization of xylose was incomplete at the two highest DDAPH concentrations after 140 h of incubation.

PA production from different DDAPH media exhibited some noteworthy differences (Fig. 3e). After an 11-h lag, the PA concentrations increase almost linearly along with the exponential cell propagation. PA production slows and ceases when the sugars were completely consumed in 30–50% DDAPH. Conversely, the PA titer continues to increase in 70 and 84% DDAPH even though cells reach the stationary phase after ~70 h. The highest PA titer of 43 g/L was obtained at 207.5 h in 84% DDAPH. Overall, the PA production rate decreases with increasing DDAPH content. The highest PA volumetric productivity (0.35 g/L h) was achieved from 30% DDAPH medium, and the lowest (0.21 g/L h) at 84% DDAPH (Fig. 2c). PA yield ranged from 0.44 to 0.48 g/g (Fig. 2d), demonstrating that the hydrolysate concentration is not as critical for PA yield as for productivity.

In terms of other byproducts, the AA production rate was higher at lower DDAPH concentrations, which is consistent with the cell growth and PA production rates

(Fig. 3g). Significant amounts of LA and SA were also formed, especially with more concentrated DDAPH (Fig. 3f, h). However, LA was consumed after glucose was depleted.

For further fermentations, although 30% DDAPH gave the highest PA productivity, the PA titer was deemed too low for economically viable downstream separations. Since 50% DDAPH led to the same PA productivity and a higher PA titer relative to 40% DDAPH, and presented the highest productivity and complete utilization of sugars when compared to less diluted media, 50% diluted DDAPH was employed for subsequent fermentations.

Effect of different nitrogen sources and pH control reagents on *P. acidipropionici*

In addition to the carbon source, efficient PA fermentation by *Propionibacterium* requires nitrogen-containing media. Therefore, with 50% DDAPH hydrolysate, several nitrogen sources including an inorganic salt (($NH_4)_2SO_4$), processed yeast products (yeast extract, YE), a less expensive soy product (soytone), and a byproduct of the corn wet milling process (corn steep liquor, CSL) were screened in serum bottles. The serum bottles contained the same total N concentration (measured in g/L) and the combination of 10 g/L YE and 5 g/L TSB was used as a control. The results show that YE and the control media produce the two highest cell densities with YE medium being marginally better (Fig. 4a). In contrast, soytone, ($NH_4)_2SO_4$, and CSL produce limited cell biomass. The same trend was observed in sugar utilization rates, in

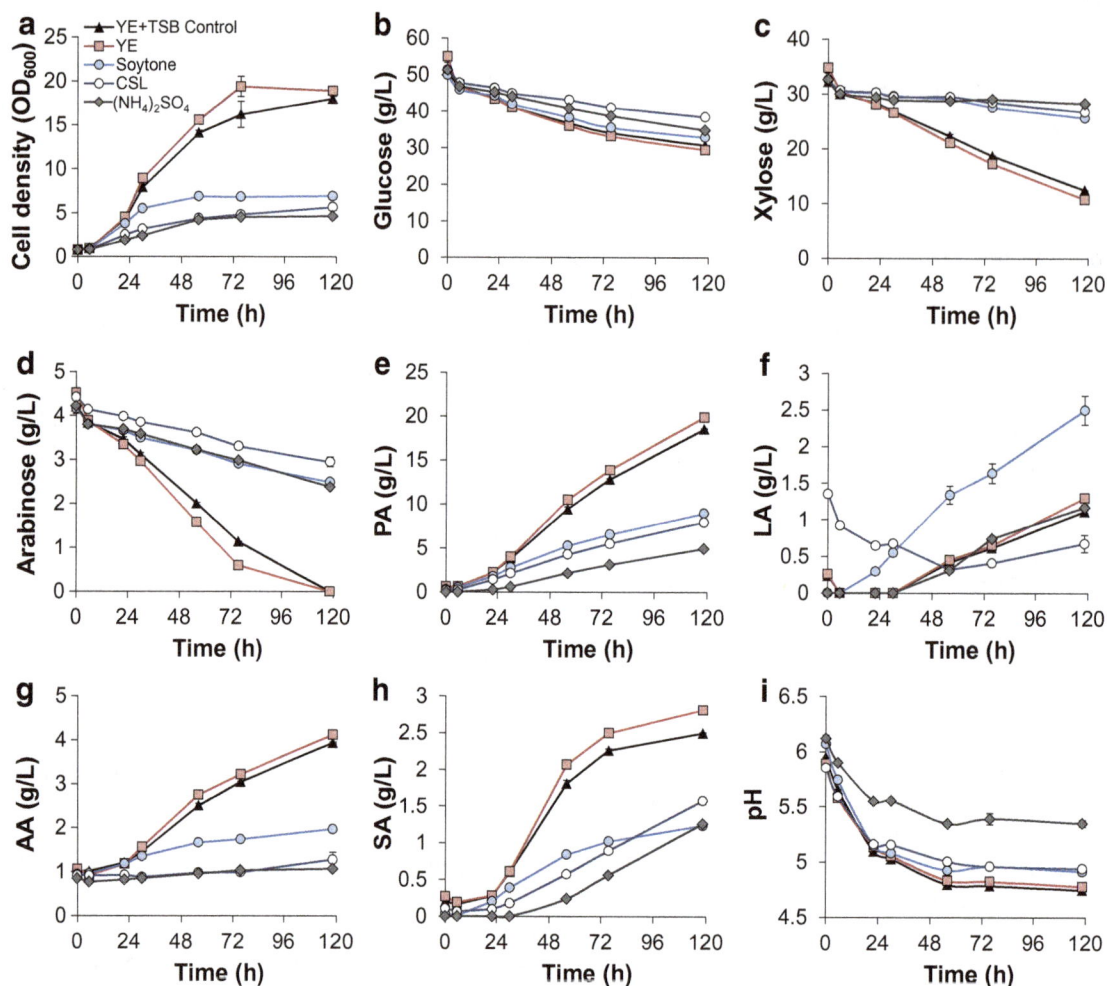

Fig. 4 Time course fermentation profiles of *P. acidipropionici* under different nitrogen sources in serum bottles. *Scatter plots* show **a** cell growth (OD$_{600}$), **b** glucose utilization, **c** xylose utilization, **d** arabinose utilization, **e** PA production, **f** LA production, **g** AA production, **h** SA production, **i** pH. The nitrogen source in the control case contains both YE and TSB

particular for xylose and arabinose (Fig. 4c, d). Figure 4c shows that over 61% xylose was consumed in YE and the control medium, while only a small amount of xylose was consumed in the other media. Moreover, arabinose was completely utilized in YE and the control media after 118.5 h, compared to the ~60% arabinose remaining in the other media (Fig. 4d). As a result, the highest acid titers were observed in YE and the control media except for LA (Fig. 4e–h). Unlike the other acids, the highest accumulation of LA was reached in soytone (Fig. 4f).

Generally, the lack of multiple nutrients, such as vitamins and/or various amino acids, can lead to poor fermentation performance, and this could be the reason for the low bacterial growth on inorganic nitrogen; however, bacterial performance in CSL media was similar. Although some references report PA production from

CSL, as both carbon and nitrogen substrates [11, 12], CSL only exhibits slight advantages over $(NH_4)_2SO_4$. Additionally, considering the high LA generation and lower PA production in soytone, this product was not further employed as a nitrogen source. As a result, YE and the control conditions (YE + TSB) were utilized as nitrogen sources going forward.

Given the lack of pH control in the serum bottle experiments, YE and control nitrogen sources were further evaluated in bioreactors at a controlled pH (6.0). NaOH [9, 12, 13, 16, 17, 19, 21, 24, 29, 33] and NH$_4$OH [8, 14, 15, 26, 28] are the most commonly used bases during PA fermentation; however, to our knowledge, no comparison of these two bases has been reported previously. As such, we investigated the effects of 4 M NaOH and 4 M NH$_4$OH on PA production in both YE and the control media. The combination of the control medium and 4 M NaOH also serves as control here, as these conditions were used in previous fermentations. The detailed fermentation data are shown in Fig. 5 and summarized in Fig. 2.

The fermentation profiles are similar under the different conditions before 29 h. Subsequently, differences became more obvious, particularly in LA production (Fig. 5f). LA accumulation occurs after 29 h and peaks at 52.5 h in all cultures. LA exhibits the highest concentrations (~7 g/L) in the control (TSB + YE) using both bases to adjust the pH. In contrast, the lowest LA accumulation (~1.5 g/L) occurs when using YE and NH_4OH.

It is worth noting that, at approximately 80 h, LA decreases to zero in all cases and thus, the yields are similar (Fig. 2). This finding is relevant for fed-batch fermentations since, if sugars are not maintained at low levels, LA will likely accumulate. Stowers et al. pressurized the reactor headspace at 150 kPa by using N_2 to prevent the LA titer from exceeding 3 g/L in a batch process, with the hypothesis that higher pressure increased the soluble CO_2 concentration in the culture, in turn enhancing CO_2 fixation by pyruvate carboxylase and converting pyruvate to PA instead of LA [28]. Although our results indicate that CO_2-sparging indeed decreases the maximum LA concentration during batch fermentation compared to N_2-sparging, the difference is less than 1 g/L (Fig. 1d). Moreover, CO_2-sparging exhibits adverse effects on PA production (Fig. 1c). Therefore, we instead decrease

LA production via use of YE as a sole nitrogen source and by replacing NaOH with NH_4OH for pH control, which results in the lowest observed LA accumulation. Finally, the maximum PA titer (30.8 g/L) and productivity (0.40 g/L h) were obtained when using YE and NH_4OH, which were 13 and 14% higher than the control results, respectively (Fig. 2).

Evaluation of PA production by *P. acidipropionici* in different feeding modes and media

The PA volumetric productivity was enhanced over 14% just by optimizing nitrogen sources and pH control reagents in batch mode (vida supra). However, high PA titer, which is critical for cost-effective recovery and purification downstream, is limited by the initial sugar concentration in batch operation. Therefore, fed-batch fermentation could increase PA titers and alleviate substrate inhibition.

One strategy to improve fed-batch fermentation performance is to optimize the nutrient profile of the feed medium. Concentrated nitrogen and sugar sources have been previously utilized as a feed for PA production in fed-batch fermentations [11, 12], whereas other studies only feed a concentrated carbon source [13, 17, 19, 24]. To our knowledge, no previous studies analyzed the influence of nitrogen concentration in the feed medium on PA production in fed-batch fermentation. Another strategy to improve fed-batch fermentation relies on the pattern of feed addition. The most common feeding

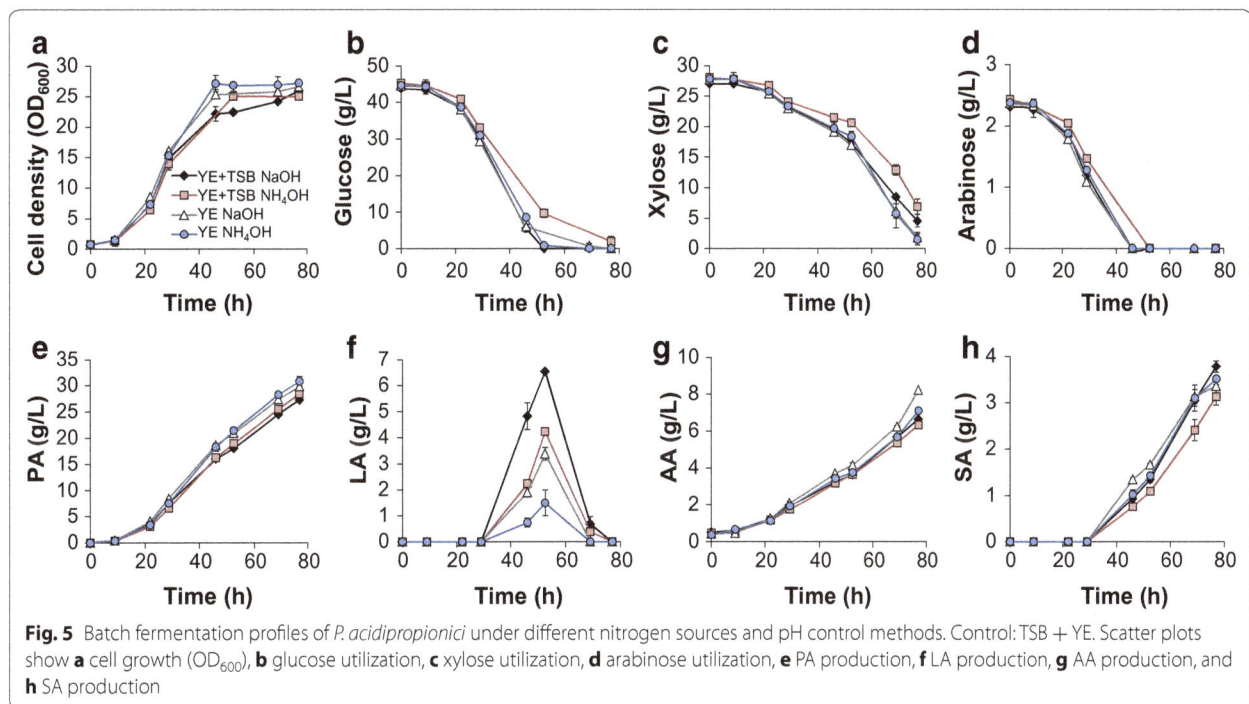

Fig. 5 Batch fermentation profiles of *P. acidipropionici* under different nitrogen sources and pH control methods. Control: TSB + YE. Scatter plots show **a** cell growth (OD$_{600}$), **b** glucose utilization, **c** xylose utilization, **d** arabinose utilization, **e** PA production, **f** LA production, **g** AA production, and **h** SA production

strategy applied in PA fermentation is pulsed feeding (PF) [11–13, 17, 19, 21, 24–26]. A more common approach for feeding control is continuous feeding (CF), wherein the carbon source is fed at similar rates as its utilization. To determine if the feed composition and/or feeding strategy could improve PA production, PF with concentrated DDAPH (500 g/L total sugars) and 13.9 g/L (low YE, or LYE) or 86.8 g/L YE (high YE, or HYE), and CF with concentrated DDAPH (500 g/L total sugar content) and 86.8 g/L YE (HYE) were compared. The fermentation data are shown in Fig. 6 and fermentation performance parameters are summarized in Fig. 2.

In the three cases, 50% DDAPH was used in the batch phase and the feeding started at 77 h when glucose was depleted. For PF, the cultures with a feed medium containing LYE and HYE media exhibit nearly identical sugar utilization profiles (Fig. 6b–d), indicating that differences in YE concentration during feeding do not affect sugar utilization. Additionally, an interesting phenomenon regarding sugar utilization was observed. Despite that the xylose consumption rate (0.25 g/L h) was much lower than the glucose consumption rate (0.63 g/L h) in the batch phase, no xylose accumulation occurred during the feeding phase. In fact, xylose consumption rates were comparable to those obtained with glucose in the three cases (Fig. 6c). It is well known that sequential utilization of different sugars is caused by carbon catabolite repression (CCR), which is generally related to a multiprotein phosphorelay system called the phosphotransferase

system (PTS) [36]. Although various sugars can be utilized by P. acidipropionici simultaneously, glucose is preferred to xylose (vida supra). Parizzi et al. suggested the presence of a CCR system by glucose in P. acidipropionici by identifying 23 genes in its genome, and also by comparing its growth rates in various carbon sources with and without hexose analogous 2-deoxy-glucose [32]. Moreover, it has been previously reported in Saccharomyces cerevisiae [37, 38] and Enterococcus mundtii [39] that glucose catabolite repression can be avoided by maintaining a low concentration of glucose. In this study, CCR was avoided by maintain the glucose concentration below 15 g/L, and as a consequence an enhanced xylose uptake rate by P. acidipropionici was observed during the feeding phase.

Although no differences in sugar utilization rates were observed between LYE and HYE media, HYE slightly enhanced the cell growth and PA production in both PF and CF. Unlike in batch cultures, in which the OD_{600} plateaued when the PA concentration was >20 g/L, the cell density continued increasing in all the fermentations after 55.5 h (when the PA concentration exceeded 21 g/L). However, cells in LYE medium reached the stationary phase earlier than cells in HYE media (91 and 167 h, respectively) and exhibit lower maximum OD_{600} (30.8 and 34, respectively) likely due to nutrient limitations (Fig. 6a).

HYE media also led to an increase of acids production, except for LA. At the end of fermentation, 62.1 and

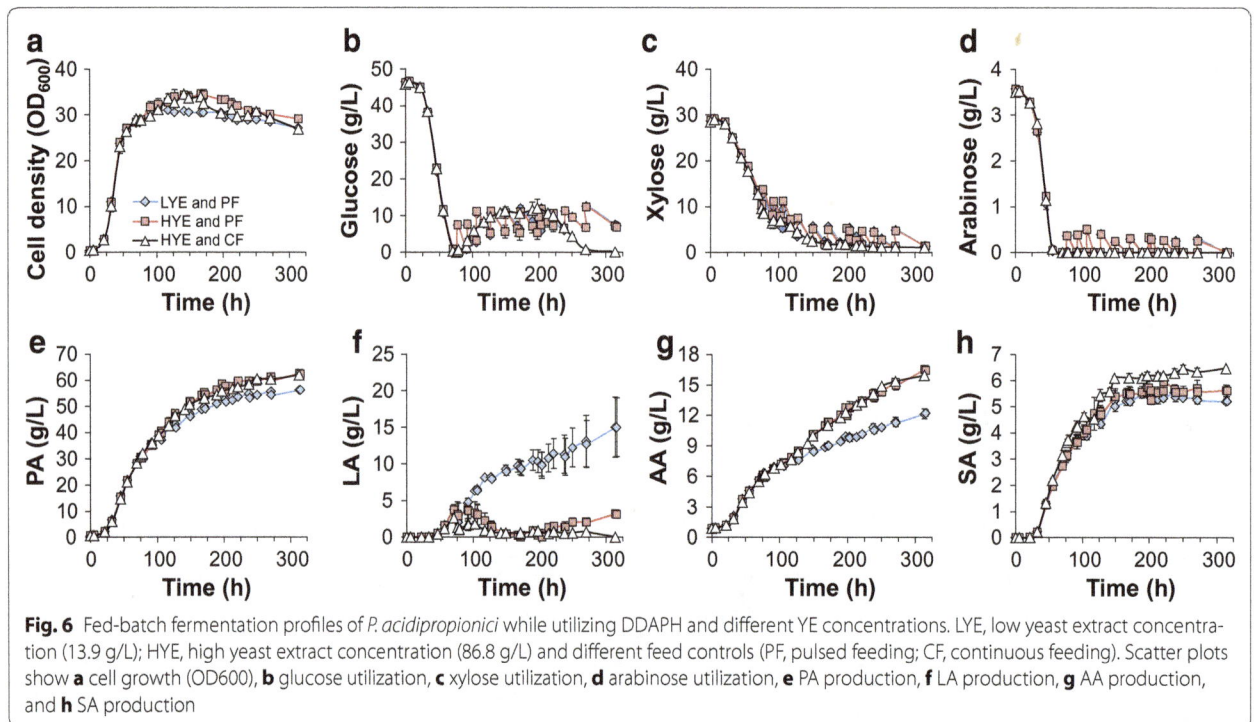

Fig. 6 Fed-batch fermentation profiles of P. acidipropionici while utilizing DDAPH and different YE concentrations. LYE, low yeast extract concentration (13.9 g/L); HYE, high yeast extract concentration (86.8 g/L) and different feed controls (PF, pulsed feeding; CF, continuous feeding). Scatter plots show **a** cell growth (OD600), **b** glucose utilization, **c** xylose utilization, **d** arabinose utilization, **e** PA production, **f** LA production, **g** AA production, and **h** SA production

62.7 g/L PA were achieved from HYE medium with PF and CF, respectively. These titers were 11% higher than that obtained in LYE medium (56.3 g/L) (Figs. 6e, 2b). Similar to PA production, both cultures grown using the HYE medium produced higher acid concentrations relative to LYE, specifically 34% for AA and 25 and 8.1% for SA (PF and CF, respectively) (Fig. 6g, h). Oppositely, LA concentrations in HYE medium were maintained below 4 g/L over the whole process (Fig. 6f), in contrast to 15 g/L titer in LYE medium. In all fermentations, LA production begun at 45.5 h and its consumption occurred after 70 h when glucose was depleted in the batch phase. At 91, 14 h after the beginning of the feeding phase, LA accumulation was observed again because of glucose feeding. However, during the following 80 h, the LA concentration in HYE medium neared zero in the presence of glucose, contrary to the increased LA accumulation in LYE medium. Moreover, by comparing LA concentrations between HYE CF and PF before 250 h, when glucose concentrations were above 4 g/L in both cultures, CF leads to a lower LA production.

Yang et al. evaluated different levels of organic nitrogen supplements (yeast extract and trypticase) on PA production from whey permeate in continuous cell-immobilized fermentation, and found that cell density, PA, and AA levels were higher when increasing nutrient levels, whereas SA exhibited the opposite trend [10]. It was also suggested that for long-term continuous fermentation, nutrient limitation could limit cell growth while maintaining cell activity and direct more carbon to target product [10]. In the current study instead, SA production was enhanced along with PA and AA in HYE medium. LA accumulation was significantly higher in LYE medium; however, this acid was not evaluated by Yang and co-workers.

Overall, by increasing the YE concentration in the feed medium and applying a CF strategy, PA titer, productivity, and yield were increased by 10, 11, and 14% compared to those from LYE PF culture, respectively (Fig. 2). Thus, HYE concentrations and CF mode were selected for the final experiments.

Fed-batch fermentation with high cell density (HCD)

Although high PA titer and limited LA accumulation were achieved via fed-batch cultivation, PA productivity was only 50% of that obtained in batch fermentation. With the purpose of increasing both titer and productivity simultaneously, fed-batch HCD fermentations were conducted. The detailed fermentation profiles are shown in Fig. 7 and results are summarized in Fig. 2. Bioreactors were inoculated at an initial OD_{600} of 91 (corresponding to a 23.5 g CDW/L). No lag was observed and the OD_{600} increases to over 100 at 20 h in the batch phase (Fig. 7a).

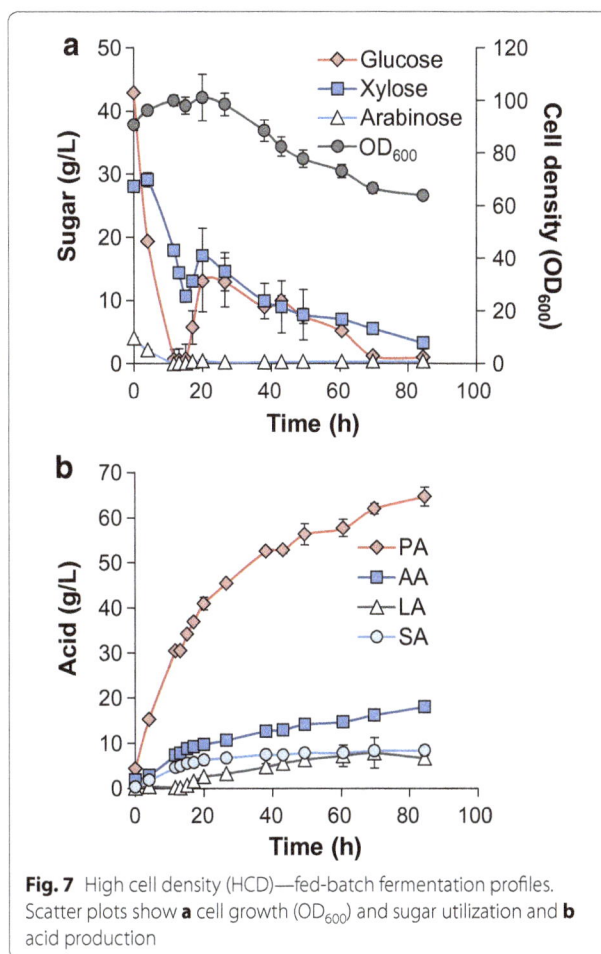

Fig. 7 High cell density (HCD)—fed-batch fermentation profiles. Scatter plots show **a** cell growth (OD_{600}) and sugar utilization and **b** acid production

Sugars were consumed in less than 20 h and, in the case of glucose and arabinose, complete consumption was observed within 11.5 h (Fig. 7a). When the feeding was initiated (at 15 h), the PA concentration was 30.5 g/L and the PA productivity was 2.35 g/L h (Fig. 2). The OD_{600} declines to 64 in the following 64.5 h, and the PA titer increases monotonically to 64.7 g/L at 84.5 h with an overall productivity of 0.77 g/L h (Fig. 2).

Additionally, compared to the previous fed-batch fermentation in HYE CF culture, a slight increase of other acids was observed in fed-batch HCD fermentation; however, PA yields were similar in both cases (0.5 g/g) (Fig. 2), likely due to the reduction of cell biomass production. Overall, the HCD fed-batch fermentation improves PA productivity 2.8-fold compared to the fed-batch process while reaching the same titer and yield (Fig. 2). Compared to the batch fermentation, fed-batch HCD fermentation improved PA titer and productivity by 110% and 92.5% (Fig. 2), respectively.

Three previous reports have employed *Propionibacterium* to produce PA from lignocellulose hydrolysate and are of interest for comparative purposes. Ramsay et al. reported a PA titer of 22.9 g/L in batch cultures from aspen with a productivity of 0.30 g/L h [20]; Liu et al. used corncob molasses to obtain a PA titer of 71.8 g/L and a productivity of 0.28 g/L h in a fed-batch fermentation [21], and Zhu et al. reported a PA concentration of 58.8 g/L at a productivity of 0.38 g/L h from cell-immobilized fed-batch fermentation in sugarcane bagasse hydrolysate [19]. Taken together, our results are among the highest PA productivity and titer reported to date from lignocellulosic feedstocks.

Conclusions

Here, we report anaerobic PA production by *P. acidipropionici* from corn stover, an abundant, industrially relevant, renewable lignocellulosic resource. Through a systematic investigation of different fermentation conditions and strategies, the PA titer and productivity were improved from 32.1 g/L and 0.31 g/L h from DDAPH batch fermentation to 64.7 g/L and 0.77 g/L h from DDAPH fed-batch HCD fermentation, respectively, roughly doubling both the titer and the productivity through changes to process conditions. Moreover, this work demonstrated the influence of nitrogen source and its concentration on LA accumulation, a major byproduct that has not been previously addressed in most of the publications describing PA fermentation.

Going forward, both the strain and the fermentation process could be further improved for industrial viability and integration into a lignocellulosic biorefinery. For instance, engineered *P. acidipropionici* with high acid tolerance [40] could be evaluated in lignocellulosic hydrolysates to enhance both PA titer and productivity. In addition, to reduce downstream process cost and increase yields, the metabolic pathway of LA production could be knocked out and industrial medium components could be used to replace the more costly laboratory-grade counterparts.

Methods

Hydrolysate preparation

Pilot-scale production of DDAPH was conducted as described by Shekiro et al. [41] with some noted modifications. Corn stover, provided by Idaho National Laboratory, was hammer-milled and filtered with a rejection screen. Milled fibers were then deacetylated using a dilute NaOH solution (0.4% w/w) at 80 °C for 2 h. Following deacetylation, remaining solids were rinsed with water and then mixed with dilute H_2SO_4 solution to achieve a 0.8% (w/w) acid concentration for dilute acid pretreatment. The slurry was thoroughly mixed for 2 h at room temperature, dewatered to approximately 40% solids, and then incubated in a horizontal pretreatment reactor at 160 °C with a residence time of 10 min. After pretreatment, the material was separated into the slurry stream with high solid content and volatile flash vent stream. Pretreated deacetylated dry slurry was neutralized using a 50% NaOH solution, and diluted with the addition of process water to 20% total solids. Novozymes Cellic® CTec2 was added to the slurry, and the mixture was incubated in a 130-L Jaygo paddle reactor at ~50 °C for 7 days under constant agitation. The sugar stream separated from lignin solids was then pH-adjusted to 6 and used as carbon substrate for PA fermentation.

Microorganism and media

Native *P. acidipropionici* ATCC 4875 was purchased from the American Type Culture Collection. The bacterium was initially revived in serum bottles under an anaerobic environment in the media detailed below. When the culture reached the exponential phase, it was preserved in 15% (v/v) glycerol stock and stored at −80 °C.

The bacterial medium contained 10 g/L yeast extract (YE) (BD Bacto™), 5 g/L tryptic soy broth (TSB) (BD™), 0.48 g/L K_2HPO_4 (Sigma-Aldrich), 0.98 g/L KH_2PO_4 (Sigma-Aldrich), and 0.05 g/L $MnSO_4$ (Sigma-Aldrich) with varying amounts and types of carbon sources (glucose, sugars from mock DDAPH, and sugars from DDAPH). Media supplemented with 40 g/L glucose were used for cell revival and the seed culture. Mock DDAPH, which is the first medium evaluated in the current study, mimics the composition of DDAPH. Mock DDAPH contained glucose, xylose, and arabinose, as major sugar components in DDAPH, at a total sugar concentration of 60 g/L but also well-known inhibitors found in DDAPH, such as furfural, HMF, and AA. The percentages of each compound in the mock DDAPH were based on the actual composition of DDAPH at 40% dilution (Table 1). Then, DDAPH at different concentrations was evaluated for PA fermentation. The sugar and inhibitor concentrations in different diluted DDAPH streams are also listed in Table 1. For fed-batch fermentations, DDAPH concentrated using a rotary evaporator (Hei-VAP value, Heidolph) was used as feed to a concentration of 500 g/L total sugars.

To optimize the nitrogen source, YE (5 g/L) and TSB (10 g/L) were directly replaced with different types of inorganic and organic nitrogen sources, including 7.1 g/L $(NH_4)_2SO_4$ (Sigma-Aldrich), 13.9 g/L yeast extract, 16.1 g/L soytone (BD Bacto™), and 16.4 g/L CSL (Sigma-Aldrich). The concentration of each nitrogen source was calculated based on its total N content reported by manufacturers to ensure that all media shared the same total N concentration (g/L). Mock-DDAPH and DDAPH substrates were sterilized using a Nalgene vacuum filtration system with a pore size of 0.2 μm. Nitrogen sources and salt components were autoclaved separately and aseptically combined to the target concentration.

Fermentation conditions for propionic acid production

Seed cultures, revived from glycerol stocks, were incubated in serum bottles at 30 °C for 48 h and then used to inoculate (7%, v/v) the fermentation medium. In order to wash metabolites and unconsumed sugars from the seed culture into the fermentation medium, cells were previously pelleted by centrifugation at 6000g for 10 min at 4 °C and then re-suspended in fresh medium to seed batch fermentations at an initial optical density at 600 nm (OD_{600}) of approximately 0.6. All fermentations (excluding seed culture propagation for HCD fermentations) were conducted in 500-mL fermentation vessels (BIOSTAT Qplus, Sartorius) containing 300 mL medium with constant agitation at 300 rpm. The bioreactor temperature was maintained at 30 °C and the pH at 6.0 by the addition of 4 M NaOH (Sigma-Aldrich) or 4 M NH_4OH (Sigma-Aldrich). Anaerobic conditions were established by sparging the medium with N_2 or CO_2 during the fermentation process at 0.2 vvm.

In fed-batch fermentation, when the initial total sugar concentration of ~80 g/L from 50% DDAPH was almost depleted, concentrated DDAPH containing 500 g/L total sugars and different concentrations of yeast extract were pulse-fed into the reactor. Continuous feeding was also performed by continuously pumping the concentrated hydrolysate into bioreactors. Sugar content was analyzed frequently and the pumping speed was adjusted to ensure that the sugar-feeding rate was equal to the sugar-consumption rate. Solution of 4 M NaOH or NH_4OH was used to maintain the pH at 6.0 and anaerobic conditions were maintained through N_2 sparging at 0.2 vvm.

For the fed-batch HCD fermentations, seed cultures were produced in 5-L bioreactors (New Brunswick) containing 3 L of bacterial medium supplemented with glucose as carbon source (40 g/L). Cultivations were performed under the same conditions as detailed before until culture reached mid to late exponential phase with OD_{600} of ~26. Cell pellets from 1.05 L seed culture were harvested by centrifugation and then resuspended in fresh DDAPH medium to seed 300 mL HCD fermentation medium, giving the initial OD_{600} values of ~91. The carbon sources in the batch and feed media were same with those used in fed-batch fermentation, and the concentrations of YE and salts were doubled. Continuous feed was applied in fed-batch HCD fermentation as detailed above. N_2 and 4 M NH_4OH were applied as previously described.

All the fermentations were performed in duplicate. Results are given as the average of the duplicate and error bars with standard error across duplicate fermentations.

Analytical methods

Cell growth was followed by measuring OD_{600} with a spectrophotometer. To eliminate color interferences from the dark DDAPH media, samples were centrifuged at 12,000 rpm for 5 min and then the cell-free supernatant was measured as the background. Dry cell weight (DCW) was determined as described by Wang et al. [26]. The cells were then collected by centrifugation, washed with distilled H_2O, and dried until constant DCW. A linear relationship between the value of OD_{600} and DCW was established by plotting OD_{600} as a function of DCW with one unit of OD_{600} equal to 0.2584 g/L DCW. The concentrations of sugars (glucose, xylose, and arabinose), organic acids (PA, AA, SA, LA), furfural and HMF were analyzed by an Agilent 1200 (Agilent Technologies, Santa Clara, CA) series HPLC system. Analytes were separated utilizing a Bio-Rad (BioRad, Hercules, CA) Aminex® HPX-87H, 9 μm × 300 mm × 7.8 mm Ion Exclusion Column (BioRad) at 55 °C with a mobile phase of 0.01 N H_2SO_4 at a flow rate of 0.6 mL/min. Individual standards were purchased from Sigma-Aldrich (Sigma-Aldrich). The levels of the calibration curve ranged from 0.05 to 50 mg/mL. A minimum of 5 calibration levels was used with an r^2 coefficient of 0.995 or better for each analyte. A check calibration standard was analyzed every 10 samples to insure the integrity of the initial calibration.

Calculation of PA yield and productivity

PA yield from the consumed sugars was calculated as the total PA produced (g) divided by total sugar consumed (g) at the end of fermentation. The PA overall volumetric productivity (g/L h) was calculated as the PA concentration (g/L) divided by the fermentation time (h).

Abbreviations

AA: acetic acid; CF: continuous feeding; CSL: corn steep liquor; DCW: dry cell weight; DDAPH: deacetylated, dilute acid pretreated and enzymatic hydrolyzed corn stover hydrolysate; HCD: high cell density; HMF: 5-hydroxymethylfurfural; HPLC: high-pressure liquid chromatography; HYE: high yeast extract concentration; LA: lactic acid; LYE: low yeast extract concentration; PA: propionic acid; PF: pulsed feeding; SA: succinic acid; TSB: tryptic soy broth; YE: yeast extract.

Authors' contributions

XW, DS, VSN, and GTB participated in the design of the study. XW conducted the experiments and analyzed the data. WEM conducted the full sugar, acids, and inhibitors analysis. XW wrote the first draft of the manuscript, and DS, VSN, JRD, and GTB revised the manuscript. ADB participated in the design, coordination, and funding support of this study. All authors read and approved the final manuscript.

Author details

[1] National Bioenergy Center, National Renewable Energy Laboratory, Golden, CO 80401, USA. [2] Chemical and Biological Engineering Department, Colorado School of Mines, Golden, CO 80401, USA.

Acknowledgements

We thank Robert Nelson and Marykate O'Brien for supplying hydrolysate and Xiaowen Chen, Brenna A. Black and Deborah Hyman for their input and help with hydrolysate component and fermentation metabolite analyses. We thank Ali Mohagheghi, Holly Smith, and Todd Vander Wall for their help during the fermentation experiments. We thank Eric Karp and Peter St. John for helpful discussions. The US Government retains and the publisher, by accepting the article for publication, acknowledges that the US Government retains a nonexclusive, paid up, irrevocable, worldwide license to publish or reproduce the published form of this work, or allow others to do so, for US Government purposes.

Competing interests

The authors declare that they have no competing interests.

Funding

This work was funded by the US Department of Energy BioEnergy Technologies Office (Grant Number DE-FOA-0000996).

References

1. Chundawat SP, Beckham GT, Himmel ME, Dale BE. Deconstruction of lignocellulosic biomass to fuels and chemicals. Chem Biomol Eng. 2011;2:121–45.
2. Gallezot P. Conversion of biomass to selected chemical products. Chem Soc Rev. 2012;41:1538–58.
3. Bozell JJ, Petersen GR. Technology development for the production of biobased products from biorefinery carbohydrates—The US Department of Energy's "top 10" revisited. Green Chem. 2010;12:539–54.
4. Salvachúa D, Mohagheghi A, Smith H, Bradfield MF, Nicol W, Black BA, Biddy MJ, Dowe N, Beckham GT. Succinic acid production on xylose-enriched biorefinery streams by Actinobacillus succinogenes in batch fermentation. Biotechnol Biofuels. 2016;9:28.
5. Salvachúa D, Smith H, St John PC, Mohagheghi A, Peterson DJ, Black BA, Dowe N, Beckham GT. Succinic acid production from lignocellulosic hydrolysate by Basfia succiniciproducens. Bioresour Technol. 2016;214:558–66.
6. Biddy MJ, Davis R, Humbird D, Tao L, Dowe N, Guarnieri MT, Linger JG, Karp EM, Salvachúa D, Vardon DR. The techno-economic basis for coproduct manufacturing to enable hydrocarbon fuel production from lignocellulosic biomass. ACS Sustain Chem Eng. 2016;4:3196–3211.
7. Davis R, Biddy MJ, Tan E, Tao L, Jones SB. Biological conversion of sugars to hydrocarbons technology pathway. Office of Scientific & Technical Information Technical Reports. 2013.
8. Blanc P, Goma G. Propionic acid fermentation: improvement of performances by coupling continuous fermentation and ultrafiltration. Bioprocess Eng. 1987;2:137–9.
9. Boyaval P, Corre C. Continuous fermentation of sweet whey permeate for propionic acid production in a CSTR with UF recycle. Biotech Lett. 1987;9:801–6.
10. Yang ST, Zhu H, Li Y, Hong G. Continuous propionate production from whey permeate using a novel fibrous bed bioreactor. Biotechnol Bioeng. 1994;43:1124–30.
11. Ozadali F, Glatz BA, Glatz CE. Fed-batch fermentation with and without on-line extraction for propionic and acetic acid production by Propionibacterium acidipropionici. Appl Microbiol Biotechnol. 1996;44:710–6.
12. Paik HD, Glatz BA. Propionic acid production by immobilized cells of

13. Huang YL, Wu Z, Zhang L, Cheung CM, Yang ST. Production of carboxylic acids from hydrolyzed corn meal by immobilized cell fermentation in a fibrous-bed bioreactor. Bioresour Technol. 2002;82:51–9.
14. Dishisha T, Alvarez MT, Hatti-Kaul R. Batch- and continuous propionic acid production from glycerol using free and immobilized cells of Propionibacterium acidipropionici. Bioresour Technol. 2012;118:553–62.
15. Dishisha T, Stahl A, Lundmark S, Hatti-Kaul R. An economical biorefinery process for propionic acid production from glycerol and potato juice using high cell density fermentation. Bioresour Technol. 2013;135:504–12.
16. Wang Z, Yang ST. Propionic acid production in glycerol/glucose co-fermentation by Propionibacterium freudenreichii subsp. shermanii. Bioresour Technol. 2013;137:116–23.
17. Liang ZX, Li L, Li S, Cai YH, Yang ST, Wang JF. Enhanced propionic acid production from Jerusalem artichoke hydrolysate by immobilized Propionibacterium acidipropionici in a fibrous-bed bioreactor. Bioprocess Biosyst Eng. 2012;35:915–21.
18. Crespo JPSG, Moura MJ, Carrondo MJT. Some engineering parameters for propionic acid fermentation coupled with ultrafiltration. Appl Biochem Biotechnol. 1990;24–25:613–25.
19. Zhu L, Wei P, Cai J, Zhu X, Wang Z, Huang L, Xu Z. Improving the productivity of propionic acid with FBB-immobilized cells of an adapted acid-tolerant Propionibacterium acidipropionici. Bioresour Technol. 2012;112:248–53.
20. Ramsay JA, Aly Hassan MC, Ramsay BA. Biological conversion of hemicellulose to propionic acid. Enzym Microb Technol. 1998;22:292–5.
21. Liu Z, Ma C, Gao C, Xu P. Efficient utilization of hemicellulose hydrolysate for propionic acid production using Propionibacterium acidipropionici. Bioresour Technol. 2012;114:711–4.
22. Wang Z, Sun J, Zhang A, Yang ST. Propionic acid fermentation. London: Wiley; 2013.
23. Choi CH, Mathews AP. Fermentation metabolism and kinetics in the production of organic acids by Propionibacterium acidipropionici. Appl Biochem Biotechnol. 1994;44:271–85.
24. Jin Z, Yang ST. Extractive fermentation for enhanced propionic acid production from lactose by Propionibacterium acidipropionici. Biotechnol Prog. 1998;14:457–65.
25. Feng XH, Chen F, Xu H, Wu B, Yao J, Ying HJ, Ouyang PK. Propionic acid fermentation by Propionibacterium freudenreichii CCTCC M207015 in a multi-point fibrous-bed bioreactor. Bioprocess Biosyst Eng. 2010;33:1077–85.
26. Wang Z, Jin Y, Yang ST. High cell density propionic acid fermentation with an acid tolerant strain of Propionibacterium acidipropionici. Biotechnol Bioeng. 2015;112:502–11.
27. Hsu ST, Yang ST. Propionic acid fermentation of lactose by Propionibacterium acidipropionici: effects of pH. Biotechnol Bioeng. 1991;38:571–8.
28. Stowers CC, Cox BM, Rodriguez BA. Development of an industrializable fermentation process for propionic acid production. J Ind Microbiol Biotechnol. 2014;41:837–52.
29. Lewis VP, Yang ST. A novel extractive fermentation for propionic production from whey lactose. Biotechnol Prog. 1992;8:104–10.
30. Yang ST, Huang H, Tay A, Qin W, Guzman L, San Nicolas EC. Extractive fermentation for the production of carboxylic acids. In: Yang ST, editor. Bioprocessing for value-added products from renewable resources. Amsterdam: Elsevier; 2007. p. 421–46.
31. Franden MA, Pilath HM, Mohagheghi A, Pienkos PT, Zhang M. Inhibition of growth of Zymomonas mobilis by model compounds found in lignocellulosic hydrolysates. Biotechnol Biofuels. 2013;6:1–15.
32. Parizzi LP, Grassi MCB, Llerena LA, Carazzolle MF, Queiroz VL, Lunardi I, Zeidler AF, Teixeira PJPL, Mieczkowski P, Rincones J, Pereira GAG. The genome sequence of Propionibacterium acidipropionici provides insights into its biotechnological and industrial potential. BMC Genom. 2012;13:562.
33. Zhang A, Sun J, Wang Z, Yang ST, Zhou H. Effects of carbon dioxide on cell growth and propionic acid production from glycerol and glucose by Propionibacterium acidipropionici. Bioresour Technol. 2014;175C:374–81.
34. Zou W, Zhu LW, Li HM, Tang YJ. Significance of CO_2 donor on the production of succinic acid by Actinobacillus succinogenes ATCC 55618. Microbial Cell Factories. 2011;10:87.

35. Woskow SA, Glatz BA. Propionic acid production by a propionic acid-tolerant strain of *Propionibacterium acidipropionici* in batch and semicontinuous fermentation. Appl Environ Microbiol. 1991;57:2821–8.

36. Wu Y, Shen X, Yuan Q, Yan Y. Metabolic engineering strategies for co-utilization of carbon sources in microbes. Bioengineering. 2016;3:10.

37. Ishola MM, Brandberg T, Taherzadeh MJ. Simultaneous glucose and xylose utilization for improved ethanol production from lignocellulosic biomass through SSFF with encapsulated yeast. Biomass Bioenerg. 2015;77:192–9.

38. Bertilsson M, Olofsson K, Lidén G. Prefermentation improves xylose utilization in simultaneous saccharification and co-fermentation of pretreated spruce. Biotechnol Biofuels. 2009;2:1–10.

39. Abdel-Rahman MA, Xiao Y, Tashiro Y, Wang Y, Zendo T, Sakai K, Sonomoto K. Fed-batch fermentation for enhanced lactic acid production from glucose/xylose mixture without carbon catabolite repression. J Biosci Bioeng. 2015;119:153–8.

40. Jiang L, Cui H, Zhu L, Hu Y, Xu X, Li S, Huang H. Enhanced propionic acid production from whey lactose with immobilized *Propionibacterium acidipropionici* and the role of trehalose synthesis in acid tolerance. Green Chem. 2015;17:250–9.

41. Shekiro J III, Kuhn EM, Nagle NJ, Tucker MP, Elander RT, Schell DJ. Characterization of pilot-scale dilute acid pretreatment performance using deacetylated corn stover. Biotechnol Biofuels. 2014;7:23.

9

Modeling and simulation of the redox regulation of the metabolism in *Escherichia coli* at different oxygen concentrations

Yu Matsuoka[1] and Hiroyuki Kurata[1,2]*

Abstract

Background: Microbial production of biofuels and biochemicals from renewable feedstocks has received considerable recent attention from environmental protection and energy production perspectives. Many biofuels and biochemicals are produced by fermentation under oxygen-limited conditions following initiation of aerobic cultivation to enhance the cell growth rate. Thus, it is of significant interest to investigate the effect of dissolved oxygen concentration on redox regulation in *Escherichia coli*, a particularly popular cellular factory due to its high growth rate and well-characterized physiology. For this, the systems biology approach such as modeling is powerful for the analysis of the metabolism and for the design of microbial cellular factories.

Results: Here, we developed a kinetic model that describes the dynamics of fermentation by taking into account transcription factors such as ArcA/B and Fnr, respiratory chain reactions and fermentative pathways, and catabolite regulation. The hallmark of the kinetic model is its ability to predict the dynamics of metabolism at different dissolved oxygen levels and facilitate the rational design of cultivation methods. The kinetic model was verified based on the experimental data for a wild-type *E. coli* strain. The model reasonably predicted the metabolic characteristics and molecular mechanisms of *fnr* and *arcA* gene-knockout mutants. Moreover, an aerobic–microaerobic dual-phase cultivation method for lactate production in a *pfl*-knockout mutant exhibited promising yield and productivity.

Conclusions: It is quite important to understand metabolic regulation mechanisms from both scientific and engineering points of view. In particular, redox regulation in response to oxygen limitation is critically important in the practical production of biofuel and biochemical compounds. The developed model can thus be used as a platform for designing microbial factories to produce a variety of biofuels and biochemicals.

Keywords: Kinetic modeling, Fermentation, Dissolved oxygen limitation, Redox regulation, ArcA, Fnr, Respiratory chain, NADH/NAD$^+$ ratio, *Escherichia coli*

Background

Microbial production of biofuels and biochemicals from renewable feedstocks has received considerable recent attention from environmental protection and energy production perspectives. A limited number of cell factory platforms have been employed for the industrial production of a wide range of fuels and chemicals. *Escherichia coli* is probably the most widely used cellular factory due

to its high growth rate and well-characterized physiology [1]. Many biofuels and biochemicals, such as ethanol and lactate, are produced by fermentation under oxygen-limited conditions. One method in particular, dual-phase cultivation method, combines the advantages afforded by aerobic and micro-aerobic (or anaerobic) conditions [2, 3]. In dual-phase processes, cultivation is initiated with an aerobic culture to increase the biomass (contributing to productivity), and it is followed by anaerobic or micro-aerobic cultivation to facilitate efficient production of the target product. It is, therefore, highly desirable to evaluate the metabolic characteristics at different dissolved

*Correspondence: kurata@bio.kyutech.ac.jp
[1] Department of Bioscience and Bioinformatics, Kyushu Institute of Technology, 680-4 Kawazu, Iizuka, Fukuoka 820-8502, Japan
Full list of author information is available at the end of the article

oxygen (DO) concentrations. For this purpose, appropriate quantitative models that can simulate such cultivations are needed.

Of the various modeling approaches currently available to cellular metabolism, flux balance analysis (FBA) approach has been extensively employed, but restricts to stoichiometric equations at the steady state, and thus it is difficult to simulate the dynamic changes in metabolic fluxes. On the other hand, a kinetic modeling approach can reproduce the dynamics of metabolite concentrations and fluxes in response to changes in genetic and environmental conditions [4], because it takes into account the mechanism of complex reactions such as allosteric modulation [5, 6], enzyme modification [7], and gene expression regulation by transcription factors (TFs) [8–11].

Of the various types of metabolic regulation, carbon catabolite regulation has been extensively modeled by a number of researchers [7, 12–19] to elucidate the mechanism of carbon uptake and metabolism. A detailed kinetic model of central carbon metabolism in *E. coli* that incorporates a constrained optimization method for parameter estimation on a supercomputer was recently developed [20]. As compared with other kinetic models, this model enabled more accurate prediction of the dynamics of wild-type (WT) cells and multiple-gene-knockout mutants in batch culture. However, from the perspective of practical applications to develop cellular factories for biofuel and biochemical production, the effect of oxygen limitation on redox regulation with carbon catabolite regulation is critically important. Considerable effort has been expended in this regard in the Systems Understanding of Microbial Oxygen Metabolism (SUMO) project [21–27].

For the proper modeling on the respiratory chain and the redox regulation, we have to consider the basic regulation mechanisms. Oxygen serves as the final electron acceptor of the respiratory chain [28]. In *E. coli*, two major oxidases, cytochrome *bo* (Cyo) and cytochrome *bd* (Cyd), transfer electrons from quinol to oxygen [29, 30]. Cyo has a low affinity for oxygen but a high reaction rate and functions primarily under aerobic conditions. By contrast, Cyd has high oxygen affinity but a lower reaction rate and functions primarily under micro-aerobic conditions. On the other hand, the dehydrogenases NADH dehydrogenase-I (Nuo) and NADH dehydrogenase-II (Ndh) oxidize electron donors such as NADH and $FADH_2$ by reducing quinone to quinol [30]. The function of the respiratory chain is the successive transport of electrons from electron donors to electron acceptors with translocation of protons from the cytoplasm to the periplasmic space via the inner membrane. The resulting proton gradient (proton motive force) drives ATP

synthesis. This series of reactions proceeds when oxygen is available, such as under aerobic or micro-aerobic conditions.

At limited oxygen concentrations, the transcription factors Fnr and ArcA/B play essential roles in metabolic regulation in *E. coli* [21]. The direct oxygen sensor Fnr regulates the expression of metabolic pathway genes under anaerobic conditions [31], whereas ArcA/B regulates these genes under both micro-aerobic and anaerobic conditions [32, 33]. The ArcA/B system is a two-component system: ArcB is a membrane-bound sensor kinase and ArcA is the cognate response regulator. ArcB autophosphorylates, and then *trans*-phosphorylates ArcA when oxygen is limited [34]. Phosphorylated ArcA in turn either activates or represses the expression of metabolic pathway genes. In addition, phosphorylated ArcA represses *cyoABCD*, which encodes Cyo, and activates *cydAB*, which encodes Cyd in the respiratory chain. Note that quinone inhibits the auto-phosphorylation of ArcB [35], which in turn represses the activity of ArcA.

The redox ratio (i.e., $NADH/NAD^+$) increases as the activity of the respiratory chain decreases in response to oxygen limitation. The excretion rates of fermentation products such as lactate, ethanol, succinate, formate (CO_2 and H_2 also), and acetate are influenced by this redox ratio. NADH is reoxidized to generate NAD^+ via these fermentative pathways to enable continuation of metabolism under micro-aerobic and anaerobic conditions. Lactate is formed by lactate dehydrogenase (LDH), whereas ethanol is formed by acetaldehyde dehydrogenase (ALDH) and alcohol dehydrogenase (ADH). Succinate is formed from phosphoenol pyruvate (PEP) via phosphoenolpyruvate carboxylase (Ppc) through the reverse pathway of the normal tricarboxylic acid (TCA) cycle from oxaloacetate to succinate, whereas the succinate dehydrogenase (SDH) pathway is reversed by fumarate reductase (Frd). Formate is formed by pyruvate formate-lyase (Pfl), and acetate is formed by phosphoacetyl transferase (Pta) and acetate kinase (Ack).

In the present study, we developed a kinetic model that describes the dynamics of the metabolism in response to different DO levels by taking into account the roles of transcription factors such as ArcA/B and Fnr, the respiratory chain reactions, the fermentative pathways as mentioned above, as well as catabolite regulation.

Methods
Modeling primary metabolism
Figure 1 shows the primary metabolic pathways of *E. coli*, including glycolysis, TCA cycle, pentose phosphate (PP), gluconeogenic, glyoxylate, and anaplerotic pathways, as well as the substrate transport system such as

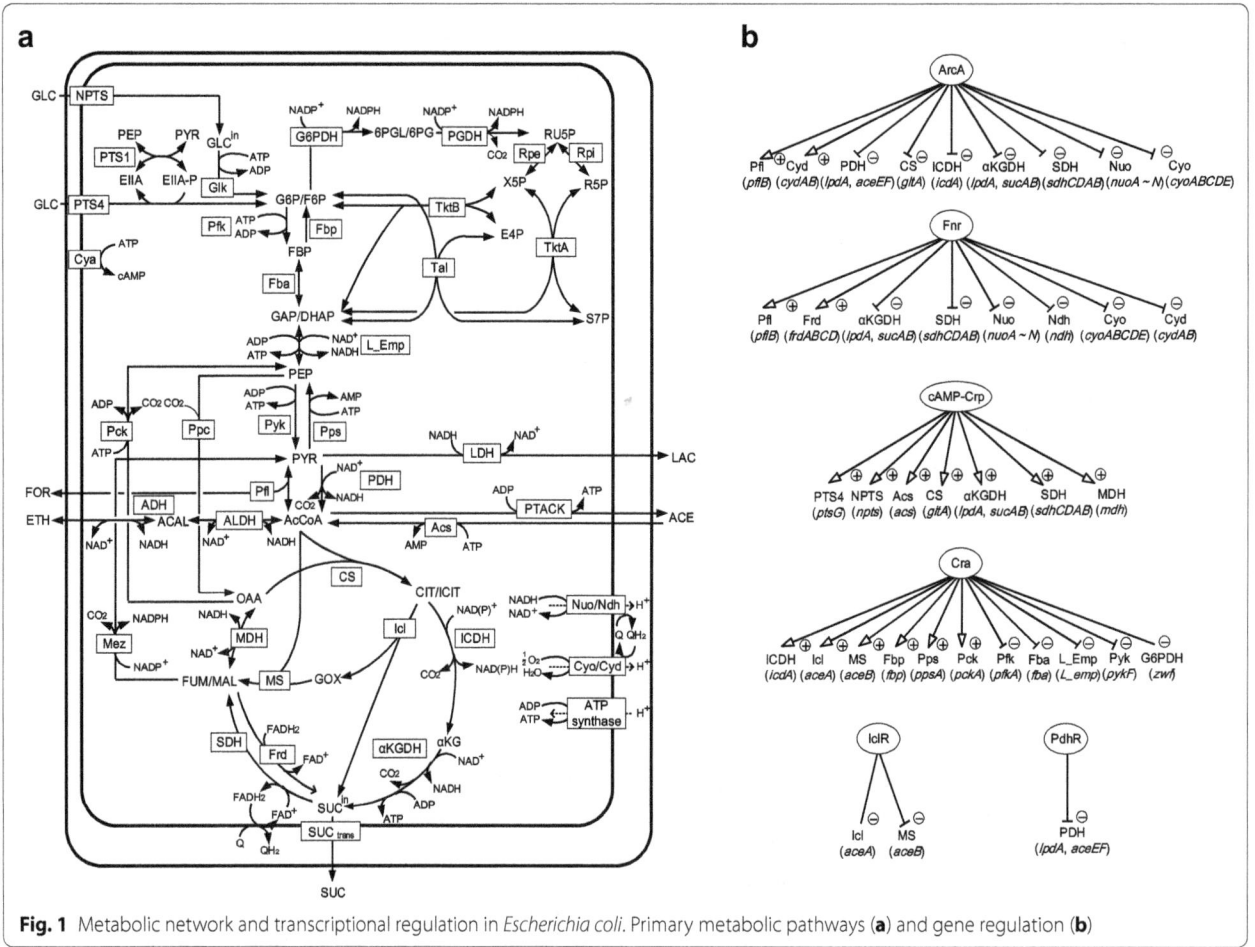

Fig. 1 Metabolic network and transcriptional regulation in *Escherichia coli*. Primary metabolic pathways (**a**) and gene regulation (**b**)

phosphotransferase system (PTS). Kinetic models of these pathways have been developed to investigate carbon uptake and metabolism under aerobic conditions [13–15, 17, 20]. Here, we constructed a kinetic model applicable under micro-aerobic (and anaerobic) conditions as well. For this purpose, the model incorporates additional fermentative pathways including such enzymes as LDH, Pfl, and ADH. We also considered respiratory chain mediators such as Nuo, Ndh, Cyo, and Cyd and redox regulation by Fnr and ArcA/B in response to changes in DO level. The detailed mass balance equations and the kinetic models are given in Additional file 1.

Once the overall metabolic fluxes of primary metabolism are calculated, the specific ATP, specific CO_2, and specific $NAD(P)H$ production rates can be estimated. ATP is produced via either substrate-level phosphorylation or oxidative phosphorylation. Referring to Fig. 1, the specific ATP production rate can be expressed as follows:

$$v_{ATP} = OP + v_{L\_Emp} + v_{Pyk} + v_{PTACK} + v_{\alpha KGDH}$$
$$- v_{Glk} - v_{Pfk} - v_{Pps} - v_{Pck} - v_{Acs}. \tag{1}$$

Note that L_Emp is the lumped pathway from glyceraldehyde-3-phosphate/dihydroxy acetone phosphate (GAP/DHAP) to PEP, and PTACK is the combined pathway for Pta and Ack (Fig. 1). In Eq. 1, OP represents the specific ATP production rate via oxidative phosphorylation, which can be estimated by introducing the H^+/ATP ratio, which indicates the ratio of proton transport-coupled ATP synthesis, where H^+/ATP = 3 [36], and calculating the proton transfer rate via Nuo, Cyo, and Cyd reactions, such that

$$OP = \frac{1}{3}\left(4 \cdot v_{Nuo} + 4 \cdot v_{Cyo} + 2 \cdot v_{Cyd}\right), \tag{2}$$

where the proton transfer efficiency, which is indicated as the number of protons delivered to the periplasmic side of the membrane per electron (H^+/e^- ratio), is taken into account for each enzyme. The H^+/e^- ratios for Nuo, Cyo, and Cyd can be estimated as 2, 2, and 1, respectively [28]. Note that NADH, which carries two electrons, is oxidized by dehydrogenases, and the electrons are subsequently transferred to cytochromes with subsequent conversion of oxygen to H_2O.

The specific growth rate (μ) was estimated based on the experimental observation that cell growth and specific ATP production rates are linearly correlated [13, 37, 38]:

$$\mu = k_{ATP} \cdot v_{ATP}, \tag{3}$$

where k_{ATP} represents the constant parameter, and v_{ATP} represents the specific ATP production rate computed using Eq. 1.

The specific CO_2 production rate can be estimated by

$$v_{CO_2} = v_{PGDH} + v_{PDH} + v_{ICDH} + v_{\alpha KGDH} + v_{Mez} + v_{Pck} - v_{Ppc}, \tag{4}$$

the specific NADPH production rate can be estimated by

$$v_{NADPH} = v_{G6PDH} + v_{PGDH} + v_{ICDH} + v_{Mez}, \tag{5}$$

and the specific production/consumption rates of NADH can be estimated by

$$v_{NADH} = v_{L\_Emp} + v_{PDH} + v_{\alpha KGDH} + v_{MDH} - v_{LDH} - v_{ALDH} - v_{ADH} - v_{Nuo} - v_{Ndh}. \tag{6}$$

To properly model primary metabolism, the metabolic regulation mechanisms must be incorporated. Enzyme-level regulation can be represented by incorporating the effectors (metabolites) into the corresponding kinetic models. For example, in *E. coli*, fructose-1,6-bisphosphate (FBP) is the feed-forward activator of pyruvate kinase (Pyk) and Ppc, whereas PEP is the feedback inhibitor of phosphofructokinase (Pfk). These effectors were incorporated in the corresponding kinetic models (Additional file 1).

Transcriptional regulation is also important and can be represented by the TFs, such that

$$v^{max} = v^{max'} \cdot f(TF_i), \tag{7}$$

where TF_i represents the activity of the ith transcription factor, and $v^{max'}$ represents the original maximum reaction rate for the corresponding pathway reaction. The detailed equations are given in Additional file 1.

In the redox regulation, Fnr and ArcA play important oxygen-dependent roles, with the activities of such TFs governed by cytoplasmic oxygen concentration $[O_2]$ and oxygenated quinone (after this, simply quinone) concentration $[Q]$, respectively. The activities of Fnr and ArcA can be expressed as Hill equation [39] as follows:

$$TF_{Fnr} = \frac{[O_2]^n}{[O_2]^n + K_{Fnr}^n}, \tag{8}$$

$$TF_{ArcA} = \frac{[Q]^n}{[Q]^n + K_{ArcA}^n}, \tag{9}$$

where K_{Fnr} and K_{ArcA} are the affinity constants and n is the negative Hill coefficient, and $[O_2]$ was defined by

$$[O_2] = k_{O_2}[DO_2], \tag{10}$$

where $[DO_2]$ is the dissolved oxygen concentration in the culture medium and k_{O_2} is the model parameter. Based on the experimental observation using biosensor [40], we assumed that $[O_2]$ is lower than $[DO_2]$. Figure 1 shows the effects of TFs on the primary metabolic pathways included in the present model. A "+" sign represents the case in which the transcription factor activates gene expression, whereas a "−" sign represents the case in which the TF represses gene expression. The gene name is written in brackets, where *npts* denotes the gene that codes for glucose transporters other than glucose-PTS, and *L_emp* denotes a hypothetical gene that codes for the lumped reactions through glyceraldehyde-3-phosphate dehydrogenase (GAPDH), phosphoglucokinase (Pgk), phosphoglucomutase (Pgm), and enolase (Eno). The DO concentration in the culture medium $[DO_2]$ was scaled from 0 to 100% and was defined as follows:

$$DO\ [\%] \equiv \frac{[DO_2]}{[DO_2]*} \times 100, \tag{11}$$

where $[DO_2]^*$ represents the saturated DO concentration at 37 °C. Thus, 0 and 100% DO levels represent the absence of oxygen (anaerobic condition) and DO saturation at 37 °C, respectively.

Model identification

Model parameters were adjusted so that the model can reproduce the experimental behavior of WT strain in the batch cultures under both micro-aerobic and aerobic conditions [41, 42], whereas other parameters, including the Michaelis–Menten and dissociation constants, were retained as those given in the references (Additional file 2). MATLAB (MathWorks) was used for all simulations. The ode15s was adopted as an ordinary differential equation solver.

Results
Experimental verification of the kinetic model

To verify the appropriateness of the kinetic model, the simulated extracellular concentrations of fermentation products were compared with the experimental data obtained from batch cultures under micro-aerobic (DO level = 1%) and aerobic (DO level = 40%) conditions [41, 42], as shown in Fig. 2. The concentrations of acetate, lactate, formate, ethanol, and succinate in the batch culture of the WT strain were plotted at the time points at which 10 g/l (micro-aerobic) and 4 g/l (aerobic) of glucose were depleted. The model reproduced most of the experimental product concentration under both micro-aerobic and aerobic conditions (Fig. 2). In addition, the model almost reproduced the experimental time courses of the WT strain under micro-aerobic (DO level = 9%) and aerobic conditions [43, 44], as shown in Additional file 3: Figure

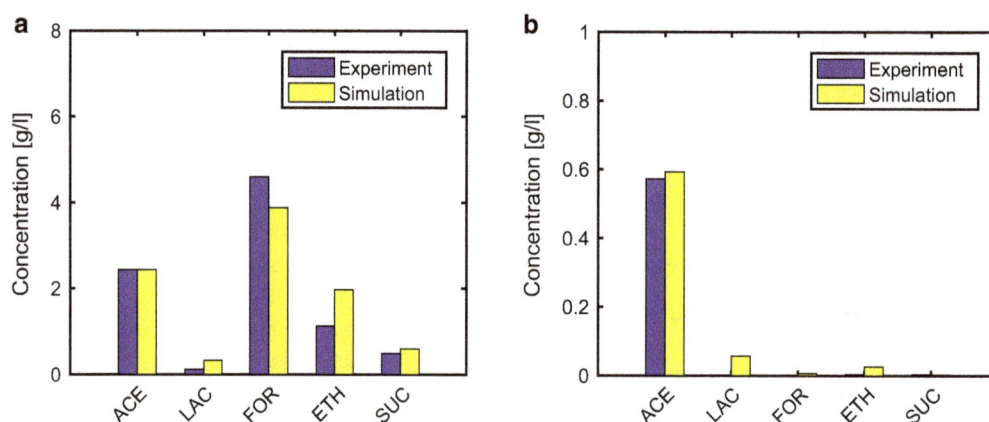

Fig. 2 Comparison of the simulation results with experimental data for wild-type *E. coli*. Micro-aerobic (**a**) and aerobic (**b**) conditions. The experimental data for micro-aerobic and aerobic conditions refer to those reported by Zhu and Shimizu [42] and Toya et al. [41], respectively. The DO levels were set to 1 and 40% for the simulation under micro-aerobic and aerobic conditions, respectively, where these conditions are comparable to the experimental conditions. The simulation results show the product concentrations at the time points at which 10 g/l (micro-aerobic) and 4 g/l (aerobic) of glucose were depleted in the batch cultures

S1. The correlation coefficient between the measured [41–44] and simulated metabolite concentrations was 0.98 ($p < 0.05$), as shown in Additional file 3: Figure S2.

Effect of DO level on the metabolic characteristics in the WT strain

Figure 3a shows the simulation result for a batch culture of the WT strain, in which the concentrations of acetate, lactate, formate, ethanol, and succinate were simulated at the time point at which 10 g/l of glucose was depleted. Acetate was the primary product at DO levels >15%. This acetate overflow was observed together with high CO_2 production at a high growth rate in *E. coli* [45, 46], as shown in Additional file 3: Figure S3. The acetate concentration increased with decrease in the DO level (3–14%), as observed experimentally [24]. As the DO level decreased, the formate, ethanol, and succinate concentrations increased; the lactate concentration increased and then steeply decreased below 5% of DO level. At DO levels <2%, the lactate concentration was lower than that of the other products as experimentally observed [42, 47].

Changes in the metabolism of the WT strain with respect to DO level

As shown in Fig. 3b, the changes in the concentrations of intracellular metabolites and fluxes were simulated with respect to DO level in the WT strain. The metabolic characteristics were evaluated by classifying the DO level into four categories: (I) anaerobic condition (DO = 0%) in which both Fnr and ArcA are active; (II) micro-aerobic conditions (0% < DO < 7%) in which both Fnr and ArcA are active; (III) micro-aerobic conditions

(7% ≤ DO < 20%) in which ArcA is primarily active and Fnr is inactive; and (IV) aerobic conditions (DO ≥ 20%) in which neither Fnr nor ArcA is active. TF_{Fnr} and TF_{ArcA} in Fig. 3b were calculated by Eqs. 8 and 9, respectively. The change in the typical carbon metabolism is illustrated for these categories in Additional file 3: Figure S4.

The specific oxygen uptake rate (qOUR), which indicates the rate of oxygen consumption via Cyo and Cyd reactions, was simulated to be high under condition IV, whereas it decreased under conditions III, II, and I (Fig. 3b). As the DO level decreased, the Cyd flux increased and then decreased at <7% DO. This up and down behavior can be attributed to the activation of Cyd synthesis by ArcA under condition III, whereas Cyd synthesis was repressed by Fnr under conditions I and II (Fig. 1). The Cyo flux was simulated to be higher than the Cyd flux under condition IV, whereas the Cyd flux was more dominant than the Cyo flux under condition II. These simulation results are supported by the experimental fact that the affinity of Cyd to oxygen is higher than that of Cyo [29]. Since quinone is produced by Cyo and Cyd, quinone decreases with a decrease in DO level. This phosphorylates ArcB and then ArcA, resulting in the increase in the ArcA activity under conditions I, II, and III.

Among the enzymes associated with consumption of pyruvate, Pfl, pyruvate dehydrogenase (PDH), and LDH, play critical roles in determining the metabolite formation pattern. As the DO level decreased, the Pfl flux was simulated to increase under condition III because ArcA activated the Pfl reaction (Fig. 1). The Pfl flux was further enhanced by Fnr and ArcA under conditions I and

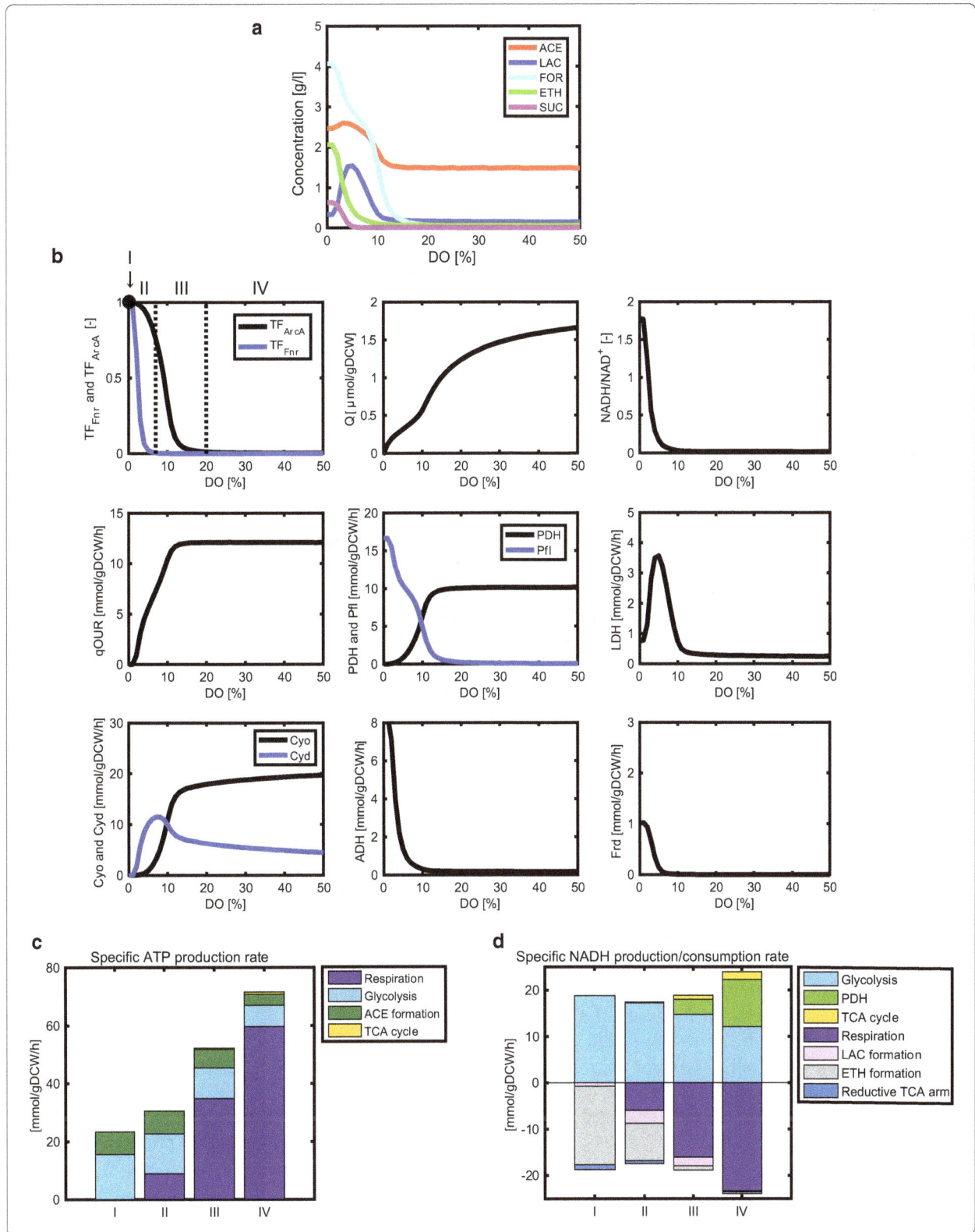

(See figure on previous page.)
Fig. 3 Simulation results of the metabolic changes with respect to DO level in wild-type *E. coli*. The changes in the metabolic (fermentation) products (**a**), and the activities of transcription factors, intracellular metabolites, and fluxes (**b**) with respect to DO level. The specific ATP production rate (**c**) and specific NADH production/consumption rates (**d**) are shown for the DO levels of 0, 3, 8, and 40% of air saturation, representatives of conditions I, II, III, and IV, respectively, where (I) anaerobic condition (DO = 0%); (II) micro-aerobic conditions under which both Fnr and ArcA are active; (III) micro-aerobic conditions under which ArcA is primarily active; and (IV) aerobic conditions under which neither Fnr nor ArcA is active. The simulation results show the product concentrations at the time point at which 10 g/l of glucose was depleted in a batch culture

II. In contrast, as the DO level decreased, the PDH flux decreased under condition III because ArcA represses the *aceE/F* genes that encode PDH (Fig. 1). The Pfl and PDH fluxes were both active under condition III, which was consistent with the experimental data [48]. The LDH flux exhibited an up and down behavior with respect to DO level. As the DO level decreased, the LDH flux increased under condition III, whereas it declined steeply under condition II.

The NADH/NAD$^+$ ratio increased steeply with decreasing DO level under conditions I and II because NADH is hard to be consumed by the NADH dehydrogenases in the respiratory chain. This simulation result was consistent with the experimental data [49]. A high NADH/NAD$^+$ ratio promoted the ADH reaction, which resulted in enhanced ethanol production under conditions I and II. Since Fnr activates the Frd flux (Fig. 1), succinate production was enhanced under conditions I and II.

To obtain a better understanding of the mechanisms by which ATP and NADH are produced or consumed under the categorized DO conditions examined, the specific production/consumption rates of ATP and NADH were simulated, as illustrated in Fig. 3c, d. DO levels of 0, 3, 8, and 40% were selected as the representatives of conditions I, II, III, and IV, respectively. The specific ATP production rate decreased in the order of conditions IV, III, II, and I (Fig. 3c). Additional file 3: Figure S5 indicates the relationship between the specific ATP production rate and the specific growth rate. Once the specific ATP production rate was calculated by Eq. 1, the specific growth rate was estimated by Eq. 3. This linear relationship between the specific ATP production rate and the specific growth rate held not only under aerobic conditions but also under micro-aerobic and anaerobic conditions with a correlation coefficient of 0.92 ($p < 0.05$), as experimentally observed [25, 37, 38, 41, 42, 47].

The DO level affected the specific ATP production rate (Fig. 3c). ATP was primarily synthesized by respiration under condition IV. By contrast, substrate-level phosphorylation by glycolysis and acetate formation became dominant under conditions I and II. NADH was consumed by the NADH dehydrogenases (Nuo and Ndh) in the respiratory chain under condition IV, whereas NADH was primarily consumed by ethanol formation under conditions I and II (Fig. 3d). This simulation result demonstrates that the ADH flux increased under conditions I and II due to a high NADH/NAD$^+$ ratio (Fig. 3b). In fact, it was experimentally shown that ethanol is produced under micro-aerobic and anaerobic conditions [42, 47, 50]. The reaction of reductive TCA arm via malate dehydrogenase (MDH) consumed NADH under conditions I and II (Fig. 3d). The resultant fumarate/malate were supplied as the substrates for the reaction of Fnr-activated Frd (Fig. 3b), producing succinate (Fig. 3a). This simulation result was consistent with the experimental observation [47, 50]. NADH was produced by glycolysis, the PDH reaction, and the TCA cycle under condition IV (Fig. 3d), whereas NADH production by the PDH flux and TCA cycle declined significantly under conditions I and II because ArcA represses the PDH flux and both ArcA and Fnr repress the TCA cycle.

Additional file 3: Figure S3A shows the carbon balances of the extracellular products, CO_2, and biomass at different DO levels (conditions I, II, III, and IV) in the WT strain. The metabolic modes changed significantly depending on DO level. Most of glucose was converted to biomass, CO_2, and acetate under condition IV. On the other hand, biomass and CO_2 production were decreased under condition I.

Prediction of the metabolic characteristics of an *fnr*-knockout mutant

As Fnr and ArcA play critical roles in redox regulation at low DO levels, it is of interest to predict the effect of

(See figure on next page.)
Fig. 4 Simulation results of the metabolic changes with respect to DO level in *fnr*-knockout mutant. The changes in the metabolic (fermentation) products (**a**), and the activities of transcription factors, intracellular metabolites, and fluxes (**b**) with respect to DO level. The specific ATP production rate (**c**) and specific NADH production/consumption rates (**d**) are shown for the DO levels of 0, 3, 8, and 40% of air saturation, representatives of conditions I, II, III, and IV, respectively. The simulation results show the product concentrations at the time point at which 10 g/l of glucose was depleted in a batch culture

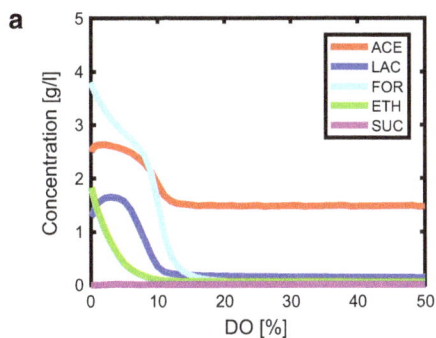

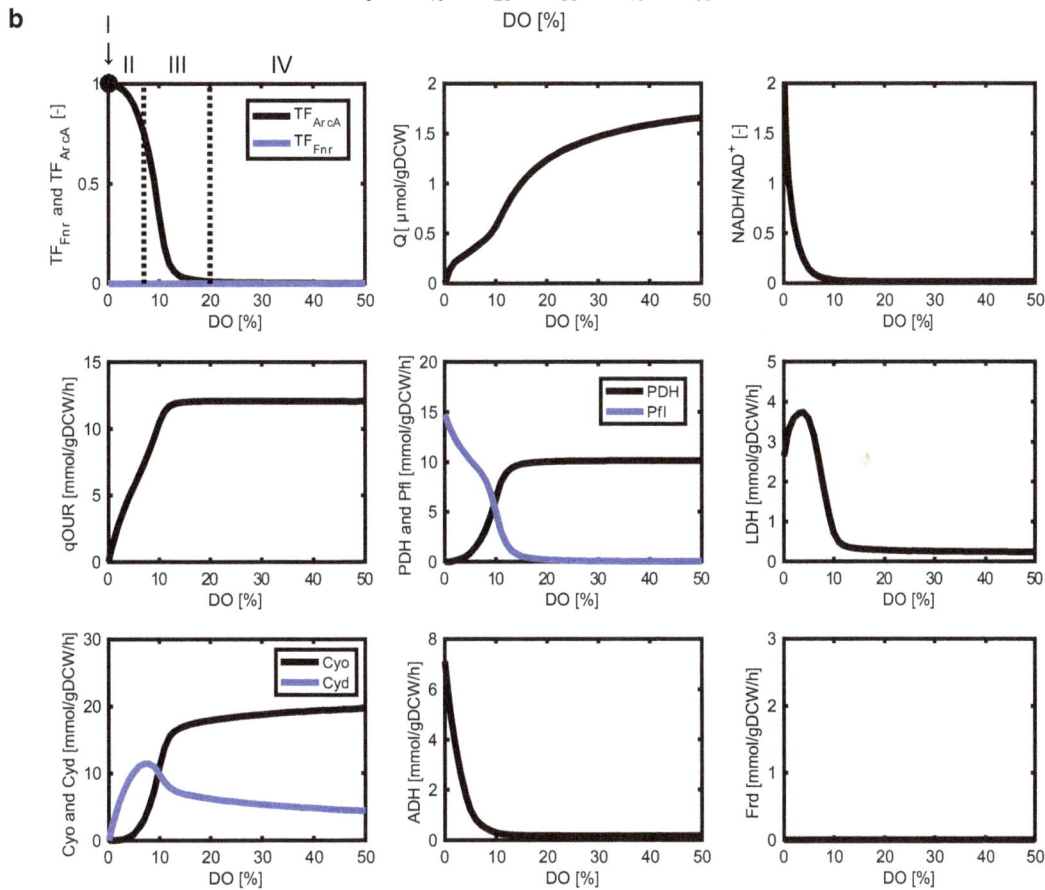

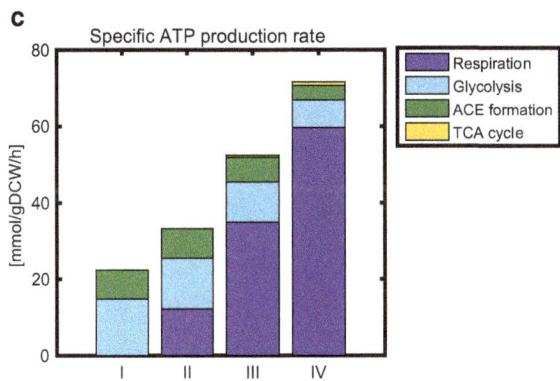

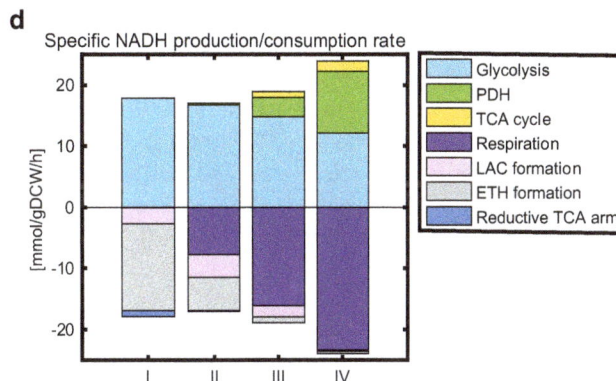

fnr or *arcA* gene knockout on the primary metabolism. Figure 4a shows the simulation results of an *fnr*-knockout mutant, in which the concentrations of acetate, lactate, formate, ethanol, and succinate were simulated at the time point at which 10 g/l of glucose was depleted. As compared to the WT strain (Fig. 3a), succinate was rarely produced at any DO level due to little activity of Frd caused by a lack of Fnr. As DO level decreased, lactate increased, peaking at 3% DO, and then slightly decreased. The lactate production in the *fnr*-knockout mutant was more enhanced than the WT strain at very low DO levels, which was supported by the experimental data [51].

Figure 4b shows the effect of DO level on the intracellular metabolic fluxes, redox status, and transcriptional activities. The Pfl flux was predicted to be lower in the *fnr*-knockout mutant than in the WT strain (Fig. 3b) under conditions I and II. The LDH flux increased under condition III and slightly decreased under conditions I and II in the *fnr*-knockout mutant, but it was higher than that of the WT strain (Fig. 3b). As the DO decreased, the NADH/NAD$^+$ ratio increased under conditions III, II, and I, which resulted in the increased ADH flux, while the Frd flux was zero due to a lack of Fnr.

The simulated specific production/consumption rates of ATP and NADH are shown in Fig. 4c, d. The profiles of the specific ATP production rates of the *fnr*-knockout mutant were almost the same as those of the WT strain (Fig. 3c), whereas the specific NADH consumption rate in the lactate and ethanol formation through LDH and ADH and the respiratory pathway somewhat differed from that of the WT strain under condition I (Figs. 3d, 4d). The NADH consumption rate through LDH in the *fnr*-knockout mutant was higher than that of the WT strain (as discussed later), whereas the NADH consumption rate through ADH was lower than that of the WT strain. As compared with the WT strain, the NADH consumption rate by the NADH dehydrogenases in the respiratory chain increased under condition II because the lack of Fnr de-repressed the NADH dehydrogenase reactions.

Additional file 3: Figure S3B shows the carbon balances of the metabolic products including CO_2 and biomass in the *fnr*-knockout mutant at different DO levels. The carbon balances differed between the WT strain and the *fnr*-knockout mutant under conditions I and II (Additional file 3: Figure S3A, B). More glucose carbon was converted into lactate in the *fnr*-knockout mutant than in the WT strain.

Prediction of the metabolic characteristics of an *arcA*-knockout mutant

As shown in Fig. 5a, the model predicted the changes in metabolic products with respect to DO level in an *arcA*-knockout mutant. Acetate production decreased slightly from 10 to 2% DO, as experimentally observed [52]. Ethanol production was predicted to be higher in the *arcA*-knockout mutant than in the WT strain and *fnr*-knockout mutant at DO <6%, which was consistent with the experimental data under micro-aerobic conditions [53, 54]. Under anaerobic condition, ethanol production was also predicted to be higher in the *arcA*-knockout mutant than in the other strains, while the ethanol production in the *arcA*-knockout mutant was reduced in the experiment [53]. This discrepancy will be discussed later. As the DO level decreased, lactate increased, peaking at 4% DO, and then decreased (Fig. 5a). The maximum concentration of lactate produced by the *arcA*-knockout mutant was higher than those of the WT strain and *fnr*-knockout mutant, as experimentally observed [53].

Figure 5b shows the effect of the DO level on the intracellular metabolic fluxes, redox status, and transcriptional activities. The Cyd flux was lower than that of the WT strain under conditions III and II (Figs. 3b, 5b), as experimentally observed [55]. As the DO level decreased under conditions III and II, the PDH flux slightly decreased. The decrease in the PDH flux was small compared to that of the WT strain and the *fnr*-knockout mutant (Figs. 3b, 4b, 5b) because the PDH flux is not repressed in the *arcA*-knockout mutant. The NADH/NAD$^+$ ratio in the *arcA*-knockout mutant was higher than that of the WT strain and the *fnr*-knockout mutant under condition II (Figs. 3b, 4b, 5b), as experimentally observed [53]. As DO level decreased, the NADH/NAD$^+$ ratio increased and then declined slightly as experimentally observed [53]. The simulated NADH/NAD$^+$ ratio of the *arcA*-knockout mutant was higher than its experimental ratio, although the simulated NADH/NAD$^+$ ratios of the WT strain and the *fnr*-knockout mutant were relatively consistent with their experimental ratios. While the activities of the PDH, citrate synthase (CS), and isocitrate dehydrogenase (ICDH) enzymes are allosterically inhibited by NADH to suppress an excess production of NADH [56, 57], the present model did not implement such allosteric inhibitions. The neglect of the allosteric inhibitions relatively reproduced the NADH/NAD$^+$ ratios of the WT strain and the *fnr*-knockout mutant because their NADH level was not so high as that of the *arcA*-knockout mutant, but would overestimate the NADH/NAD$^+$ ratio of the *arcA*-knockout mutant. As the DO level decreased, the Frd flux steeply increased and then declined (Fig. 5b). The simulation result of the Frd flux showed the similar trend as that of the NADH/NAD$^+$ ratio due to the fact that NADH is oxidized at MDH with Frd.

The specific production/consumption rates of ATP and NADH were simulated as shown in Fig. 5c, d. The specific ATP production rate increased in the *arcA*-knockout

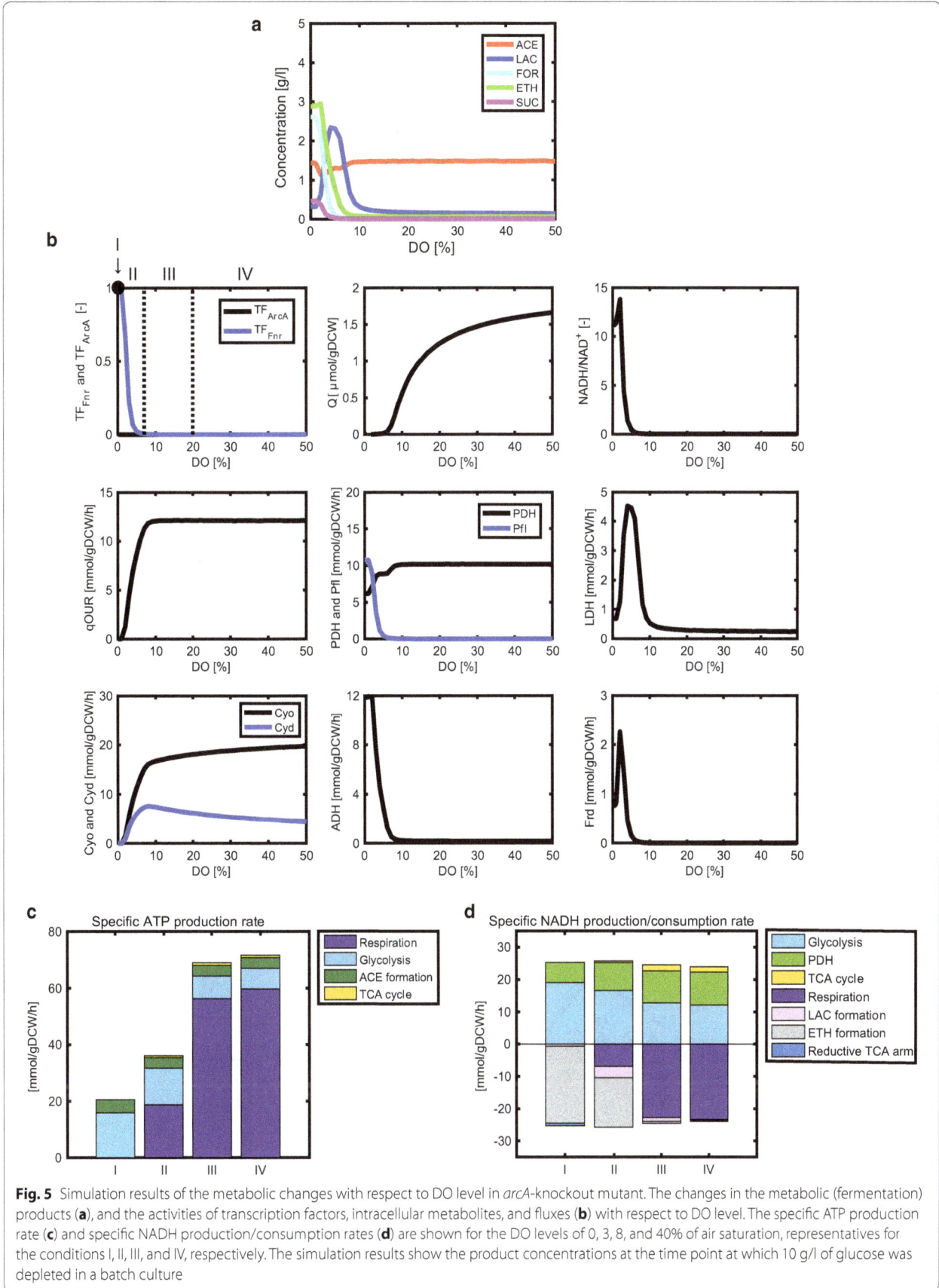

Fig. 5 Simulation results of the metabolic changes with respect to DO level in *arcA*-knockout mutant. The changes in the metabolic (fermentation) products (**a**), and the activities of transcription factors, intracellular metabolites, and fluxes (**b**) with respect to DO level. The specific ATP production rate (**c**) and specific NADH production/consumption rates (**d**) are shown for the DO levels of 0, 3, 8, and 40% of air saturation, representatives for the conditions I, II, III, and IV, respectively. The simulation results show the product concentrations at the time point at which 10 g/l of glucose was depleted in a batch culture

mutant under condition III as compared with the WT strain (Figs. 3c, 5c) because Cyo is activated in the *arcA*-knockout mutant (Fig. 1). For this, the qOUR for the *arcA*-knockout mutant was higher than that of the WT strain under condition III (Figs. 3b, 5b), as experimentally observed [32]. The specific NADH production rate in the TCA cycle was slightly higher than that of the WT strain under condition III (Figs. 3d, 5d), as experimentally observed [32], because the TCA cycle is not repressed in this mutant. The specific NADH consumption rate through ethanol formation (by ALDH and ADH) was higher than that of the WT strain under conditions I and II (Figs. 3d, 5d).

Additional file 3: Figure S3C shows the carbon balances of the metabolic products including CO_2 and biomass in the *arcA*-knockout mutant at different DO levels. More glucose carbon was converted to biomass, and more CO_2 was produced in the *arcA*-knockout mutant than the WT strain due to de-repression of the PDH and TCA cycle under condition III (Additional file 3: Figure S3A, C).

Rational design of a method for lactate production by a *pfl*-knockout mutant

The present model was utilized for the rational design of microbial cellular factories and optimization of target metabolite production. While a *pfl*-knockout mutant was experimentally reported to exclusively produce lactate [44], the effect of the DO level on the energy generation, biomass formation, and productivity has been rarely investigated. Simulated time course data of a batch culture of the *pfl*-knockout mutant reasonably predicted

the experimental data (Fig. 6a) [44]. Here, we considered operation strategies for the efficient production of lactate by the *pfl*-knockout mutant.

Dual-phase cultivation was designed to enhance the target metabolite production, starting with an aerobic cultivation to promote the cell growth, followed by an anaerobic or micro-aerobic condition to facilitate the target metabolite production. The switching time when the culture condition is changed from aerobic to micro-aerobic condition is generally a key parameter for enhanced productivity. The effect of the switching time on lactate yield (g of product/g of substrate consumed) and productivity (g/l of product concentration/h of cultivation time) was simulated for the *pfl*-knockout mutant when 10 g/l glucose was supplied as a carbon source (Fig. 6b). The DO levels of 40 and 1% were set to the aerobic and micro-aerobic conditions, respectively. The symbols in Fig. 6b represent the productivity of lactate obtained from the experiments [44, 58]. As expected, the yield was the highest when the cells were cultured consistently under the micro-aerobic condition, although the productivity was low. The productivity was improved to 0.81 g/l/h (at a switching time of 4.5 h) by the dual-phase cultivation, as compared to 0.38 g/l/h under the micro-aerobic condition throughout the cultivation (Fig. 6b).

Discussion

Advantages of the proposed kinetic model

There are several modeling approaches to simulate the fermentation characteristics. Khodayari et al. [59] simulated the succinate overproduction by *E. coli* under

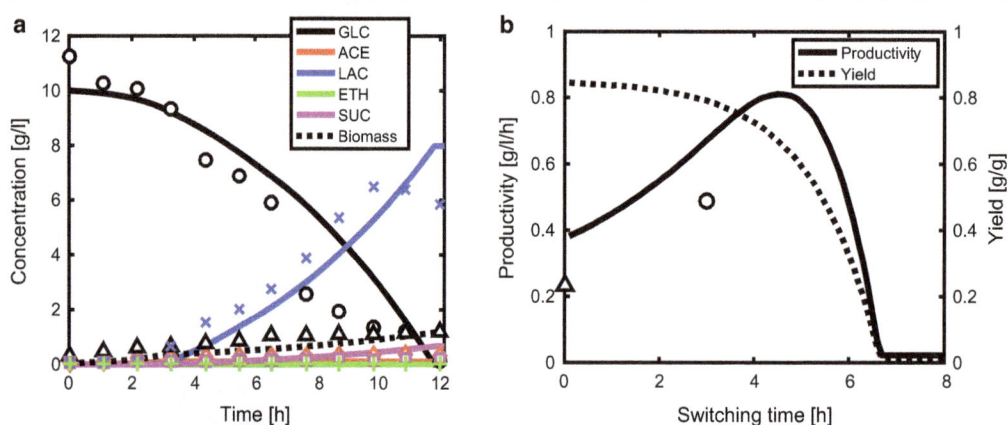

Fig. 6 Batch cultivation of *pfl*-knockout mutant in the dual-phase cultivation starting with aerobic cultivation followed by micro-aerobic cultivation. The DO levels of 40 and 1% were set for the aerobic and micro-aerobic conditions, respectively. 10 g/l glucose was used as a carbon source. **a** Time course data of extracellular product and biomass cultured for 3 h under aerobic condition followed by micro-aerobic condition, where the lines show simulation results and the symbols represent experimental data [44]: *open circle* glucose; *open diamond* acetate; *multiplication sign* lactate; *full width plus sign* ethanol; *open square* succinate; *open up-pointing triangle* biomass. **b** Effect of the switching time on the yield and productivity. *Yield* represents g of lactate/g of glucose consumed. *Productivity* represents g/l of lactate concentration/h of cultivation time. *The lines* represent the simulation results and the *symbols* represent the experimental data of the productivity (*open up-pointing triangle* Liu et al. [58]; *open circle* Zhu and Shimizu [44])

both aerobic and anaerobic conditions using the kinetic model-based k-OptForce method with ensemble modeling approach and parameterization based on the data obtained from multiple mutant strains. Their model was able to predict the metabolism that improves the succinate yield under aerobic condition but failed to predict it under anaerobic condition. It is essential to predict the dynamics of the cell growth and metabolite production over a broad range of DO levels and to understand the metabolic regulation mechanisms for the rational design of useful metabolite production. To meet these requirements, we have developed a kinetic model that implements redox regulation by Fnr and ArcA into central carbon metabolism [15, 20]. An advantage in the proposed model is to accurately simulate metabolisms under anaerobic, micro-aerobic, and aerobic conditions.

The model was constructed and verified using available experimental data for the WT strain [24, 29, 41–44, 47–51, 54]. The model-predicted behaviors were validated by the experimental data of the *fnr*-knockout mutant [51], the *arcA*-knockout mutant [32, 52, 53, 55, 60], and the *pfl*-knockout mutant [44].

To achieve efficient production of a target metabolite, the dual-phase cultivation method was investigated to improve the lactate production using the *pfl*-knockout mutant. This investigation revealed the importance of the optimal switching time from aerobic to micro-aerobic conditions to maximize the productivity (Fig. 6b). In addition, the trade-off between yield and productivity must be considered in practice, because the yield decreases with increased duration of the aerobic period (Fig. 6b).

Regulation mechanisms underlying the metabolic changes in response to DO level

In the simulation of the WT strain, the LDH flux exhibited up and down changes with respect to DO level (Fig. 3b). The LDH flux increased more under condition III than under condition IV. Under condition III, ArcA repressed the PDH flux while increasing the Pfl flux. Although the total flux from pyruvate to acetyl-CoA (PDH flux + Pfl flux) was almost the same between under conditions III and IV, the increased NADH/NAD$^+$ ratio increased the LDH flux under condition III. On the other hand, the LDH flux decreased under condition II because pyruvate, the substrate of the LDH reaction, was consumed by the Fnr-enhanced Pfl reaction.

Lactate production increased in the *fnr*-knockout mutant under conditions I and II (Figs. 3a, 4a) as compared with the WT strain, as experimentally observed [51]. Since the Pfl flux was reduced in the *fnr*-knockout mutant under conditions I and II (Figs. 3b, 4b), the total flux from pyruvate to acetyl-CoA was also reduced as

compared to the WT strain, resulting in the accumulation of pyruvate. Pyruvate was converted to lactate, accompanied by NADH consumption. On the other hand, the *arcA*-knockout mutant also exhibited higher lactate production than the WT strain around 4% DO under condition II (Figs. 3a, 5a) because the total flux from pyruvate to acetyl-CoA was reduced as compared to the WT strain as experimentally observed [60].

At lower oxygen levels under conditions I and II, the *arcA*-knockout mutant exhibited a marked increase in ethanol production (Figs. 3a, 5a). Although the total flux from pyruvate to acetyl-CoA was almost the same as that of the WT strain under conditions I and II, the NADH/NAD$^+$ ratio in the *arcA*-knockout mutant was much higher than those in the WT strain and the *fnr*-knockout mutant due to the high flux of PDH (Figs. 3b, 4b, 5b). This resulted in enhanced ethanol production. These simulation results were consistent with the experimental observation except for condition I (anaerobic condition) [53]. While the simulated ethanol production flux of the *arcA*-knockout mutant was higher than that of the WT strain, the experimental ethanol production flux in the *arcA*-knockout mutant was comparable to that in the WT strain under anaerobic condition. The discrepancy in the ethanol production under anaerobic condition would be due to the overestimation of the NADH/NAD$^+$ ratio in the *arcA*-knockout mutant. This overestimation results from the fact that the simulated reductive pathway flux through Ppc-MDH/Fum-Frd in the *arcA*-knockout mutant was lower than the experimental flux under anaerobic condition. Such underestimation of the reductive flux may be caused by the neglect of some effectors [61, 62] on the Ppc reaction responsible for the MDH and Frd fluxes. The present model includes the effect of FBP on the Ppc activity, but did not include the effects of acetyl-CoA, malate, and aspartate.

Toward virtual metabolism

Synthetic biology aims to understand the mechanisms governing the dynamic behaviors of biochemical networks in response to environmental stresses or genetic variations and facilitate the rational design or engineering of cells at the gene-regulation level. Synthetic biology approaches consist of the construction of a rigorously defined biochemical network map, development of mathematical models, experimental validation of these models, and analysis and rational design of biological systems, ultimately leading to computer-aided design of cells [63–65]. The proposed kinetic model was constructed according to this synthetic approach to provide a platform for the rational design of biofuel and biochemical production by *E. coli* and for further modeling efforts, including extension to amino acid, nucleotide, lipid, and polysaccharide metabolisms, as well as cell physiology. A comprehensive dynamic model, called 'virtual *E. coli*,' is expected to reproduce the complex dynamics of a series of genetic mutants under different conditions, such as consumption of multiple sugars, nitrogen, and phosphate starvation, osmotic pressure, and changes in pH. In addition, the kinetic model of the *E. coli* central carbon metabolism would be a feasible reference model for constructing the kinetic models of a variety of microbes, because central carbon metabolisms are relatively conserved across them.

On the other hand, another characteristic of microbes is their metabolic variety due to evolution under various growth conditions on earth. For example, yeast produces ethanol via pyruvate decarboxylase (PDC) and ADH, *Clostridia* employs acetone–butanol–ethanol (ABE) pathway, and *Zymomonas* has the Entner–Doudoroff (ED) pathway. Since the detailed metabolic pathways depend on the microbes, it is essential to take into account their differences to construct their kinetic models, while using the *E. coli* kinetic model as a reference model.

Conclusions

It is quite important to understand metabolic regulation mechanisms from both scientific and engineering points of view. In particular, redox regulation in response to oxygen limitation is critically important in the practical production of biofuel and biochemical compounds. Therefore, we developed a kinetic model with enzymatic and transcriptional regulations to predict the dynamics of metabolism at different DO levels. Transcription factor activities, metabolite concentrations, and fluxes of the WT strain and *fnr*- and *arcA*-knockout mutants were simulated to validate the model. Using this kinetic model, a rational operation strategy for the *pfl*-knockout mutant was designed to enhance lactate production. A dual-phase strategy was considered that involves initial cultivation under aerobic condition to enhance the cell growth rate, with subsequent cultivation under anaerobic or micro-aerobic condition to enhance the lactate production.

Abbreviations

Primary metabolic pathway and transport system

EI enzyme I, *EIIA* enzyme IIA, *ED* pathway, Entner–Doudoroff pathway, *HPr* histidine-phosphorylatable protein, *PP* pathway, pentose phosphate pathway, *PTS* phosphotransferase system, *TCA cycle* tricarboxylic acid cycle.

Metabolites

ACAL acetaldehyde, *AcCoA* acetyl-CoA, *CIT* citrate, *DHAP* dihydroxy acetone phosphate, *E4P*

erythrose-4-phosphate, *ETH* ethanol, *FBP* fructose-1,6-bisphosphate, *FOR* formate, *F6P* fructose-6-phosphate, *FUM* fumarate, *G6P* glucose-6-phosphate, *GAP* glyceraldehyde-3-phosphate, *GLC* glucose, *GOX* glyoxylate, *ICI* isocitrate, *αKG* α-ketoglutarate, *LAC* lactate, *MAL* malate, *OAA* oxaloacetate, *PEP* phosphoenol pyruvate, *6PG* 6-phosphogluconate, *6PGL* 6-phosphoglucono-lactone, *PYR* pyruvate, *Q* quinone, *QH$_2$* quinol, *R5P* ribose-5-phosphate, *RU5P* ribulose-5-phosphate, *S7P* sedoheptulose-7-phosphate, *SUC* succinate.

Enzymes

Ack acetate kinase, *Acs* acetyl coenzyme A synthetase, *ADH* alcohol dehydrogenase, *ALDH* acetaldehyde dehydrogenase, *Cya* adenylate cyclase, *Cyd* cytochrome *bd*, *Cyo* cytochrome *bo*, *CS* citrate synthase, *Eno* enolase, *Fba* fructose-1,6-bisphosphate aldolase, *Fbp* fructose bisphosphatase, *Frd* fumarate reductase, *Fum* fumarase, *G6PDH* glucose-6-phosphate dehydrogenase, *GAPDH* glyceraldehyde-3-phosphate dehydrogenase, *Glk* glucokinase, *ICDH* isocitrate dehydrogenase, *Icl* isocitrate lyase, *αKGDH* α-ketoglutarate dehydrogenase, *LDH* lactate dehydrogenase, *MDH* malate dehydrogenase, *Mez* malic enzyme, *MS* malate synthase, *Ndh* NADH dehydrogenase-II, *Nuo* NADH dehydrogenase-I, *Pck* phosphoenolpyruvate carboxykinase, *PDH* pyruvate dehydrogenase, *Pfk* phosphofructokinase, *Pfl* pyruvate formate-lyase, *PGDH* 6-phosphogluconate dehydrogenase, *Pgk* phosphoglucokinase, *Pgm* phosphoglucomutase, *Ppc* phosphoenolpyruvate carboxylase, *Pps* phosphoenolpyruvate synthase, *Pta* phosphotransacetylase, *Pyk* pyruvate kinase, *Rpe* ribulose phosphate epimerase, *Rpi* ribose phosphate isomerase, *SDH* succinate dehydrogenase, *Tal* transaldolase, *TktA* transketolase I, *TktB* transketolase II.

Authors' contributions

YM designed the research, developed the kinetic model, performed the simulations, analyzed the data, and wrote the manuscript. HK analyzed the data and wrote the manuscript. Both authors read and approved the final manuscript.

Author details

[1] Department of Bioscience and Bioinformatics, Kyushu Institute of Technology, 680-4 Kawazu, Iizuka, Fukuoka 820-8502, Japan. [2] Biomedical Informatics R&D Center, Kyushu Institute of Technology, 680-4 Kawazu, Iizuka, Fukuoka 820-8502, Japan.

Acknowledgements

Not applicable.

Competing interests

The authors declare that they have no competing interests.

Funding

This work was supported by a Grant-in-Aid for Scientific Research (B) (16H02898) from the Japan Society for the Promotion of Science (JSPS) and was partially supported by the developing key technologies for discovering and manufacturing pharmaceuticals used for next-generation treatments and diagnoses, both from the Ministry of Economy, Trade and Industry, Japan (METI), and from the Japan Agency for Medical Research and Development (AMED).

References

1. Huffer S, Roche CM, Blanch HW, Clark DS. *Escherichia coli* for biofuel production: bridging the gap from promise to practice. Trends Biotechnol. 2012;30(10):538–45.
2. Lange J, Takors R, Blombach B. Zero-growth bioprocesses—a challenge for microbial production strains and bioprocess engineering. Eng Life Sci. 2016. doi:10.1002/elsc.201600108.
3. Vemuri GN, Eiteman MA, Altman E. Succinate production in dual-phase *Escherichia coli* fermentations depends on the time of transition from aerobic to anaerobic conditions. J Ind Microbiol Biotechnol. 2002;28(6):325–32.
4. Matsuoka Y, Shimizu K. Current status and future perspectives of kinetic modeling for the cell metabolism with incorporation of the metabolic regulation mechanism. Bioresour Bioprocess. 2015;2:4.
5. Link H, Kochanowski K, Sauer U. Systematic identification of allosteric protein-metabolite interactions that control enzyme activity *in vivo*. Nat Biotechnol. 2013;31(4):357–61.
6. Machado D, Herrgård MJ, Rocha I. Modeling the contribution of allosteric regulation for flux control in the central carbon metabolism of *E. coli*. Front Bioeng Biotechnol. 2015;3:154.
7. Kremling A, Bettenbrock K, Laube B, Jahreis K, Lengeler JW, Gilles ED. The organization of metabolic reaction networks. III. Application for diauxic growth on glucose and lactose. Metab Eng. 2001;3(4):362–79.
8. Hardiman T, Lemuth K, Keller MA, Reuss M, Siemann-Herzberg M. Topology of the global regulatory network of carbon limitation in *Escherichia coli*. J Biotechnol. 2007;132(4):359–74.
9. Matsuoka Y, Shimizu K. Metabolic regulation in *Escherichia coli* in response to culture environments via global regulators. Biotechnol J. 2011;6(11):1330–41.
10. Perrenoud A, Sauer U. Impact of global transcriptional regulation by ArcA, ArcB, Cra, Crp, Cya, Fnr, and Mlc on glucose catabolism in *Escherichia coli*. J Bacteriol. 2005;187(9):3171–9.
11. Shimizu K. Toward systematic metabolic engineering based on the analysis of metabolic regulation by the integration of different levels of information. Biochem Eng J. 2009;46(3):235–51.
12. Bettenbrock K, Fischer S, Kremling A, Jahreis K, Sauter T, Gilles ED. A quantitative approach to catabolite repression in *Escherichia coli*. J Biol Chem. 2006;281(5):2578–84.
13. Kadir TAA, Mannan AA, Kierzek AM, McFadden J, Shimizu K. Modeling and simulation of the main metabolism in *Escherichia coli* and its several single-gene knockout mutants with experimental verification. Microb Cell Fact. 2010;9:88.
14. Kotte O, Zaugg JB, Heinemann M. Bacterial adaptation through distributed sensing of metabolic fluxes. Mol Syst Biol. 2010;6:355.
15. Matsuoka Y, Shimizu K. Catabolite regulation analysis of *Escherichia coli* for acetate overflow mechanism and co-consumption of multiple sugars based on systems biology approach using computer simulation. J Biotechnol. 2013;168(2):155–73.
16. Nishio Y, Usuda Y, Matsui K, Kurata H. Computer-aided rational design of the phosphotransferase system for enhanced glucose uptake in *Escherichia coli*. Mol Syst Biol. 2008;4:160.
17. Usuda Y, Nishio Y, Iwatani S, Van Dien SJ, Imaizumi A, Shimbo K, Kageyama N, Iwahata D, Miyano H, Matsui K. Dynamic modeling of *Escherichia coli* metabolic and regulatory systems for amino-acid production. J Biotechnol. 2010;147(1):17–30.
18. Kurata H, Maeda K, Matsuoka Y. Dynamic modeling of metabolic and gene regulatory systems toward developing virtual microbes. J Chem Eng Jpn. 2014;47(1):1–9.
19. Kremling A, Geiselmann J, Ropers D, de Jong H. Understanding carbon catabolite repression in *Escherichia coli* using quantitative models. Trends Microbiol. 2015;23(2):99–109.
20. Jahan N, Maeda K, Matsuoka Y, Sugimoto Y, Kurata H. Development of an accurate kinetic model for the central carbon metabolism of *Escherichia coli*. Microb Cell Fact. 2016;15(1):112.
21. Bettenbrock K, Bai H, Ederer M, Green J, Hellingwerf KJ, Holcombe M, Kunz S, Rolfe MD, Sanguinetti G, Sawodny O, et al. Towards a systems level understanding of the oxygen response of *Escherichia coli*. Adv Microb Physiol. 2014;64:65–114.
22. Ederer M, Steinsiek S, Stagge S, Rolfe MD, Ter Beek A, Knies D, de Mattos MJT, Sauter T, Green J, Poole RK, et al. A mathematical model of metabolism and regulation provides a systems-level view of how *Escherichia coli* responds to oxygen. Front Microbiol. 2014;5:124.
23. Rolfe MD, Ocone A, Stapleton MR, Hall S, Trotter EW, Poole RK, Sanguinetti G, Green J. Systems analysis of transcription factor activities in environments with stable and dynamic oxygen concentrations. Open Biol. 2012;2(7):120091.
24. Rolfe MD, Ter Beek A, Graham AI, Trotter EW, Asif HMS, Sanguinetti G, de Mattos JT, Poole RK, Green J. Transcript profiling and inference of *Escherichia coli* K-12 ArcA activity across the range of physiologically relevant oxygen concentrations. J Biol Chem. 2011;286(12):10147–54.
25. Steinsiek S, Frixel S, Stagge S, Bettenbrock K. Sumo: characterization of *E. coli* MG1655 and *frdA* and *sdhC* mutants at various aerobiosis levels. J Biotechnol. 2011;154(1):35–45.
26. Steinsiek S, Stagge S, Bettenbrock K. Analysis of *Escherichia coli* mutants with a linear respiratory chain. PLoS ONE. 2014;9(1):e87307.
27. Trotter EW, Rolfe MD, Hounslow AM, Craven CJ, Williamson MP, Sanguinetti G, Poole RK, Green J. Reprogramming of *Escherichia coli* K-12 metabolism during the initial phase of transition from an anaerobic to a micro-aerobic environment. PLoS ONE. 2011;6(9):e25501.
28. Borisov VB, Verkhovsky MI. Oxygen as acceptor. EcoSal Plus. 2009. doi:10.1128/ecosalplus.3.2.7.
29. Alexeeva S, Hellingwerf KJ, Teixeira de Mattos MJ. Quantitative assessment of oxygen availability: perceived aerobiosis and its effect on flux distribution in the respiratory chain of *Escherichia coli*. J Bacteriol. 2002;184(5):1402–6.
30. Unden G, Bongaerts J. Alternative respiratory pathways of *Escherichia coli*: energetics and transcriptional regulation in response to electron acceptors. Biochim Biophys Acta. 1997;1320(3):217–34.
31. Kang Y, Weber KD, Qiu Y, Kiley PJ, Blattner FR. Genome-wide expression analysis indicates that FNR of *Escherichia coli* K-12 regulates a large number of genes of unknown function. J Bacteriol. 2005;187(3):1135–60.
32. Alexeeva S, Hellingwerf KJ, Teixeira de Mattos MJ. Requirement of ArcA for redox regulation in *Escherichia coli* under microaerobic but not anaerobic or aerobic conditions. J Bacteriol. 2003;185(1):204–9.
33. Gunsalus RP. Control of electron flow in *Escherichia coli*: coordinated transcription of respiratory pathway genes. J Bacteriol. 1992;174(22):7069–74.
34. Kwon O, Georgellis D, Lin ECC. Phosphorelay as the sole physiological route of signal transmission by the arc two-component system of *Escherichia coli*. J Bacteriol. 2000;182(13):3858–62.
35. Georgellis D, Kwon O, Lin EC. Quinones as the redox signal for the *arc* two-component system of bacteria. Science. 2001;292(5525):2314–6.
36. Tomashek JJ, Brusilow WSA. Stoichiometry of energy coupling by proton-translocating ATPases: a history of variability. J Bioenerg Biomembr. 2000;32(5):493–500.
37. Nanchen A, Schicker A, Sauer U. Nonlinear dependency of intracellular fluxes on growth rate in miniaturized continuous cultures of *Escherichia coli*. Appl Environ Microbiol. 2006;72(2):1164–72.
38. Yao R, Hirose Y, Sarkar D, Nakahigashi K, Ye Q, Shimizu K. Catabolic regulation analysis of *Escherichia coli* and its *crp, mlc, mgsA, pgi* and *ptsG* mutants. Microb Cell Fact. 2011;10:67.
39. Henkel SG, Ter Beek A, Steinsiek S, Stagge S, Bettenbrock K, de Mattos MJT, Sauter T, Sawodny O, Ederer M. Basic regulatory principles of *Escherichia coli*'s electron transport chain for varying oxygen conditions. PLoS ONE. 2014;9(9):e107640.
40. Potzkei J, Kunze M, Drepper T, Gensch T, Jaeger KE, Büchs J. Real-time determination of intracellular oxygen in bacteria using a genetically encoded FRET-based biosensor. BMC Biol. 2012;10:28.

41. Toya Y, Nakahigashi K, Tomita M, Shimizu K. Metabolic regulation analysis of wild-type and *arcA* mutant *Escherichia coli* under nitrate conditions using different levels of omics data. Mol BioSyst. 2012;8(10):2593–604.

42. Zhu J, Shimizu K. Effect of a single-gene knockout on the metabolic regulation in *Escherichia coli* for D-lactate production under microaerobic condition. Metab Eng. 2005;7(2):104–15.

43. Toya Y, Ishii N, Nakahigashi K, Hirasawa T, Soga T, Tomita M, Shimizu K. ^{13}C-metabolic flux analysis for batch culture of *Escherichia coli* and its *pyk* and *pgi* gene knockout mutants based on mass isotopomer distribution of intracellular metabolites. Biotechnol Prog. 2010;26(4):975–92.

44. Zhu J, Shimizu K. The effect of *pfl* gene knockout on the metabolism for optically pure D-lactate production by *Escherichia coli*. Appl Microbiol Biotechnol. 2004;64(3):367–75.

45. Valgepea K, Adamberg K, Nahku R, Lahtvee PJ, Arike L, Vilu R. Systems biology approach reveals that overflow metabolism of acetate in *Escherichia coli* is triggered by carbon catabolite repression of acetyl-CoA synthetase. BMC Syst Biol. 2010;4:166.

46. Bernal V, Castano-Cerezo S, Canovas M. Acetate metabolism regulation in *Escherichia coli*: carbon overflow, pathogenicity, and beyond. Appl Microbiol Biotechnol. 2016;100(21):8985–9001.

47. Gonzalez JE, Long CP, Antoniewicz MR. Comprehensive analysis of glucose and xylose metabolism in *Escherichia coli* under aerobic and anaerobic conditions by ^{13}C metabolic flux analysis. Metab Eng. 2016;39:9–18.

48. Alexeeva S, de Kort B, Sawers G, Hellingwerf KJ, de Mattos MJ. Effects of limited aeration and of the ArcAB system on intermediary pyruvate catabolism in *Escherichia coli*. J Bacteriol. 2000;182(17):4934–40.

49. de Graef MR, Alexeeva S, Snoep JL, Teixeira de Mattos MJ. The steady-state internal redox state (NADH/NAD) reflects the external redox state and is correlated with catabolic adaptation in *Escherichia coli*. J Bacteriol. 1999;181(8):2351–7.

50. Chen XW, Alonso AP, Allen DK, Reed JL, Shachar-Hill Y. Synergy between ^{13}C-metabolic flux analysis and flux balance analysis for understanding metabolic adaption to anaerobiosis in *E. coli*. Metab Eng. 2011;13(1):38–48.

51. Kim HJ, Hou BK, Lee SG, Kim JS, Lee DW, Lee SJ. Genome-wide analysis of redox reactions reveals metabolic engineering targets for D-lactate overproduction in *Escherichia coli*. Metab Eng. 2013;18:44–52.

52. Waegeman H, Beauprez J, Moens H, Maertens J, De Mey M, Foulquie-Moreno MR, Heijnen JJ, Charlier D, Soetaert W. Effect of *iclR* and *arcA* knockouts on biomass formation and metabolic fluxes in *Escherichia coli* K12 and its implications on understanding the metabolism of *Escherichia coli* BL21 (DE3). BMC Microbiol. 2011;11:70.

53. Levanon SS, San KY, Bennett GN. Effect of oxygen on the *Escherichia coli* ArcA and FNR regulation systems and metabolic responses. Biotechnol Bioeng. 2005;89(5):556–64.

54. Zhu J, Shalel-Levanon S, Bennett G, San KY. Effect of the global redox sensing/regulation networks on *Escherichia coli* and metabolic flux distribution based on C-13 labeling experiments. Metab Eng. 2006;8(6):619–27.

55. Govantes F, Orjalo AV, Gunsalus RP. Interplay between three global regulatory proteins mediates oxygen regulation of the *Escherichia coli* cytochrome d oxidase (*cydAB*) operon. Mol Microbiol. 2000;38(5):1061–73.

56. Kim Y, Ingram LO, Shanmugam KT. Dihydrolipoamide dehydrogenase mutation alters the NADH sensitivity of pyruvate dehydrogenase complex of *Escherichia coli* K-12. J Bacteriol. 2008;190(11):3851–8.

57. Molgat GF, Donald LJ, Duckworth HW. Chimeric allosteric citrate synthases: construction and properties of citrate synthases containing domains from two different enzymes. Arch Biochem Biophys. 1992;298(1):238–46.

58. Liu HM, Kang JH, Qi QS, Chen GJ. Production of lactate in *Escherichia coli* by redox regulation genetically and physiologically. Appl Biochem Biotech. 2011;164(2):162–9.

59. Khodayari A, Chowdhury A, Maranas CD. Succinate overproduction: a case study of computational strain design using a comprehensive *Escherichia coli* kinetic model. Front Bioeng Biotechnol. 2015;2:76.

60. Nikel PI, Zhu J, San KY, Mendez BS, Bennett GN. Metabolic flux analysis of *Escherichia coli* creB and *arcA* mutants reveals shared control of carbon catabolism under microaerobic growth conditions. J Bacteriol. 2009;191(17):5538–48.

61. Yang C, Hua Q, Baba T, Mori H, Shimizu K. Analysis of *Escherichia coli* anaplerotic metabolism and its regulation mechanisms from the metabolic responses to altered dilution rates and phosphoenolpyruvate carboxykinase knockout. Biotechnol Bioeng. 2003;84(2):129–44.

62. Kai Y, Matsumura H, Inoue T, Terada K, Nagara Y, Yoshinaga T, Kihara A, Tsumura K, Izui K. Three-dimensional structure of phosphoenolpyruvate carboxylase: a proposed mechanism for allosteric inhibition. Proc Natl Acad Sci USA. 1999;96(3):823–8.

63. Kurata H, Maeda K, Onaka T, Takata T. BioFNet: biological functional network database for analysis and synthesis of biological systems. Brief Bioinform. 2013;15(5):699–709.

64. Kurata H, Masaki K, Sumida Y, Iwasaki R. CADLIVE dynamic simulator: direct link of biochemical networks to dynamic models. Genome Res. 2005;15(4):590–600.

65. Kurata H, Matoba N, Shimizu N. CADLIVE for constructing a large-scale biochemical network based on a simulation-directed notation and its application to yeast cell cycle. Nucleic Acids Res. 2003;31(14):4071–84.

10

Metagenomic mining pectinolytic microbes and enzymes from an apple pomace-adapted compost microbial community

Man Zhou[1], Peng Guo[2], Tao Wang[1], Lina Gao[1], Huijun Yin[1], Cheng Cai[2], Jie Gu[3] and Xin Lü[1*]

Abstract

Background: Degradation of pectin in lignocellulosic materials is one of the key steps for biofuel production. Biological hydrolysis of pectin, i.e., degradation by pectinolytic microbes and enzymes, is an attractive paradigm because of its obvious advantages, such as environmentally friendly procedures, low in energy demand for lignin removal, and the possibility to be integrated in consolidated process. In this study, a metagenomics sequence-guided strategy coupled with enrichment culture technique was used to facilitate targeted discovery of pectinolytic microbes and enzymes. An apple pomace-adapted compost (APAC) habitat was constructed to boost the enrichment of pectinolytic microorganisms.

Results: Analyses of 16S rDNA high-throughput sequencing revealed that microbial communities changed dramatically during composting with some bacterial populations being greatly enriched. Metagenomics data showed that apple pomace-adapted compost microbial community (APACMC) was dominated by *Proteobacteria* and *Bacteroidetes*. Functional analysis and carbohydrate-active enzyme profiles confirmed that APACMC had been successfully enriched for the targeted functions. Among the 1756 putative genes encoding pectinolytic enzymes, 129 were predicted as novel (with an identity <30% to any CAZy database entry) and only 1.92% were more than 75% identical with proteins in NCBI environmental database, demonstrating that they have not been observed in previous metagenome projects. Phylogenetic analysis showed that APACMC harbored a broad range of pectinolytic bacteria and many of them were previously unrecognized.

Conclusions: The immensely diverse pectinolytic microbes and enzymes found in our study will expand the arsenal of proficient degraders and enzymes for lignocellulosic biofuel production. Our study provides a powerful approach for targeted mining microbes and enzymes in numerous industries.

Keywords: Lignocellulosic biofuel, Pectin, Metagenomic, Pectinolytic microbes and enzymes, Compost habitat

Background

High worldwide demand for energy and increasing concerns over global climate change have prompted the development of sustainable and environmentally friendly energy [16, 45]. Lignocellulosic biofuel, which derived from the most abundant renewable organic material on our planet, represents a promising alternative to fossil fuels [12, 15]. However, the major obstacles to industrial-scale production of biofuel from lignocellulosic feedstocks lie in the recalcitrant nature of biomass toward enzymatic breakdown and the relatively low activity of currently available hydrolytic enzymes [15, 44].

Pectin is one of the plant cell wall components. It is abundant in the middle lamella and primary cell walls, though presents at low levels in secondary walls [8]. For the cell walls of pectin-rich biomass, for example apple

*Correspondence: xinlu@nwsuaf.edu.cn
[1] College of Food Science and Engineering, Northwest A&F University, Yangling, Shaanxi Province, China
Full list of author information is available at the end of the article

pomace, it contains 12–35% pectin on a dry weight basis [11]. In plant biomass, pectin embeds in the cellulose–hemicellulose network of the cell wall and regulates intercellular adhesion like glues [16]. It is the complex matrix of pectin that masks cellulose and/or hemicellulose through hydrogen bonding interactions [53], and blocks their accessibility to degradative enzymes [8], thus resulting in plant biomass that is less susceptible to degradation and more recalcitrant to deconstruction [23]. As a result, degradation of pectin in lignocellulosic materials has been established as essential for efficient bioconversion of lignocellulose [47]. Recently, the reduction of bulk percentage of pectin through genetic manipulation or enzymatic means has been proved to reduce the recalcitrance and accelerate the lignocellulose saccharification of herbaceous plants [23], *Arabidopsis* [12], switchgrass [8], and woody biomass [3, 4].

Removal of pectin can be achieved in physical, chemical, or biological manner. Biological hydrolysis of pectin by pectinolytic microbes and enzymes is favored as it is environmentally benign and energy efficient [26]. Pectinolytic enzymes have multiple benefits in the efficient hydrolysis of lignocellulosic materials: first, yield of fermentable sugars by hydrolysis of pectin itself [4]; second, facilitation of sugar release by disrupting the pectin network around cellulose and lignin [23], and exposure of other polymers to degradation by hemicellulases and cellulases [30]; third, improvement of cell wall porosity [3] and reduction of mechanical strength because of its crosslinking and water complexation features [47]. Especially for pectin-rich lignocellulosic biomass which also could serve as the feedstock for lignocellulosic biofuel [30], for instance apple pomace, pectinolytic enzymes will play a more prominent role.

Despite pectinolytic enzymes playing a crucial part in the lignocellulosic biofuel production, most of the currently available pectinolytic enzymes are costly, inefficient, and susceptive to fluctuations in feedstock [5]. In consequence, search for microbes and enzymes from naturally evolved pectinolytic microbial communities offers a promising strategy for the discovery of new pectinolytic enzymes. Given the unique features of compost habitat, there is tremendous potential to discover robust organisms and novel enzymes which tolerate harsh pretreatment scenarios under industrial conditions [2]. Thus, compost is considered as one of the most attractive DNA pools for target gene discovery [2, 24].

Metagenomics, which directly analyzes the total DNA from environmental samples, provides a powerful strategy in unveiling the novel microbes and enzymes in microbial communities without the technical challenges of cultivation [15, 37]. However, as environmental samples generally hold a huge reservoir of extensive microbes

and enzymes, it is unfeasible to characterize them accurately. Hence, to reduce the complexity of metagenomic datasets, render further assembly more amenable, and more importantly, improve the specificity of the sample's DNA, the oriented enrichment culture technique is essential to be employed for the establishment of microbial consortia with desired functionality [28]. In this manner, the enzyme repertoire of enriched consortia can be tailored to degrade specific feedstock [52]. Since apple pomace is an pectin-rich lignocellulosic biomass [30], pectinolytic enzymes could be exploited from the established pectinolytic microbes which are selectively enriched in abundance from compost communities by cultivation with apple pomace as the sole carbon source.

In this study, a metagenomic sequence-guided strategy combined with enrichment culture technique was used to targetedly discover the pectinolytic microbes and enzymes from an apple pomace-adapted compost microbial community (APACMC). The pipeline of this strategy is shown in Fig. 1. Firstly, the unique APACMC was constructed from the cow manure compost habitat to boost the enrichment of pectinolytic microorganisms. The dynamic microbial changes of APACMC were characterized by 16S rDNA high-throughput sequencing. Secondly, a targeted metagenomic approach was applied to facilitate the identification of pectinolytic microbes and enzymes. A more accurate microbial taxonomic analysis and function characterization were conducted. Thirdly, the metagenome sequences were annotated and phylogenetically affiliated against carbohydrate-active enzymes (CAZymes) database. Finally, after the specific investigation of genes related to pectinolytic CAZymes and their taxonomic affiliations, the robust microorganisms and novel enzymes processing the degradation of pectin were identified.

Results and discussion
Changes in physicochemical properties during composting
The variations of physicochemical properties during composting are strongly associated with the biological reactions involving organic matter, and thus, these changes reflect the microbial activity and progress of the composting process [41]. The dynamic changes in physicochemical properties (i.e., temperature, water content, and pH) during the apple pomace-adapted compost (APAC) process are illustrated in Fig. 2. During the 30-day enrichment period (Fig. 2a), the temperature of the APAC pile maintained at 25–35 °C for 24 h to allow the compost microbes to establish, then it rapidly reached 60–70 °C to trigger the thermophilic phase. After the temperature reached 68 °C at day 15, the temperature declined gradually back to the ambient temperature over the rest 15 days to trigger the cooling and maturation

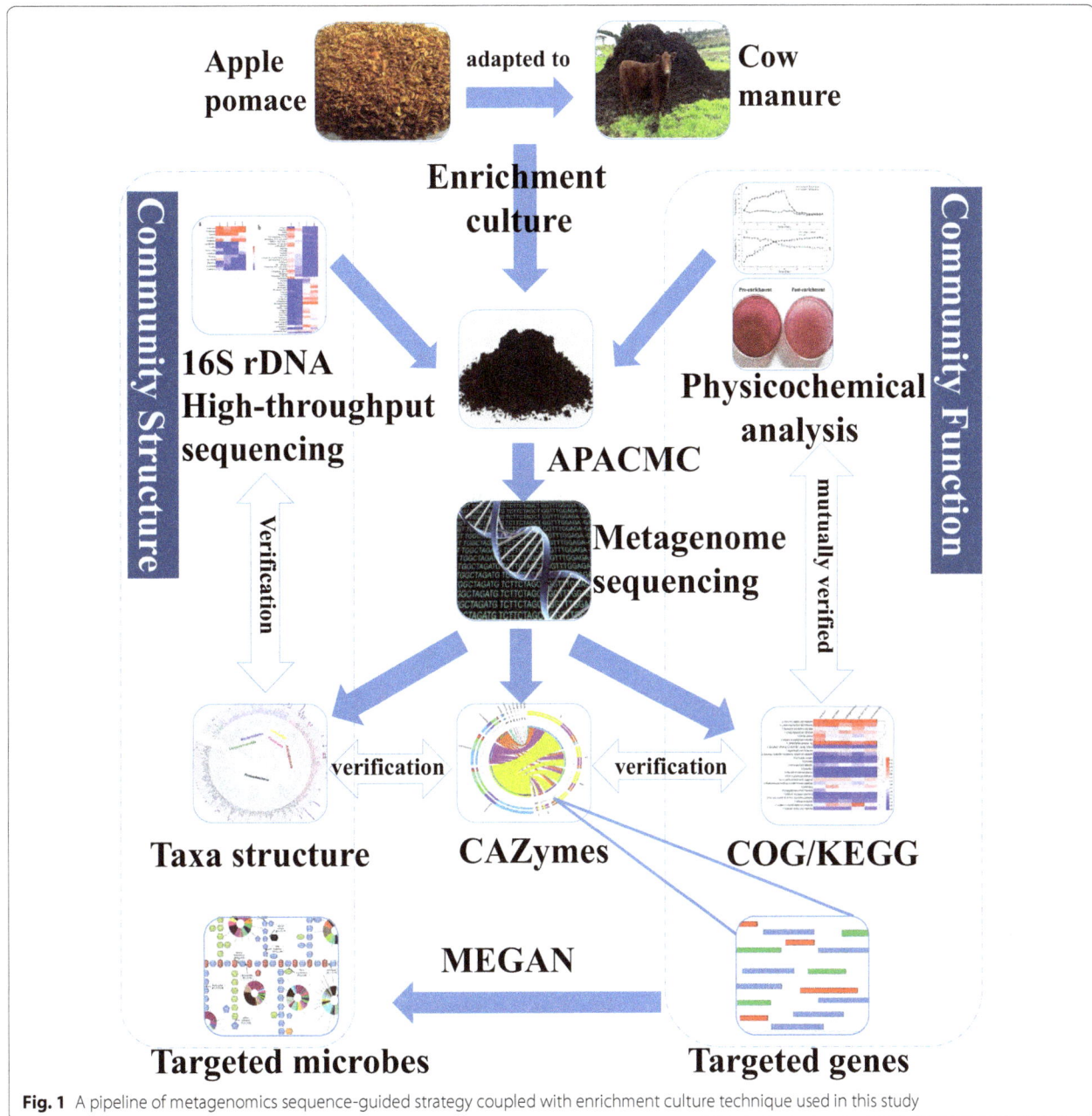

Fig. 1 A pipeline of metagenomics sequence-guided strategy coupled with enrichment culture technique used in this study

phase. As shown in Fig. 2b, the initial pH of APAC was in the range of 3.8–4.0 as the acid–base nature of apple pomace. Eventually, the pH value of APAC gradually rose to approximately 8.5. The escalating pH during composting may be attributed to the release of ammonia, methanol, and the decomposition of organic acids of apple pomace [51]. The water content of APAC dropped fast at the early stages and then declined slowly. The small variations of pH and water content at the end of enrichment indicated that the microorganisms were still active and the degradation of apple pomace continued.

Changes in bacterial community structure during composting

To characterize the changes of microbiota structure during composting, 16S rDNA sequencing on the representative samples of different phases, i.e., CM0 (day 0), Mes5 (day 5 of Mesophilic), The15 (day 15 of Thermophilic), and Mat30 (day 30 of Maturation), was performed. As expected, the sample CM0 had the highest α diversity (OUT numbers and Chao1 estimator) while The15 had the lowest (Additional file 1: Table S1). Although the main phyla throughout the entire composting process

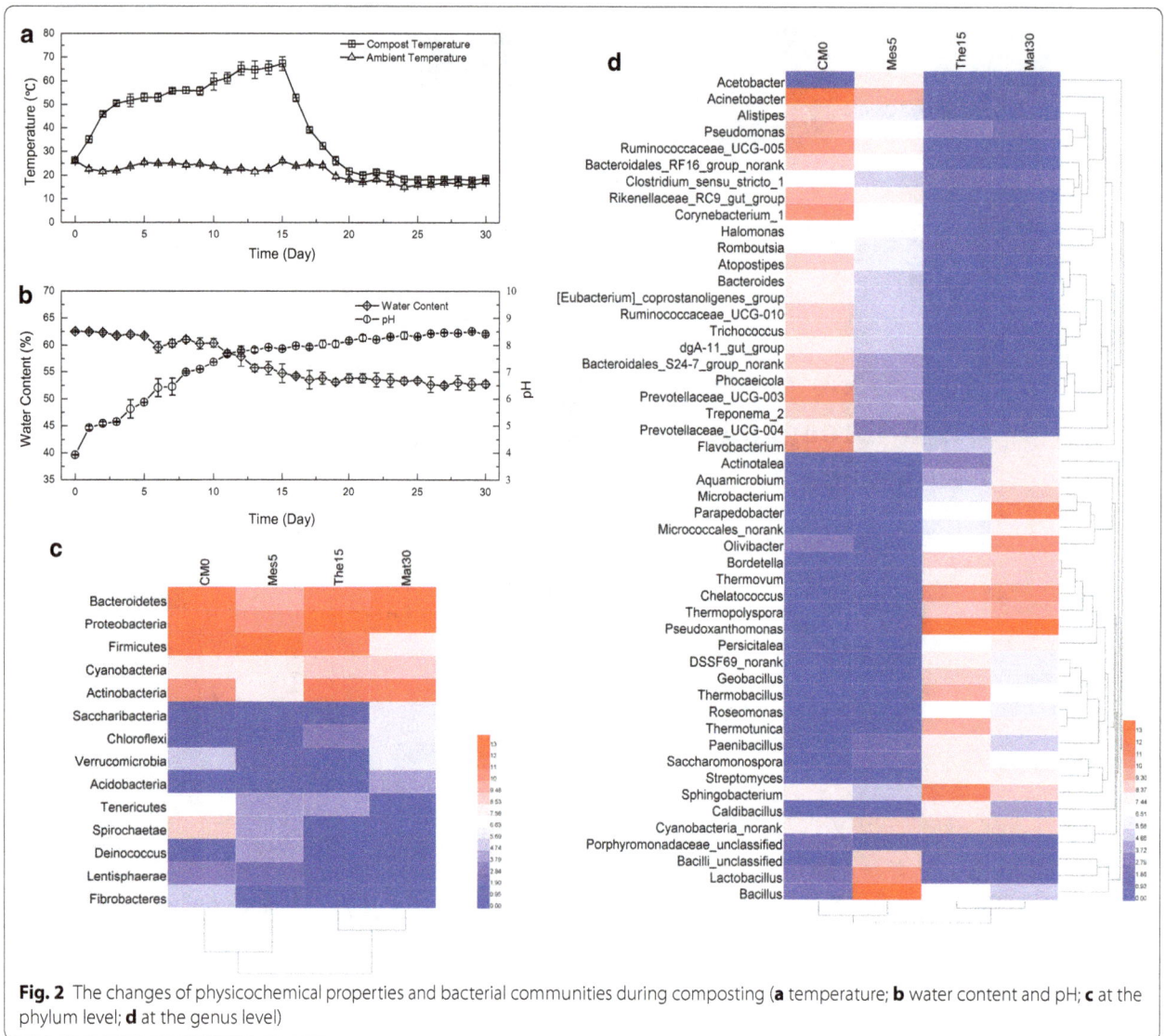

Fig. 2 The changes of physicochemical properties and bacterial communities during composting (**a** temperature; **b** water content and pH; **c** at the phylum level; **d** at the genus level)

did not change greatly (Fig. 2c), namely *Actinobacteria*, *Bacteroidetes*, *Proteobacteria*, *Firmicutes*, and *Cyanobacteria*, the main genera varied dramatically with some bacterial populations being greatly enriched (Fig. 2d). At the genus level, the bacterial community profiles of the main genera were clustered into two groups, which the bacterial community structures in The15 and Mat30 differed remarkably from CM0. The genera *Acinetobacter* (21.9%), *Planococcaceae unclassified* (8.0%), and *Ruminococcaceae* UCG-005 (6.9%) were the dominant in the CM0, whereas they declined to a very low level or completely disappeared after day 15 (Fig. 2d). By contrast, the genera *Pseudoxanthomonas* (36.7%), *Parapedobacter* (16.8%), *Chelatococcus* (7.2%), *Olivibacter* (4.7%), and *Sphingobacterium* (2.4%) were enriched in the composting process. The evolution of specific populations reflected that

APAC had adapted to apple pomace degradation. Furthermore, most of the species abundant in Mat30 have been detected showing highly positive correlations on lignocellulose/pectin-degrading activities [9, 36, 44, 52]. The predominance of lignocellulolytic or pectinolytic species suggests that APACMC has potential to degrade lignocellulose and pectin effectively.

Moreover, the pectinolytic activities of APAC increased dramatically after the composting process, which preliminarily proved the effectiveness of the microbial enrichment (Additional file 2: Fig. S1). Further, it was showed that 83.25% pectin was degraded through 30-day composting. Consequently, APACMC was supposed to be successfully established with pectinolytic capability by means of the enrichment culture technique we adopted.

Microbial diversity in APACMC metagenome

To obtain more detailed information on the diversities of pectinolytic microbes and genes encoding pectinolytic enzymes in APACMC, shotgun sequence of Mat30 was performed by using Illumina HiSeq4000 platform. Metagenomic sequencing of APACMC yielded 89,623,103 reads after quality filtering. After assembly, 272,516 predicted ORFs with the average length of 668 bp were obtained (Additional file 3: Table S2). The analysis of metagenomic datasets showed that APACMC was predominately composed of bacterial members (~99.7%), along with very few archaea, eukarya, and uncharacterized organisms. This could be explained by the fact that fungal activities were precluded as APACMC sustained high temperatures between 55 and 68 °C (Fig. 2a).

To estimate the microbial diversity more accurately, various taxonomic protocols such as MEGAN, MetaPhlAn, and MG-RAST were used, while minor differences in the rank abundance order were observed. Taxonomic analysis revealed that APACMC was primarily consisted of members from phyla *Proteobacteria*, *Bacteroidetes*, *Actinobacteria*, and *Firmicutes* (Fig. 3a), which agreed with the result of 16S rDNA sequencing (Mat30 in Fig. 2c). Several previous studies have reported that thermophilic compost communities contain high-abundance genera within these phyla [18, 36]. Other phyla, such as *Verrucomicrobia*, *Cyanobacteria*, and *Planctomycetes* were presented at very low abundances, which together accounted for only 1.04% of the total sequences. Meanwhile, around 623 predicted genes could not be assigned to a definite bacteria phylum, which may belong to yet uncharacterized bacteria.

The prevalence of genus *Sphingobacterium* (17.11%), *Pseudoxanthomonas* (13.46%), *Bordetella* (4.83%), *Microbacterium* (2.23%), *Pedobacter* (2.13%), *Niabella* (2.10%), and *Thermobacillus* (1.33%) was in accordance with previous studies, which have found these species to be major components of bacterial consortia with high cellulolytic or pectinolytic activity in compost habitats [9, 18, 36, 40]. The taxonomic compositions at genus level were also consistent with the result of 16S rDNA sequencing (Mat30 in Fig. 2d). In addition, it was found some *Enterobacteriaceae* species also had high capacity for metabolizing pectin, such as genera *Yersinia*, *Klebsiella*, *Dickeya*, and *Pectobacterium* [1]. Moreover, the thermophilic pectinolytic bacteria, such as *Actinomadura* [38], *Bacillus* [25], *Geobacillus* [43], *Streptomyces* [50], *Thermomonospora*, and *Thermobifida* [44] were also detected.

To gain further insight into the diversity of APACMC, the metagenomic dataset was taxonomically profiled at the species level (Fig. 3b). The result showed that *Nannocystis exedens* (30.44%, marked with the highest bar, see Fig. 3b) was the most prevalent species accompanying with *Sphingobacterium* sp. 21 (23.33%), *Parapedobacter composti* (18.97%), *Cytophagaceae bacterium SCN 52-12* (13.69%), *Thermobacillus composti* (18.24%) and *Saccharomonospora glauca* (11.91%). It was reported that most of them (marked with asterisk, see Fig. 3b) are proficient degraders of lignocellulose [18, 36, 40]. Simultaneously, a number of thermophilic pectinolytic bacteria

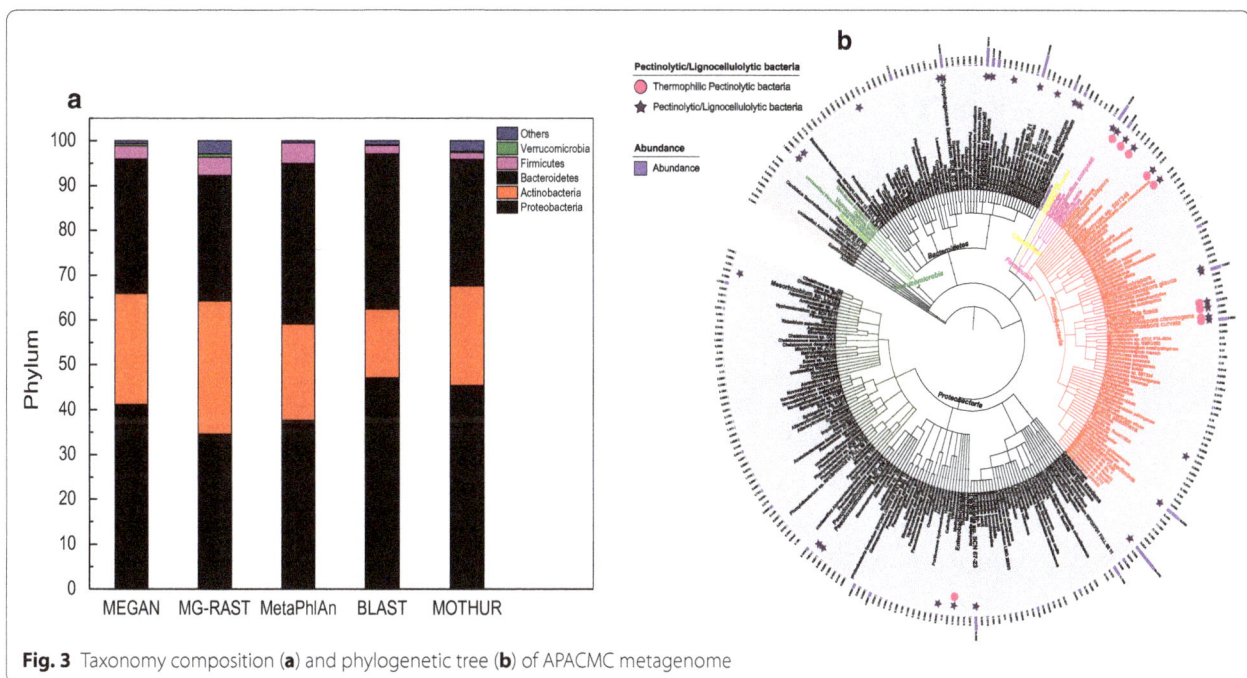

Fig. 3 Taxonomy composition (**a**) and phylogenetic tree (**b**) of APACMC metagenome

(marked with solid circle, see Fig. 3b) including *Thermobispora bispora* [44], *Thermomonospora curvata* [50], and *Thermobifida fusca* [52] were found. Collectively, large proportion of lignocellulolytic and pectinolytic microorganisms further confirms that APACMC has possessed the targeted functions.

Functional profiles of predicated genes in APACMC metagenome

Annotations by the MG-RAST pipeline revealed that 99.98% of the predicted genes were protein coding, among which 81.01% had been assigned a putative function. The COG and KEGG repertoire of the predicted genes was analyzed to assess the primary functions of these genes in APACMC. The COG categories analysis (Fig. 4a) showed that APACMC was enriched for amino acid metabolism (8.8% in all COG functional categories), general function (8.7%), inorganic ion metabolism (7.9%), carbohydrate transport and metabolism (7.1%), energy production and conversion (6.6%), and cell wall/membrane/envelope biogenesis (6.1%). The comparative COG analysis of the APACMC with another four well-known lignocellulose-degrading consortia from rain forest compost, switchgrass-adapted compost [2], Sao Paulo zoo park compost [24], and rice straw-adapted compost [36] revealed that they shared similar metabolic patterns,

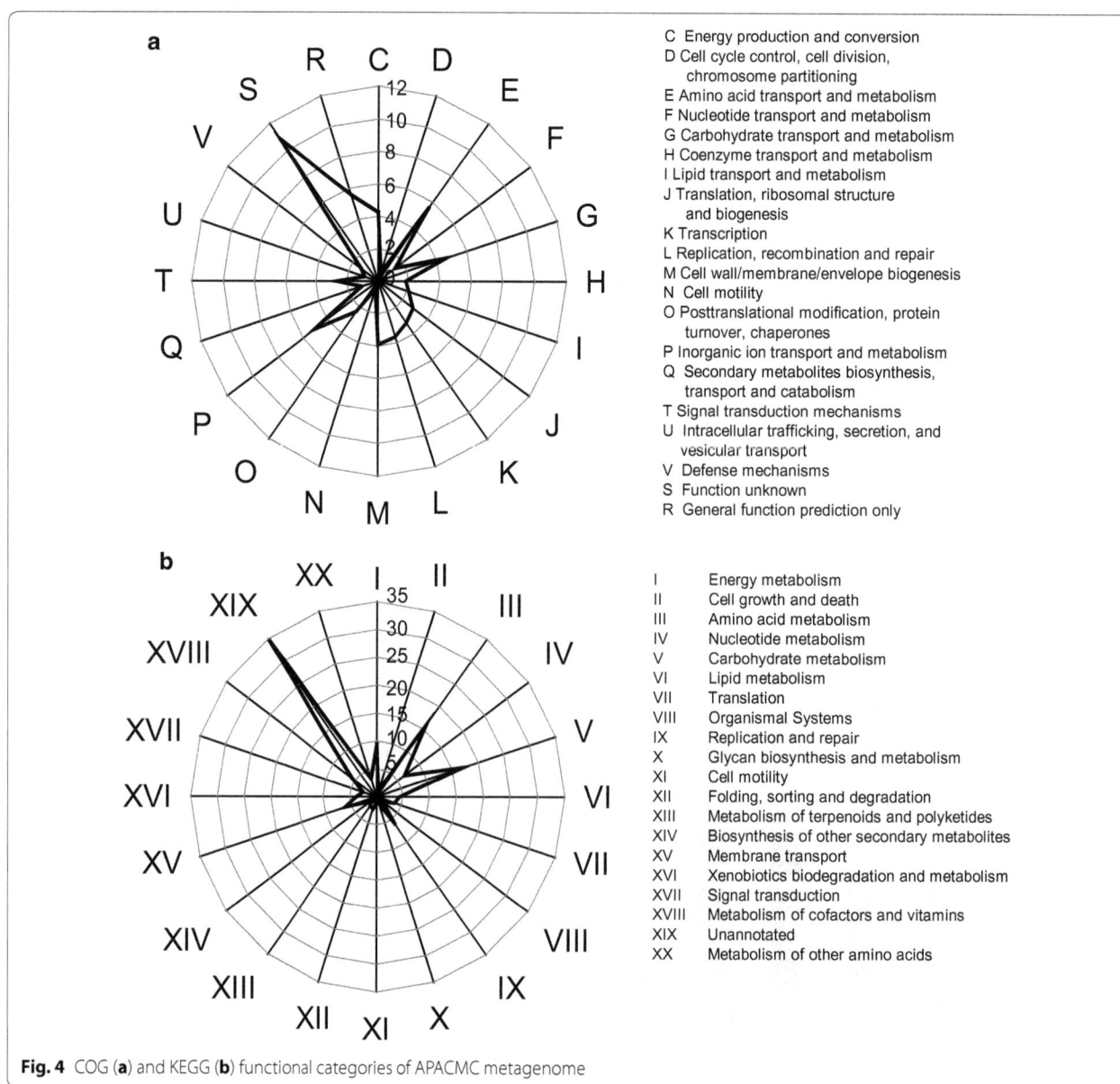

C Energy production and conversion
D Cell cycle control, cell division, chromosome partitioning
E Amino acid transport and metabolism
F Nucleotide transport and metabolism
G Carbohydrate transport and metabolism
H Coenzyme transport and metabolism
I Lipid transport and metabolism
J Translation, ribosomal structure and biogenesis
K Transcription
L Replication, recombination and repair
M Cell wall/membrane/envelope biogenesis
N Cell motility
O Posttranslational modification, protein turnover, chaperones
P Inorganic ion transport and metabolism
Q Secondary metabolites biosynthesis, transport and catabolism
T Signal transduction mechanisms
U Intracellular trafficking, secretion, and vesicular transport
V Defense mechanisms
S Function unknown
R General function prediction only

I   Energy metabolism
II   Cell growth and death
III   Amino acid metabolism
IV   Nucleotide metabolism
V   Carbohydrate metabolism
VI   Lipid metabolism
VII   Translation
VIII   Organismal Systems
IX   Replication and repair
X   Glycan biosynthesis and metabolism
XI   Cell motility
XII   Folding, sorting and degradation
XIII   Metabolism of terpenoids and polyketides
XIV   Biosynthesis of other secondary metabolites
XV   Membrane transport
XVI   Xenobiotics biodegradation and metabolism
XVII   Signal transduction
XVIII Metabolism of cofactors and vitamins
XIX   Unannotated
XX   Metabolism of other amino acids

Fig. 4 COG (**a**) and KEGG (**b**) functional categories of APACMC metagenome

particularly associated with carbohydrates and amino acids transport and metabolism (Additional file 4: Fig. S2). The KEGG ontology exhibited analogous patterns (Fig. 4b), where carbohydrate metabolism (17.0%), amino acid metabolism (16.0%), energy metabolism (9.4%), nucleotide metabolism (6.6%), and membrane transport (6.2%) were abundant. Generally, these observations indicate that APACMC has successfully enriched several desired functional capacities, especially, for carbohydrate metabolism.

In order to get more detailed information about the decomposition of pectin, specific COGs involved in pectin transport and metabolism were further analyzed (Additional file 5: Table S3). APACMC harbored a broad spectrum of genes involved in the metabolism of different monosaccharide building blocks of pectin (e.g., arabinose, fucose, galactose, mannose, rhamnose, xylose, etc.), all of which accounted for 25.8% of the COG subcategory G (Carbohydrate transport and metabolism). Additionally, the genes associated with carbohydrate transporters and phosphotransferase systems were also very plentiful. For example, ABC-type sugar transport system, permease, TonB, and phosphotransferase system, which are responsible for the uptake, transport, and phosphorylation of sugars [18, 44], took up 7.2, 10.6, 2.6, and 1.0% of the COG subcategory G, respectively. In summary, the rich diversity of gene functions in carbohydrate transport and metabolism indicates that APACMC has enriched a great potential for the degradation of pectin.

The diversity, abundance, and phylogenetic distribution of CAZymes in APACMC metagenome

It is well established that the plant biomass-degrading capacities of microbial consortia are closely related to genes encoding CAZymes [52]. To gain an overview of microbial degradation of main polymers in apple pomace, we screened APACMC metagenome for the discovery of microorganisms and genes encoding CAZymes. All the candidate genes of APACMC metagenome were searched against the CAZy database using dbCAN [49] for the presence of at least one relevant catalytic domain or carbohydrate-binding module, rather than overall sequence similarity to known CAZymes. The results showed that APACMC harbored a total of 9274 different CAZyme genes, which distributed heterogeneously among glycoside hydrolases (GHs, 35.6%), glycosyltransferases (GTs, 26.9%), carbohydrate esterases (CEs, 17.5%), carbohydrate-binding modules (CBMs, 13.3%), auxiliary activities (AAs, 4.6%), and polysaccharide lyases (PLs, 2.1%) (Additional file 6: Table S4).

To link metabolic functions to abundant consortia members, all the CAZyme genes of APACMC were

taxonomically classified. The phylogenetic distribution of CAZyme genes showed that the abundance of CAZyme genes varied across bacterial phyla and most of them were derived from *Bacteroidetes* (51.86%), *Proteobacteria* (30.82%), *Actinobacteria* (13.52%), and *Firmicutes* (2.31%) (Fig. 5a). Notably, members of *Bacteroidetes* were predominant in the CAZymes classes (GHs, GTs, CEs, PLs, and CBMs) of APACMC, which was remarkably different from those of rice straw-adapted compost (*Actinobacteria*) [44] and corn stover-adapted compost (*Proteobacteria*) [52].

As the CAZyme genes were unevenly distributed within each phylum, the extensive phylogenetic distributions of CAZyme genes at lower taxonomic levels were further investigated. In addition to *Sphingobacterium* and *Niabella* of the phylum *Bacteroidetes*, CAZyme genes were also abundant in *Pseudoxanthomonas* and *Chelatococcus* of the phylum *Proteobacteria*, *Microbacterium* of the phylum *Actinobacteria*, as well as *Thermobacillus* of the phylum *Firmicutes* (Fig. 5b). At the species level, six different species of phylum *Bacteroidetes*, which accounted for 25.06% of the total CAZyme genes, were present in the top-10 richest members harboring CAZyme genes (Fig. 5c). This finding indicates that members of *Bacteroidetes* possess a much abundant and wider range of CAZyme catalog in APACMC. Besides, the CAZyme genes were also detected in uncharacterized species of *Chelatococcus* (5.04%), *Pseudoxanthomonas* sp. GW2 (3.62%), and *Pseudoxanthomonas* sp. J31 (4.48%) of the phylum *Proteobacteria* and *Thermobacillus composti* (1.57%) of the phylum of *Firmicutes*.

The phylogenetic distributions of CAZyme genes corresponded well to the structure of the ecologically dominant species in APACMC, which confirms the assumption that the functional traits of consortia have a direct correlation with their taxonomic profiles [44]. In conclusion, the CAZymes profile reveals that polysaccharides of apple pomace are decomposed by the predominant *Bacteroidetes* in cooperation with *Proteobacteria*, *Actinobacteria*, and *Firmicutes*. Together with the COG profiles for glycan degradation, the diverse repertoire of CAZymes provides a basis for a collaborative system tailored to the processing and metabolizing of apple pomace in the compost habitat.

Mining for pectinolytic enzymes

Pectin is the major composition of apple pomace and is an extremely structurally complex polysaccharide, which is constituted of as many as 17 different monosaccharides and more than 20 different linkages [5]. The representative structure of pectin is schematically shown in Fig. 6. It is basically composed of homogalacturonan (HG), rhamnogalacturonan I (RG-I), the substituted galacturonans

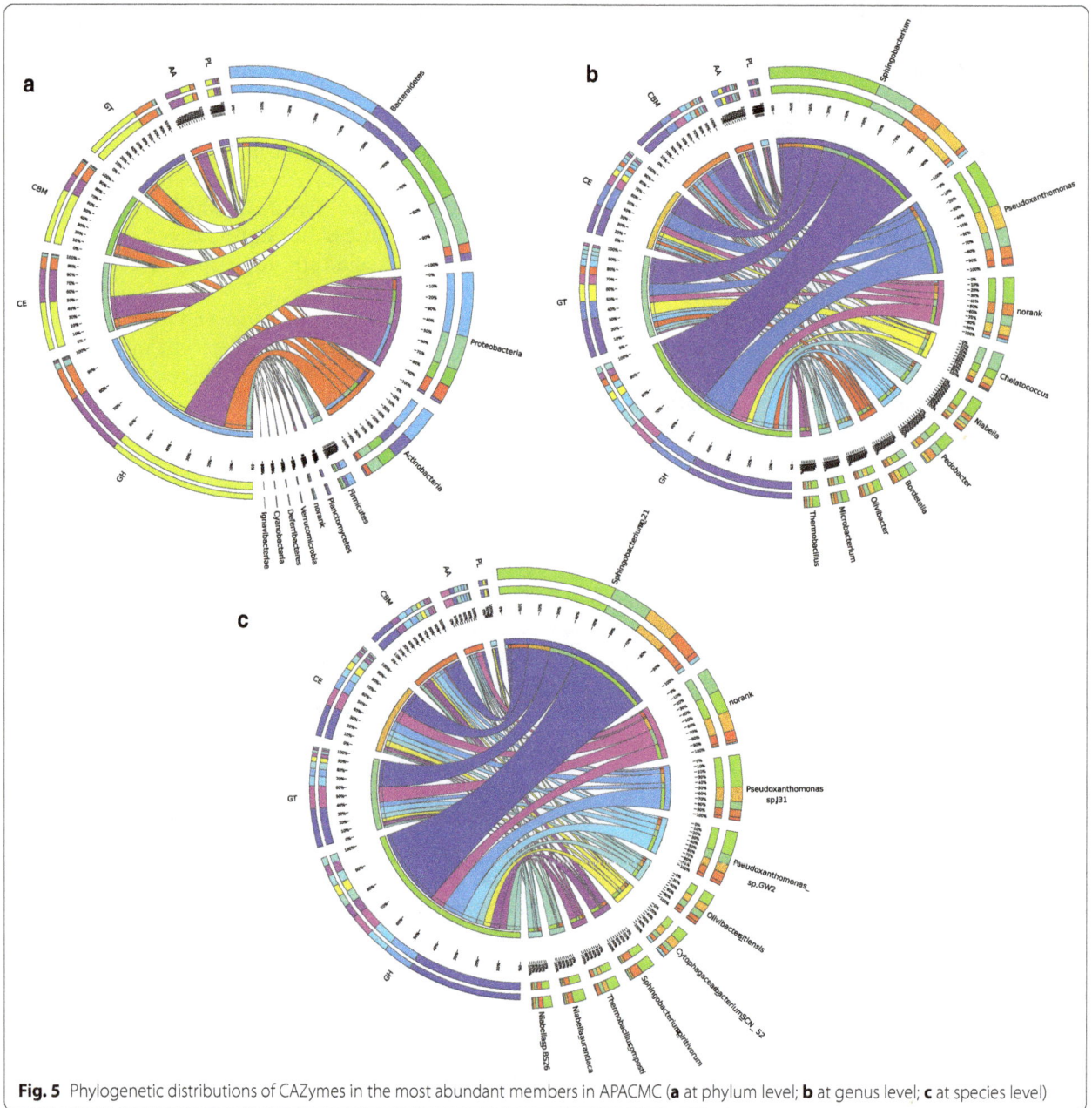

Fig. 5 Phylogenetic distributions of CAZymes in the most abundant members in APACMC (**a** at phylum level; **b** at genus level; **c** at species level)

rhamnogalacturonan II (RG-II), and xylogalacturonan (XGA) [19]. Due to its complex and heterogeneous structure, the efficient and complete degradation of pectin involves a battery of enzymes which act specifically and synergistically. These pectin-degrading enzymes are classified as de-polymerases (hydrolases and lyases), pectinesterases, and de-branching enzymes based on the action mode and site.

The different types of pectinolytic enzymes and their cleavage sites are depicted in Fig. 6. The degradation of pectin is caused by the de-esterification of methoxyl groups, affecting the texture and rigidity of the cell wall [1]. Pectin methylesterases (PMEs, EC 3.1.1.11) remove the methyl groups from the HG backbone to give access to de-polymerases, while pectin acetylesterases (PAEs, EC 3.1.1.6) remove acetyl groups from acetylated HG and RG [32]. Hydrolases (polygalacturonases PGs, EC 3.2.1.15, 67 and 82) and lyases (pectin lyases: PNLs, EC 4.2.2.10 and pectate lyases: PELs, EC 4.2.2.2 and 9) preferentially degrade the α-1,4-glycosidic bonds of HG/XGA backbones by hydrolysis and β-elimination, respectively. Similarly, in the initial deconstruction of pectin, PNLs

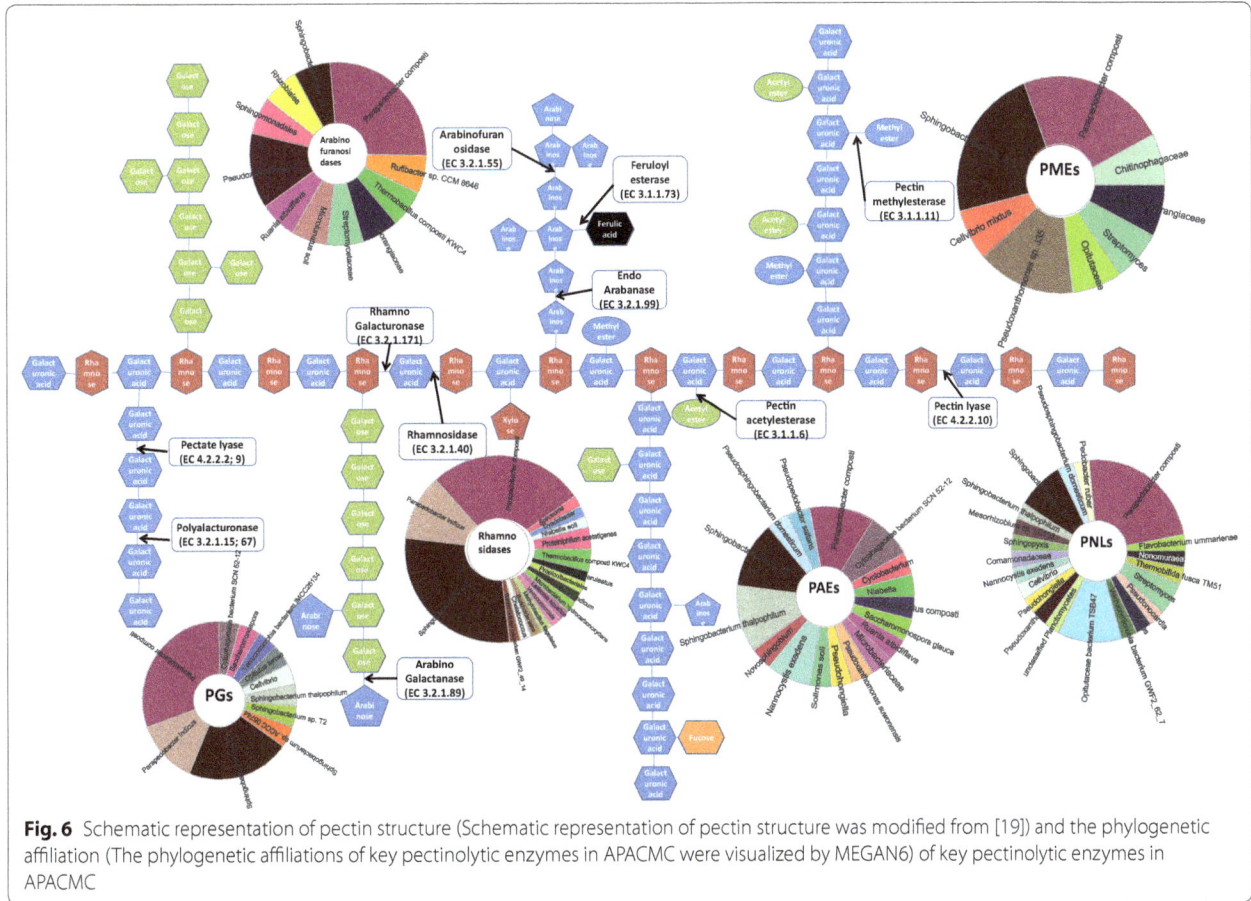

Fig. 6 Schematic representation of pectin structure (Schematic representation of pectin structure was modified from [19]) and the phylogenetic affiliation (The phylogenetic affiliations of key pectinolytic enzymes in APACMC were visualized by MEGAN6) of key pectinolytic enzymes in APACMC

also play an essential role since it is the only enzyme that can cleave the α-1, 4 bonds of highly esterified pectin without prior actions of other enzymes. De-branching enzymes are responsible for the cleavage of the backbone or lateral chains of RG-I and RG-II [5]. Rhamnogalacturonases (EC 3.2.1.171, 173 and 174) and rhamnogalacturonan lyases (RGLs: EC 4.2.2.23 and 24) are involved in the cleavage of the RG-I backbone; α-L-rhamnosidases (EC 3.2.1.40) act on hydrolytic cleavage of the RG chain at non-reducing end releasing rhamnose; arabinofuranosidases (EC 3.2.1.55) attack on α-L-arabinofuranosides, α-L-arabinans, arabinoxylans, and arabinogalactans; arabinogalactanases (EC 3.2.1.89) and β-galactosidases (EC 3.2.1.23) act randomly on the galactan core of AGs; and feruloyl esterases (EC 3.2.1.73) are needed to release the ferulic acid attached to C-2 of arabinose and C-6 of galactose.

According to the results of CAZymes annotation, we found an extremely abundant of genes associated with the complete degradation of pectin (Additional file 7: Table S5). A total of 1756 entries were identified as encoding pectinolytic enzymes, which took up to 18.93% of the total CAZy genes. As summarized in Table 1, these

entries contained 105 PLs candidates from 6 families, 881 GHs mainly from 17 families, 537 CEs from 5 families, and 233 CBMs from 5 families. Compared to another two compost habitats (Table 1), the catalog of pectinolytic enzymes in APACMC was much more abundant and diverse than RSA (843 candidates from 31 families) and EMSD5 (398 candidates from 28 families), indicating that APACMC has a better potential for pectin degradation based on CAZyme inventory. As shown in Table 1, a large panel of pectinolytic enzymes was found, such as PGs, PNLs, PELs, RGLs, PMEs, PAEs, α-L-rhamnosidases, arabinofuranosidases, arabinogalactanases, and β-galactosidases. Furthermore, we also detected a wealth of CBMs which possibly associate to pectin degradation. For example, some members of family CBM32 have been found to bind oligogalacturonides to counteract the loss of binding affinity between thermophilic pectinases and their substrates at elevated temperature [1, 52]. These findings indicate that APACMC exhibits a collaborative enzymatic system efficient in the complete degradation of pectin.

To assess the identity of these possible pectinolytic enzymes with known proteins, these amino acid

Table 1 Summary of pectinolytic enzymes in three metagenomes

Pectinolytic enzymes	CAZy family		Predominant activity	This study	RSA[a]	EMSD5[b]
Depolymerizing enzymes	Glycoside hydrolases (GHs)	GH4	Exo-polygalacturonase (EC 3.2.1.67)	14	21	24
		GH28	Polygalacturonase (EC 3.2.1.15); exo-polygalacturonase (EC 3.2.1.67); exo-polygalacturonosidase (EC 3.2.1.82); rhamnogalacturonase (EC 3.2.1.171); rhamnogalacturonan α-1,2-galacturonohydrolase (EC 3.2.1.173); rhamnogalacturonan α-L-rhamnopyranohydrolase (EC 3.2.1.174)	32	11	3
	Polysaccharide lyases (PLs)	PL1	Pectate lyase (EC 4.2.2.2); exo-pectate lyase (EC 4.2.2.9); pectin lyase (EC 4.2.2.10)	30	8	4
		PL3	Pectate lyase (EC 4.2.2.2)	2	4	0
		PL4	Rhamnogalacturonan lyase (EC 4.2.2.-)	0	0	1
		PL9	Pectate lyase (EC 4.2.2.2); exopolygalacturonate lyase (EC 4.2.2.9)	12	15	4
		PL10	pectate lyase (EC 4.2.2.2)	13	8	0
		PL11	Rhamnogalacturonan endolyase (EC 4.2.2.23); rhamnogalacturonan exolyase (EC 4.2.2.24)	18	10	1
		PL22	Oligogalacturonate lyase/oligogalacturonide lyase (EC 4.2.2.6, 9)	30	21	12
Pectinesterases	Carbohydrate esterases (CEs)	CE1	Feruloyl esterase (EC 3.1.1.73)	482	223	82
		CE8	Pectin methylesterase (EC 3.1.1.11)	17	8	6
		CE12	Pectin acetylesterase (EC 3.1.1.6); rhamnogalacturonan acetylesterase (EC 3.1.1.86)	34	4	8
		CE13	Pectin acetylesterase (EC 3.1.1.6)	2	0	0
		CE16	Pectin acetylesterase (EC 3.1.1.6)	4	0	0
De-branching enzymes	Glycoside hydrolases (GHs)	GH1	β-Glucosidase (EC 3.2.1.23)	60	63	44
		GH2	β-Glucosidase (EC 3.2.1.23); α-L-arabinofuranosidase (EC 3.2.1.55)	76	27	19
		GH3	β-Glucosidase (EC 3.2.1.23); α-L-arabinofuranosidase (EC 3.2.1.55)	143	83	32
		GH10	Endo-1,4-β-xylanase (EC 3.2.1.8); endo-1,3-β-xylanase (EC 3.2.1.32)	50	37	6
		GH35	β-Glucosidase (EC 3.2.1.23)	9	7	1
		GH42	β-Glucosidase (EC 3.2.1.23)	33	14	9
		GH43	α-L-Arabinofuranosidase (EC 3.2.1.55); arabinanase (EC 3.2.1.99)	205	51	33
		GH51	α-L-Arabinofuranosidase (EC 3.2.1.55)	40	24	4
		GH53	Arabinogalactanase (EC 3.2.1.89)	8	6	6
		GH59	β-Glucosidases (EC 3.2.1.23)	2	2	0
		GH62	α-L-Arabinofuranosidase (EC 3.2.1.55)	2	6	0
		GH78	α-L-Rhamnosidase (EC 3.2.1.40)	92	39	8
		GH105	Unsaturated rhamnogalacturonyl hydrolase (EC 3.2.1.172)	35	4	5
		GH106	α-L-Rhamnosidase (EC 3.2.1.40)	30	6	3
		GH127	β-L-Arabinofuranosidase (EC 3.2.1.185)	51	17	11
	Carbohydrate-binding modules (CBMs)	CBM13	Arabinanase (GH43D;RUM_09280); feruloyl esterase I (FaeI;CGSCsYakCAS_18248); pectate lyase B (PelB)	25	14	24
		CBM32	Binding to galactose, lactose, polygalacturonic acid and LacNAc	137	63	23
		CBM35	Arabinanase; feruloyl esterase D (Fae1;XylD;XynD;CJA_3282); pectate lyase (PelA;CJA_3104)	27	17	10
		CBM61	Modules of approx. 150 residues found appended to GH43, GH53 catalytic domains	3	3	2
		CBM66	Pectate lyase (PecB;Athe_1854;Cbes_1854;Cbes1854)	63	27	13
Total			Families	33	31	28
			ORFs	1756	843	398

[a] RSA (Rice Straw-Adapted) microbial consortia adapted to rice straw from Ref. [44]

[b] EMSD5 microbial consortia adapted to corn stover from reference of [52]

sequences of 1756 putative pectinolytic genes were searched against NCBI non-redundant (NCBI-NR), CAZy, NCBI environmental (NCBI-ENV), and Swiss-Prot databases by DIAMOND [6] (Fig. 7; Additional file 8: Table S6). Firstly, the results based on NCBI-NR showed that the amino acid sequence identity of these 1756 genes ranged from 25 to 100%, with an average of 76.95% (Additional file 8: Table S6). And 23.29% of these sequences were most similar to proteins annotated as "hypothetical/predicted protein" or "proteins of unknown function" in NCBI-NR. Secondly, only 9.83% of these sequences were highly similar (>95% sequence identity) to any CAZy database entry, indicating that most of these sequences had not been previously deposited in CAZy [15]. And 129 sequences were considered as novel with less than 30% identity [31]. Thirdly, only 1.92% of these putative pectinolytic genes are more than 75% identical to sequences deposited in the NCBI-ENV database, demonstrating that these enzymes also have not been observed in previous metagenome projects [15]. Lastly, 145 sequences had less than 30% identity to any known proteins deposited in Swiss-Prot, indicating that their assigned activity has not been verified biochemically. Summarily, the large amount of relatively low identity sequences indicate that the strategy we adopted has great potential in mining novel enzymes from environmental sources.

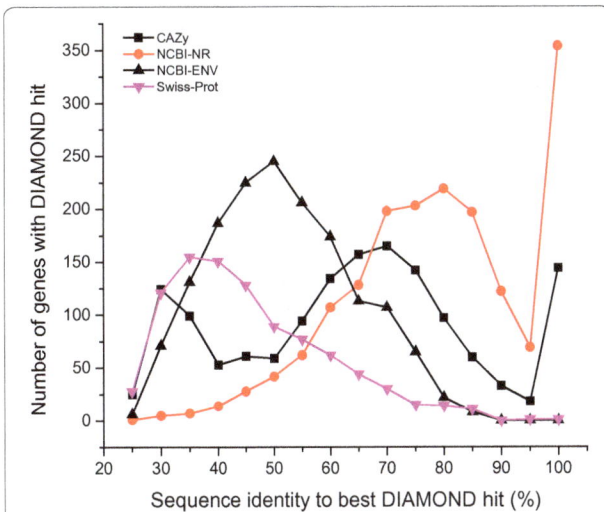

Fig. 7 Similarity distribution of putative pectinolytic candidates (*n* = 1756) containing a catalytic domain (CD) or a carbohydrate-binding module (CBM) associated with pectinolytic activity. Sequences were compared to the NCBI-NR (*red* 1756 hits), CAZy (*black* 1464 hits), NCBI-ENV (*blue* 1560 hits), and Swiss-Prot (*pink* 927 hits) databases (best BLAST hit, *E* value $\leq 1e^{-5}$); 26 genes contained both a CD and CBM, whereas 1498 and 232 genes contained only a CD or CBM, respectively

Mining for pectinolytic microbes

To explore the phylogenetic origins of these pectinolytic enzymes, we examined the top BLASTX hit organism of each identified enzyme at species level, deciphered the role of individual pectinolytic microbe and their potential synergistic action in the process of pectin degradation. Of the 1756 sequences encoding pectinolytic enzymes, most of their phylogenetic affiliations predicted by BLASTX were consistent with the predicted source organisms of APACMC metagenomic bins (Additional file 7: Table S5). Many of these genes are homologous to those found in the top-10 abundant community members (Fig. 5c), such as *Sphingobacterium* sp. *21*, *Sphingobacterium spiritivorum*, *Thermobacillus composti*, and *Cytophagaceae bacterium SCN 52-12*, which further verified that APACMC was successfully target-enriched for pectin degradation.

To provide a systematic overview of pectin degradation by individual member of APACMC, the specific taxonomic assignments of key pectinolytic enzymes, i.e., PGs, PMEs, PNLs, PELs, α-ʟ-rhamnosidases, and arabinofuranosidases, were illustrated in Fig. 6. Clearly, pectinolytic species in APACMC were considerably diverse. The majority of candidate PGs was mainly originated from a variety of *Bacteroidetes* species, which consisted of *Parapedobacter composti*, *Sphingobacterium* sp. *21*, *Parapedobacter indicus*, *Sphingobacterium thalpophilum*, as well as *Opitutus terrae* and *Verrucomicrobia bacterium IMCC26134* from the *Verrucomicrobia*. It is generally known that the bacterial sources of PELs and PNLs are some specific bacteria such as *Bacillus* sp. and *Pseudomonas* sp. [5]. However, in our study, a broad range of other bacteria were the major producers, including *Parapedobacter composti*, *Sphingobacterium* sp. *21* and *Sphingobacterium thalpophilum*, *Opitutaceae bacterium TSB47*, *Thermobifida fusca TM 51*, *Saccharomonospora glauca*, *Nannocystis exedens*, and *Mesorhizobium* sp. *LC 103*. The taxonomic classifications of genes encoding PMEs and PAEs revealed that they were dominantly from *Sphingobacterium* sp. *21*, *Parapedobacter composti*, *Sphingobacterium thalpophilum*, *Cytophagaceae bacterium SCN 52-12* of *Bacteroidetes*, *Pseudoxanthomonas* sp. *J35*, *Cellvibrio mixtus*, *Nannocystis exedens* of *Proteobacteria*, *Saccharomonospora glauca*, and *Ruania albidiflava* of *Actinobacteria* along with *Thermobacillus composti* of *Firmicutes*. The sequences encoding α-ʟ-rhamnosidases and arabinofuranosidases were predominantly affiliated with organisms from *Parapedobacter composti*, *Sphingobacterium* sp. *21* and *Parapedobacter indicus* of *Bacteroidetes*, *Streptomyces caeruleatus* and *Ruania albidiflava* of *Actinobacteria*, and *Thermobacillus composti* of *Firmicutes*. These collective data indicate a synergistic action of multiple members derived from

Bacteroidetes, *Proteobacteria*, *Actinobacteria*, *Firmicutes*, and *Verrucomicrobia* in the degradation of pectin.

Most of the currently available pectinolytic enzymes are reported from filamentous fungal species (e.g., *Aspergillus* sp. and *Penicillium* sp.). However, as the wide functional diversity, broad array of terminal electron acceptors, high ability to degrade lignin [37], as well as more amenable to genetic manipulation, pectinolytic bacteria are likely to play important roles in future biotechnology strategies. Surprisingly, a variety of bacteria were identified as the major producers of various pectinolytic enzymes, such as *Cellvibrio mixtus*, *Cytophagaceae bacterium SCN 52-12*, *Nannocystis exedens*, *Opitutaceae bacterium TSB47*, *Parapedobacter composti*, *Parapedobacter indicus*, *Ruania albidiflava*, *Saccharomonospora glauca*, *Sphingobacterium* sp. *21*, *Sphingobacterium thalpophilum*, and *Thermobacillus composti*. Many of these species were initially described to degrade pectin. Strikingly, *Parapedobacter composti*, *Sphingobacterium* sp. *21*, and *Sphingobacterium thalpophilum*, which were identified as the top-10 richest members, each of them harbors a great number of genes encoding various pectinolytic enzymes, indicating that they are well equipped with systematic pectinolytic enzymes. This make them to be promising bacterial sources of pectinolytic enzymes and potential efficient degraders of pectin in the future. Although putative pectinolytic enzymes have been annotated in the genomes of these type-strains [14], so far the information available on the pectinolytic enzymes of these strains is very limited. Our data provide insight into their potentials and highlight their importance in the complex degradation of pectin.

Conclusions

Novel pectinolytic microbes and enzymes have potential application in numerous industrial processes. Here, we adopted a strategy which combined metagenomics sequencing with enrichment culture technique to rapidly discover efficient pectin degraders and novel pectinolytic enzyme sequences. The immensely diverse pectinolytic microbes and enzymes found in our study will not only shed light on the current understanding of microbial interaction and enzymatic synergism in pectin degradation, but also expand the arsenal of proficient degraders and enzymes for lignocellulosic biofuel production. When combined with high-throughput strategies, such as cell-free protein expression system, droplet-based microfluidics, fluorescence-activated cell sorting (FCAS), and nanostructure-initiator mass spectrometry (NIMS), the efficiency of this strategy for obtaining novel enzymes may meet the ever-growing demand from various industries.

Methods

Enrichment of apple pomace-adapted microbial community in compost habitat

The composting materials were composed of apple pomace and fresh cattle manure. Apple pomace was kindly provided by Shaanxi Haisheng fresh fruit juice Co. Ltd., China [46]. The cattle manure was collected from the Northwest A&F University farm located in Yangling, China. The composting experiment was conducted from September 15 to October 29, 2016 according to Sun et al. [41] with slight modifications. Briefly, the cattle manure was mixed with apple pomace to adjust the C/N ratio to 30:1 and the moisture content to around 60% and then the mixture was placed in rectangular foam containers as described in Sun et al. [41]. The piles of compost were turned and sampled daily. Samples were pooled from the top, middle, and bottom of the composting, and then mixed completely. When the pile temperature dropped to ambient temperature at the end of the maturing stage, the composting process was considered as completed. The each sample was split into two parts: one part was stored at 4 °C for subsequent physicochemical analysis and the other was stored at −80 °C for high-throughput sequencing.

Physicochemical analysis

The pile temperature was monitored every 24 h by inserting a mercury thermometer in the center of the composting material. The moisture content was measured gravimetrically after drying samples at 105 °C for 24 h. The pH values of the samples were tested in water (solid-to-water ratio of 1:10, w/v) with a pH meter [51]. The contents of pectin were determined by modified carbazole method [46].

16S rDNA sequencing and phylogenetic classification

The 16S rDNA sequencing was performed at the Frasergen Genoimcs Institute (Wuhan, China) using the Illumina MiSeq platform [33]. The 16S V3-V4 region was amplified using the primers 338F and 806R. After the processing of raw data, the high-quality sequences were subjected to filter singletons, remove chimeras, and cluster into operational taxonomic units (OTUs) at a 97% identity using UPARSE [10]. A representative sequence of each OTU was assigned to a taxonomic level in the SILVA database [35] using the RDP classifier. Microbial diversity and richness measurements were performed using MOTHUR [39]. The microbial diversity was estimated by Shannon and Simpson, and the richness was determined by Chao and Ace estimators.

Metagenome sequencing, de novo assembly, and Open Reading Frames (ORFs) prediction

Metagenome sequencing, de novo assembly, and ORFs prediction were performed by Frasergen Genoimcs Institute (Wuhan, China) according to Qin with slight modifications [34]. Briefly, a library with 400-bp clone insert size was constructed and sequenced on Illumina HiSeq4000 platform. Sequence reads were quality trimmed to an accuracy of 98.0% and duplicate reads were identified and removed prior to assembly. Nearly 89.6 million high-quality reads were generated (16.2 Gb). High-quality short reads of the DNA sample were assembled by the SOAPdenovo assembler [22] with a k-mer length of 39–47. The assembled contigs longer than 500 bp were subject to ORFs prediction using the Meta-Gene [29] with default parameters. The redundant ORFs were removed by CD-HIT [13] from the non-redundant gene catalog and the abundances were annotated by SOAPaligner [22].

Taxonomic assignment and functional classification in metagenomic database

Taxonomic annotation of predicated genes was performed by BLASTP against the NCBI-NR database with an E value of $1e^{-5}$. The APACMC metagenomic dataset was also taxonomically profiled at species level by MetaPhlAn2 [42], MG-RAST [27] and MEGAN 6 [17]. The phylogenetic tree was generated using iTOL software [21]. Functional classification of predicted gene was performed by BLASTP against eggNOG database and by KOBAS 2.0 (a Orthology Based Annotation System) [48] against KEGG database. The "function comparison" module of integrated microbial genomes with microbiome samples (IMG/M) [7] were applied to compare the COG category of APACMC against another four well-known lignocelluloses-degrading microbiomes available on IMG/M, including rain forest compost (IMG Submission ID 5968), switchgrass-adapted compost [2], sao paulo zoo park compost [24], and rice straw-adapted compost [36].

Carbohydrate-active enzymes (CAZymes): annotation and phylogenetic analysis

Searches for CAZymes were performed as described by Wang and coworkers [44]. Briefly, the amino acid sequences of the predicted ORFs in the APACMC metagenome were annotated by dbCAN, an automated CAZyme signature domain-based annotation method based on family-specific HMMs [49] by MAFFT and HMMER. After identification, these sequences were searched against NCBI non-redundant (NCBI-NR), CAZy database, NCBI environmental database (NCBI-ENV), and Swiss-Prot database by DIAMOND [6] with

a cutoff of E value $<1e^{-5}$. The phylogenetic distributions in the top ten abundant members possessing CAZymes were visualized via software Circos [20] at the level of phylum, genus, and specie.

Specific pectin-degrading genes: annotation and phylogenetic analysis

The predicted sequences encoding pectinolytic enzymes were re-annotated and verified using DIAMOND using a sensitive setting [6] against the proteins deposited in NCBI-NR database. The phylogenetic origins of candidate genes were determined by MEGAN 6 [17].

Sequence data submission

The assembled metagenome datasets were submitted to IMG/M and Metagenomics RAST server (MG-RAST) under the project ID 117466 and mgs566360, respectively.

Additional files

> **Additional file 1: Figure S1.** Red agar test.tif. The effectiveness of enrichment cultures (post- and pre-) by grown on ruthenium red agar plates.
>
> **Additional file 2: Table S1.** Diversity and OUT distributions.xlsx. Diversity and OTU distribution of CM0, Mes5, The15 and Mat30.
>
> **Additional file 3: Table S2.** De novo assembly results.docx. Illumnia reads and de novo assembly results of APACMC metagenome.
>
> **Additional file 4: Figure S2.** The comparison of COG category.tiff. The COG comparison of APAMC with other well-known lignocellulosic metagenomes.
>
> **Additional file 5: Table S3.** Specific COGs.xlsx. Specific COGs envolved in apple pomace deconstruction.
>
> **Additional file 6: Table S4.** Annotation of CAZymes genes.xlsx. Annotation of CAZymes genes by dbCAN.
>
> **Additional file 7: Table S5.** Pectolytic enzyme genes.xlsx. Catalog of genes encoding pectinolytic enzymes.
>
> **Additional file 8: Table S6.** Distributions of identities.xlsx. Distributions of identities of predicated pectinolytic enzyme sequences.

Abbreviations
AAs: auxiliary activities; APAC: apple pomace-adapted compost; APACMC: apple pomace-adapted compost microbial community; CAZymes: carbohydrate-active enzymes; CBMs: carbohydrate-binding modules; CEs: carbohydrate esterases; GHs: glycoside hydrolases; COGs: clusters of orthologous groups; GTs: glycosyltransferases; HG: homogalacturonan; PAEs: pectin acetylesterases; PELs: pectate lyases; PGs: polygalacturonases; PLs: polysaccharide lyases; PMEs: pectin methylesterases; PNLs: pectin lyases; MG-RAST: metagenomics RAST; ORFs: open reading frames; OTU: operational taxonomic unit; RGLs: rhamnogalacturonan lyases; RG-I: rhamnogalacturonan I,; RG-II: galacturonans rhamnogalacturonan II; XGA: xylogalacturonan.

Authors' contributions
MZ conceived and designed the experiments. MZ performed the majority of experimental work, analyzed the results, and drafted this manuscript. PG participated in the analysis of bioinformatics. TW, LG, and HY assisted in the enrichment of APACMC and the chemical composition test. JG and CC helped to guide the composting process and process bioinformatics, respectively. MZ performed all data analyses. XL led and coordinated the overall project. All authors read and approved the final manuscript.

Author details
[1] College of Food Science and Engineering, Northwest A&F University, Yangling, Shaanxi Province, China. [2] College of Information Engineering, Northwest A&F University, Yangling, Shaanxi Province, China. [3] College of Natural Resources and Environment, Northwest A&F University, Yangling, Shaanxi Province, China.

Acknowledgements
Not applicable.

Competing interests
The authors declare that they have no competing interests.

Funding
This study was supported by the Special Fund for Agro-scientific Research in the Public Interest (No. 201503135), Ministry of Agriculture, China.

References
1. Abbott DW, Boraston AB. Structural biology of pectin degradation by Enterobacteriaceae. Microbiol Mol Biol Rev. 2008;72(2):301–16.
2. Allgaier M, Reddy A, Park JI, Ivanova N, D'Haeseleer P, Lowry S, Sapra R, Hazen TC, Simmons BA, VanderGheynst JS, Hugenholtz P. Targeted discovery of glycoside hydrolases from a switchgrass-adapted compost community. PLoS ONE. 2010;5(1):e8812.
3. Biswal AK, Hao Z, Pattathil S, Yang X, Winkeler K, Collins C, Mohanty SS, Richardson EA, Gelineo-Albersheim I, Hunt K, Ryno D, Sykes RW, Turner GB, Ziebell A, Gjersing E, Lukowitz W, Davis MF, Decker SR, Hahn MG, Mohnen D. Downregulation of GAUT12 in Populus deltoides by RNA silencing results in reduced recalcitrance, increased growth and reduced xylan and pectin in a woody biofuel feedstock. Biotechnol Biofuels. 2015;8:41.
4. Biswal AK, Soeno K, Gandla ML, Immerzeel P, Pattathil S, Lucenius J, Serimaa R, Hahn MG, Moritz T, Jönsson LJ. Aspen pectate lyase Ptxt PL1-27 mobilizes matrix polysaccharides from woody tissues and improves saccharification yield. Biotechnol Biofuels. 2014;7(1):1.
5. Bonnin E, Garnier C, Ralet MC. Pectin-modifying enzymes and pectin-derived materials: applications and impacts. Appl Microbiol Biotechnol. 2014;98(2):519–32.
6. Buchfink B, Xie C, Huson DH. Fast and sensitive protein alignment using DIAMOND. Nat Methods. 2015;12(1):59–60.
7. Chen IA, Markowitz VM, Chu K, Palaniappan K, Szeto E, Pillay M, Ratner A, Huang J, Andersen E, Huntemann M, Varghese N, Hadjithomas M, Tennessen K, Nielsen T, Ivanova NN, Kyrpides NC. IMG/M: integrated genome and metagenome comparative data analysis system. Nucleic Acids Res. 2017;45(D1):D507–16.
8. Chung Daehwan, Pattathil Sivakumar, Biswal Ajaya K, Hahn Michael G, Mohnen Debra, Westpheling J. Deletion of a gene cluster encoding pectin degrading enzymes in Caldicellulosiruptor bescii reveals an important role for pectin in plant biomass recalcitrance. Biotechnol Biofuels. 2014;7(1):0.
9. DeAngelis KM, Gladden JM, Allgaier M, D'haeseleer P, Fortney JL, Reddy A, Hugenholtz P, Singer SW, Vander Gheynst JS, Silver WL, Simmons BA, Hazen TC. Strategies for Enhancing the Effectiveness of Metagenomic-based Enzyme Discovery in Lignocellulolytic Microbial Communities. Bioenergy Research. 2010;3(2):146–58.
10. Edgar RC. UPARSE: highly accurate OTU sequences from microbial amplicon reads. Nat Methods. 2013;10(10):996–8.
11. Edwards MC, Doran-Peterson J. Pectin-rich biomass as feedstock for fuel ethanol production. Appl Microbiol Biotechnol. 2012;95(3):565–75.
12. Francocci F, Bastianelli E, Lionetti V, Ferrari S, De Lorenzo G, Bellincampi D, Cervone F. Analysis of pectin mutants and natural accessions of Arabidopsis highlights the impact of de-methyl-esterified homogalacturonan on tissue saccharification. Biotechnol Biofuels. 2013;6(1):163.
13. Fu L, Niu B, Zhu Z, Wu S, Li W. CD-HIT: accelerated for clustering the next-generation sequencing data. Bioinformatics. 2012;28(23):3150–2.
14. Hahnke RL, Meier-Kolthoff JP, Garcia-Lopez M, Mukherjee S, Huntemann M, Ivanova NN, Woyke T, Kyrpides NC, Klenk HP, Goker M. Genome-Based Taxonomic Classification of Bacteroidetes. Front Microbiol. 2016;7:2003.
15. Hess M, Sczyrba A, Egan R, Kim T-W, Chokhawala H, Schroth G, Luo S, Clark DS, Chen F, Zhang T. Metagenomic discovery of biomass-degrading genes and genomes from cow rumen. Science. 2011;331(6016):463–7.
16. Himmel ME, Ding SY, Johnson DK, Adney WS, Nimlos MR, Brady JW, Foust TD. Biomass recalcitrance: engineering plants and enzymes for biofuels production. Science. 2007;315(5813):804–7.
17. Huson DH, Beier S, Flade I, Gorska A, El-Hadidi M, Mitra S, Ruscheweyh HJ, Tappu R. MEGAN community edition—interactive exploration and analysis of large-scale microbiome sequencing data. PLoS Comput Biol. 2016;12(6):e1004957.
18. Jimenez DJ, Chaves-Moreno D, van Elsas JD. Unveiling the metabolic potential of two soil-derived microbial consortia selected on wheat straw. Sci Rep. 2015;5:13845.
19. Khan M, Nakkeeran E, Umesh-Kumar S. Potential application of pectinase in developing functional foods. Annu Rev Food Sci Technol. 2013;4:21–34.
20. Krzywinski M, Schein J, Birol I, Connors J, Gascoyne R, Horsman D, Jones SJ, Marra MA. Circos: an information aesthetic for comparative genomics. Genome Res. 2009;19(9):1639–45.
21. Letunic I, Bork P. Interactive tree of life (iTOL) v3: an online tool for the display and annotation of phylogenetic and other trees. Nucleic Acids Res. 2016;44(W1):W242–5.
22. Li R, Zhu H, Ruan J, Qian W, Fang X, Shi Z, Li Y, Li S, Shan G, Kristiansen K, Li S, Yang H, Wang J, Wang J. De novo assembly of human genomes with massively parallel short read sequencing. Genome Res. 2010;20(2):265–72.
23. Lionetti V, Francocci F, Ferrari S, Volpi C, Bellincampi D, Galletti R, D'Ovidio R, De Lorenzo G, Cervone F. Engineering the cell wall by reducing de-methyl-esterified homogalacturonan improves saccharification of plant tissues for bioconversion. Proc Natl Acad Sci. 2010;107(2):616–21.
24. Martins LF, Antunes LP, Pascon RC, de Oliveira JC, Digiampietri LA, Barbosa D, Peixoto BM, Vallim MA, Viana-Niero C, Ostroski EH, Telles GP, Dias Z, da Cruz JB, Juliano L, Verjovski-Almeida S, da Silva AM, Setubal JC. Metagenomic analysis of a tropical composting operation at the sao paulo zoo park reveals diversity of biomass degradation functions and organisms. PLoS ONE. 2013;8(4):e61928.
25. Mei Y, Chen Y, Zhai R, Liu Y. Cloning, purification and biochemical properties of a thermostable pectinase from Bacillus halodurans M29. J Mol Catal B Enzym. 2013;94:77–81.
26. Menon V, Rao M. Trends in bioconversion of lignocellulose: biofuels, platform chemicals & biorefinery concept. Prog Energy Combust Sci. 2012;38(4):522–50.
27. Meyer F, Paarmann D, D'Souza M, Olson R, Glass EM, Kubal M, Paczian T, Rodriguez A, Stevens R, Wilke A. The metagenomics RAST server–a public resource for the automatic phylogenetic and functional analysis of metagenomes. BMC Bioinform. 2008;9(1):386.
28. Montella S, Amore A, Faraco V. Metagenomics for the development of new biocatalysts to advance lignocellulose saccharification for bioeconomic development. Crit Rev Biotechnol 2015: 1–12.
29. Noguchi H, Park J, Takagi T. MetaGene: prokaryotic gene finding from environmental genome shotgun sequences. Nucleic Acids Res. 2006;34(19):5623–30.
30. Parmar I, Rupasinghe HP. Bio-conversion of apple pomace into ethanol and acetic acid: enzymatic hydrolysis and fermentation. Bioresour Technol. 2013;130(2013):613–20.
31. Pearson WR. An introduction to sequence similarity ("homology") searching. Curr Protoc Bioinform. 2013:3.1.1–3.1.8.
32. Pelloux J, Rusterucci C, J ME. New insights into pectin methylesterase structure and function. Trends Plant Sci. 2007;12(6):267–77.
33. Qian X, Sun W, Gu J, Wang XJ, Zhang YJ, Duan ML, Li HC, Zhang RR. Reducing antibiotic resistance genes, integrons, and pathogens in dairy manure by continuous thermophilic composting. Bioresour Technol. 2016;220:425–32.
34. Qin J, Li R, Raes J, Et A. A human gut microbial gene catalogue established by metagenomic sequencing. Nature. 2010;464(7285):59–65.
35. Quast C, Pruesse E, Yilmaz P, Gerken J, Schweer T, Yarza P, Peplies J, Glockner FO. The SILVA ribosomal RNA gene database project: improved data processing and web-based tools. Nucleic Acids Res. 2013;41:D590–5966.

36. Reddy AP, Simmons CW, D'haeseleer P, Khudyakov J, Burd H, Hadi M, Simmons BA, Singer SW, Thelen MP, VanderGheynst JS. Discovery of microorganisms and enzymes involved in high-solids decomposition of rice straw using metagenomic analyses. PLoS ONE. 2013;8(10):e77985.

37. Rosnow JJ, Anderson LN, Nair RN, Baker ES, Wright AT. Profiling microbial lignocellulose degradation and utilization by emergent omics technologies. Crit Rev Biotechnol. 2016:1–15.

38. Saoudi B, Habbeche A, Kerouaz B, Haberra S, Ben Romdhane Z, Tichati L, Boudelaa M, Belghith H, Gargouri A, Ladjama A. Purification and characterization of a new thermoalkaliphilic pectate lyase from Actinomadura keratinilytica Cpt20. Process Biochem. 2015;50(12):2259–66.

39. Schloss PD, Gevers D, Westcott SL. Reducing the effects of PCR amplification and sequencing artifacts on 16S rRNA-based studies. PLoS ONE. 2011;6(12):e27310.

40. Simmons CW, Reddy AP, D'haeseleer P, Khudyakov J, Billis K, Pati A, Simmons BA, Singer SW, Thelen MP, VanderGheynst JS. Metatranscriptomic analysis of lignocellulolytic microbial communities involved in high-solids decomposition of rice straw. Biotechnol Biofuels. 2014;7(1):495.

41. Sun J, Qian X, Gu J, Wang X, Gao H. Effects of oxytetracycline on the abundance and community structure of nitrogen-fixing bacteria during cattle manure composting. Bioresour Technol. 2016;216:801–7.

42. Truong DT, Franzosa EA, Tickle TL, Scholz M, Weingart G, Pasolli E, Tett A, Huttenhower C, Segata N. MetaPhlAn2 for enhanced metagenomic taxonomic profiling. Nat Methods. 2015;12(10):902–3.

43. Valladares Juarez AG, Dreyer J, Gopel PK, Koschke N, Frank D, Markl H, Muller R. Characterisation of a new thermoalkaliphilic bacterium for the production of high-quality hemp fibres, *Geobacillus thermoglucosidasius* strain PB94A. Appl Microbiol Biotechnol. 2009;83(3):521–7.

44. Wang C, Dong D, Wang H, Muller K, Qin Y, Wang H, Wu W. Metagenomic analysis of microbial consortia enriched from compost: new insights into the role of Actinobacteria in lignocellulose decomposition. Biotechnol Biofuels. 2016;9:22.

45. Wang W, Yang H, Zhang Y, Xu J. IoT-enabled real-time energy efficiency optimisation method for energy-intensive manufacturing enterprises. Int J Computer Inter Manuf. 2017;1–18.

46. Wang X, Chen Q, Lü X. Pectin extracted from apple pomace and citrus peel by subcritical water. Food Hydrocolloids. 2014;38:129–37.

47. Xiao C, Anderson CT. Roles of pectin in biomass yield and processing for biofuels. Front Plant Sci. 2013;4:67.

48. Xie C, Mao X, Huang J, Ding Y, Wu J, Dong S, Kong L, Gao G, Li CY, Wei L. KOBAS 2.0: a web server for annotation and identification of enriched pathways and diseases. Nucleic Acids Res. 2011;39:W316–22.

49. Yin Y, Mao X, Yang J, Chen X, Mao F, Xu Y. dbCAN: a web resource for automated carbohydrate-active enzyme annotation. Nucleic Acids Res. 2012;40:W445–51.

50. Yuan P, Meng K, Shi P, Luo H, Huang H, Tu T, Yang P, Yao B. An alkaline-active and alkali-stable pectate lyase from Streptomyces sp. S27 with potential in textile industry. J Ind Microbiol Biotechnol. 2012;39(6):909–15.

51. Zhang Y, Li H, Gu J, Qian X, Yin Y, Li Y, Zhang R, Wang X. Effects of adding different surfactants on antibiotic resistance genes and intl1 during chicken manure composting. Bioresour Technol. 2016;219:545–51.

52. Zhu N, Yang J, Ji L, Liu J, Yang Y, Yuan H. Metagenomic and metaproteomic analyses of a corn stover-adapted microbial consortium EMSD5 reveal its taxonomic and enzymatic basis for degrading lignocellulose. Biotechnol Biofuels. 2016;9:243.

53. Zykwinska A, Thibault JF, Ralet MC. Organization of pectic arabinan and galactan side chains in association with cellulose microfibrils in primary cell walls and related models envisaged. J Exp Bot. 2007;58(7):1795–802.

11

Saccharomyces cerevisiae strain comparison in glucose–xylose fermentations on defined substrates and in high-gravity SSCF: convergence in strain performance despite differences in genetic and evolutionary engineering history

Vera Novy[1], Ruifei Wang[2], Johan O. Westman[2], Carl Johan Franzén[2] and Bernd Nidetzky[1*]

Abstract

Background: The most advanced strains of xylose-fermenting *Saccharomyces cerevisiae* still utilize xylose far less efficiently than glucose, despite the extensive metabolic and evolutionary engineering applied in their development. Systematic comparison of strains across literature is difficult due to widely varying conditions used for determining key physiological parameters. Here, we evaluate an industrial and a laboratory *S. cerevisiae* strain, which has the assimilation of xylose via xylitol in common, but differ fundamentally in the history of their adaptive laboratory evolution development, and in the cofactor specificity of the xylose reductase (XR) and xylitol dehydrogenase (XDH).

Results: In xylose and mixed glucose–xylose shaken bottle fermentations, with and without addition of inhibitor-rich wheat straw hydrolyzate, the specific xylose uptake rate of KE6-12.A (0.27–1.08 g g_{CDW}^{-1} h^{-1}) was 1.1 to twofold higher than that of IBB10B05 (0.10–0.82 g g_{CDW}^{-1} h^{-1}). KE6-12.A further showed a 1.1 to ninefold higher glycerol yield (0.08–0.15 g g^{-1}) than IBB10B05 (0.01–0.09 g g^{-1}). However, the ethanol yield (0.30–0.40 g g^{-1}), xylitol yield (0.08–0.26 g g^{-1}), and maximum specific growth rate (0.04–0.27 h^{-1}) were in close range for both strains. The robustness of flocculating variants of KE6-12.A (KE-Flow) and IBB10B05 (B-Flow) was analyzed in high-gravity simultaneous saccharification and co-fermentation. As in shaken bottles, KE-Flow showed faster xylose conversion and higher glycerol formation than B-Flow, but final ethanol titres (61 g L^{-1}) and cell viability were again comparable for both strains.

Conclusions: Individual specific traits, elicited by the engineering strategy, can affect global physiological parameters of *S. cerevisiae* in different and, sometimes, unpredictable ways. The industrial strain background and prolonged evolution history in KE6-12.A improved the specific xylose uptake rate more substantially than the superior XR, XDH, and xylulokinase activities were able to elicit in IBB10B05. Use of an engineered XR/XDH pathway in IBB10B05 resulted in a lower glycerol rather than a lower xylitol yield. However, the strain development programs were remarkably convergent in terms of the achieved overall strain performance. This highlights the importance of comparative strain evaluation to advance the engineering strategies for next-generation *S. cerevisiae* strain development.

*Correspondence: bernd.nidetzky@tugraz.at
[1] Institute of Biotechnology and Biochemical Engineering, Graz University of Technology, Graz, Austria
Full list of author information is available at the end of the article

Background

Bioethanol, produced from lignocellulosic feedstock, is one of the most promising fossil fuel substitutes and it can help to mitigate climate change and secure energy supply chains [1, 2]. However, there are still major obstacles in the bioethanol production process, which have to be overcome to realize the full potential for commercialization [1, 3].

A main challenge is to find, or engineer, a fermentation organism that performs well in the difficult substrate presented by the lignocellulosic hydrolyzates [4, 5]. During the pretreatment step, high levels of inhibitory compounds (e.g., aromatic aldehydes and organic acids) are formed by secondary decomposition processes [6, 7]. The lignocellulosic hydrolyzates further contain significant concentrations of hemicellulose-derived pentoses, mainly xylose, besides the cellulose-derived glucose [3]. Realization of the full potential of the feedstock requires conversion of all provided sugars [8].

To target this, extensive research effort has been spent on enabling *Saccharomyces cerevisiae* to ferment xylose [9–11]. Based on its inherent robustness and process stability, this yeast is the preferred organism of the industries and a promising candidate for lignocellulose-to-ethanol processes [9–11]. *S. cerevisiae*, however, is naturally unable to ferment xylose [12, 13], necessitating the introduction of a heterologous xylose assimilation pathway into the yeast's genome. Two different pathways are available; the bacterial direct isomerization of xylose-to-xylulose, catalyzed by xylose isomerase (XI) [14], and the fungal "net" isomerization in two oxidoreductive steps via xylitol, catalyzed by xylose reductase (XR) and xylitol dehydrogenase (XDH) [9–11]. Both strategies have resulted in strains with the desired xylose-converting phenotype [9–11, 14]. Despite the recent success of strains harboring the XI [15–17], the XR/XDH pathway remains a strong option for development [9–11, 18].

Irrespective of the basic engineering strategy applied, however, the resultant strains display specific xylose uptake rates (q_{Xylose}) considerably lower than the corresponding glucose uptake rates [9–11]. A substrate uptake rate is a complex manifestation of the microbial physiology and may be limited by the actual uptake into the cell, metabolic integration, or both. It is often more convenient to try to evolve a complex physiological parameter rather than engineer it rationally [19]. Strategies applied to improve q_{Xylose} in *S. cerevisiae* include evolution in repetitive batch cultivations [20–22], continuous chemostat experiments [23, 24], or a combination of the two [25]. Strain selection has mainly been based on aerobic [21, 22] or anaerobic growth on xylose [20, 23]. Laboratory evolution has been further applied to increase the yeast's tolerance against the stressors and inhibitors present in the lignocellulosic substrates [24–26].

The main difficulty of evolutionary engineering lies in the proper choice of both selection pressure and screening parameter [19, 27]. According to the slogan "you only get what you screen for," strains evolved for improved aerobic growth on xylose, might not actually show an improved anaerobic specific rate of ethanol production ($q_{Ethanol}$), and an accelerated q_{Xylose} might result in decreased ethanol yields ($Y_{Ethanol}$) [27]. Furthermore, strains are often characterized only under a few cultivation conditions [10, 27]. Because the maximum specific growth rate (μ_{max}), q_{Xylose}, and $Y_{Ethanol}$ are highly dependent on the experimental set-up (e.g., sugar substrate concentrations, pH, inhibitor content, cell density), broad variation in the experimental conditions across literature makes a rigorous comparison of the different strains difficult.

Another challenge in advancing large-scale bioethanol production from lignocellulosic feedstock is to achieve the high final ethanol titers necessary to render the process cost-effective (40–50 g L^{-1}, e.g., [8]). This requires the processing of high solid loadings which is associated with problems such as high concentrations of inhibitors [6, 28], mass and heat transfer limitations due to high viscosities [29], and insufficient xylose fermentation caused by high glucose-to-xylose ratios [30, 31]. Fed-batch simultaneous saccharification and co-fermentation (SSCF), with substrate, enzyme, and cell feeding, or a combination thereof, has been shown to be useful to overcome these problems [24, 28–31].

In this study, we compare two xylose-fermenting strains of *S. cerevisiae*, IBB10B05 [20] and KE6-12.A ([25], Albers et al., unpublished), that were established independently through completely different development programs. Both strains harbor the XR/XDH pathway and were evolved for growth on xylose and accelerated xylose conversion ([20, 25], Albers et al., unpublished) but they differ fundamentally in their metabolic and evolutionary engineering history. Strain characterization was conducted in anaerobic shaken bottle experiments on synthetic sugar substrates with and without addition of inhibitor-rich wheat straw hydrolyzate. This allowed for precise determination of the metabolite yields, the growth rates, and the specific substrate uptake rates. To further compare the strains in a process set-up closer to industrial applications, the severity of the fermentation conditions was increased and flocculating variants of IBB10B05 (B-Flow) and KE6-12.A (KE-Flow) were applied in high-gravity multi-feed SSCFs. This study will give insights into how the specific traits of the two strains, which were elicited by different metabolic and evolutionary engineering strategies, can affect the global fermentation performance under laboratory conditions and in industrially relevant experimental set-ups.

Methods
Strains

The genetically and evolutionary engineered *S. cerevisiae* strains IBB10B05 (Graz University of Technology, Austria) and KE6-12.A (Chalmers University of Technology, Sweden) were used. IBB10B05 is a descendant of BP10001, which was enabled to xylose fermentation by the genomic integration of a mutated (K274R; N276D) XR variant from *Candida tenuis*, the wild-type XDH from *Galactocandida mastotermitis* and an additional copy of the endogenous xylulose kinase 1 [32]. Evolutionary engineering of BP10001 was described before [20], and will be only briefly summarized in the following. Throughout the evolution procedure, mineral medium was utilized with xylose as sole carbon source (XM). The pH was stabilized at 6.5 with K_2HPO_4 buffer and incubation was under strictly anaerobic conditions at 30 °C. BP10001 was firstly cultivated in a batch culture for 91 days. Subsequently, cells were transferred to XM-agar plates. The fastest growing colony was subjected to further engineering by repetitive batches. After several rounds, the clone showing the highest μ_{max} and q_{Xylose} was IBB10B05. In total IBB10B05 was evolved from BP10001 in 61 generations [20].

KE6-12.A is a non-commercial strain derived from TMB3400 by evolutionary engineering [25]. TMB3400 was generated by genomic integration of *Pichia stipitis* XR and XDH genes, and a combination of chemical mutagenesis and laboratory evolution was then used [21]. TMB3400 was further evolved resulting in KE6-12.A, and a detailed description of the secondary evolution procedure will be published elsewhere (Albers et al., unpublished). In short, the parent strain (obtained after initial evolutions with heat treatment and high xylose levels for 15 and 77 generations) was cultivated in a continuous culture at pH 5.0 and 35 °C. The cultivation was started with a batch phase without any air inflow using glucose and xylose-based mineral medium. Subsequently, the continuous phase was initiated by feeding xylose with increasing levels of inhibitor-rich bagasse hydrolyzate. The cultivation was run as a turbidostat with low aeration. During the continuous phase, the last strain saved as frozen stock contained a mixed population (denoted KE6-12), generated after 120 generations. In a later study, the best performing single cell line was singled out and denoted KE6-12.A [25].

In SSCF experiments, flocculating variants of IBB10B05 and KE6-12.A were used. The strains were made flocculating by genomic integration of the *FLOw* gene at the *HO* locus [33]. The resulting flocculating IBB10B05 and KE6-12.A were denoted B-Flow and KE-Flow, respectively.

Raw materials

The liquid and the solid fractions of pretreated wheat straw were obtained from SEKAB E-technology (Örnsköldsvik, Sweden). The wheat straw was pretreated by acid-catalyzed (0.2% (w/v) H_2SO_4) steam explosion. After pretreatment, the slurry was separated by press filtration into a xylose- and inhibitor-rich liquid (denoted herein hydrolyzate) and a cellulose-rich solid fraction. The two fractions were used independently in this study. The pretreatment strategy will be published in detail in a separate publication [33]. The compositions of both fractions are summarized in Additional file 1: Table S1. Prior to use, the pH of the liquid fraction was adjusted to 6.5 with NaOH, after which it was sterilized using 0.45 μm filters (Klari-Flex, Whatman, Maidstone, United Kingdom).

Shaken bottle fermentations
Media

Unless otherwise stated, all chemicals were from Carl Roth + Co KG (Karlsruhe, Germany). YPD medium contained 10 g L^{-1} yeast extract, 20 g L^{-1} casein peptone, and 20 g L^{-1} glucose. YPD agar plates additionally contained 20 g L^{-1} agar. YX, YG, and YGX media contained yeast extract (10 g L^{-1}) and the carbon sources xylose (40 g L^{-1}), glucose (40 g L^{-1}), and a combination thereof (40 g L^{-1} xylose, 40 g L^{-1} glucose), respectively. Fermentations conducted in a hydrolyzate matrix contained 70 vol% hydrolyzate (H), 10 vol% yeast extract solution (10 g L^{-1}), 10 vol% sugar solution, and 10 vol% inoculum. Xylose was added to the H-YX medium to reach a final concentration of 30 g L^{-1}. Glucose and xylose were added to the H-YGX medium to reach final concentrations of 40 and 30 g L^{-1}, respectively. Because of the low concentration of glucose in the hydrolyzate (Additional file 1: Table S1), H-YX media additionally contained ~2 g L^{-1} glucose. Low cell density fermentations were additionally supplemented with 0.1 vol% ergosterol solution (10 g L^{-1} ergosterol, 420 g L^{-1} Tween-80, both Sigma-Aldrich, St. Louis, MO, USA, boiled in 96 vol% ethanol).

Fermentations

Cells were stored in glycerol stocks and initially plated on YPD agar plates. Incubation was at 30 °C for 48 h. Cells were then used to inoculate 50 mL of YPD medium in 300 mL baffled shake flasks. Incubation was at 30 °C overnight. Cells were transferred to 300 mL of YPD medium in 1000 mL baffled shake flasks to a starting OD_{600} of 0.05, and incubated at 30 °C. Cells were harvested within the exponential growth phase (OD_{600} < 2.5) by centrifugation (4420g, 4 °C, 20 min, Sorvall RC-5B) and the cell pellet was washed and resuspended in 0.9% (w/v) NaCl solution. Reactions were performed anaerobically at 30 °C in glass bottles, tightly sealed with rubber septa (90 mL working volume). The bottles were sparged with N_2 prior to and shortly after inoculation. Starting OD_{600} was either 5 (high cell density fermentations) or 0.1 (low cell density fermentations). Incubation was performed at 180 rpm in a CERTOMAT BS-1 incubator shaker (Sartorius AG, Göttingen, Germany).

Analysis of cell growth, cell viability, sugars, and metabolites

Samples of 1.5 mL were frequently removed from shaken bottle fermentations and immediately put on ice. One milliliter of the sample volume was then centrifuged (15,700g, 4 °C, 10 min, Centrifuge 5415 R, Eppendorf, Hamburg, Germany) and the supernatant stored at −20 °C prior to HPLC analysis. The cell growth was recorded as increase in OD_{600}. The cell dry weight (CDW) was determined by filtering 1 mL of cell suspension through pre-weighed cellulose-acetate filter papers. After washing thoroughly with water, the filter paper was dried for 15 min in a microwave, cooled down in a desiccator, and weighed. Cell dry weights were recorded for YX, YG, and YGX fermentations and determined in triplicates. For analysis of colony forming units (CFU), the cell suspension was diluted with 0.9% (w/v) NaCl solution, and 1 mL of the appropriately diluted cell suspension was plated on YPD agar plates. Incubation was at 30 °C for 48 h. Extracellular fermentation products (ethanol, glycerol, xylitol, and acetic acid) and sugars (xylose and glucose) were analyzed by HPLC (Merck-Hitachi LaChrom system, L-7250 autosampler, L-7490 RI detector, L-7400 UV detector; Merck, Whitehouse Station, NJ). The system was equipped with an Aminex HPX-87H column and an Aminex Cation H guard column (both Bio-Rad, Hercules, CA). The operating temperature was 65 °C, and the flow rate of the mobile phase (5 mM sulfuric acid) was 0.6 mL/min.

Data processing and evaluations

The maximal specific growth rate (μ_{max}; h^{-1}) was determined as the slope of the linear region of the $\ln(OD_{600})$ vs time trajectory. Carbon balances were calculated with the assumption that 1 mol CO_2 was formed per mol acetate and ethanol. For biomass yields, a C-molar weight of 26.4 g $Cmol^{-1}$ was applied [34]. The specific uptake rates $q_{Glucose}$ and q_{Xylose} were calculated by first plotting glucose and xylose concentrations against fermentation time. The resulting scatter plots were fitted with suitable equations, and the first derivatives of the fitted equations were used to calculate the volumetric uptake rates Q (g L^{-1} h^{-1}). To calculate $q_{Glucose}$ and q_{Xylose} (g g_{CDW}^{-1} h^{-1}), Q was further

normalized to the CDW. Similar to previously published studies, both $q_{Glucose}$ and q_{Xylose} decreased with reaction time. Thus, reported values herein represent arithmetic means of the first four determinations made within the initial phase of the reaction. Please note: In fermentations containing glucose *and* xylose (YGX, H-YG and H-YGX), both strains showed an initial phase where only glucose was consumed ("glucose phase") and only subsequently xylose uptake started ("xylose phase"). q_{Xylose} therefore represents the arithmetic mean of the first four sampling points of the xylose phase. Based on the improved co-fermentation capacity of both evolved strains, however, it was not possible to separate the phases completely, resulting in residual glucose being present in the time frame when q_{Xylose} was determined.

High-gravity SSCF

The SSCF fermentation strategy will be published in full detail in another publication [33], and will be only briefly summarized here. Seed cultures were prepared in shake flask cultures containing YPD medium. Subsequently, cell propagation was accomplished in batch followed by fed-batch cultivation in 3.6 L bioreactors (INFORS HT, Switzerland). The batch and the feed media contained molasses, hydrolyzate, and media supplements, and propagation was run at 35 °C under aerobic conditions. For the SSCF, the solid fraction of the pretreated wheat straw was utilized as substrate and the desired dry mass loading was adjusted with hydrolyzate to reduce water consumption. The SSCF was run in a multi-feed approach, feeding both the wheat straw solids and cells from the cell propagation reactor at predetermined time points [33, 35]. In total, 20% (w/w) water insoluble solids (WIS) were loaded to the reactor. The enzyme (Ctec2, Novozymes, Denmark) loading was 10 Filter Paper Units (FPU) per g WIS. Cells were added to maintain a CDW/WIS ratio of 0.02 g g^{-1}. The SSCF was run at pH 5. A temperature profile was utilized, where the first 24 h were run at 35 °C after which the temperature was lowered to 30 °C. In total, the SSCF was run for 120 h. Samples were taken to measure external metabolites by HPLC, cell growth by total cell count, and cell viability by CFU [33].

Results

Shaken bottle fermentations

The strains IBB10B05 and KE6-12.A were compared in xylose and mixed glucose–xylose fermentations conducted in complex media or a hydrolyzate matrix. In this first part of the study, the fermentation performance of the strains was evaluated in anaerobic shaken bottle experiments. Yeast extract (10 g L^{-1}) was the sole medium additive. As shown by us [36, 37] and others

[38], yeast extract is sufficient for fermentations of pure sugar substrates as well as lignocellulosic hydrolyzates. It can replace mineral medium and expensive vitamin and trace element additives [36–38]. The hydrolyzate matrix represented the liquid fraction after dilute acid-catalyzed steam explosion, during which significant amounts of the hemicellulose were hydrolyzed into xylose (Additional file 1: Table S1). The hydrolyzate further contained inhibitory compounds including acetic acid, 5-hydroxymethyl-furfural (HMF), and furfural (Additional file 1: Table S1). These experiments were, hence, designed to evaluate the robustness of the strains. Fermentations were either run at high cell density (starting OD_{600} ~5) or low cell density (starting OD_{600} ~0.1). High cell density was used to analyze the conversion capacity of the yeast strains. Because of the high starting OD_{600} and the limited nutrients in shaken bottle experiments, only marginal cell growth was observed and the OD_{600} doubled maximally once within the fermentation time. Variations in growth are reflected in the biomass yields ($Y_{Biomass}$). To still be able to analyze the ability of the strains to grow anaerobically on the sugar substrates under the provided conditions, low cell density fermentations were additionally conducted.

Comparison of KE6-12.A and IBB10B05 in high cell density fermentations

IBB10B05 and KE6-12.A were first analyzed in high cell density fermentations of xylose (YX) and glucose and xylose (YGX). The resulting time courses are depicted in Fig. 1. The physiological parameters calculated from the data are summarized in Table 1. In fermentations of xylose only, KE6-12.A was faster in metabolizing xylose than IBB10B05 (Fig. 1a, b), resulting in an almost twice as high q_{Xylose} (Table 1). The $Y_{Ethanol}$ and $Y_{Xylitol}$ were similar for both strains at ~0.30 and ~0.25 g g^{-1}, respectively. In mixed glucose–xylose fermentations, KE6-12.A also showed faster sugar uptake (Fig. 1c, d) and the $q_{Glucose}$ and q_{Xylose} were 1.3-fold and a 2.7-fold higher, respectively, than they were in IBB10B05. The $Y_{Ethanol}$ was 0.40 g g^{-1} for both strains and the by-product distribution was also similar for the two strains (Table 1). In fermentations of the mixed sugar substrates, both strains showed some degree of true co-fermentation of glucose and xylose between 5 and 15 h fermentation time (Fig. 1).

In the next step, strain performance was compared in an inhibitor-rich hydrolyzate. The hydrolyzate was supplemented with xylose (H-YX) and glucose and xylose

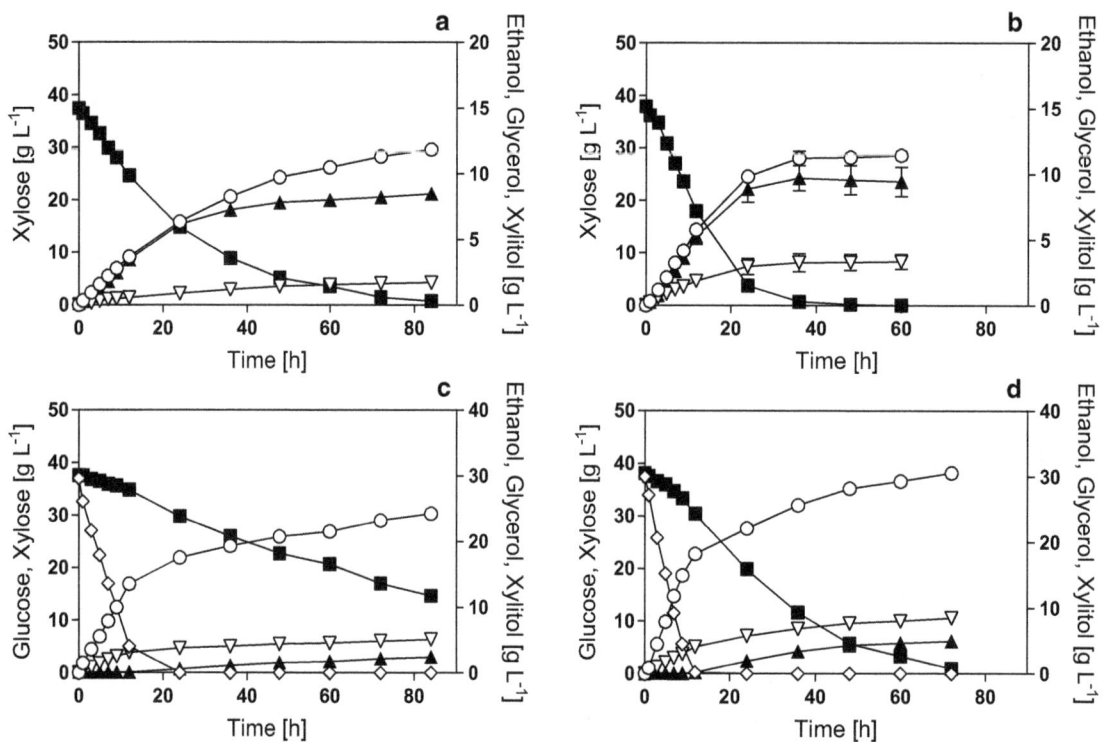

Fig. 1 Time courses of shaken bottle fermentations in complex media supplemented with xylose or with glucose and xylose. Fermentations were performed in YX (**a**, **b**) and YGX (**c**, **d**) media using strains IBB10B05 (**a**, **c**) and KE6-12.A (**b**, **d**). The starting OD_{600} was 5. Data points are mean values from biological duplicates. *Error bars* indicate the spread. Symbols: Xylose (*filled squares*), glucose (*empty diamonds*), ethanol (*empty circles*), glycerol (*empty triangles*), and xylitol (*filled triangles*)

Table 1 The physiological parameters of strains IBB10B05 and KE6-12.A in high cell density fermentations (starting OD_{600} 5) of xylose (YX) and glucose and xylose (YGX) in complex media

	YX		YGX[a]	
	IBB10B05	KE6-12A	IBB10B05	KE6-12A
$q_{Glucose}$ [g g_{CDW}^{-1} h^{-1}]	–	–	0.92 ± 0.02	1.23 ± 0.03
q_{Xylose} [g g_{CDW}^{-1} h^{-1}]	0.34 ± 0.00	0.66 ± 0.04	0.10 ± 0.01	0.27 ± 0.02
$Y_{Ethanol}$ [g g^{-1}]	0.31 ± 0.00	0.30 ± 0.01	0.40 ± 0.00	0.40 ± 0.00
$Y_{Glycerol}$ [g g^{-1}]	0.04 ± 0.00	0.09 ± 0.01	0.09 ± 0.00	0.11 ± 0.00
$Y_{Xylitol}$ [g g^{-1}]	0.24 ± 0.00	0.25 ± 0.03	0.04 ± 0.00	0.07 ± 0.01
$Y_{Acetate}$ [g g^{-1}]	0.04 ± 0.00	0.01 ± 0.00	0.02 ± 0.00	0.01 ± 0.00
$Y_{Biomass}$ [g g^{-1}]	0.06 ± 0.01	0.04 ± 0.00	0.05 ± 0.00	0.02 ± 0.00
C-recovery [%]	100.8 ± 0.7	96.4 ± 2.3	97.5 ± 1.0	99.7 ± 0.2

Data represent the mean values and the spread between biological duplicates

[a] Yields are based on consumed xylose and glucose

(H-YGX). The time courses using IBB10B05 and KE6-12.A are depicted in Fig. 2. The corresponding physiological parameters are summarized in Table 2. For clarity reasons, glucose is not depicted in Fig. 2a and b,

but the hydrolyzate contained a small amount of glucose (~2 g L^{-1}, Additional file 1: Table S1), which was consumed by both strains at equal speed and was depleted within the first 2.5 h of fermentation. In fermentations of xylose, KE6-12.A showed a 1.4-fold higher q_{Xylose} than IBB10B05 did. The $Y_{Ethanol}$ and $Y_{Xylitol}$ were similar at ~0.31 and ~0.25 g g^{-1}, respectively. The two strains, however, varied significantly in the formation of glycerol and acetate. KE6-12.A produced 0.15 g g^{-1} glycerol but no acetate. IBB10B05 produced 0.03 g g^{-1} glycerol but 0.04 g g^{-1} acetate. In fermentations of H-YGX, q_{Xylose} was 1.5-fold higher for KE6-12.A than for IBB10B05. $q_{Glucose}$ (~2.1 g g_{CDW}^{-1} h^{-1}) and $Y_{Ethanol}$ (0.39 g g^{-1}) were similar for both strains. As in the fermentations of H-YX, KE6-12.A produced more glycerol and less acetate than IBB10B05, but the differences were smaller (Table 2). Glucose and xylose co-consumption was less pronounced for both strains in H-YGX as compared to YGX fermentations.

Addition of hydrolyzate affected the specific substrate uptake rates differently in the various experimental setups. In fermentations of xylose only (YX and H-YX, Tables 1, 2), the addition of hydrolyzate slowed down the xylose conversion in both strains, and q_{Xylose} was

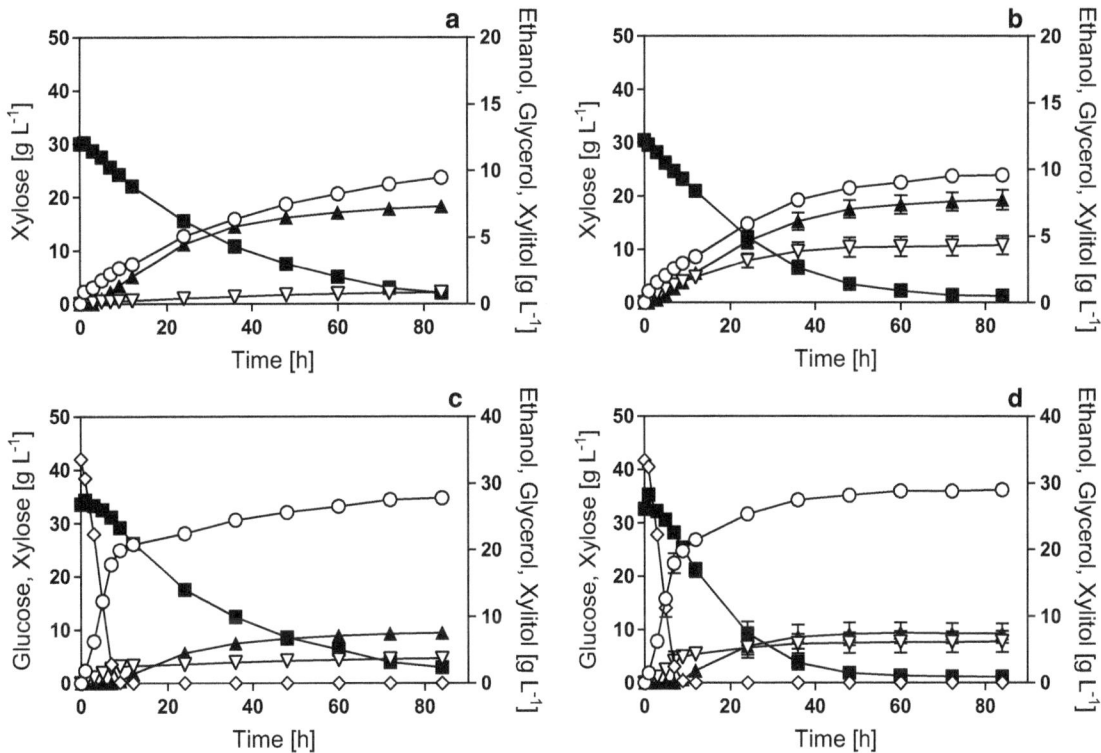

Fig. 2 Time courses of shaken bottle fermentations of xylose or with glucose and xylose in a hydrolyzate matrix. Fermentations were performed in H-YX (**a**, **b**) and H-YGX (**c**, **d**) media using strains IBB10B05 (**a**, **c**) and KE6-12.A (**b**, **d**). The starting OD_{600} was 5. Data points are mean values from biological duplicates. *Error bars* indicate the spread. Symbols: Xylose (*filled squares*), glucose (*empty diamonds*), ethanol (*empty circles*), glycerol (*empty triangles*), and xylitol (*filled triangles*)

Table 2 The physiological parameters of strains IBB10B05 and KE6-12.A in high cell density fermentations (starting OD$_{600}$ 5) of xylose (YX) and glucose and xylose (YGX) in a hydrolyzate matrix

	H-YX		H-YGX	
	IBB10B05	KE6-12.A	IBB10B05	KE6-12.A
$q_{Glucose}$ [g g$_{CDW}^{-1}$ h^{-1}]			2.10 ± 0.11	2.13 ± 0.22
q_{Xylose} [g g$_{CDW}^{-1}$ h^{-1}]	0.30 ± 0.02	0.43 ± 0.03	0.24 ± 0.01	0.36 ± 0.03
$Y_{Ethanol}$ [g g^{-1}]	0.32 ± 0.02	0.31 ± 0.01	0.39 ± 0.00	0.40 ± 0.00
$Y_{Glycerol}$ [g g^{-1}]	0.03 ± 0.00	0.15 ± 0.03	0.05 ± 0.00	0.08 ± 0.01
$Y_{Xylitol}$ [g g^{-1}]	0.26 ± 0.01	0.24 ± 0.02	0.08 ± 0.00	0.08 ± 0.02
$Y_{Acetate}$ [g g^{-1}]	0.04 ± 0.00	0.00 ± 0.00	0.03 ± 0.00	0.01 ± 0.00
$Y_{Biomass}$ [g g^{-1}]	0.03 ± 0.00	0.03 ± 0.01	0.04 ± 0.00	0.03 ± 0.01
C-recovery [%]	100.1 ± 4.5	102.2 ± 1.9	96.7 ± 1.5	97.1 ± 0.5

Data represent the mean values and the spread between biological duplicates

Yields are based on consumed xylose and glucose

reduced 1.1- and 1.5-fold in IBB10B05 and KE6-12.A, respectively. When fermentations were conducted with mixed sugar substrates, addition of hydrolyzate instead enhanced q_{Xylose} as well as $q_{Glucose}$ (Tables 1, 2). Thus, IBB10B05 showed a 2.3- and 2.4-fold increase in $q_{Glucose}$ and q_{Xylose}, respectively, in H-YGX as compared to YGX fermentations. In KE6-12.A, the difference was 1.7-fold ($q_{Glucose}$) and 1.3-fold (q_{Xylose}).

Comparison of KE6-12.A and IBB10B05 in low cell density fermentations

IBB10B05 and KE6-12.A were also compared in low cell density fermentations. The results are displayed in Table 3, which summarizes the maximal specific growth rate and the corresponding specific sugar uptake rates. The time courses and a summary of the metabolite yields of fermentations conducted without added hydrolyzate can be found in the Additional file 2: Figure S1 and Additional file 3: Table S2, respectively. Here, the two strains exhibited similar growth rates of 0.04 h^{-1} (YX)

and 0.27 h^{-1} (YGX). In YX media, the q_{Xylose} was slightly higher for IBB10B05 than KE6-12.A (Table 3). In YGX media, both strains metabolized glucose at equal rate, but q_{Xylose} was 1.5-fold lower in IBB10B05 than in KE6-12.A (Table 3).

The time courses of fermentations conducted in a hydrolyzate matrix are depicted in the Additional file 4: Figure S2 and the metabolic yields are summarized in Additional file 5: Table S3. Under these conditions the μ_{max} of both strains was similar at ~0.20 h^{-1} when mixed sugar substrates were used (H-YGX). In fermentations of xylose only (H-YX), IBB10B05 showed a 1.3-fold lower μ_{max} as compared to KE6-12.A. The specific glucose and xylose uptake rates in H-YGX fermentations varied only insignificantly, but IBB10B05 tended to convert glucose faster and xylose slower than KE6-12.A (Table 3). In H-YX fermentations, the q_{Xylose} of KE6-12.A was 1.3-fold higher as compared to IBB10B05.

In contrast to high cell density fermentations, addition of the hydrolyzate affected the specific sugar uptake rates positively in all experimental set-ups, irrespective of the sugar substrate or strain used (Table 3). It was further observed, that inoculation with low cell densities tended to result in higher specific sugar conversion rates than in fermentations started with large inocula (Tables 1, 2, and 3). This effect was stronger in IBB10B05, which showed an up to 2.7-fold higher q_{Xylose} in low cell density fermentations compared to the corresponding high cell density fermentation (Tables 1, 2, and 3).

High-gravity multi-feed SSCF

To compare the strains under more realistic process conditions, we conducted a high-gravity SSCF experiment with the flocculating variants of IBB10B05 (B-Flow) and KE6-12.A (KE-Flow). The process was operated with solids and cell feeding. This is a result of a series of development studies, which included the modeling and optimization of the cell and solids feeding strategy [35, 39], and the use of the flocculating yeast strains to

Table 3 Comparison of the maximal growth rates and specific substrate uptake rates of strains IBB10B05 and KE6-12A in low cell density fermentations (starting OD$_{600}$ 0.1) in complex media and a hydrolyzate matrix containing xylose (YX and H-YX) or a combination of glucose and xylose (YGX and H-YGX)

	IBB10B05			KE6-12.A		
	μ_{max} [h^{-1}]	$q_{Glucose}$ [g g$_{CDW}^{-1}$ h^{-1}]	q_{Xylose} [g g$_{CDW}^{-1}$ h^{-1}]	μ_{max} [h^{-1}]	$q_{Glucose}$ [g g$_{CDW}^{-1}$ h^{-1}]	q_{Xylose} [g g$_{CDW}^{-1}$ h^{-1}]
YX	0.05 ± 0.00		0.77 ± 0.03	0.04 ± 0.00		0.68 ± 0.13
YGX	0.27 ± 0.01	1.35 ± 0.35	0.11 ± 0.02	0.27 ± 0.01	1.28 ± 0.20	0.17 ± 0.20
H-YX	0.13 ± 0.01		0.82 ± 0.06	0.17 ± 0.01		1.08 ± 0.04
H-YGX	0.21 ± 0.00	1.84 ± 0.22	0.52 ± 0.12	0.20 ± 0.01	1.58 ± 0.56	0.60 ± 0.03

Data represent the mean values and the spread between biological duplicates

simplify harvesting and handling of the yeast cells and potentially improving their inhibitor tolerance [33, 40]. The process has been designed for strain KE6-12.A with the aims of (a) maximizing the solids loading while controlling the apparent viscosity to reduce mass and heat transfer limitations, (b) controlling the amount of inhibitors added to the reactor, (c) keeping favorable glucose/xylose ratios to promote xylose fermentations, and (d) maintaining cell viability throughout the fermentations [35, 39]. The process was initially run at 35 °C to promote enzymatic hydrolysis. However, it has been clearly shown for KE-Flow (flocculating KE6-12.A) as well as for B-Flow (flocculating IBB10B05) that the combined stresses of the SSCF, i.e., high inhibitor and ethanol concentrations, have a much more severe impact on biomass growth at 35 °C than at 30 °C [33]. To accommodate both aspects of process efficiency, namely enzymatic hydrolysis rates and cell viability, the process was run at 35 °C for the first 24 h after which the temperature was lowered to 30 °C [33]. The resulting time courses of the SSCFs are depicted in Fig. 3. Table 4 shows the measured and calculated values for the xylose and the glucose uptake. Ethanol production and by-product formation are also summarized in the table. Both strains utilized almost all the available glucose monomers within the first 10 h of SSCF (Fig. 3). From that period on, the glucose concentration varied only slightly between 0.2 and 1.0 g L^{-1}. This indicates that the cell viability, which was maintained above 40% for the major part of the fermentation (Additional file 6: Figure S3), was sufficient to continuously consume the glucose released by enzymatic hydrolysis. It further shows that the rate of the enzymatic hydrolysis and the rate of glucose consumption by the yeast cells were well matched.

Table 4 Sugar uptake and product formation in 120 h of SSCF fermentations using the flocculating strains B-Flow (IBB10B05) and KE-Flow (KE6-12.A)

	B-Flow	KE-Flow
Xylose consumption [g L^{-1}]^a	12.2	18.5
Glucose consumption [g L^{-1}]^b	136.5	139.3
Ethanol production [g L^{-1}] ($Y_{Ethanol}$ [g g^{-1}])^c	69.6 (0.47)	71.0 (0.45)
Glycerol production [g L^{-1}] ($Y_{Glycerol}$ [g g^{-1}])^c	5.2 (0.03)	7.8 (0.05)
Xylitol production [g L^{-1}] ($Y_{Xylitol}$ [g g^{-1}])^c	1.8 (0.01)	4.0 (0.03)
Acetate production [g L^{-1}] ($Y_{Acetate}$ [g g^{-1}])^c	1.7 (0.01)	0.1 (0.00)

^{a} Consumed xylose under the assumption that no additional xylose was released by enzymatic hydrolysis

^{b} Consumed glucose calculated based on the produced ethanol using the theoretical ethanol on glucose yield of 0.51 g g^{-1}

^{c} Metabolic yields based on the xylose and glucose consumption calculated as described in a and b

This validates the cell and solids feeding scheme not only for strain KE-Flow, for which it was developed, but also for B-Flow. Moreover, it demonstrates that the model-based feeding design [35, 39] is not strain specific, thus offering flexible application. In 120 h of fermentation, both strains produced approximately 60 g L^{-1} of ethanol. B-Flow consumed 12.2 g L^{-1} of the initially available xylose and produced only minor amounts of glycerol, xylitol, and acetate (Fig. 3; Table 4). KE-Flow converted almost all the initially available xylose (~19 g L^{-1}) and produced slightly more glycerol and xylitol but less acetate than B-Flow.

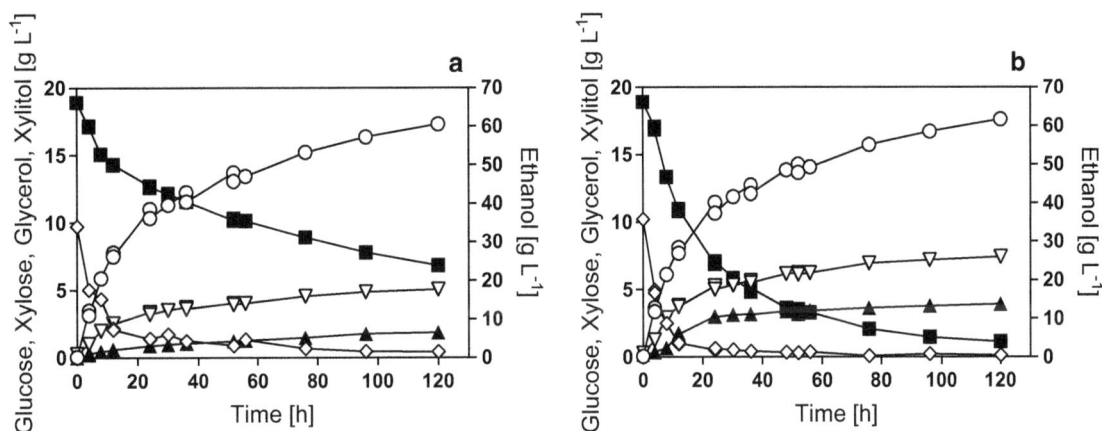

Fig. 3 Time courses of mixed glucose–xylose fermentation in high-gravity SSCF fermentations with cell and substrate feed. Fermentations were performed using the strains B-Flow (flocculating IBB10B05; **a**) and KE-Flow (flocculating KE6-12.A; **b**). Solids were added after 0, 4, 12, 24, 36, and 52 h. Cells were fed after 0, 24, 36, and 52 h. Data of KE-Flow were taken from [33]. Symbols: Xylose (*filled squares*), glucose (*empty diamonds*), ethanol (*empty circles*), glycerol (*empty triangles*), and xylitol (*filled triangles*)

Discussion

Laboratory evolution is an extremely powerful tool to enhance xylose-to-ethanol fermentation in yeasts. In this study, we compared two xylose-fermenting *S. cerevisiae* strains, IBB10B05 and KE6-12.A, which differ fundamentally in their metabolic and evolutionary history. IBB10B05 is based on the CEN.PK 113-5D genomic background and harbors an engineered NADH-preferring XR and a wild-type XDH [32, 41, 42]. It was evolved on mineral media with xylose as sole carbon source under strictly anaerobic conditions [20]. IBB10B05 was previously characterized in synthetic media [20, 37], in spent sulfite liquor [36] and in wheat straw hydrolyzates [37]. KE6-12.A harbors wild-type versions of *P. stipitis* XR and XDH and was evolved in a multitude of rounds, including chemostat evolution on xylose with increasing amounts of inhibitor-rich bagasse hydrolyzate under aerobic conditions ([25], Albers et al., unpublished). KE6-12.A was previously analyzed in fermentations of dilute lignocellulosic hydrolyzates [25] and high-gravity SSCFs [35, 39]. In this study, IBB10B05 and KE6-12.A were characterized and compared in identical experimental set-ups, with the aim of generating more information of how the strain background and the metabolic and the evolutionary engineering strategy affects the respective strain performance.

KE6-12.A and IBB10B05 show similar $Y_{Ethanol}$

To facilitate comparison, the physiological parameters from high cell density shaken bottle fermentations are summarized in Fig. 4. Independent of the fermentation media, both strains showed approximately the same $Y_{Ethanol}$, which was ~0.30 g g^{-1} in fermentations of xylose, and ~0.40 g g^{-1} in mixed glucose–xylose fermentations (Fig. 4b). In low cell density fermentations and SSCFs, the $Y_{Ethanol}$ was also similar for both strains (Table 4; Additional file 3: Table S2, Additional file 5: Table S3). These results indicate that $Y_{Ethanol}$ alone is not sufficient for detailed strain comparison. As described before, it is rather the biomass growth, the sugar consumption rates, the by-product formation, and the ethanol productivities that differ between various strains and which are sensitive to changes in the experimental set-up [6, 43, 44].

Development of xylose uptake in KE6-12.A and IBB10B05

The q_{Xylose} of KE6-12.A exceeded that of IBB10B05 regardless of the fermentation medium used (Fig. 4a). KE6-12.A further showed a faster xylose conversion in low cell density fermentations (Table 3) and in SSCFs (Fig. 3). This is in accordance with evidence from previously published studies, that industrial strains are preferred progenitor strains to realize high substrate conversion rates [45–47]. However, in this study, the

two strains do not only vary in their strain background. IBB10B05 and KE6-12.A also differ in their metabolic and evolutionary engineering strategy, individually designed to increase q_{Xylose}.

Strain IBB10B05 incorporates XR, XDH, and XK enzymes with reported activities of ~1.2, ~0.9, and 1.9 U/mg$_{crude\ cell\ protein}$, respectively [20]. These activities are significantly higher than corresponding activities reported for strain TMB3400 (XR ~ 0.08 U/mg$_{crude\ cell\ protein}$, XDH ~ 0.22 U/mg$_{crude\ cell\ protein}$, and XK ~ 0.08 U/mg$_{crude\ cell\ protein}$ [21]), the parent strain of KE6-12.A. In accordance to flux control theory [48], accelerated xylose conversion in IBB10B05 was suggested to be mainly caused by high levels of XR, XDH, and XK activity [20, 49, 50]. One would expect therefore that IBB10B05 exhibit higher q_{Xylose} than KE6-12.A. TMB3400, in contrast, was shown to contain significantly enhanced levels of transporter proteins as result of evolution [21]. Furthermore, evolution for increased inhibitor resistance, as represented here by strain KE6-12.A, was associated with an increased expression of genes involved in the pentose phosphate pathway [51]. Increased flux through the pentose phosphate pathway could create a kinetic pull effect through the xylose assimilation pathway involving the XR, XDH, and XK catalyzed reactions. KE6-12.A further has a much longer laboratory evolution history than IBB10B05, including chemical mutagenesis, evolution for improved xylose conversion on pure sugar substrate [21], and chemostat experiments on xylose with inhibitor-rich bagasse hydrolyzate (Albers et al., unpublished). The last step alone already involved 102 generations, whereas IBB10B05 was evolved in just 61 generations in total [20].

We therefore speculate that the industrial strain background in combination with the prolonged laboratory evolution history, and the resulting traits elicited in KE6-12.A, overcompensated the effect of the high XR, XDH, and XK activities in IBB10B05.

Impact of the lignocellulosic substrates on q_{Xylose}

Lignocellulosic hydrolyzates contain inhibitors such as acetic acid, HMF, and furfural, all of which negatively impact cell viability, biomass growth, and ethanol productivity. The physiological parameter, which is most susceptible to inhibition in engineered *S. cerevisiae*, is q_{Xylose} [6, 43, 44]. In this study, the inhibitor tolerance of IBB10B05 and KE6-12.A was firstly evaluated by comparing the fermentation performance in shaken bottle fermentations with and without added hydrolyzate.

In high cell density fermentations of xylose only, addition of hydrolyzate reduced q_{Xylose} in both strains (YX and H-YX, Fig. 4a), but to different extents. In IBB10B05 this effect was much less pronounced than in KE6-12.A (Tables 1, 2). The likely reason for the strongly decreased

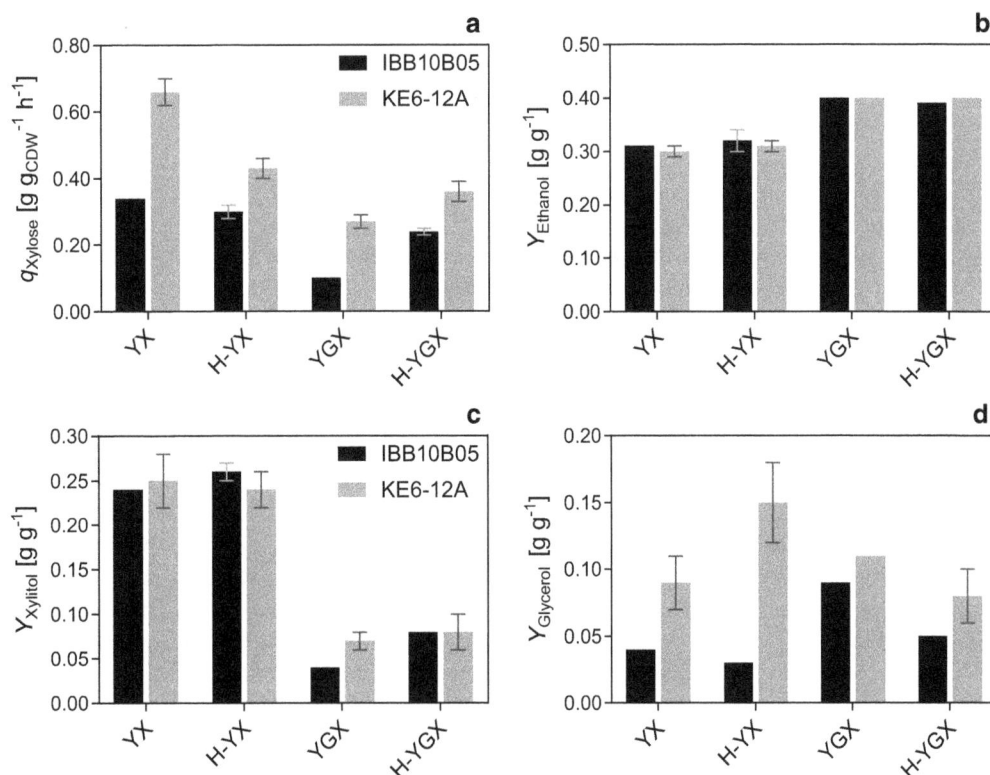

Fig. 4 Comparison of xylose uptake rates and product yields of strains IBB10B05 and KEG-12.A in high cell density shaken bottle fermentations. Depicted are q_{Xylose} (**a**), $Y_{Ethanol}$ (**b**), $Y_{Xylitol}$ (**c**), and $Y_{Glycerol}$ (**d**) using strains IBB10B05 (*black bars*) and KE6-12.A (*gray bars*). Fermentations were conducted in complex media (YX and YGX) and a hydrolyzate matrix (H-YX and H-YGX) as indicated. Data were taken from Tables 1 and 2 and represent the mean values and the spread of biological duplicates

q_{Xylose} in KE6-12.A is a loss of viability (measured in colony forming units, Additional file 7: Figure S4). In fermentations of H-YX, the cell viability decreased rapidly within the first 50 h to ~40% of the original value. In fermentation of YX only, no drop in viability was observed (Additional file 7: Figure S4). In contrast to KE6-12.A, the viability of IBB10B05 stayed equally constant at almost 100% over fermentation time, independent of the addition of hydrolyzate (Additional file 7: Figure S4).

However, it seems unlikely that the lignocellulose-derived inhibitors had a stronger inhibiting effect on KE6-12.A than they had on IBB10B05. Industrial strains were shown to be more inhibitor tolerant than laboratory strains [45–47], and KE6-12.A was evolved for increased inhibitor resistance ([25], Albers et al., unpublished). Moreover, KE6-12.A did not show a decrease in q_{Xylose} when hydrolyzate was added to low cell density fermentations of xylose (YX and H-YX, Table 3), which are more prone to inhibition by toxic compounds than are fermentations using large inocula [52].

It is likely that the observed differences are a result of the respective evolution strategy in combination with the

experimental set-up used. In high cell density fermentations, which were designed to resemble larger scale applications, expensive media additives such as ergosterol or oleic acid were avoided. Both compounds are essential for anaerobic growth [53]. Thus, low cell density fermentations, designed to analyze differences in μ_{max}, were supplemented with an ergosterol solution additionally containing Tween-80. The lack of these essential compounds in high cell density fermentations in combination with the strictly anaerobic conditions represents a significant stress on the yeast cell [53]. This stress was targeted by the evolution strategy of IBB10B05, which was kept anaerobic during the entire evolution procedure [20]. KE6-12.A, in contrast, was evolved under aerobic conditions [21, 25]. We would like to suggest, therefore, that the drop in both viability and q_{Xylose} of KE6-12.A was brought about by the lack of ergosterol and/or oleic acid under conditions of lignocellulose-derived stressors in the hydrolyzate.

In mixed glucose–xylose fermentation, addition of the hydrolyzate had a beneficial impact on the glucose and xylose uptake rates (Fig. 4; Tables 1, 2). This effect was even more pronounced in low cell density fermentations (Table 3; Additional file 2: Figure S1, Additional file 4: Figure S2). In these fermentations q_{Xylose} was affected positively by the hydrolyzate, even when just the fermentation of xylose was analyzed (H-YX and YX, Table 3). This "boosting" impact of the hydrolyzate can have several reasons. The low amounts of acetic acid and salts, which are present in the hydrolyzate (Additional file 1: Table S1), can exercise moderate stress on the yeast cells [54–56]. The resulting enhanced need for energy and redox equivalents can trigger an increase in the glycolytic flux, which in turn results in higher fermentation rates [56]. The hydrolyzate further contained small amounts of furfural and HMF (Additional file 1: Table S1), which can both act as electron acceptors, facilitating NADH re-oxidation [57, 58]. This also renders higher glycolytic rates possible. The increased glycolytic flux and the corresponding kinetic pull through the pathways upstream of glycolysis may have further positively affected the q_{Xylose} in IBB10B05 and KE6-12.A in fermentations with added hydrolyzate (Fig. 4a; Table 3).

The inhibitor tolerance of the two strains was further evaluated under the high severity conditions of the SSCF, where B-Flow (IBB10B05) and KE-Flow (KE6-12.A) showed a comparable glucose uptake, produced a similar amount of ethanol, and displayed comparable viability over fermentation time (Fig. 3; Additional file 6: Figure S3). This indicates that the two strains are equally tolerant against the high severity conditions, even though KE6-12.A, in contrast to IBB10B05, was evolved for increased inhibitor tolerance.

The inhibitor tolerance of yeast cells has been shown to depend on the overexpression of enzymes, which can reduce the lignocellulose-derived furaldehydes (e.g., HMF and furfural) into their less harmful corresponding alcohols [43, 58]. Responsible for the furaldehyde reductions are native enzymes, e.g., the alcohol dehydrogenase ADH6, and also the heterologous XR in engineered *S. cerevisiae* [43, 58, 59]. Overexpression of the XR has been further suggested to play a role in the stress response of xylose-fermenting *S. cerevisiae*, similar to native aldose reductases [18, 60]. Thus, the high XR activity in IBB10B05 might have increased the inhibitor tolerance to a similar extent in IBB10B05 as the metabolic alterations caused by evolutionary engineering did in KE6-12.A.

A common trait of both evolved yeast strains is the strongly accelerated xylose metabolism [20, 21, 25]. This increase in q_{Xylose} results in significantly improved ATP generation rates, which, in turn, not only increase the ethanol productivity, but also provide the means to cope with lignocellulose-derived stressors, e.g., organic acids [20, 43].

Co-enzyme specificity of the XR and its impact on by-product formation

Another difference between IBB10B05 and KE6-12.A is the type of XR and XDH the strains have incorporated. Whereas IBB10B05 harbors an engineered NADH-preferring XR, which renders the xylose assimilation pathway redox neutral [32], KE6-12.A contains the wild-type enzymes. The mismatched co-enzyme usage of the latter is widely accepted to be the main reason for excessive xylitol formation [32, 41, 61, 62]. As summarized in Fig. 4c, however, no difference in xylitol yields was detected between the two strains. Instead, the main strain-dependent difference was found in the $Y_{Glycerol}$ (Fig. 4d). Thus, in all high cell density shaken bottle fermentations (Fig. 4d), as well as in low cell density fermentations (Additional file 3: Table S2, Additional file 5: Table S3), and SSCFs (Table 4), KE6-12.A produced significantly more glycerol than IBB10B05. Glycerol, like xylitol, functions as a "redox-sink"; its formation serves to remove excess NADH [63]. It therefore seems likely that the comparably high glycerol formation in KE6-12.A is an indicator for redox imbalances caused by the unequal co-enzyme specificity of the XR and the XDH. This is supported by the fact that the largest difference between the $Y_{Glycerol}$ of IBB10B05 and KE6-12.A was found in fermentations of xylose as sole sugar substrate (Fig. 4d).

Figure 4 further indicates that both strains showed an increased $Y_{Xylitol}$ at high q_{Xylose} (Fig. 4a, c). This is in accordance with a previously published study on strain IBB10B05, in which $Y_{Xylitol}$ was demonstrated to increase with the q_{Xylose} [36]. The underlying reason for this effect is probably a kinetic bottleneck at the level of the XDH [42, 49].

q_{Xylose} in both strains is dependent on the glucose concentration

In all presented experiments, the q_{Xylose} was lower in mixed glucose–xylose fermentations (YGX, H-YGX) than in xylose fermentations only (YX, H-YX), irrespective of fermentation matrix, cell density, or strain used (Fig. 4a; Table 3). It is well known that glucose can inhibit q_{Xylose} in engineered *S. cerevisiae* (e.g., [64–66]), which natively does not harbor specific xylose transporter proteins [67, 68]. Although the homologous hexose transporters (e.g., Hxt1-7p) can facilitate xylose uptake, their affinity for glucose is so much higher that xylose uptake is inhibited even at high xylose-to-glucose ratios [67, 68]. In contrast, basal amounts of glucose (<2 g L^{-1}, e.g., [66]) have been shown to positively affect q_{Xylose}. Upregulation of transporter gene expression and an increase in glycolytic flux, which can create a kinetic pull through the xylose catabolism upstream of glycolysis, are likely the reasons for this [65–67]. The dependence of q_{Xylose} on the glucose concentration in engineered *S. cerevisiae* has been exemplified for the progenitor strain of IBB10B05, BP10001, which showed an increase of q_{Xylose} from 0.15 g g_{CDW}^{-1} h^{-1} (no glucose) to 0.30 g g_{CDW}^{-1} h^{-1} at glucose concentrations below 0.3 g L^{-1}. At glucose concentrations above 1 g L^{-1}, however, xylose uptake decreased rapidly and ceased completely at >5 g L^{-1} [66].

In low cell density fermentations conducted in this study, both strains exhibited a higher q_{Xylose} in H-YX than in YX media (Table 3). In line with previous evidence [36, 66], this was probably caused by the basal glucose concentration in the hydrolyzate (~2 g L^{-1}, Additional file 1: Table S1), which stayed in the medium for ~12 h of low cell density fermentations and thus, likely positively affected q_{Xylose}.

Figure 4a and Table 3 further indicate that inhibition by glucose on q_{Xylose} was stronger in complex media than in fermentations conducted in a hydrolyzate matrix. Interpretation of the effect is difficult. However, the result supports the notion that specific sugar conversion rates are complex manifestations of yeast physiology strongly dependent on the fermentation conditions.

Inhibition of xylose transport by glucose is also the reason for the sequential sugar uptake by engineered strains of *S. cerevisiae* [10, 11]. In this study, both strains showed a short phase of true sugar co-consumption in fermentations of complex media (Figs. 1, 2). In the SSCFs, the phase of glucose and xylose co-fermentation was even extended to the whole process, judging from the continued increase in ethanol and decrease in xylose concentrations (Fig. 3). The increased ability of evolved *S. cerevisiae* strains to co-consume glucose and xylose was described before [15, 16, 69] and has been ascribed to the overexpression of transporter proteins, xylose pathway enzymes, and enzymes of the pentose phosphate pathway [16, 20, 69]. This has been also demonstrated for the parent strain of KE6-12.A, TMB3400, and for IBB10B05 [20, 21].

The higher sugar co-consumption in SSCFs as compared to shaken bottle fermentations was probably a result of the presence of basal amounts of glucose (see "q_{Xylose} in both strains is dependent on the glucose concentration"), released by enzymatic hydrolysis, and the cell propagation strategy (see "Methods"). Continuous cultivation of yeast on inhibitor-rich medium containing high amounts of xylose can promote the xylose fermentation capacity, inhibitor tolerance, and sugar co-consumption by short-term adaptation [70, 71].

Conclusion

In this study, key physiological parameters of KE6-12.A and IBB10B05 were compared to evaluate the influence of the metabolic and evolutionary engineering strategies

on strain performance. Despite minor differences in the physiological characteristics of the two strains, the global fermentation performance was remarkably convergent. These results indicate that the individual specific traits of the two strains, which were elicited by the respective metabolic or evolutionary engineering strategies, affected the physiological parameters in different ways and to varying extents. They furthermore highlight the importance of comparative strain evaluation across laboratories to dissect the benefits of individual specific traits brought about by strain engineering on the global fermentation performance.

Additional files

Additional file 1: Table S1. Composition of the solid and the liquid fraction (here denoted hydrolyzate) of the pretreated wheat straw. Data were taken from [33].

Additional file 2: Figure S1. Time courses of low cell density shaken bottle fermentations in complex media supplemented with xylose or with glucose and xylose. Fermentations were performed in YX (a, b) and YGX (c, d) media using strains IBB10B05 (a, c) and KE6-12.A (b, d). The starting OD_{600} was 0.1. Data points are mean values from biological replicates. Error bars indicate the spread. Symbols: Xylose (filled squares), glucose (empty diamonds), ethanol (empty circles), glycerol (empty triangles), xylitol (filled triangles), and OD_{600} (crosses and dashed lines).

Additional file 3: Table S2. Comparison of the physiological parameters of strains IBB10B05 and KE6-12.A in low cell density fermentations (starting OD_{600} 0.1) of xylose (YX) and glucose and xylose (YGX).

Additional file 4: Figure S2. Time courses of low cell density shaken bottle fermentations with xylose or with glucose and xylose in a hydrolyzate matrix. Fermentations were performed in H-YX (a, b) and H-YGX (c, d) media using strains IBB10B05 (a, c) and KE6-12.A (b, d). The starting OD_{600} was 0.1. Data points are mean values from biological replicates. Error bars indicate the spread. Symbols: Xylose (filled squares), glucose (empty diamonds), ethanol (empty circles), glycerol (empty triangles), xylitol (filled triangles), and OD_{600} (crosses and dashed lines).

Additional file 5: Table S3. Comparison of the physiological parameters of strains IBB10B05 and KE6-12.A in low cell density fermentations (starting OD_{600} 0.1) of xylose (H-YX) and mixed glucose-xylose (H-YGX) in a hydrolyzate matrix.

Additional file 6: Figure S3. Comparison of the change in viability over time in high gravity SSCF fermentations. Depicted are the colony forming units (CFU) per total cell count using B-Flow (IBB10B05; panel a) and KE-Flow (KE6-12A; panel b). The starting OD_{600} was 5. Data represent mean values of 3 counted plates. Data for KE-Flow were taken from [33]. Error bars indicate the spread.

Additional file 7: Figure S4. The change in viability over time in shaken bottle fermentations. Depicted are the colony forming units (CFU) per OD_{600} value, relative to the value at t = 0 h. Fermentations were conducted in complex media (a) and a hydrolyzate matrix (b) supplemented with xylose (circles) or glucose and xylose (triangles) using strains IBB10B05 (filled symbols) and KE6-12.A (empty symbols). The starting OD_{600} was 5. Data points are mean values from biological replicates. Error bars indicate the spread.

Authors' contributions
All authors contributed to the design of the research. VN, RW, and JOW planned and performed the experiments and analyzed the data. VN drafted the manuscript from contributions by all authors. VN, CJF, and BN edited the final version. All authors read and approved the final manuscript.

Author details
[1] Institute of Biotechnology and Biochemical Engineering, Graz University of Technology, Graz, Austria. [2] Division of Industrial Biotechnology, Department of Biology and Biological Engineering, Chalmers University of Technology, Gothenburg, Sweden.

Acknowledgements
We like to thank Karin Longus for her indispensable help in performing the shaken bottle fermentations.

Competing interests
The authors declare that they have no competing interests.

Funding
Funding by the Swedish Energy Agency (Grant No. P P37353-1) and the Chalmers Energy Initiative (http://www.chalmers.se/en/areas-of-advance/energy/cei/) is gratefully acknowledged.

References
1. Sims REH, Mabee W, Saddler JN, Taylor M. An overview of second generation biofuel technologies. Bioresour Technol. 2010;101(6):1570–80.
2. von Blottnitz H, Curran MA. A review of assessments conducted on bioethanol as a transportation fuel from a net energy, greenhouse gas, and environmental life cycle perspective. J Clean Prod. 2007;15(7):607–19.
3. Chundawat SPS, Beckham GT, Himmel ME, Dale BE. Deconstruction of lignocellulosic biomass to fuels and chemicals. Annu Rev Chem Biomol Eng. 2011;2(1):121–45.
4. Mosier N, Wyman C, Dale B, Elander R, Lee YY, Holtzapple M, Ladisch M. Features of promising technologies for pretreatment of lignocellulosic biomass. Bioresour Technol. 2005;96(6):673–86.
5. Alvira P, Tomás-Pejó E, Ballesteros M, Negro MJ. Pretreatment technologies for an efficient bioethanol production process based on enzymatic hydrolysis: a review. Bioresour Technol. 2010;101(13):4851–61.
6. Palmqvist E, Hahn-Hagerdal B. Fermentation of lignocellulosic hydrolysates. II: inhibitors and mechanisms of inhibition. Bioresour Technol. 2000;1:25–33.
7. Klinke HB, Thomsen AB, Ahring BK. Inhibition of ethanol-producing yeast and bacteria by degradation products produced during pre-treatment of biomass. Appl Microbiol Biotechnol. 2004;66(1):10–26.
8. Viikari L, Vehmaanperä J, Koivula A. Lignocellulosic ethanol: from science to industry. Biomass Bioenergy. 2012;46:13–24.
9. Chu BCH, Lee H. Genetic improvement of Saccharomyces cerevisiae for xylose fermentation. Biotechnol Adv. 2007;25(5):425–41.
10. Kim SR, Park YC, Jin YS, Seo JH. Strain engineering of Saccharomyces cerevisiae for enhanced xylose metabolism. Biotechnol Adv. 2013;31(6):851–61.
11. Moysés D, Reis V, Almeida J, Moraes L, Torres F. Xylose fermentation by Saccharomyces cerevisiae: challenges and prospects. Int J Mol Sci. 2016;17(3):207.
12. Toivari MH, Salusjarvi L, Ruohonen L, Penttila M. Endogenous xylose pathway in Saccharomyces cerevisiae. Appl Environ Microbiol. 2004;70(6):3681–6.
13. Jeffries TW, Jin YS. Metabolic engineering for improved fermentation of pentoses by yeasts. Appl Environ Microbiol. 2004;63(5):495–509.
14. van Maris A, Winkler A, Kuyper M, de Laat W, van Dijken J, Pronk J. Development of efficient xylose fermentation in Saccharomyces cerevisiae: xylose isomerase as a key component. Adv Biochem Eng Biotechnol. 2007;108:179–204.
15. Demeke M, Dietz H, Li Y, Foulquie-Moreno M, Mutturi S, Deprez S, Den Abt T, Bonini B, Liden G, Dumortier F, et al. Development of a D-xylose fermenting and inhibitor tolerant industrial Saccharomyces cerevisiae strain with high performance in lignocellulose hydrolyzates using metabolic and evolutionary engineering. Biotechnol Biofuels. 2013;6(1):89.
16. Kuyper M, Toirkens MJ, Diderich JA, Winkler AA, van Dijken JP, Pronk JT. Evolutionary engineering of mixed-sugar utilization by a xylose-fermenting Saccharomyces cerevisiae strain. FEMS Yeast Res. 2005;5(10):925–34.
17. Brat D, Boles E, Wiedemann B. Functional expression of a bacterial xylose isomerase in Saccharomyces cerevisiae. Appl Environ Microbiol. 2009;75(8):2304–11.
18. Karhumaa K, Sanchez RG, Hahn-Hagerdal B, Gorwa-Grauslund M. Comparison of the xylose reductase-xylitol dehydrogenase and the xylose isomerase pathways for xylose fermentation by recombinant Saccharomyces cerevisiae. Microb Cell Fact. 2007;6:5.
19. Sauer U. Evolutionary engineering of industrially important microbial phenotypes. In: Nielsen J, Eggeling L, Dynesen J, Gárdonyi M, Gill R, de Graaf A, Hahn-Hägerdal B, Jönsson L, Khosla C, Licari R, et al., editors. Metabolic engineering, vol. 73. Heidelberg: Springer Berlin; 2001. p. 129–69.
20. Klimacek M, Kirl E, Krahulec S, Longus K, Novy V, Nidetzky B. Stepwise metabolic adaption from pure metabolization to balanced anaerobic growth on xylose explored for recombinant Saccharomyces cerevisiae. Microb Cell Fact. 2014;13:37.
21. Wahlbom CF, van Zyl WH, Jonsson LJ, Hahn-Hagerdal B, Otero RR. Generation of the improved recombinant xylose-utilizing Saccharomyces cerevisiae TMB 3400 by random mutagenesis and physiological comparison with Pichia stipitis CBS 6054. FEMS Yeast Res. 2003;3(3):319–26.
22. Liu G, Liu J, Cui X, Cai L. Sequence-dependent prediction of recombination hotspots in Saccharomyces cerevisiae. J Theor Biol. 2012;293:49–54.
23. Sonderegger M, Sauer U. Evolutionary engineering of Saccharomyces cerevisiae for anaerobic growth on xylose. Appl Environ Microbiol. 2003;69(4):1990–8.
24. Koppram R, Albers E, Olsson L. Evolutionary engineering strategies to enhance tolerance of xylose utilizing recombinant yeast to inhibitors derived from spruce biomass. Biotechnol Biofuels. 2012;5(1):32.
25. Tomás-Pejó E, Bonander N, Olsson L. Industrial yeasts strains for biorefinery solutions: constructing and selecting efficient barcoded xylose fermenting strains for ethanol. Biofuels Bioprod Biorefin. 2014;8(5):626–34.
26. Cakar ZP, Turanli-Yildiz B, Alkim C, Yilmaz U. Evolutionary engineering of Saccharomyces cerevisiae for improved industrially important properties. FEMS Yeast Res. 2012;12(2):171–82.
27. Cai Z, Zhang B, Li Y. Engineering Saccharomyces cerevisiae for efficient anaerobic xylose fermentation: reflections and perspectives. Biotechnol J. 2012;7(1):34–46.
28. Almeida JRM, Modig T, Petersson A, Hähn-Hägerdal B, Lidén G, Gorwa-Grauslund MF. Increased tolerance and conversion of inhibitors in lignocellulosic hydrolysates by Saccharomyces cerevisiae. J Chem Technol Biotechnol. 2007;82(4):340–9.
29. Hodge DB, Karim MN, Schell DJ, McMillan JD. Model-based fed-batch for high-solids enzymatic cellulose hydrolysis. Appl Biochem Biotechnol. 2009;152(1):88–107.
30. Olofsson K, Palmqvist B, Lidén G. Improving simultaneous saccharification and co-fermentation of pretreated wheat straw using both enzyme and substrate feeding. Biotechnol Biofuels. 2010;3:17.
31. Olofsson K, Rudolf A, Liden G. Designing simultaneous saccharification and fermentation for improved xylose conversion by a recombinant strain of Saccharomyces cerevisiae. J Biotechnol. 2008;134:112–20.
32. Petschacher B, Nidetzky B. Altering the coenzyme preference of xylose reductase to favor utilization of NADH enhances ethanol yield from xylose in a metabolically engineered strain of Saccharomyces cerevisiae. Microb Cell Fact. 2008;7(1):9.
33. Westman JO, Wang R, Novy V, Franzén CJ. Sustaining fermentation in high-gravity ethanol production by feeding yeast to a temperature-profiled multi-feed simultaneous saccharification and co-fermentation of wheat straw. Biotechnol Biofuels. 2017. doi:10.1186/s13068-017-0893-y.
34. Lange HC, Heijnen JJ. Statistical reconciliation of the elemental and molecular biomass composition of Saccharomyces cerevisiae. Biotechnol Bioeng. 2001;75(3):334–44.
35. Wang R, Unrean P, Franzén CJ. Model-based optimization and scale-up of multi-feed simultaneous saccharification and co-fermentation of steam pre-treated lignocellulose enables high gravity ethanol production. Biotechnol Biofuels. 2016;9(1):88.
36. Novy V, Krahulec S, Longus K, Klimacek M, Nidetzky B. Co-fermentation of hexose and pentose sugars in a spent sulfite liquor matrix with genetically modified Saccharomyces cerevisiae. Bioresour Technol. 2013;130:439–48.
37. Novy V, Krahulec S, Wegleiter M, Muller G, Longus K, Klimacek M, Nidetzky B. Process intensification through microbial strain evolution: mixed glu-

cose-xylose fermentation in wheat straw hydrolyzates by three generations of recombinant *Saccharomyces cerevisiae*. Biotechnol Biofuels. 2014;7(1):49.

38. Jorgensen H. Effect of nutrients on fermentation of pretreated wheat straw at very high dry matter content by *Saccharomyces cerevisiae*. Appl Biochem Biotechnol. 2009;153(1–3):44–57.

39. Wang R, Koppram R, Olsson L, Franzén CJ. Kinetic modeling of multi-feed simultaneous saccharification and co-fermentation of pretreated birch to ethanol. Bioresour Technol. 2014;172:303–11.

40. Westman JO, Mapelli V, Taherzadeh MJ, Franzén CJ. Flocculation causes inhibitor tolerance in *Saccharomyces cerevisiae* for second-generation bioethanol production. Appl Environ Microbiol. 2014;80(22):6908–18.

41. Krahulec S, Klimacek M, Nidetzky B. Analysis and prediction of the physiological effects of altered coenzyme specificity in xylose reductase and xylitol dehydrogenase during xylose fermentation by *Saccharomyces cerevisiae*. J Biotechnol. 2012;158(4):192–202.

42. Krahulec S, Klimacek M, Nidetzky B. Engineering of a matched pair of xylose reductase and xylitol dehydrogenase for xylose fermentation by *Saccharomyces cerevisiae*. Biotechnol J. 2009;4(5):684–94.

43. Almeida JR, Runquist D, Sanchez i Nogue V, Liden G, Gorwa-Grauslund MF. Stress-related challenges in pentose fermentation to ethanol by the yeast *Saccharomyces cerevisiae*. Biotechnol J. 2011;6(3):286–99.

44. Casey E, Sedlak M, Ho NW, Mosier NS. Effect of acetic acid and pH on the co-fermentation of glucose and xylose to ethanol by a genetically engineered strain of *Saccharomyces cerevisiae*. FEMS Yeast Res. 2010;10(4):385–93.

45. Matsushika A, Inoue H, Murakami K, Takimura O, Sawayama S. Bioethanol production performance of five recombinant strains of laboratory and industrial xylose-fermenting *Saccharomyces cerevisiae*. Bioresour Technol. 2009;100(8):2392–8.

46. Sonderegger M, Jeppsson M, Larsson C, Gorwa-Grauslund MF, Boles E, Olsson L, Spencer-Martins I, Hahn-Hägerdal B, Sauer U. Fermentation performance of engineered and evolved xylose-fermenting *Saccharomyces cerevisiae* strains. Biotechnol Bioeng. 2004;87(1):90–8.

47. Hahn-Hägerdal B, Karhumaa K, Fonseca C, Spencer-Martins I, Gorwa-Grauslund M. Towards industrial pentose-fermenting yeast strains. Appl Microbiol Biotechnol. 2007;74(5):937–53.

48. Stephanopoulos G, Aristidou AA, Nielsen J. Metabolic engineering: principles and methodologies. Cambridge: Academic press; 1998.

49. Eliasson A, Hofmeyr J-HS, Pedler S, Hahn-Hägerdal B. The xylose reductase/xylitol dehydrogenase/xylulokinase ratio affects product formation in recombinant xylose-utilising *Saccharomyces cerevisiae*. Enzym Microb Technol. 2001;29(4–5):288–97.

50. Sonderegger M, Jeppsson M, Hahn-Hagerdal B, Sauer U. Molecular basis for anaerobic growth of *Saccharomyces cerevisiae* on xylose, investigated by global gene expression and metabolic flux analysis. Appl Environ Microbiol. 2004;70(4):2307–17.

51. Gorsich SW, Dien BS, Nichols NN, Slininger PJ, Liu ZL, Skory CD. Tolerance to furfural-induced stress is associated with pentose phosphate pathway genes ZWF1, GND1, RPE1, and TKL1 in *Saccharomyces cerevisiae*. Appl Microbiol Biotechnol. 2006;71(3):339–49.

52. Pienkos PT, Zhang M. Role of pretreatment and conditioning processes on toxicity of lignocellulosic biomass hydrolysates. Cellulose. 2009;16(4):743–62.

53. Andreasen AA, Stier TJB. Anaerobic nutrition of *Saccharomyces cerevisiae*. I. Ergosterol requirement for growth in a defined medium. J Cell Physiol. 1953;41(1):23–36.

54. Bellissimi E, van Dijken JP, Pronk JT, van Maris AJA. Effects of acetic acid on the kinetics of xylose fermentation by an engineered, xylose-isomerase-based *Saccharomyces cerevisiae* strain. FEMS Yeast Res. 2009;9(3):358–64.

55. Guo Z, Olsson L. Physiological response of *Saccharomyces cerevisiae*
to weak acids present in lignocellulosic hydrolysate. FEMS Yeast Res. 2014;14(8):1234–48.

56. Olz R, Larsson K, Adler L, Gustafsson L. Energy flux and osmoregulation of *Saccharomyces cerevisiae* grown in chemostats under NaCl stress. J Bacteriol. 1993;175(8):2205–13.

57. Wahlbom CF, Hahn-Hagerdal B. Furfural, 5-hydroxymethyl furfural, and acetoin act as external electron acceptors during anaerobic fermentation of xylose in recombinant *Saccharomyces cerevisiae*. Biotechnol Bioeng. 2002;78(2):172–8.

58. Almeida JRM, Bertilsson M, Gorwa-Grauslund MF, Gorsich S, Lidén G. Metabolic effects of furaldehydes and impacts on biotechnological processes. Appl Microbiol Biotechnol. 2009;82(4):625–38.

59. Almeida JR, Modig T, Roder A, Liden G, Gorwa-Grauslund MF. *Pichia stipitis* xylose reductase helps detoxifying lignocellulosic hydrolysate by reducing 5-hydroxymethyl-furfural (HMF). Biotechnol Biofuels. 2008;1(1):12.

60. Chang Q, Harter TM, Rikimaru LT, Petrash JM. Aldo–keto reductases as modulators of stress response. Chem Biol Interact. 2003;143–144:325–32.

61. Matsushika A, Watanabe S, Kodaki T, Makino K, Inoue H, Murakami K, Takimura O, Sawayama S. Expression of protein engineered NADP+-dependent xylitol dehydrogenase increases ethanol production from xylose in recombinant *Saccharomyces cerevisiae*. Appl Microbiol Biotechnol. 2008;81(2):243–55.

62. Klimacek M, Krahulec S, Sauer U, Nidetzky B. Limitations in xylose-fermenting *Saccharomyces cerevisiae*, made evident through comprehensive metabolite profiling and thermodynamic analysis. Appl Environ Microbiol. 2010;76(22):7566–74.

63. Bakker BM, Overkamp KM, van Maris AJ, Kotter P, Luttik MA, van Dijken JP, Pronk JT. Stoichiometry and compartmentation of NADH metabolism in *Saccharomyces cerevisiae*. FEMS Microbiol Rev. 2001;25(1):15–37.

64. Madhavan A, Tamalampudi S, Srivastava A, Fukuda H, Bisaria VS, Kondo A. Alcoholic fermentation of xylose and mixed sugars using recombinant *Saccharomyces cerevisiae* engineered for xylose utilization. Appl Microbiol Biotechnol. 2009;82(6):1037–47.

65. Meinander NQ, Boels I, Hahn-Hägerdal B. Fermentation of xylose/glucose mixtures by metabolically engineered *Saccharomyces cerevisiae* strains expressing XYL1 and XYL2 from *Pichia stipitis* with and without overexpression of TAL1. Bioresour Technol. 1999;68(1):79–87.

66. Krahulec S, Petschacher B, Wallner M, Longus K, Klimacek M, Nidetzky B. Fermentation of mixed glucose-xylose substrates by engineered strains of *Saccharomyces cerevisiae*: role of the coenzyme specificity of xylose reductase, and effect of glucose on xylose utilization. Microb Cell Fact. 2010;9(1):16.

67. Sedlak M, Ho NWY. Characterization of the effectiveness of hexose transporters for transporting xylose during glucose and xylose co-fermentation by a recombinant *Saccharomyces* yeast. Yeast. 2004;21(8):671–84.

68. Hamacher T, Becker J, Gardonyi M, Hahn-Hägerdal B, Boles E. Characterization of the xylose-transporting properties of yeast hexose transporters and their influence on xylose utilization. Microbiology (Reading, England). 2002;148(Pt 9):2783–8.

69. Ho NWY, Chen Z, Brainard AP. Genetically engineered *Saccharomyces* yeast capable of effective cofermentation of glucose and xylose. Appl Environ Microbiol. 1998;64(5):1852–9.

70. Nielsen F, Tomas-Pejo E, Olsson L, Wallberg O. Short-term adaptation during propagation improves the performance of xylose-fermenting *Saccharomyces cerevisiae* in simultaneous saccharification and co-fermentation. Biotechnol Biofuels. 2015;8:219.

71. Tomas-Pejo E, Olsson L. Influence of the propagation strategy for obtaining robust *Saccharomyces cerevisiae* cells that efficiently co-ferment xylose and glucose in lignocellulosic hydrolysates. Microb Biotechnol. 2015;8(6):999–1005.

12

Thermotolerant genes essential for survival at a critical high temperature in thermotolerant ethanologenic *Zymomonas mobilis* TISTR 548

Kannikar Charoensuk[1], Tomoko Sakurada[2], Amina Tokiyama[3], Masayuki Murata[2], Tomoyuki Kosaka[2,3,4], Pornthap Thanonkeo[5] and Mamoru Yamada[2,3,4*]

Abstract

Background: High-temperature fermentation (HTF) technology is expected to reduce the cost of bioconversion of biomass to fuels or chemicals. For stable HTF, the development of a thermotolerant microbe is indispensable. Elucidation of the molecular mechanism of thermotolerance would enable the thermal stability of microbes to be improved.

Results: Thermotolerant genes that are essential for survival at a critical high temperature (CHT) were identified via transposon mutagenesis in ethanologenic, thermotolerant *Zymomonas mobilis* TISTR 548. Surprisingly, no genes for general heat shock proteins except for *degP* were included. Cells with transposon insertion in these genes showed a defect in growth at around 39 °C but grew normally at 30 °C. Of those, more than 60% were found to be sensitive to ethanol at 30 °C, indicating that the mechanism of thermotolerance partially overlaps with that of ethanol tolerance in the organism. Products of these genes were classified into nine categories of metabolism, membrane stabilization, transporter, DNA repair, tRNA modification, protein quality control, translation control, cell division, and transcriptional regulation.

Conclusions: The thermotolerant genes of *Escherichia coli* and *Acetobacter tropicalis* that had been identified can be functionally classified into 9 categories according to the classification of those of *Z. mobilis*, and the ratio of thermotolerant genes to total genomic genes in *Z. mobilis* is nearly the same as that in *E. coli,* though the ratio in *A. tropicalis* is relatively low. There are 7 conserved thermotolerant genes that are shared by these three or two microbes. These findings suggest that *Z. mobilis* possesses molecular mechanisms for its survival at a CHT that are similar to those in *E. coli* and *A. tropicalis*. The mechanisms may mainly contribute to membrane stabilization, protection and repair of damage of macromolecules and maintenance of cellular metabolism at a CHT. Notably, the contribution of heat shock proteins to such survival seems to be very low.

Keywords: *Zymomonas mobilis*, Ethanologenic microbe, Transposon mutagenesis, Thermotolerant gene, Ethanol-tolerant

Background

Zymomonas mobilis is an efficient ethanologenic microbe that has been isolated from sugarcane or alcoholic beverages such as African palm wine, and it causes cider sickness and spoiling of beer [1]. The organism bears an anaerobic catabolism via the Entner–Doudoroff pathway [2], which utilizes 1 mol of glucose to yield 2 mol of pyruvate, which is then decarboxylated to acetaldehyde and reduced to ethanol. Due to its strong metabolic activity and low ATP productivity compared to those of the Emden–Meyerhof pathway in the conventional ethanol

*Correspondence: m-yamada@yamaguchi-u.ac.jp
[3] Department of Biological Chemistry, Faculty of Agriculture, Yamaguchi University, 1677-1 Yoshida, Yamaguchi 753-8515, Japan
Full list of author information is available at the end of the article

producer yeast and high-yield ethanol production as a result of the Entner–Doudoroff pathway [1, 3] as well as the fact that the organism is generally regarded as being safe (GRAS) [4], *Z. mobilis* has been focused for its applications to production of useful materials including ethanol as a biofuel, oligosaccharides as food additives, and levan as a medicine [5, 6].

Since the ethanol fermentation process is exothermic [7, 8], ethanologenic microorganisms are exposed to heat stress in addition to other stresses including ethanol [9, 10]. Heat stress has an impact on their growth or viability [11, 12] to prevent fermentation, and the impact is enhanced in the presence of other inhibiting factors, i.e., low pH, high ethanol concentration, and high osmolarity [13–18]. Thus, thermotolerant *Z. mobilis* is thought to be beneficial for the production of useful materials. *Z. mobilis* TISTR 548 is a thermotolerant strain that can grow even at 39 °C [19–21], which is 5–10 °C higher than the optimum temperature for the same genus [22] and the same species [1, 23], and it can efficiently produce ethanol to an extent similar to that of ZM4 [3]. However, information on the molecular mechanism of the thermotolerance of thermotolerant *Z. mobilis* is limited, though some heat shock proteins have been analyzed [24, 25].

Elucidation of the molecular mechanism of microbial survival at a critical high temperature (CHT) may be useful for the development of high-temperature fermentation systems, which have several advantages including reduction in cooling cost, saving of enzyme cost in simultaneous saccharification and fermentation or prevention of contamination of unfavorable microbes [26, 27]. We thus performed transposon mutagenesis of the thermotolerant *Z. mobilis* TISTR 548 to isolate thermosensitive mutants, each of which is defective of one of the so-called thermotolerant genes. The physiological functions of these genes allow us to decipher the molecular mechanism of its survival at a CHT. Moreover, we may be able to understand the general strategy of Gram-negative bacteria to cope with thermal stresses at their individual CHTs by comparison of the mechanism in *Z. mobilis* as α-proteobacteria with those of other bacteria, *Escherichia coli* as γ-proteobacteria and *Acetobacter tropicalis* as α-proteobacteria, that have been investigated [28, 29]. *E. coli* is intrinsically thermotolerant compared to general mesophilic microbes and used for production of useful materials like amino acids, hormones, or vaccines. *Z. mobilis* TISTR548 and *A. tropicalis* are thermotolerant and efficiently produces ethanol and acetic acid, respectively, at relatively high temperatures [19, 29]. Thus, the knowledge of the general strategy might be applicable for relatively thermosensitive mesophilic microbes that have been utilized for production of useful materials in fermentation companies.

Results

Isolation of thermosensitive mutants by transposon mutagenesis in thermotolerant *Z. mobilis*

Thermotolerant *Z. mobilis* strain TISTR 548 was subjected to transposon mutagenesis via *E. coli* S17-1 harboring pSUP2021Tn*10* as a donor strain for conjugal mating [30]. The growth levels of about 8000 transconjugants obtained were compared on YPD plates at 30 and 39.5 °C, and thermosensitive ones that exhibited no or almost no growth at the high temperature were selected. They were subjected to repeated examination on YPD plates as a second screening and resultantly obtained 123 thermosensitive isolates were further subjected to the final screening in a YPD liquid medium under a static condition at 30 and 39.5 °C. Eventually, 38 isolates that exhibited defective or very weak growth in the liquid culture at the high temperatures were selected as thermosensitive mutants and were used for the following experiments.

The insertion site of Tn*10* in the genome of each mutant was determined by thermal asymmetric interlaced (TAIL)-PCR followed by nucleotide sequencing. The genomic sequences flanking Tn*10* were analyzed by using public databases to identify a disrupted gene. As a result, out of the 38 thermosensitive mutants, only 26 were found to have a Tn*10* insertion in independent genes and 12 were overlapped (Additional file 1: Table S1). This overlapping suggests that the isolation of thermosensitive mutants was nearly saturated. The 26 thermosensitive mutants including 14 representatives showed impaired growth at 39 or 39.5 °C but a similar level of growth to that of the parental strain at 30 °C (Additional file 1: Figure S1).

The gene organization around each Tn*10*-inserted gene might cause a polar effect of the insertion on the transcription of a downstream gene(s) that is intrinsically transcribed by read-through from an upstream promoter(s). Such an organization was found in 12 of the 26 mutants (Additional file 1: Figure S2). The possibility of such polar effects was thus examined by RT-PCR with total RNA that had been prepared from cells grown at 30 and 39.5 °C (Additional file 1: Figure S3). The data suggest that all genes located downstream of the transposon-inserted genes are expressed at the same levels of expression as those in the parental strain. Therefore, it is thought that the thermosensitive phenotype of the 26 thermosensitive mutants is due to the disruption of each gene inserted by Tn*10*, not due to a polar effect on its downstream gene(s). Taken together, 26 independent thermosensitive mutants were obtained and thus 26 thermotolerant genes were identified in thermotolerant *Z. mobilis* TISTR 548.

Function and classification of thermotolerant genes in thermotolerant *Z. mobilis*

In order to know the physiological functions of thermotolerant genes, database searching was performed. As a result, out of the 26 thermotolerant genes, 24 genes were functionally annotated and classified into 9 categories of general metabolism, membrane stabilization, transporter, DNA repair, tRNA/rRNA modification, protein quality control, translation control, cell division, and transcriptional regulation (Table 1). The remaining 2 genes encode unknown proteins.

Group A consists of two genes related to general metabolism, ZZ6_0707 and ZZ6_1376, that encode glucose sorbosone dehydrogenase and 5, 10-methylene-tetrahydrofolate reductase, respectively. The former oxidizes glucose or sorbosone and belongs to a family that possesses a beta-propeller fold. The best characterized in the family is soluble glucose dehydrogenase from *Acinetobacter calcoaceticus*, which oxidizes glucose to glucono-δ-lactone [31]. The latter catalyzes the conversion of 5,10-methylenetetrahydrofolate, which is used for de novo thymidylate biosynthesis, to 5-methyltetrahydrofolate [32], which is used for methionine biosynthesis [32].

Group B is the largest group that consists of 12 genes related to membrane stabilization or membrane formation. Of these, ZZ6_1146 encodes glucosamine/fructose 6-phosphate aminotransferase, which is the first and rate-limiting enzyme in the hexosamine biosynthetic pathway and catalyzes the formation of glucosamine-6-phosphate using glutamine as an ammonia donor. This amino sugar is essential for the formation of a plethora of glycoconjugates for the peptidoglycan macromolecule in prokaryotes [33]. ZZ6_0929 encodes glycosyltransferase group 1, which is involved in biosynthesis of the lipopolysaccharide (LPS) core [34]. This enzyme has two putative conserved domains: one domain covering 94% of the protein is named GT1_mtfB_like. MtfB (mannosyltransferase B) in *E. coli* has been shown to direct growth of the O9-specific polysaccharide chain [35]. The other covering 53% of the protein is named RfaB and is involved in assembly of the lipopolysaccharide core in *E. coli* [36]. ZZ6_0923 encodes phospholipase D/transphosphatidylase possessing the domain of cardiolipin synthase, which catalyzes phosphatidyl group transfer from one phosphatidylglycerol molecule to another to form cardiolipin and glycerol [37]. The *cls*⁻ for a defective cardiolipin synthase that shows a low level of cardiolipin in phospholipid composition has been reported [38], and the *cls* gene may be related to membrane stabilization. ZZ6_1551 encodes squalene hopene cyclase, which is a key enzyme for hopanoid biosynthesis and cyclizes squalene to hopene [39]. Hopanoids belong to a triterpene series widespread

among prokaryotes and play roles in membrane stabilization. Several different hopanoid derivatives are present in *Z. mobilis* [40]. ZZ6_1046 and ZZ6_1043 encode TolQ and TolB, respectively. Both proteins are components of the Tol–Pal (peptidoglycan-associated lipoprotein) system, which is involved in the maintenance of outer membrane stability [41]. Tol proteins are located in the cell envelope and are thought to be involved in the integration of some outer membrane components such as porins and lipopolysaccharides [42]. ZZ6_1254 encodes a protein-export membrane protein, SecD, in the Sec system, and mutations of the gene exhibit pleiotropic defects in protein export in *E. coli* [43]. ZZ6_1477 encodes a preprotein import (inner membrane) translocase subunit, Tim44. In mitochondria, Tim44 is a component to anchor mHsp70 to the TIM23 channel and associates transiently with the TIM23 complex for import of matrix-localized proteins in mitochondria [44]. ZZ6_0158 encodes an autotransporter secretion inner membrane protein, TamB, that forms a complex of the translocation and assembly module with the outer membrane protein, TamA. The complex functions in translocation of autotransporters across the outer membrane [45]. ZZ6_1210 encodes a competence protein, ComEC, that is a DNA transformation transporter (DNA-T) core component (KEGG). Competent cells generally possess a DNA transport complex that is most likely composed of surface-exposed DNA receptors, which facilitate DNA translocation through the cell wall, membrane pores, and motor molecules that power DNA transport [46]. ZZ6_0840 encodes a hypothetical transmembrane protein that possesses a zinc finger domain at its N-terminal portion and a Hid1 superfamily domain at its middle portion as putative conserved protein domains. Hid1 is a high-temperature-induced dauer-forming protein 1 with many putative transmembrane segments in *Caenorhabditis elegans* [47]. ZZ6_0541 encodes a protein bearing an SH3-like domain (COG3807). There are many SH3-like domain-containing proteins [48], but the function of the domain has not been clarified yet except for SH3-like domain-dependent interaction between CheA and CheW [49].

Group C as transporter includes a single gene, ZZ6_1289, that encodes a putative Fe^{2+}/Mn^{2+} transporter, which shares 58% identity to Fe^{2+}/Mn^{2+} transporter pcl1 in *Acetobacter pasteurianus*.

Group D consists of genes for DNA repair. ZZ6_0616 encodes the DNA repair protein RadC. RadC functions specifically in recombination repair that is associated with a replication fork and is required for growth-medium-dependent repair of DNA double strand breaks in *E. coli* [50]. ZZ6_0934 encodes XseA, a large subunit of exonuclease VII that is implicated in the

Table 1 Classification of thermotolerant genes and characterization of their Tn*10*-inserted mutants in *Z. mobilis* TISTR 548

Category	Tn*10*-inserted gene	Function	Protein type[a]	Growth at high temperature compared with that of parental strain[b]			Sensitivity to ethanol[c]		Effect of MgCl$_2$[d]
				38 °C	39 °C	39.5 °C	2.0% (v/v)	2.5% (v/v)	
	(WT, TISTR548)			++++	+++++	+++	++++	++++	−
General	ZZ6_0707	Glucose sorbosone dehydrogenase	S	+	+	−	++++	++++	−
Metabolism (Group A)	ZZ6_1376	5,10-methylenetetrahydrofolate reductase	S	++++	++++	+	+++	+++	+++
Membrane	ZZ6_1146	Glucosamine/fructose 6-phosphate aminotransferase	M	+	+		++	++	+++
Stabilization (Group B)	ZZ6_0929	Glycosyltransferase group 1	S	+	−	−	++++	+	++++
	ZZ6_0923	Phospholipase D/transphosphatidylase	M	−	−	−	−	−	−
	ZZ6_1551	Squalene hopene cyclase (Shc)	S	−	−	−	+	−	+++
	ZZ6_1046	Tol/Pal system component TolQ	M	+	+	−	++	++	−
	ZZ6_1043	Tol/Pal system component TolB	S	+	+	+	++++	++++	−
	ZZ6_1254	Protein export membrane protein SecD	M	−	−	−	++	+	++
	ZZ6_1477	Preprotein translocase subunit Tim44	M	−	−	−	++++	++++	−
	ZZ6_0158	Autotransporter secretion inner membrane protein TamB	M	+	−		++	++	++++
	ZZ6_1210	Competence protein ComEC	M	−	−	−	+	+	+++
	ZZ6_0840	Hypothetical transmembrane protein	M	−	−	−	++++	+++	++
	ZZ6_0541	Hypothetical transmembrane protein	M	++++	+++	+	++	++	++
Transporter (Group C)	ZZ6_1289	Putative Fe^{2+}/Mn^{2+} transporter	M	−	−	−	+++	+++	−
DNA repair (Group D)	ZZ6_0616	DNA repair protein RadC	S	++++	+++	+	+++	+++	−
	ZZ6_0934	Exonuclease VII (XseA)	S	−	−	−	+++	+++	−
	ZZ6_0681	DNA repair protein RadA	S	+	+	−	+++	+++	++
tRNA/rRNA modification (Group E)	ZZ6_0023	tRNA/rRNA methyltransferase (SpoU)	S	+++	++	++	++	++	++

Table 1 continued

Category	Tn*10*-inserted gene	Function	Protein type[a]	Growth at high temperature compared with that of parental strain[b]			Sensitivity to ethanol[c]		Effect of MgCl$_2$[d]
				38 °C	39 °C	39.5 °C	2.0% (v/v)	2.5% (v/v)	
Protein quality control (Group F)	ZZ6_1659	Zn-dependent peptidase	S	++++	+++	++	++++	++++	−
	ZZ6_0980	Serin protease DegP	S	−	−	−	−	−	+
Translation control (Group G)	ZZ6_0702	ATP-dependent helicase HrpB	S	−	−	−	+	−	−
Cell division (Group H)	ZZ6_0979	ParA/MinD-like ATPase	S	−	−	−	++	++	−
Transcriptional regulation (Group I)	ZZ6_0019	Trp repressor-binding protein WrbA	S	−	−	−	+++	++	−
Others	ZZ6_0962	Pseudogene	(S)	+	+	−	++++	++++	++
	ZZ6_0861	Hypothetical protein	S	+	+	−	++	++	−

[a] Protein type was described as described in "Methods" sections."S" and "M" mean soluble protein and membrane protein, respectively

[b] The growth of representative of isolated mutants was compared to that of the parental strain on 3% YPD plates at 38, 39, and 39.5 °C. The number of "+" indicates the degree of cell growth at high temperature compared to that of the parental strain, while "−" indicates no growth

[c] The tolerance of representative of isolated mutants to ethanol was determined by comparison of growth on 3% YPD plates containing 2.0 and 2.5% (v/v) ethanol. The number of "+" indicates the degree of cell growth at 30 °C under the ethanol stress condition compared to that of the parental strain, while "−" indicates no growth

[d] The effect of MgCl$_2$ on the growth of representative of isolated mutants was determined by comparison of growth in 3% YPD liquid medium containing 20 mM MgCl$_2$ at 39.5 °C. The number of "+" indicates the following degree of cell growth compared to that of the growth in the absence of MgCl$_2$: ++, $P < 0.05$; +++, $P < 0.01$; ++++, $P < 0.001$. "−" indicates no significant improvement of growth by the addition of MgCl$_2$

resection of a nicked mismatched strand in a methyl-directed mismatch repair pathway [51]. ZZ6_0681 encodes the DNA repair protein RadA. In *E. coli*, RadA is involved in recombination and recombination repair and is likely involved in the stabilization or processing of branched DNA molecules or blocked replication forks [52]. *radA* mutants show a modest decrease in survival after UV or X-irradiation exposure [53].

Group E consists of one gene for tRNA/rRNA modification. ZZ6_0023 encodes SpoU, which is a tRNA/rRNA methyltransferase. This enzyme may contribute to stabilization of the structure of tRNA or ribosome [54]. Analysis of the nucleoside modification pattern of tRNA, 16S rRNA, and 23S rRNA in *E. coli* has shown that the modified nucleoside 2′-*O*-methylguanosine, present in a subset of tRNAs at residue 18, is completely absent in the *spoU* mutant [55].

Group F genes are related to protein quality control. ZZ6_1659 encodes a Zn-dependent peptidase (peptidase with a M16 domain) (KEGG). The M16 family of zinc peptidases comprises a pair of homologous domains that form two halves of a "clam-shell" surrounding the active site, and closure of the clam-shell is required for proteolytic activity [56]. ZZ6_0980 encodes the serine protease DegP, and the orthologue gene has been identified as a thermotolerant gene in *E. coli* and *A. tropicalis* [28, 29].

DegP is a chaperone/serine protease located in the periplasm and acts to remove damaged proteins [57, 58].

Group G consists of one gene for translation control. ZZ6_0702 encodes the ATP-dependent helicase HrpB, that acts as an RNA helicase. Some in this helicase group unwind RNA molecules with a 3′ to 5′ polarity [59]. HrpA is an orthologue of HrpB involved in mRNA processing in *E. coli*. *hrpA* mutations in regions for predicted binding and hydrolysis of nucleotide triphosphate abolish the ability for mRNA processing [60].

Group H as cell division includes ZZ6_0979 for ParA/MinD-like ATPase. In *E. coli*, MinD activates a MinC-dependent mechanism responsible for the inactivation of potential division sites and renders the division inhibition system sensitive to MinE, which are required for correct placement of a division site [61]. MinD binds ATP and bears ATPase activity. On the other hand, ParA is required for the equipartition of P1 plasmids during cell division [62].

Group I consists of one gene related to transcriptional regulation. ZZ6_0019 encodes the flavoprotein WrbA, that binds to the tryptophan repressor TrpR and functions as an accessory element in blocking the TrpR-specific transcriptional process [63]. WrbA enhances the formation and/or stabilization of noncovalent complexes between TrpR holorepressor and its primary operator targets [64]. WrbA also functions as an NAD(P)H/quinone oxidoreductase [64] and belongs to the family of multimeric flavodoxin-like proteins [65] as a new type (type IV) of NAD(P)H:quinone oxidoreductase, which protects cells against oxidative stress [64] and may prepare cells for long-term maintenance under stress conditions [66].

There are two genes that deviate from the 9 categories. ZZ6_0962 is named as a pseudogene but should have a crucial function at a high temperature as observed in this study. The pseudogene has an inserted transposon in the gene, but the contribution of the transposon to thermotolerance is unknown. ZZ6_0861 encodes a hypothetical small protein consisting of 82 amino acid residues.

Effect of supplemented MgCl₂ on growth of thermosensitive mutants

Mg^{2+} is known to stabilize the outer membrane structure in cells by binding extracellularly [67] and the thermosensitive phenotype of mutants due to the disruption of genes for membrane stabilization is suppressed by the addition of $MgCl_2$ at a CHT in *E. coli* [28]. Thus, the effect of $MgCl_2$ on growth of thermosensitive mutants in *Z. mobilis* was tested at its CHT.

Thermosensitive mutants and the parental strain were grown in YPD medium with or without 20 mM $MgCl_2$ at 39.5 °C for 24 h under a static condition (Additional file 1: Figure S4; Table 1). The growth of 13 thermosensitive mutants was significantly improved by the supplementation of $MgCl_2$, 120–260% of that of the parental strain. Eight of them were in Group B and have disrupted genes for membrane stabilization or membrane formation. These results suggest that Mg^{2+} stabilizes the membrane structure at a CHT and protects cells from heat, as has been proposed in *E. coli*.

Effect of ethanol stress on growth of thermosensitive mutants

Zymomonas mobilis as an efficient ethanol producer is often exposed to ethanol stress under fermentation conditions. The effect of exogenous ethanol on thermosensitive mutants was thus examined on YPD plates containing 2.0 or 2.5% ethanol at 30 °C. In consequence, about half of the thermosensitive mutants exhibited repressed growth in the presence of ethanol, less than 50% growth compared to that in the absence of ethanol (Table 1). Interestingly, most of the thermosensitive mutants that were classified into the membrane stabilization group exhibited sensitivity to ethanol stress, and most of the ethanol-sensitive mutants were classified into the group in which the thermosensitive growth phenotype was suppressed by the addition of $MgCl_2$. Therefore, these results suggest that the mechanism of

thermotolerance at a CHT partially overlaps with that of ethanol stress resistance and allows us to speculate that stabilization of the membrane structure is one of crucial points for ethanol tolerance.

Discussion

In this study, we isolated 38 thermosensitive mutants by transposon mutagenesis and finally identified 26 thermotolerant genes that are required for survival at a CHT in thermotolerant *Z. mobilis* TISTR 548. Physiological functions and classification of these gene products may allow us to obtain a clue regarding the thermotolerance mechanism of this organism. The gene products were classified into 9 categories (Table 1). About half of them are related to membrane stabilization or membrane formation including enzymes for peptidoglycan or lipid biosynthesis and proteins for protein secretion systems. Most of these, genes for glucosamine/fructose 6-phosphate aminotransferase (ZZ6_1146), glycosyltransferase (ZZ6_0929), squalene hopene cyclase (ZZ6_1551), protein export membrane protein SecD (ZZ6_1254), autotransporter secretion inner membrane protein TamB (ZZ6_0158), competence protein ComEC (ZZ6_1210), hypothetical transmembrane protein (ZZ6_0840), and hypothetical transmembrane protein (ZZ6_0541) were found to be required for ethanol tolerance. Therefore, it is thought that membrane stabilization and maintenance are essential for survival at a CHT. Surprisingly, as found in *E. coli* [28], there was no heat shock protein in these thermotolerant gene products except for DegP, suggesting that not all heat shock proteins may be essential for survival under high temperatures. DegP, which functions in the periplasm as a chaperone at low temperatures and as a protease at high temperatures [68], is thought to play a role in the maintenance of homeostasis of the periplasm or membranes. In *E. coli*, *groEL* as an essential gene was induced at a CHT [28] and thus some heat shock proteins may be required under such an extreme condition.

Thermotolerant genes have also been identified in *E. coli* BW25113 and *A. tropicalis* SKU1100: 72 and 24 genes, respectively [28, 29; unpublished data]. The thermotolerant genes of the two microbes can be classified into 9 categories according to the classification of those of *Z. mobilis*, and the number and distribution of these genes are shown in Table 2. The ratios of thermotolerant genes to total genomic genes in *Z. mobilis*, *E. coli*, and *A. tropicalis* are 1.47, 1.68, and 0.70%, respectively. We do not know the reason why the ratio in *A. tropicalis* is relatively low. In the case of *E. coli*, a single-gene knockout library was used for screening thermosensitive mutants and thus almost all of the genes except for essential genes were examined. On the other hand, in the case of *Z. mobilis* and *A. tropicalis*, transposon mutagenesis

Table 2 Comparison of thermotolerant genes among *Z. mobilis* TISTR 548, *E. coli* BW25113, and *A. tropicalis* SKU1100

Category	No. of thermotolerant gene (ratio %[a])		
	Z. mobilis	*E. coli*[b]	*A. tropicalis*[c]
General metabolism	2 (0.11%)	22 (0.51%)	1 (0.03%)
Membrane stabilization	12 (0.68%)	18 (0.42%)	5 (0.15%)
Transporter	1 (0.06%)	3 (0.07%)	3 (0.09%)
DNA repair and DNA modification	3 (0.17%)	6 (0.14%)	1 (0.03%)
tRNA and rRNA modification	1 (0.06%)	9 (0.21%)	0 (0%)
Protein quality control and stress response	2 (0.11%)	4 (0.09%)	5 (0.15%)
Translational control	1 (0.06%)	3 (0.07%)	2 (0.06%)
Cell division	1 (0.06%)	3 (0.07%)	2 (0.06%)
Transcriptional regulation	1 (0.06%)	0 (0%)	2 (0.06%)
Others	2 (0.11%)	3 (0.07%)	4 (0.12%)
Sum of thermotolerant gene	26 (1.47%)	72 (1.68%)	24 (0.70%)
Total genomic genes	1765	4288	3412

[a] Ratio was estimated using the number of total genomic genes

[b] Data of Murata et al. [28] and unpublished data

[c] Data of Soemphol et al. [29]

was applied for screening thermosensitive mutants, and the ratios of the number of thermotolerant genes, for each of which two or more transposon-inserted mutants were isolated, to the total number of thermotolerant genes (Additional file 1: Table S1) [29] were 35 and 21%, respectively. Therefore, the low ratio of multiple mutants for the same gene in *A. tropicalis* suggests the possibility that there are still unidentified thermotolerant genes in *A. tropicalis* SKU1100. In all categories except for general metabolism, ratios of thermotolerant genes in *Z. mobilis* are closer to those in *E. coli* than those in *A. tropicalis*. Notably, *Z. mobilis* has a higher ratio of thermotolerant genes for membrane stabilization than the ratios in other two microbes: 46, 25, and 20% in *Z. mobilis*, *E. coli*, and *A. tropicalis*, respectively.

On the other hand, *E. coli* possesses several discriminating sets of thermotolerant genes, which are absent in the other two microbes: 4 genes (*aceE*, *aceF*, *lpd*, and *lipA*) for pyruvate metabolism, 3 genes (*atpA*, *atpD*, and *atpG*) for ATPase, 3 genes (*cydB*, *yhcB*, and *cydD*) for ubiquinol oxidase or its formation, and 3 genes (*ubiE*, *ubiH*, and *ubiX*) for ubiquinone biosynthesis in the category of general metabolism, 8 genes (*gmhB*, *lpcA*, *rfaC*, *rfaD*, *afaE*, *rfaF*, *rfaG*, and *lpxL*) for lipopolysaccharide biosynthesis and 5 genes (*ydcL*, *yfdL*, *ynbE*, *nlpI*, and *ycdO*) for peptidoglycan-associated lipoproteins or predicted lipoproteins in the category of membrane stability, 5 genes (*dnaQ*, *holC*, *priA*, *ruvA*, and *ruvC*) for

DNA double-strand break repair in the category of DNA repair, and 6 genes (*iscS*, *yheL*, *yheM*, *yheN*, *yhhP*, and *yccM*) for a sulfur relay system in the category of tRNA modification [28; unpublished data]. Of these sets, genes for the lipopolysaccharide biosynthesis and the sulfur relay system are postulated to have been acquired by horizontal gene transfer [28]. The genes in the 4 categories described above seem to contribute to specific strategies for thermotolerance in *E. coli* [28; some thermotolerant genes will be described elsewhere].

There are common thermotolerant genes or thermotolerant genes related to the same physiological function or pathway among the three microbes. In the category of protein quality control, the three microbes share *degP* and both *Z. mobilis* and *A. tropicalis* have a gene for Zn-dependent protease (ZZ6_1659 and ATPR_0429, respectively). In membrane stabilization, one gene related to hopanoid biosynthesis is present in *Z. mobilis* and *A. tropicalis* (*shc* and ATPR_1188, respectively) and two to three genes for the Tol-Pal system are present in *Z. mobilis* (*tolQ* and *tolB*) and *E. coli* (*pal*, *tolQ* and *tolR*). One gene related to MinC-dependent cell division inhibition in cell division is present in *Z. mobilis* and *A. tropicalis* (*minD* and *minC*, respectively), and *wrbA* in transcriptional regulation and *nhaA* for the Na^+/H^+ antiporter in transporters are shared by *Z. mobilis* and *A. tropicalis*. On the basis of the functions of these genes and combinations of other thermotolerant genes in each category, some common strategies for thermotolerance have emerged: in the category of membrane stabilization, synthesis or modification of peptidoglycan and maintenance of integrity for all three microbes, and hopanoid or lipid synthesis for *Z. mobilis* and *A. tropicalis*; in DNA repair, double-strand DNA repair, which may be accumulated at a CHT, for *Z. mobilis* and *E. coli*; tRNA modification, probably for a stable structure at such a high temperature, for *Z. mobilis* and *E. coli*; in chaperone and protease, removal of damaged proteins, especially by periplasmic serine protease DegP, for all three microbes; control of chromosome segregation for *E. coli* and *A. tropicalis*, and control of cell division for all three microbes; and in transcriptional regulation, Trp repressor-binding protein WrbA (still unclear why necessary) for *Z. mobilis* and *A. tropicalis*. In addition, import or export of some metal ions may be important probably for keeping homeostasis of some ions, export of toxic ions or maintenance of membrane potential.

At a CHT, several problems including protein unfolding or increase in membrane fluidity occur. Reactive oxygen species increase as the temperature increases [69], causing the damage of macromolecules including DNA [70, 71]. The requirement of genes for the 9 categories allows us to make speculations about various types of damage of membrane and proteins or about the abnormal structures of macromolecules including proteins, DNAs and RNAs at a CHT. Microbes would have thus acquired thermotolerant genes to overcome these problems. Moreover, it is assumed that these genes are involved in the response of cells to other stresses including osmotic stress or oxygen stress. In fact, *Z. mobilis* increases thermotolerance by the addition of sorbitol [72] and exhibits faster growth and higher ethanol production under a static condition than that under a shaking condition [19, unpublished]. Further experiments are required for clarifying this assumption.

Conclusions

The thermotolerant genes of thermotolerant ethanologenic *Z. mobilis* TISTR 548 have been identified. Comparison with thermotolerant genes in *E. coli* and *A. tropicalis* reveal that these genes of the three microbes can be classified into 9 categories and that there are common thermotolerant genes or thermotolerant genes related to the same physiological function or pathway among the three microbes, which suggest several common strategies, including membrane stabilization, protection and repair of macromolecules of proteins, DNAs and RNAs, and maintenance of cellular metabolism-like cell division, transcription or translation, for the three microbes to survive at CHT. Considering the genetic conversion of non-thermotolerant to thermotolerant bacteria, such strategies might be applicable.

Methods

Materials

A DNA sequencing Kit (ABI PRISM ® Terminator v 3.1 Cycle sequencing Kit) was obtained from Applied Biosystem Japan. Oligonucleotide primers were synthesized by Proligo Japan K.K. (Tokyo, Japan). Other chemicals were all of analytical grade and obtained from commercial sources.

Microorganisms and media

Zymomonas mobilis TISTR 548 [19, 20] and its derivatives were grown in YPD (3% glucose, 0.5% peptone, and 0.3% yeast extract) medium. *E. coli* S17-1 harboring pSUP2021 Tn*10* [30] was grown in LB (0.5% yeast extract, 1% NaCl, and 1% Bactotryptone) medium supplemented with 12.5 μg/ml of tetracycline.

Conjugation and transposon mutagenesis

Escherichia coli S17-1 harboring pSUP2021 Tn*10* as a donor for conjugal mating was grown in LB medium containing 12.5 μg/ml of tetracycline under a shaking condition at 100 rpm at 37 °C. The recipient *Z. mobilis* TISTR 548 was grown in YPD medium under a static

condition at 30 °C. Cells of both strains were grown to the mid-log phase, washed three times with LB medium, recovered by centrifugation at 5000 rpm for 1 min, and suspended in a small volume of LB medium. Both cell suspensions were then mixed at a ratio of donor and recipient of 3:2 and stood for 3 h at 30 °C. The suspensions were spotted on the surfaces of LB agar plates and incubated at 30 °C for 5 h. After the mating steps, cells were recovered, resuspended in a small volume of YPD medium, and spread on YPD agar plates containing 0.15% acetic acid and 12.5 μg/ml of tetracycline. Transconjugants (transposon-inserted mutants) that appeared on the plates after 3-day incubation at 30 °C were subjected to the following screening.

Screening of thermosensitive mutants

About 8000 transconjugants were subjected to the first screening in which they were grown at 30 and 39.5 °C on YPD agar plates. Transposon-inserted mutants that showed no or almost no growth on the plates at 39.5 °C were selected for the next screening. The second screening was performed under the same condition as that in the first screening. Selected mutants were then subjected to the last screening in which their thermosensitivity was examined in 2-ml liquid culture of YPD medium at 30 and 39.5 °C for 24 h under a static condition. Cell growth was determined by measuring cell turbidity at OD_{550}. Mutants that showed a value at OD_{550} significantly less than that of the parent strain were selected and defined as thermosensitive mutants.

Examination of the effects of heat and ethanol stresses on growth of thermosensitive mutants

Thermosensitive mutants and the parental strain were pre-cultured in YPD medium under a static condition at 30 °C until a mid-log phase. For the heat stress experiment, the pre-cultured cells were serially diluted with YPD medium, spotted on YPD agar plates, and incubated at 30, 38, 39, and 39.5 °C for 40 h. For the ethanol stress experiment, the pre-cultured cells were serially diluted, spotted on YPD plates supplemented with 2.0 or 2.5% ethanol, and incubated at 30 °C for 40 h. Growth ability was examined in triplicate.

Effect of Mg²⁺ on growth of thermosensitive mutants

Thermosensitive mutants and the parental strain were pre-cultured in YPD medium under a static condition at 30 °C until a mid-log phase. The pre-cultured cells were inoculated in YPD medium with or without 20 mM $MgCl_2$ and incubated at 39.5 °C for 24 h under a static condition. The experiments were performed more than 3 times. The significance of the effect of $MgCl_2$ on cell growth was evaluated by a *t* test.

Identification of the transposon (Tn10)-inserted site in a thermosensitive mutant genome by TAIL-PCR followed by nucleotide sequencing

The Tn10-inserted site in the genome of each thermosensitive mutant was determined by TAIL-PCR [73] followed by nucleotide sequencing. The genomic DNA from thermosensitive mutants was isolated as described previously [74]. The concentration of isolated genomic DNA was measured by using Nanodrop (Nanodrop Technologies, Wilmington, DE). TAIL-PCR was performed by using TaKaRa PCR Thermal Cycler Dice® mini (TaKaRa). Three specific primers for TAIL-PCR were TnISR-1 (GATCCTCTCGTTTGTTGCGGTCAGGCC) [30], TnISR-1.5 (AGGGCTGCTAAAGGAAGCGG) (this work) and TnISR-2 (ACGAAGCGCAAAGAGGAA-GCAGG) [29], and an arbitrary degenerated primer was AD2 (GTNCGASWCANAWGTT) [73]. The first PCR was carried out in a 50-μl mixture containing 10 ng of chromosome DNA, 5.0 μM TnISR-1, 25 μM AD2 primer, 500 μM each of dNTPs, 0.5 U PrimeSTAR (TaKaRa) and 1× buffer supplied for the enzyme. Two percent of the first PCR product was used as a template for the second PCR, which was performed using the same reaction mixture as that used for the first PCR except that TnISR-1.5 was used as a specific primer. The third PCR was also performed using the same reaction mixture as that used for the first PCR except that TnISR2 was used as a specific primer and the concentration of AD2 was reduced to 12.5 μM [25]. The second or third PCR product was purified by using a PCR product purification kit (Qiagen) and subjected to nucleotide sequencing on an ABI PRISM 310 Genetic Analyzer (Applied Biosystems) or DNA Sequencer GenomeLab GeXP (Beckman Coulter). The sequencing reaction was performed with a BigDye® Terminator v3.1 Cycle Sequencing Kit (Applied Biosystems) or a GenomeLab Dye Terminator Cycle Sequencing with Quick Start Kit (Beckman Coulter).

RT-PCR

Zymomonas mobilis cells were grown in 50 ml of YPD medium under a static condition at 30 °C until exponential phase, and then the temperature was increased to 39.5 °C and the cultivation was continued for 8 min. As a control, the cultivation was continued for 8 min at 30 °C. Total RNA was prepared from these heat-stressed or not heat-stressed cells by the hot phenol method [75]. RT-PCR analysis was performed using an mRNA-selective RT-PCR kit (TaKaRa) and primers (Additional file 1: Table S2) to examine the expression of immediate downstream genes of Tn10-inserted genes as described previously [28]. The reverse transcription reaction was carried out at 42 °C for 15 min, followed by PCR at 85 °C for 1 min, 45 °C for 1 min, and extension at 72 °C for 1 min, using the two specific primers for each gene. After the completion of 15, 20, 25, and 30 cycles, the PCR products were analyzed by 0.9% agarose gel electrophoresis and stained with ethidium bromide [76]. The relative amounts of RT-PCR products on the gel were compared by measuring the density of bands on the gel by using image J (https://imagej.nih.gov/ij/). Under our conditions, the RNA-selective RT-PCR was able to specifically detect mRNA because no band was observed when reverse transcriptase was omitted.

Bioinformatics analysis

The intrinsic gene that was inserted by Tn10 in each thermotolerant mutant was confirmed to be a thermotolerant gene after analyses of the gene organization and/or expression of its downstream gene. Thermotolerant genes were then subjected to functional classification by bioinformatics analysis mainly according to the instructions of KEGG (http://www.genome.jp/kegg/), NCBI (http://www.ncbi.nlm.nih.gov/), Inter Pro (http://www.ebi.ac.uk/interpro/), and Uniprot (http://www.uniprot.org/). Protein type was analyzed by TMHMM (http://www.cbs.dtu.dk/services/TMHMM/). Homology searching and alignment were performed using BLAST [77]. The Z. mobilis TISTR 548 thermotolerant genes were designed as ZZ6_XXXX according to Z. mobilis subsp. mobilis ATCC29191 because the genome sequence of TISTR 548 was found to be almost identical to that of ATCC29191 after draft sequencing (unpublished).

Abbreviations
HTF: high-temperature fermentation; TISTR: Thailand Institute of Scientific and Technological Research; GRAS: generally regarded as being safe; CHT: critical high temperature; TAIL-PCR: thermal asymmetric interlaced PCR; LPS: lipopolysaccharide; DNA-T: DNA transformation transporter; NADH: reduced form of nicotinamide adenine dinucleotide; NADPH: reduced form of nicotinamide adenine dinucleotide phosphate; TnISR: transposon-inserted region; AD: arbitrary degenerate.

Authors' contributions
Conceived and designed the experiments: PT, MM, MY. Performed the experiments: KC, TS, AT, MM. Analyzed the data: KC, TS, AT, MM, TK, PT, MY. Wrote the paper: KC, MM, MY. All authors read and approved the final manuscript.

Author details
[1] Division of Product Development and Management Technology, Faculty of Agro-Industrial Technology, Rajamangala University of Technology Tawan-ok, Chanthaburi Campus, Chanthaburi 22100, Thailand. [2] Life Science, Graduate School of Science and Technology for Innovation, Yamaguchi University, Ube 755-8505, Japan. [3] Department of Biological Chemistry, Faculty of Agriculture, Yamaguchi University, 1677-1 Yoshida, Yamaguchi 753-8515, Japan. [4] Research Center for Thermotolerant Microbial Resources, Yamaguchi University, Yamaguchi 753-8315, Japan. [5] Department of Biotechnology, Faculty of Technology, Khon Kaen University, Khon Kaen 40002, Thailand.

Acknowledgements

We thank K. Matsushita, T. Yakushi, W. Soemphol, and N. Lertwattanasakul for their helpful discussion. This work was supported by The Core to Core Program A. Advanced Research Networks, which was granted by the Japan Society for the Promotion of Science, the National Research Council of Thailand, Ministry of Science and Technology in Vietnam, National Univ. of Laos, Univ. of Brawijaya and Beuth Univ. of Applied Science Berlin, and supported by the Program for Promotion of Basic Research Activities for Innovative Biosciences, which was granted by Japan Science and Technology Agency.

Competing interests

The authors declare that they have no competing interests.

Funding

Current Funding Sources is 16H02485. Recipient person is Mamoru Yamada, Ph. D.

References

1. Swings J, De Ley J. The biology of Zymomonas. Bacteriol Rev. 1977;41:1–46.
2. Gibbs M, De Moss RD. Anaerobic dissimilation of C14. Labelled glucose fructose by Pseudomonas lindneri. J Biol Chem. 1954;207:689–94.
3. Seo J, Chong H, Park HS, Yoon K, Jung C, Kim JJ, et al. The genome sequence of the ethanologenic Bacterium Zymomonas mobilis ZM4. Nat Biotechnol. 2005;23:63–8.
4. Yang S, Pelletier DA, Lu TYS, Brown SD. The Zymomonas mobilis regulator hfq contributes to tolerance against multiple lignocellulosic pretreatment inhibitors. BMC Microbiol. 2010;135:1–11.
5. Calazans GMT, Lopes CE, Lima RMOC, Defranca FP. Antitumor activities of levans produced by Zymomonas mobilis strains. Biotechnol Lett. 1997;19:19–21.
6. Yoo SH, Yoon EJ, Cha J, Lee HG. Antitumor activity of levan polysaccharides from selected microorganisms. Int J Biol Macromol. 2004;34:37–41.
7. Uden VN, Duarte HDC. Effects of ethanol on the temperature profile of Saccharomyces cerevisiae. Z Allg Mikrobiol. 1981;21:743–50.
8. Ghose TK, Bandyopadhyay KK. Studies on immobilized Saccharomyces cerevisiae. II. Effect of temperature distribution on continuous rapid ethanol formation in molasses fermentation. Biotechnol Bioeng. 1982;24:797–804.
9. Attfield PV. Stress tolerance: the key to effective strains of industrial baker's yeast. Nat Biotechnol. 1997;15:1351–7.
10. Wang Y, Gong L, Liang J, Zhang Y. Effects of alcohol on expressions of apoE in mice livers and brains. Wei Sheng Yan Jiu. 2007;36:737–40.
11. Basso LC, de Amorim HV, de Oliveira AJ, Lopes ML. Yeast selection for fuel ethanol production in Brazil. FEMS Yeast Res. 2008;8:1155–63.
12. Babiker MA, Banat A, Hoshida H, Ano A, Nonklang S, Akada R. High-temperature fermentation: how can processes for ethanol production at high temperatures become superior to the traditional process using mesophilic yeast? Appl Microbiol Biotechnol. 2010;85:861–7.
13. Piper PW. The heat shock and ethanol stress responses of yeast exhibit extensive similarity and functional overlap. FEMS Microbiol Lett. 1995;134:121–7.
14. Carmelo V, Santos R, Viegas CA, Sa´-Correia I. Modification of Saccharomyces cerevisiae thermotolerance following rapid exposure to acid stress. Int J Food Microbiol. 1998;42:225–30.
15. Ciani M, Beco L, Comitini F. Fermentation behaviour and metabolic interactions of multistarter wine yeast fermentations. Int J Food Microbiol. 2006;108:239–45.
16. Pizarro F, Varela C, Martabit C, Bruno C, Pe´rez-Correa JR, Agosin E. Coupling kinetic expressions and metabolic networks for predicting wine fermentations. Biotechnol Bioeng. 2007;98:986–98.
17. Coleman MC, Fish R, Block DE. Temperature-dependent kinetic model for nitrogen-limited wine fermentations. Appl Environ Microb. 2007;73:5875–84.
18. Gibson BR, Lawrence SJ, Leclaire JP, Powell CD, Smart KA. Yeast responses to stresses associated with industrial brewery handling. FEMS Microbiol Rev. 2007;31:535–69.
19. Sootsuwan K, Irie A, Murata M, Lertwattanasakul N, Thanonkeo P, Yamada M. Thermotolerant Zymomonas mobilis: comparison of ethanol fermentation capability with that of an efficient type strain. Open Biotechnol J. 2007;1:59–65.
20. Charoensuk K, Irie A, Lertwattanasakul N, Sootsuwan K, Thanonkeo P, Yamada M. Physiological importance of cytochrome c peroxidase in ethanologenic Thermotolerant Zymomonas mobilis. J Mol Microbiol Biotechnol. 2011;20:70–82.
21. Thanonkeo P, Thanonkeo S, Charoensuk K, Yamada M. Ethanol production from Jerusalem artichoke (Helianthus tuberosus L.) by Zymomonas mobilis TISTR 548. Afr J Biotechnol. 2011;10:10691–7.
22. Manaia CM, Moore ERB. Pseudomonas thermotolerans sp. nov., a thermotolerant species of the genus Pseudomonas sensustricto. Int J Syst Evol Microbiol. 2002;52:2203–9.
23. Saeki A, Theeragool G, Matsushita K, Toyama H, Lotong N, Adachi O. Development of thermotolerant acetic acid bacteria useful for vinegar fermentation at higher temperatures. Biosci Biotechnol Biochem. 1997;61:138–45.
24. Michel GPF, Starka J. Effect of ethanol and heat stresses on the protein pattern of Zymomonas mobilis. J Bacteriol. 1986;165:1040–2.
25. Thanonkeo P, Sootsuwan K, Leelavacharamas V, Yamada M. Cloning and transcriptional analysis of groES and groEL in ethanol-producing bacterium Zymomonas mobilis TISTR 548. Pak J Biol Sci. 2007;10:13–22.
26. Rodrussamee N, Lertwattanasakul N, Hirata K, Suprayogi, Limtong S, Kosaka T, Yamada M. Growth and ethanol fermentation ability on hexose and pentose sugars and glucose effect under various conditions in thermotolerant yeast Kluyveromyces marxianus. Appl Microbiol Biotechnol. 2011;90:1573–86.
27. Murata M, Nitiyon S, Lertwattanasakul N, Sootsuwan K, Kosaka T, Thanonkeo P, Limtong S, Yamada M. High-temperature fermentation technology for low-cost bioethanol. J Jpn Inst Energy. 2015;94:1154–212.
28. Murata M, Fujimoto H, Nishimura K, Charoensuk K, Nagamitsu H, Raina S, et al. Molecular strategy for survival at a critical high temperature in Escherichia coli. PLoS ONE. 2011;6:e20063.
29. Soemphol W, Deeraksa A, Matsutani M, Yakushi T, Toyama H, Adachi O, et al. Global analysis of the genes involved in the thermotolerance mechanism of thermotolerant Acetobacter tropicalis SKU1100. Biosci Biotechnol Biochem. 2011;75:1921–8.
30. Deeraksa A, Moonmangmee S, Toyama HMY, Adachi O, Matsushita K. Characterization and spontaneous mutation of a novel gene, polE, involved in pellicle formation in Acetobacter tropicalis SKU1100. Microbiology. 2005;151:4111–20.
31. Oubrie A, Rozeboom HJ, Dijkstra BW. Active-site structure of the soluble quinoprotein glucose dehydrogenase complexed with methylhydrazine: a covalent cofactor-inhibitor complex. Proc Natl Acad Sci USA. 1999;96:11787–91.
32. Kim SH, Lee BR, Kim JN, Kim BG. NdgR, a common transcriptional activator for methionine and leucine biosynthesis in Streptomyces coelicolor. J Bacteriol. 2012;94:6837–46.
33. Badet-Denisot MA, Fernandez-Herrero LA, Berenguer J, Ooi T, Badet B. Characterization of L-glutamine:D-fructose-6-phosphate amidotransferase from an extreme thermophile Thermus thermophilus HB8. Arch Biochem Biophys. 1997;337:129–36.
34. Roncero C, Casadaban MJ. Genetic analysis of the genes involved in synthesis of the lipopolysaccharide core in Escherichia coli K-12: three operons in the rfa locus. J Bacteriol. 1992;174:3250–60.
35. Kido N, Torgov VI, Sugiyama T, Uchiya K, Sugihara H, Komatsu T, et al. Expression of the O9 polysaccharide of Escherichia coli: sequencing of the E. coli O9 rfb gene cluster, characterization of mannosyl transferases, and evidence for an ATP-binding cassette transport system. J Bacteriol. 1995;177:2178–87.
36. Pradel E, Parker CT, Schnaitman CA. Structures of the rfaB, rfaI, rfaJ, and rfaS genes of Escherichia coli K-12 and their roles in assembly of the lipopolysaccharide core. J Bacteriol. 1992;174:4736–45.
37. Tropp BE. Cardiolipin synthase from Escherichia coli. Biochim Biophys Acta. 1997;1348:192–200.
38. Shibuya I, Miyazaki C, Ohta A. Alteration of phospholipid composition

by combined defects in phosphatidylserine and cardiolipin synthases and physiological consequences in *Escherichia coli*. J Bacteriol. 1985;161:1086–92.

39. Siedenburg G, Jendrossek D. Squalene-hopene cyclases. Appl Environ Microbiol. 2011;77:3905–15.

40. Hermans MA, Neuss B, Sahm H. Content and composition of hopanoids in *Zymomonas mobilis* under various growth conditions. J Bacteriol. 1991;173:5592–5.

41. Kampfenkel K, Braun V. Membrane topologies of the TolQ and TolR proteins of *Escherichia coli*: inactivation of TolQ by a missense mutation in the proposed first transmembrane segment. J Bacteriol. 1993;75:4485–91.

42. Ray MC, Germon P, Vianney A, Portalier R, Lazzaroni JC. Identification by genetic suppression of *Escherichia coli* TolB residues important for TolB–Pal interaction. J Bacteriol. 2000;182:821–4.

43. Gardel C, Benson S, Hunt J, Michaelis S, Beckwith J. *secD*, a new gene involved in protein export in *Escherichia coli*. J Bacteriol. 1987;169:1286–90.

44. Slutsky-Leiderman O, Marom M, Iosefson O, Levy R, Maoz S, Azem A. The interplay between components of the mitochondrial protein translocation motor studied using purified components. J Biol Chem. 2007;282:33935–44942.

45. Selkrig J, Mosbahi K, Webb CT, Belousoff MJ, Perry AJ, Wells TJ, et al. Discovery of an archetypal protein transport system in bacterial outer membranes. Nat Struct Mol Biol. 2012;19:506–10.

46. Chen I, Dubnau D. DNA uptake during bacterial transformation. Nat Rev Microbiol. 2004;2:241–9.

47. Ailion M, Thomas JH. Isolation and characterization of high-temperature-induced dauer formation mutants in *Caenorhabiditis elegans*. Genetics. 2003;165:127–44.

48. Whisstock JC, Lesk AM. SH3 domains in prokaryotes. Trends Biochem Sci. 1999;24:132–3.

49. Bilwes AM, Alex LA, Crane BR, Simon MI. Structure of CheA, a signal-transducing histidine kinase. Cell. 1999;96:131–41.

50. Saveson CJ, Lovett ST. Tandem repeat recombination induced by replication fork defects in *Escherichia coli* requires a novel factor, RadC. Genetics. 1999;152:5–13.

51. Harris RS, Ross KJ, Lombardo MJ, Rosenberg SM. Mismatch repair in *Escherichia coli* cells lacking single-strand exonucleases ExoI, ExoVII, and RecJ. J Bacteriol. 1998;180:989–93.

52. Beam CE, Saveson CJ, Lovett ST. Role for *radA/sms* in recombination intermediate processing in *Escherichia coli*. J Bacteriol. 2002;184:6836–44.

53. Sargentini NJ, Smith KC. Quantitation of the involvement of the *recA*, *recB*, *recC*, *recF*, *recJ*, *recN*, *lexA*, *radA*, *radB*, *uvrD*, and *umuC* genes in the repair of X-ray-induced DNA double-strand breaks in *Escherichia coli*. Radiat Res. 1986;107:58–72.

54. Decatur WA, Fournier MJ. rRNA modifications and ribosome function. Trends Biochem Sci. 2002;27:344–51.

55. Persson BC, Jäger G, Gustafsson C. The *spoU* gene of *Escherichia coli*, the fourth gene of the *spoT* operon, is essential for tRNA (Gm18) 2'-*O*-methyl-transferase activity. Nucleic Acids Res. 1997;25:4093–147.

56. Aleshin AE, Gramatikova S, Hura GL, Bobkov A, Strongin AY, Stec B, et al. Crystal and solution structures of a prokaryotic M16B peptidase: an open and shut case. Structure. 2009;17:1465–75.

57. Lipinska B, Zylicz M, Georgopoulos C. The HtrA (DegP) protein, essential for *Escherichia coli* survival at high temperatures, is an endopeptidase. J Bacteriol. 1990;172:1791–7.

58. Meltzer M, Hasenbein S, Mamant N, Merdanovic M, Poepsel S, Hauske P, et al. Structure, function and regulation of the conserved serine proteases DegP and DegS of *Escherichia coli*. Res Microbiol. 2009;160:660–6.

59. Lee CG, Hurwitz J. A new RNA helicase isolated from HeLa cells that catalytically translocates in the 3' to 5' direction. J Biol Chem. 1992;267:4398–407.

60. Koo JT, Choe J, Moseley SL. HrpA, a DEAH-box RNA helicase, is involved in mRNA processing of a fimbrial operon in *Escherichia coli*. Mol Microbiol. 2004;52:1813–26.

61. de Boer PA, Crossley RE, Hand AR, Rothfield LI. The MinD protein is a membrane ATPase required for the correct placement of the *Escherichia coli* division site. EMBO J. 1991;10:4371–80.

62. Nordström K, Austin SJ. Mechanisms that contribute to the stable segregation of plasmids. Annu Rev Genet. 1989;23:37–69.

63. Yang W, Ni L, Somerville RL. A stationary-phase protein of *Escherichia coli* that affects the mode of association between the *trp* repressor protein and operator-bearing DNA. Proc Natl Acad Sci USA. 1993;90:5796–800.

64. Patridge EV, Ferry JG. WrbA from *Escherichia coli* and *Archaeoglobus fulgidus* is an NAD(P)H: quinone oxidoreductase. J Bacteriol. 2006;188:3498–506.

65. Grandori R, Khalifah P, Boice JA, Fairman R, Giovanielli K, Carey J. Biochemical characterization of WrbA, founding member of a new family of multimeric flavodoxin-like proteins. J Biol Chem. 1998;273:20960–6.

66. Chang DE, Smalley DJ, Conway T. Gene expression profiling of *Escherichia coli* growth transitions: an expanded stringent response model. Mol Microbiol. 2002;45:289–306.

67. Nikaido H. Molecular basis of bacterial outer membrane permeability revisited. Microbiol Mol Biol Rev. 2003;67:593–656.

68. Spiess C, Beil A, Ehrmann M. A temperature-dependent switch from chaperone to protease in a widely conserved heat shock protein. Cell. 1999;97:339–47.

69. Noor R, Murata M, Yamada M. Oxidative stress as a trigger for growth phase-specific sigmaE-dependent cell lysis in *Escherichia coli*. J Mol Microbiol Biotechnol. 2009;17:177–87.

70. Condon S. Responses of lactic acid bacteria to oxygen. FEMS Microbiol Rev. 1987;46:269–80.

71. Zagorski N, Imlay J. The chemistry behind oxidative damage. ASBMB Today. 2009;4:33–6.

72. Sootsuwan K, Thanonkeo P, Keeratirakha N, Thanonkeo S, Jaisil P, Yamada M. Sorbitol required for cell growth and ethanol production by Zymomonas mobilis under heat, ethanol, and osmotic stresses. Biotechnol Biofuels. 2013;6(1):180

73. Lui YG, Mitsukawa N, Oosumi T, Whittier R. Efficient isolation and mapping of Arabidopsis Thaliana T-DNA insert junctions by thermal asymmetric interlaced PCR. Plant J. 1995;8:457–63.

74. Sambrook J, Russell DW. Molecular cloning: a laboratory manual. 3rd ed. Cold Spring Harbour: Cold Spring Harbour Laboratory Press; 2001.

75. Aiba H, Adhya S, de Cromburgghe B. Evidence for two functional gal promoters in intact Escherichia coli cells. J Biol Chem. 1981;256:11905–10.

76. Tsunedomi R, Izu H, Kawai T, Matsushita K, Ferenci T, Yamada M. The activator of GntII genes for gluconate metabolism, GntH, exerts negative control of GntR-regulated GntI genes in Escherichia coli. J Bacteriol. 2003;185:1783–95.

77. Altschul SF, Gish W, Miller W, Myers EW, Lipman DJ. Basic local alignment search tool. J Mol Biol. 1990;215:403–10.

78. Desiniotis A, Kouvelis VN, Davenport K, Bruce D, Detter C, Tapia R, et al. Complete genome sequence of the ethanol-producing Zymomonas mobilis subsp. mobilis centrotype ATCC 29191. J Bacteriol. 2012;194:5966–7.

13

Single cell oil production by *Trichosporon cutaneum* from steam-exploded corn stover and its upgradation for production of long-chain α,ω-dicarboxylic acids

Chen Zhao[1], Hao Fang[1,2]* and Shaolin Chen[1]

Abstract

Background: Single cell oil (SCO) production from lignocelluloses by oleaginous microorganisms is still high in production cost, making the subsequent production of biofuels inviable economically in such an era of low oil prices. Therefore, how to upgrade the final products of lignocellulose-based bioprocess to more valuable ones is becoming a more and more important issue.

Results: Differently sourced cellulases were compared in the enzymatic hydrolysis of the steam-exploded corn stover (SECS) and the cellulase from the mixed culture of *Trichoderma reesei* and *Aspergillus niger* was found to have the highest enzymatic hydrolysis yield 86.67 ± 4.06%. Three-stage enzymatic hydrolysis could greatly improve the efficiency of the enzymatic hydrolysis of SECS, achieving a yield of 74.24 ± 2.69% within 30 h. Different bioprocesses from SECS to SCO were compared and the bioprocess C with the three-stage enzymatic hydrolysis was the most efficient, producing 57.15 g dry cell biomass containing 31.80 g SCO from 327.63 g SECS. An efficient and comprehensive process from corn stover to long-chain α,ω-dicarboxylic acids (DCAs) was established by employing self-metathesis, capable of producing 6.02 g long-chain DCAs from 409.54 g corn stover and 6.02 g alkenes as byproducts.

Conclusions: On-site cellulase production by the mixed culture of *T. reesei* and *A. niger* is proven the most efficient in providing cellulase to the lignocellulose-based bioprocess. Three-stage enzymatic hydrolysis was found to have very good application value in SCO production by *Trichosporon cutaneum* from SECS. A whole process from corn stover to long-chain DCAs via a combination of biological and chemical approaches was successfully established and it is an enlightening example of the comprehensive utilization of agricultural wastes.

Keywords: Single cell oil, *Trichosporon cutaneum*, Steam-exploded corn stover, On-site cellulase production, Three-stage enzymatic hydrolysis, α,ω-dicarboxylic acids, Self-metathesis

Background

As an agricultural country with more than 1.3 billion people to feed, China has plenty of agricultural wastes or residues that should be utilized in an economic and environmentally friendly way. This is good for rural economy and environment protection because the reality is that those agricultural residues are always treated improperly, e.g., combusted directly, which causes serious environmental problem [1–3]. Bioconversion of them to fuels or chemicals, therefore, is promising and beneficial to the rural economy and environment. This kind of scientific research and relevant industry should be encouraged and supported as the growth rate of Chinese economy decreases but environmental burden increases.

Single cell oil (SCO) from microorganism is thought to be a desirable alternative oil source to plant oil or animal fat due to the high productivity, the low land requirement, as well as their particular and precise biochemical

*Correspondence: fanghao@nwsuaf.edu.cn
[1] College of Life Sciences, Northwest A&F University, 22 Xinong Road, Yangling 712100, Shaanxi, China
Full list of author information is available at the end of the article

and physicochemical properties [4–6]. Many oleaginous microorganisms can accumulate high lipid content, some up to 80% dry cell weight or even higher [7]. Among them, *Trichosporon cutaneum* is a promising producer of SCO from lignocelluloses because of its high lipid yield and strong tolerance to inhibitors ubiquitously existing in the pretreated lignocellulosic materials [8, 9].

Pretreatment is still the most expensive single-unit operation in the lignocellulose-based bioprocesses [4, 10]. Steam explosion, which is the most commonly used pretreatment method and opined to be close to commercialization, can greatly improve the enzymatic digestibility of lignocellulosic materials [11–13]. Steam explosion produces lots of inhibitors that are resulted from the decomposition of hemicellulose and lignin [11, 14], thus making *T. cutaneum* a desirable candidate for SCO production from steam-exploded lignocellulosic materials.

The concept of on-site enzyme production has many advantages such as saving costs of separation, concentration, storage, and transportation [15, 16]. In addition, use of lignocellulosic biomass as substrate to induce cellulase production has an increased enzymatic hydrolysis specificity for the substrate itself than others [4, 17]. Mixed culture of *Trichoderma reesei* and *Aspergillus niger* is advantageous over the monoculture of *T. reesei* or *A. niger* in cellulase production and the enzymatic hydrolysis of steam-exploded corn stover (SECS) [12, 13].

Multi-stage enzymatic hydrolysis was found to be capable of improving the efficiency of enzymatic hydrolysis and the whole bioprocess, especially when high solid loading was used [4, 18, 19]. The volume of the multi-stage enzymatic hydrolysate is several times of one-stage enzymatic hydrolysate, leading to lower sugar concentration. The application value of multi-stage enzymatic hydrolysis, however, was proven in SCO production where high initial concentrations of fermentable sugars were unfavorable, unlike bioethanol production which prefers high sugar concentration [4, 18].

Single cell oil is a good starting material for biodiesel [6]. However, biodiesel is an imperfect final product nowadays because of the low oil prices, making it economically uncompetitive. Thus, transforming SCO to value-added chemicals is important to the commercialization of the lignocellulose-based bioprocess. Long-chain α,ω-dicarboxylic acids (DCAs), which are important platform chemicals and building blocks for biodegradable polymers [20], are much more valuable than biodiesel and other biofuels.

In this work, mixed culture of *T. reesei* and *A. niger* was established to produce and supply cellulase in the context of on-site enzyme production. Then SECS was enzymatically hydrolyzed by the cellulase from the mixed culture of *T. reesei* and *A. niger*. Differently sourced cellulases

were compared in the enzymatic hydrolysis of SECS. Subsequently, the enzymatic hydrolysates of SECS were fermented by *T. cutaneum* for SCO production. The bioprocesses using different cellulases were compared to select the most efficient one. Moreover, the three-stage enzymatic hydrolysis was adopted to further enhance the efficiency of the bioprocess from SECS to SCO. Furthermore, the unsaturated fatty acids hydrolyzed from SCO were upgraded to long-chain DCAs via self-metathesis. We tried to seek the most efficient process of SCO production and the comprehensive utilization and conversion of corn stover to value-added chemicals.

Methods
Steam-exploded corn stover

The lignocellulosic material corn stover was from Kaifeng City, Henan Province, China. It was air-dried and stored at room temperature before use. Before steam explosion, it was sliced to a proper size (5–10 cm). The pretreatment of steam explosion was carried out in a 3.5 L reactor (Wuhai Gerun Environmental Protection Equipment Co., Ltd., China) under the following conditions: temperature 200 °C, pressure 1.6 MPa, pressure maintained duration 7 min, and substrate loading 100 g (dry material). The SECS was collected and washed 3 times using distilled water with a ratio of solid to liquid 1:10 (g:mL). Then the washed SECS residues were kept in refrigerator at 4 °C until further use. The composition of the washed SECS was as follows (dry material): glucan 52.5%, xylan 7.2%, lignin 22.8%, ash 11.4%, and others 6.1%.

Microorganisms and media

Trichoderma reesei Rut-C30 and *A. niger* NL02 used for the mixed culture for cellulase production from SECS were obtained from the strain collection of the Department of Biochemical Engineering, Nanjing Forestry University. *T. cutaneum* ACCC20271, purchased from Agricultural Culture Collection of China, were used for SCO production from SECS enzymatic hydrolysate.

As for the seed medium for the preparation of *T. reesei* and *A. niger* inoculums and the fermentation medium for the cellulase production, please refer to our previous papers [4, 13]. These media were autoclaved at 121 °C for 20 or 30 min (20 min for the medium without SECS and 30 min for the medium with SECS).

Yeast peptone dextrose was used as seed medium for *T. cutaneum* and its composition was as follows (g/L): glucose 20, peptone 20, yeast extract 10. The fermentation medium for SCO production by *T. cutaneum* had the following composition: SECS enzymatic hydrolysate, $(NH_4)_2SO_4$ in accordance with C/N ratio, KH_2PO_4 3 g/L, $MgSO_4$ 0.5 g/L, trace element solution 1%(v/v) and vitamin solution 0.1%(v/v). The composition of the trace element solution was as follows (g/L): EDTA 15, $MnCl_2 \cdot 4H_2O$ 1.0, $CuSO_4 \cdot 5H_2O$ 0.3, $CaCl_2 \cdot 2H_2O$ 4.5, $NaMoO_4 \cdot 2H_2O$ 0.4, H_3BO_3 1.0, KI 0.1, $CoCl_2 \cdot 6H_2O$ 0.3, $ZnSO_4 \cdot 7H_2O$ 4.5 and $FeSO_4 \cdot 7H_2O$ 3.0. The components of the vitamin solution were 0.05 g/L D-biotin, 1 g/L calcium pantothenate, 1 g/L nicotinic acid, 1 g/L thiamine hydrochloride, 1 g/L pyridoxine hydrochloride, 0.2 g/L para-aminobenzonic acid, and 25 g/L (myo)inositol. The enzymatic hydrolysate was autoclaved at 121 °C for 30 min and the other solutions were sterilized by filtering through 0.22-μm membrane (Millipore, MA, USA). They were blended before use.

Mixed culture of *T. reesei* and *A. niger*

For the mixed culture for cellulase production, 10% (v inoculum/v total volume) *T. reesei* and 10 or 2% (v inoculum/v total volume) *A. niger* inoculums were inoculated. The delay time of *A. niger* inoculation was 0 h (inoculated simultaneously), 24, or 48 h. These two conditions derived 6 mixed culture forms, denoted as 0 h/1:1, 0 h/5:1, 24 h/1:1, 24 h/5:1, 48 h/1:1, and 48 h/5:1. As to the details about the mixed culture, please refer to our previous work [12]. At least three parallel samples ($n \geq 3$) were used in the analysis and data are shown in the form of means ± standard deviations.

Enzymatic hydrolysis of SECS

One-stage enzymatic hydrolysis of SECS

Both one- and three-stage enzymatic hydrolysis of SECS were conducted in 250-mL Erlenmeyer flasks with a working volume 50 mL containing 2.5 mL 1 M citrate buffer solution (pH 4.8), SECS, cellulase (added finally), and a supplementary amount of water to make up 50 mL. Once cellulase was added, flasks were incubated in an orbital shaker (140 rpm) at 50 °C for 48 or 72 h. Periodic sampling was done for analysis. At least three parallel samples ($n \geq 3$) were used in the analysis and data are shown in the form of means ± standard deviations.

Three-stage enzymatic hydrolysis of SECS

As for the three-stage enzymatic hydrolysis, the first, second, and third stages were conducted for 9, 9, and 12 h, respectively. Initial cellulase dosage was 15 FPIU/g glucan. At the end of each stage, the solid residue was separated by centrifugation (3000 rpm, 10 min). Fresh water and buffer were then added to the solid residue for the next stage enzymatic hydrolysis. To compensate the enzyme activity lost, 3 and 2 FPIU/g glucan fresh cellulase were added at the beginning of the second and the third stage, respectively. Thus the total cellulase dosage was 20 FPIU/g glucan. Periodic sampling was implemented for analysis. At least three parallel samples

($n \geq 3$) were used in the analysis and data are shown in the form of means ± standard deviations.

The yield of enzymatic hydrolysis of SECS was calculated according to the following equation:

$$\text{Yield (\%)} = (\text{glucose} + \text{xylose}) \, (\text{g}) \times 0.9$$
$$\times 100 / (\text{glucan} + \text{xylan}) \text{ in substrate (g)}. \tag{1}$$

A conversion factor of 0.9 was used to eliminate the interfering effect because of the molecular weight changes of sugars before and after hydrolysis (a molecule of sugar without and with a molecule of H_2O) so as to assure the accuracy. At least three parallel samples ($n \geq 3$) were used in the analysis and data are shown in the form of means ± standard deviations.

Fermentation of SECS enzymatic hydrolysate by *T. cutaneum*

The pre-culture was performed on the YPD medium at 28 °C and 150 rpm for 24 h. Then the pre-cultured *T. cutaneum* was inoculated into 250-mL Erlenmeyer flasks containing 50 mL the fermentation medium and incubated at 28 °C and 180 rpm with initial pH 6.0. Or the fermentation was carried out under the conditions we mentioned elsewhere. Sampling was conducted periodically for analysis to monitor the growth of *T. cutaneum* and the lipid accumulation during the fermentation process. At least three parallel samples ($n \geq 3$) were used in the analysis and data are shown in the form of means ± standard deviations.

SCO extraction from *T. cutaneum*

Trichosporon cutaneum cells were harvested by centrifugation and mixed thoroughly with 4 M HCl with a ratio of 6 mL 4 M HCl versus 1 g dry cell weight (DCW) by vortex. The mixtures were kept at room temperature for 30 min and then maintained in water bath at 100 °C for 3 min. Subsequently, they were cooled down quickly at −20 °C and added with a double volume of chloroform and methanol mixture (1:1 in volume ratio). Then, the mixtures were shaken completely at 5000 rpm for 5 min. The chloroform layer was collected, blended with an equal volume of 0.1% (w/v) NaCl solution, and vortexed for 5 min. The chloroform layer was collected and SCO was extracted by volatilizing chloroform. At least three parallel samples ($n \geq 3$) were used in the analysis and data are shown in the form of means ± standard deviations.

Self-metathesis reaction

The free fatty acids, hydrolyzed from SCO produced by *T. cutaneum* using acid-hydrolysis reaction and separated using physical method, were transferred into a three-necked round-bottomed flask. Then the flask was outgassed by purging with nitrogen gas for 0.5 h and added with the first-generation Grubbs catalyst or the second-generation Grubbs catalyst at a certain dosage so as to start the metathesis reaction. The catalysts were purchased from Sigma-Aldrich Co. LLC. The subsequent operations were performed according to the references [20, 21].

The unit of mol% was defined as 1 molar catalyst per 100 molar fatty acids. The conversion was calculated as follows:

$$
\begin{aligned}
\text{Conversion (\%)} = & \left(\text{initial amount of fatty acids} \right. \\
& \left. -\text{residual amount of fatty acids}\right)(g) \\
& \times 100/\text{initial amount of fatty acids (g)}.
\end{aligned}
$$
(2)

The reaction was repeated at least three times. At least three parallel samples ($n \geq 3$) were used in the analysis and data are shown in the form of means $\pm$ standard deviations.

Analytical methods

Determination of enzymatic activities of cellulase

Filter paper activity (FPA), beta-glucosidase activity (BGA), Avicelase activity, and CMCase activity were assayed in accordance with the standard method recommended by the International Union of Pure and Applied Chemistry (IUPAC) [22] with some modifications. The substrate used for assaying FPA was 50 mg (1 × 6 cm strip) Whatman No.1 filter paper (Kent, UK). The Unit (FPIU) of FPA was defined as the amount of enzyme needed for releasing 1 μmol of reducing sugars in 1 min. The substrate used for assaying BGA was pNPG (p-nitrophenyl-β-d-1,4-glucopyranoside) (Sigma-Aldrich, St. Louis, MO, USA) and the Unit (IU) of FPA was defined as the amount of enzyme required to release 1 μmol of p-nitrophenol in 1 min. The substrates used for measuring Avicelase activity and CMCase activity were microcrystalline cellulose PH101 and carboxymethyl cellulose, respectively, purchased from Sigma-Aldrich, Co. LLC. The Unit (U) of Avicelase activity or CMCase activity was defined as the amount of enzyme required for generating 1 mg of reducing sugars in 1 h. For more details about the determination of enzymatic activities, consult our previous papers [1, 4]. At least three parallel samples ($n \geq 3$) were used in the analysis and data are shown in the form of means $\pm$ standard deviations.

Determination of monomeric sugars

Monomeric sugars were analyzed by Agilent 1100 (Agilent Technologies, Santa Clara, CA, USA)

high-performance liquid chromatography (HPLC) (Bio-rad Aminex HPX-87P ion exclusion column). Deionized and degassed water was employed as the mobile phase at a flow rate of 0.6 mL/min. The column temperature was fixed at 55 °C. The eluate was detected by a refractive index detector. At least three parallel samples ($n \geq 3$) were used in the analysis and data are shown in the form of means $\pm$ standard deviations.

Determinations of dry cell biomass and SCO

Cell biomass was dried at 105 °C to a constant weight and measured by electronic balance. For SCO determination, cells (~30 mg biomass in 1 mL water solution), 4.5 mL of methanol (Sinopharm Chemical Reagent Co. Ltd., Shanghai, China), and 1 mL of tridecanoic acid (Sigma-Aldrich Co. LLC) as internal standard (approximately 0.5 mg/mL) were added into a tube. The tube was capped and vortexed for 30 s. Then a volume of 0.2 mL 12 M H_2SO_4 (Sinopharm Chemical Reagent Co. Ltd., Shanghai, China) was added and mixed by vortex. The tube was heated in water bath at 85 °C for 15 min for esterification. Then the tube was cooled down with tap water. Add 2 mL H_2O and mix by vortex. Add 2 mL hexane (Sinopharm Chemical Reagent Co. Ltd., Shanghai, China) and mix again for fatty acid methyl esters (FAME) extraction. The hexane layer was collected and moved into vial for analysis.

The fatty acid composition was determined using capillary gas chromatography (GC). SP-2560 (100 m × 0.25 mm × 0.20 μm) capillary column (Supelco) was installed on a Hewlett Packard 5890 gas chromatograph equipped with a Hewlett Packard 3396 Series II integrator and 7673 controller, a flame ionization detector, and split injection (Agilent Technologies Inc., Santa Clara, CA, USA). The injector was kept at 260 °C, with an injection volume of 1 μL by split injection mode (ratio of 30:1). The initial oven temperature was set at 120 °C, then heated at a increasing rate of 3 °C/min to 240 °C and held for 20 min. The detector temperature was set at 250 °C. Helium was used as the carrier gas at a flow rate of 0.5 mL/min, and the column head pressure was 280 kPa. At least three parallel samples ($n \geq 3$) were used in the analysis and data are shown in the form of means $\pm$ standard deviations.

Determination of DCAs and other chemicals

The DCAs and byproducts resulted from the self-metathesis reaction were quantified with GC and characterized with GC–MS. DCAs and fatty acids were analyzed in the form of methyl ester using the method described by Miao et al. [23]. The equipment and conditions used for GC analysis were the same as described above. The GC–MS analysis employed Agilent 6890N Gas Chromatograph coupled with 5975B Mass Selective

Detector and CDS Analytical Pyroprobe 500 Pyrolysis Injection Probe. The column was capillary column HP-5MS (30 m × 250 μm × 0.25 μm). The injector was kept at 260 °C, and 1 μL sample was loaded by split injection mode (ratio of 30:1). The initial oven temperature was set at 120 °C, then increased at a heating rate of 3 °C/min to 240 °C and held for 20 min. The detector temperature was 250 °C. Helium was used as the carrier gas at a flow rate of 0.5 mL/min. At least three parallel samples ($n \geq 3$) were used in the analysis and data are shown in the form of means ± standard deviations.

Results and discussion
Mixed culture

Different mixed culture forms of *T. reesei* and *A. niger* were tried to identify the optimal form of cellulase production with the highest FPA and the bettered composition. Figure 1a shows the monoculture of *T. reesei* or *A. niger* and mixed cultures of *T. reesei* and *A. niger* after 5 days of cellulase production using SECS as the substrate and the inducer. It was found that the monoculture of *T. reesei* led to high FPA but extremely low BGA. This is because *T. reesei* is relatively complete in the composition of cellulase mixture but deficient in β-glucosidase [13, 24, 25]. Nonetheless, the monoculture of *A. niger* resulted in high BGA but very low FPA, mainly because *A. niger* is famous for its high BGA but not as robust as *T. reesei* in secreting a complete cellulase mixture that can degrade cellulose to monomeric sugars [12, 24].

The mixed culture forms 0 h/1:1 and 0 h/5:1 had the similar pattern to the monoculture of *A. niger*, except that the FPA values were slightly increased and BGA values were obviously enhanced. The BGAs of the mixture form 0 h/1:1 and 0 h/5:1 were higher than the monoculture of *A. niger*, indicating that the mix culture form 0 h/1:1 and 0 h/5:1 facilitated the β-glucosidase production. *A. niger* dominated in the mixed culture form 0 h/1:1 and 0 h/5:1. The mixed culture forms 24 h/1:1, 24 h/5:1, and 48 h/1:1 derived lower FPAs than the monoculture of *T. reesei* and BGAs than the monoculture of *A. niger*. This indicates that the competition between *T. reesei* and *A. niger* was too fierce to develop synergism. Consequently, these three mixed culture forms, 24 h/1:1, 24 h/5:1, and 48 h/1:1, are not suitable for cellulase production because the FPA was not adequately high.

The mixed culture form 48 h/5:1 had the highest FPA and relatively high BGA (Fig. 1a), indicating that the deficiency of *T. reesei* was overcome by mixed culture and the productivity was enhanced. The time course of the mixed culture form 48 h/5:1 is shown in Fig. 1b. All enzymatic activities increased as the fermentation process was underway and peaked after 5 days of fermentation, except Avicelase activity which continued increasing

during the whole time course. The continuous increase in Avicelase activity may be because the main component of exoglucanases cellobiohydrolase I was driven by a strong promotor *Pcbh1* which manages cellobiohydrolase gene expression continuously [26]. Other enzymatic activities declined on Day 6 because *T. reesei* entered the phase of decline. Thus, the cellulase (FPA 3.42 ± 0.25 FPIU/mL, BGA 1.16 ± 0.15 IU/mL, Avicelase 15.39 ± 0.50 U/mL, and CMCase 117.51 ± 4.12 U/mL) was harvested after 5-d fermentation and applied to the subsequent experiment.

Enzymatic hydrolysis of SECS

Steam-exploded corn stover was hydrolyzed for the production of monomeric sugars using the cellulase from the mixed culture form 48 h/5:1. The commercial cellulase, Celluclast, and the cellulase from the monoculture of *T. reesei* were used as control. The results of the enzymatic hydrolysis of SECS are shown in Fig. 2a. After 48-h enzymatic hydrolysis, the yields of the cellulase from the mixed culture form 48 h/5:1, the cellulase from the monoculture of *T. reesei*, and the commercial cellulase were 86.67 ± 4.06%, 73.63 ± 3.46, and 61.73 ± 2.55 g/L, respectively. It was found that the cellulase from mixed culture 48 h/5:1 had better performance in the enzymatic hydrolysis of SECS than the cellulase from the monoculture of *T. reesei* when they were at the same dosage, 25 FPIU/g glucan. This is the same as the observation in our previous work [12, 13] and indicates that the composition of the cellulase was ameliorated by the mixed culture of *T. reesei* and *A. niger*. Moreover, both the cellulase from the mixed culture form 48 h/5:1 and the cellulase from the monoculture of *T. reesei* released higher concentration of glucose from SECS than the commercial cellulase, Celluclase, which was purchased from Sigma-Aldrich (St. Louis, MO, USA). This is because the use of lignocellulosic biomass as substrate to induce cellulase production has an increased enzymatic hydrolysis specificity for the substrate itself than others [15, 17, 27]. Therefore, the on-site cellulase production by mixed culture of *T. reesei* and *A. niger* is applicable and promising in the lignocellulose-based bioprocesses.

An enhancement in cellulase productivity and activities by the mixed culture of *T. reesei* and *A. niger* was reported by Ahamed and Vermette [28], but they did not test its performance in enzymatic hydrolysis. Some report found that the mixed culture of *T. reesei* and *A. phoenicis* had a lower FPA than the monoculture of *T. reesei*, but its cellulase had better performance than the cellulase from the monoculture of *T. reesei* and commercial cellulase in the enzymatic hydrolysis of lignocellulose [29]. Here, we obtained a mixed culture system with higher productivity of cellulase and better enzymatic hydrolysis performance.

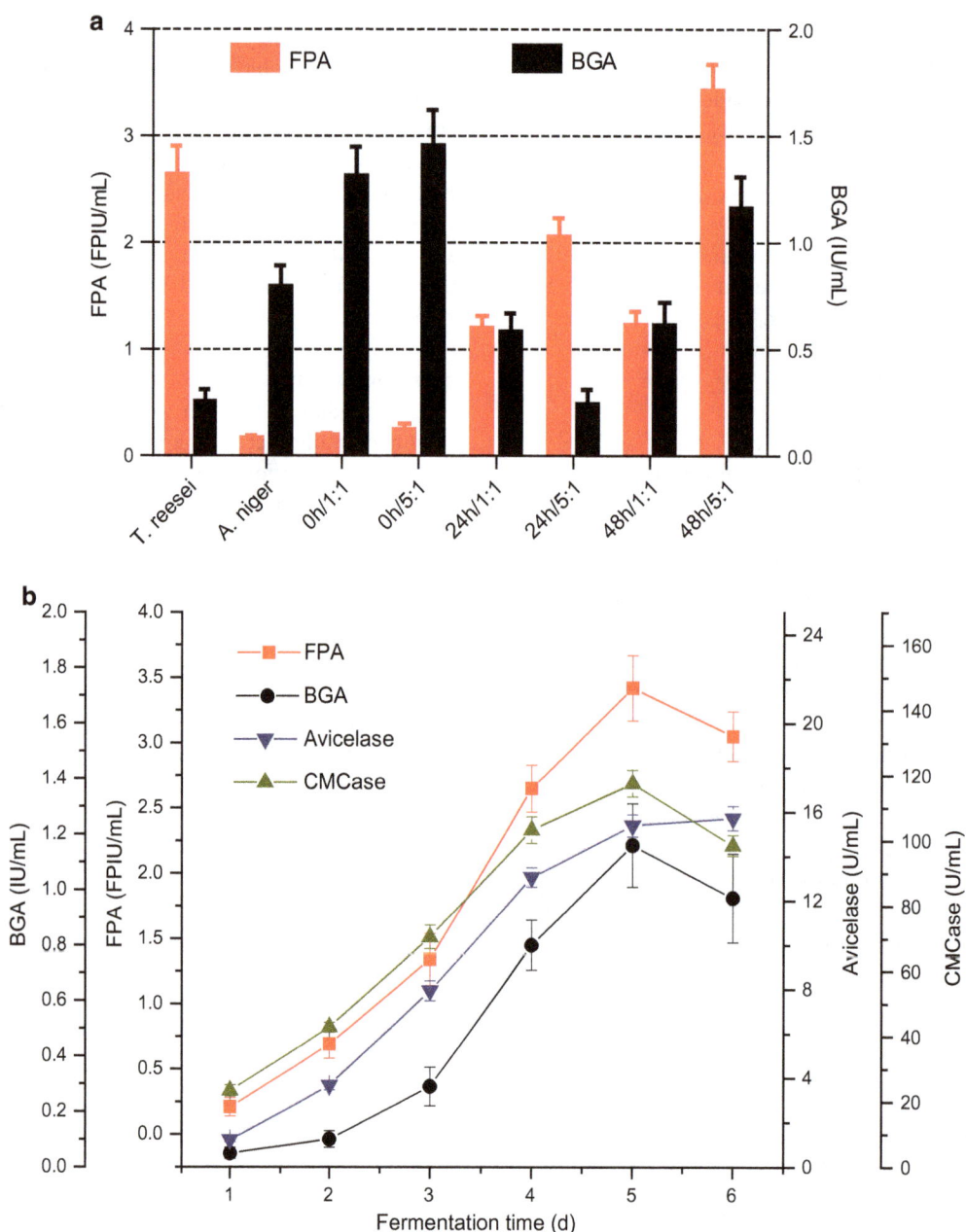

Fig. 1 a FPAs and BGAs in the fermented broth obtained after 5 days fermentation from the monoculture of *T. reesei* or *A. niger* and the mixed cultures of *T. reesei* and *A. niger* grown in the medium containing steam-exploded corn stover (SECS). **b** Time course of cellulase production by the mixed culture (48 h/5:1) of *T. reesei* and *A. niger* induced by SECS. Data shown are means of at least three parallel samples ($n \geq 3$) and *error bars* are standard deviations (mean ± SD)

In some lignocellulose-based bioprocesses such as bioethanol production, high solid loading is preferred because it could increase the product concentration and decrease the operating costs [18, 30]. The solid loading of 300 g SECS (dry material) was used as the substrate to produce higher concentration of glucose. The cellulase from the mixed culture form 48 h/5:1, the cellulase from the monoculture of *T. reesei* and the commercial cellulase were compared, and the results are shown in Fig. 2b. The cellulase from the mixed culture 48 h/5:1 still outperformed the other two cellulases when the solid loading was increased from 100 to 300 g/L (dry material). This also showed the superiority of the cellulase from the mixed culture over the cellulase from the

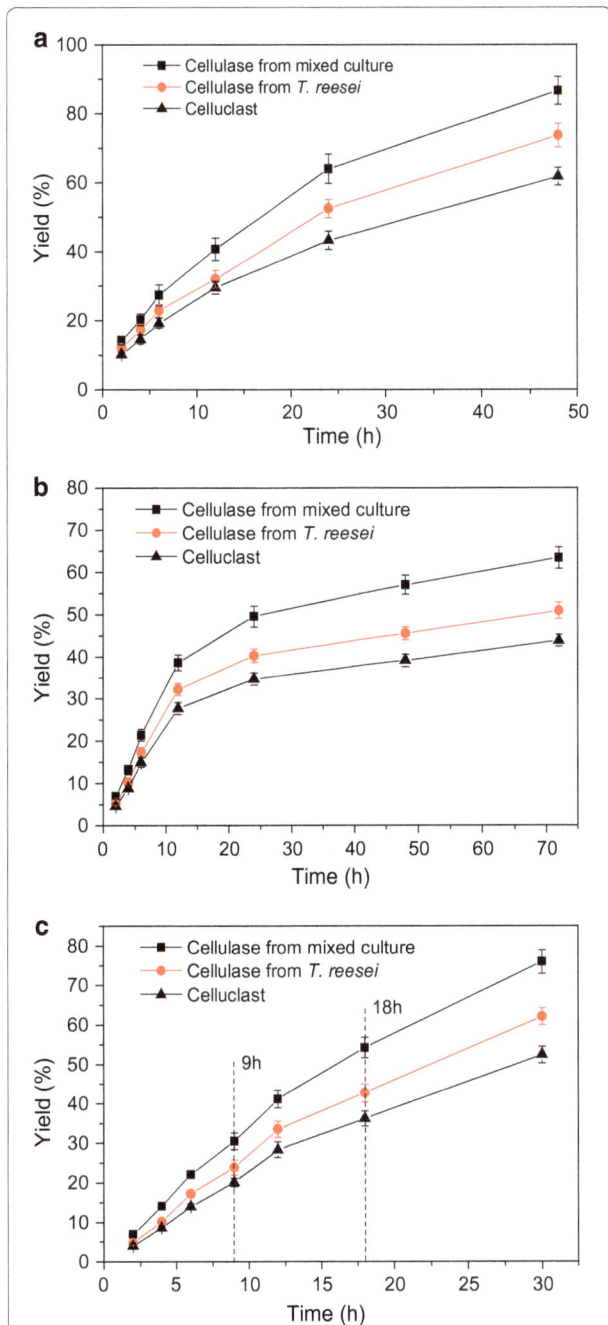

Fig. 2 a One-stage enzymatic hydrolysis of SECS. The dosage of cellulase and the concentration of SECS (dry material) were 25 FPIU/g glucan and 100 g/L, respectively. **b** One-stage enzymatic hydrolysis of SECS. The dosage of cellulase and the concentration of SECS (dry material) were 30 FPIU/g glucan and 300 g/L, respectively.
c Three-stage (9 + 9 + 12 h) enzymatic hydrolysis of SECS. The total dosage of cellulase and the concentration of SECS (dry material) were 20 FPIU/g glucan and 300 g/L, respectively. The initial cellulase loading was 15, and 3 and 2 FPIU/g of glucan fresh cellulase were added for the second and the third stage, respectively. All enzymatic hydrolysis experiments were conducted in 250-mL Erlenmeyer flasks with a volume of 50 mL. Data shown are means of at least three parallel samples ($n \geq 3$) and *error bars* are standard deviations (mean $\pm$ SD)

monoculture of *T. reesei* and Celluclast. The yields after 48- and 72-h enzymatic hydrolysis by the cellulase from the mixed culture form 48 h/5:1 were only 57.03 $\pm$ 2.23 and 63.35 $\pm$ 2.53%, respectively, much lower than the previous enzymatic hydrolysis even when the cellulase dosage was 30 FPIU/g glucan which was higher than the cellulase dosage of 25 FPIU/g glucan used in the enzymatic hydrolysis of 100 g/L SECS.

Multi-stage enzymatic hydrolysis such as three-stage enzymatic hydrolysis is able to improve yield, shorten enzymatic hydrolysis time, and lessen cellulase dosage [18, 19]. Hence, it was used in this work for the enzymatic saccharification of 300 g SECS and the time course is shown in Fig. 2c. It was found that the efficiency of enzymatic hydrolysis was improved substantially. The time of the three-stage enzymatic hydrolysis was shortened from 72 to 30 h but the yield is higher than the one-stage enzymatic hydrolysis of SECS (Fig. 2b). The yield was able to reach 74.24 $\pm$ 2.69% by three-stage enzymatic hydrolysis just using 30 h, and a cellulase dosage of 20 FPIU/g. Three-stage enzymatic hydrolysis can to a great degree improve the efficiency of enzymatic hydrolysis, saving cellulase, and accelerating the enzymatic hydrolysis [18, 19].

In addition, the cellulase produced by the mixed culture of *T. reesei* and *A. niger* was superior to the cellulase produced by the monoculture of *T. reesei* and the commercial cellulase Celluclast in all the enzymatic hydrolysis processes in this work, suggesting that the composition of the cellulase was ameliorated, the synergism was enhanced and the degradation ability of the cellulase was strengthened. This work demonstrates that the mixed culture of *T. reesei* and *A. niger* we established is a good approach to realize on-site cellulase production and cellulase autarky. The enzymatic hydrolysates resulted from 100 and 300 g SECS were used as feedstock for the SCO production by *T. cutaneum*.

Effects of culture conditions on SCO production by *T. cutaneum*

Different C/N ratios (carbon–nitrogen ratios) were compared to seek the most suitable one for SCO production by *T. cutaneum* in the enzymatic hydrolysate of SECS. The results of the fermentation for 8 days in the enzymatic hydrolysate containing 50.84 $\pm$ 2.37 g/L glucose and 6.65 $\pm$ 0.32 g/L xylose with the C/N ratios ranging from 20:1 to "∞" (without addition of nitrogen source) are presented in Fig. 3a. The molar C/N ratio 80:1 was found to be the best C/N ratio equal to the concentration of ammonium sulfate 1.58 g/L. If the C/N ratio was higher than 80:1, less cell biomass, lipid, and lipid content were produced. As the concentration of ammonium sulfate increased over 1.58 g/L, i.e., the C/N ratio

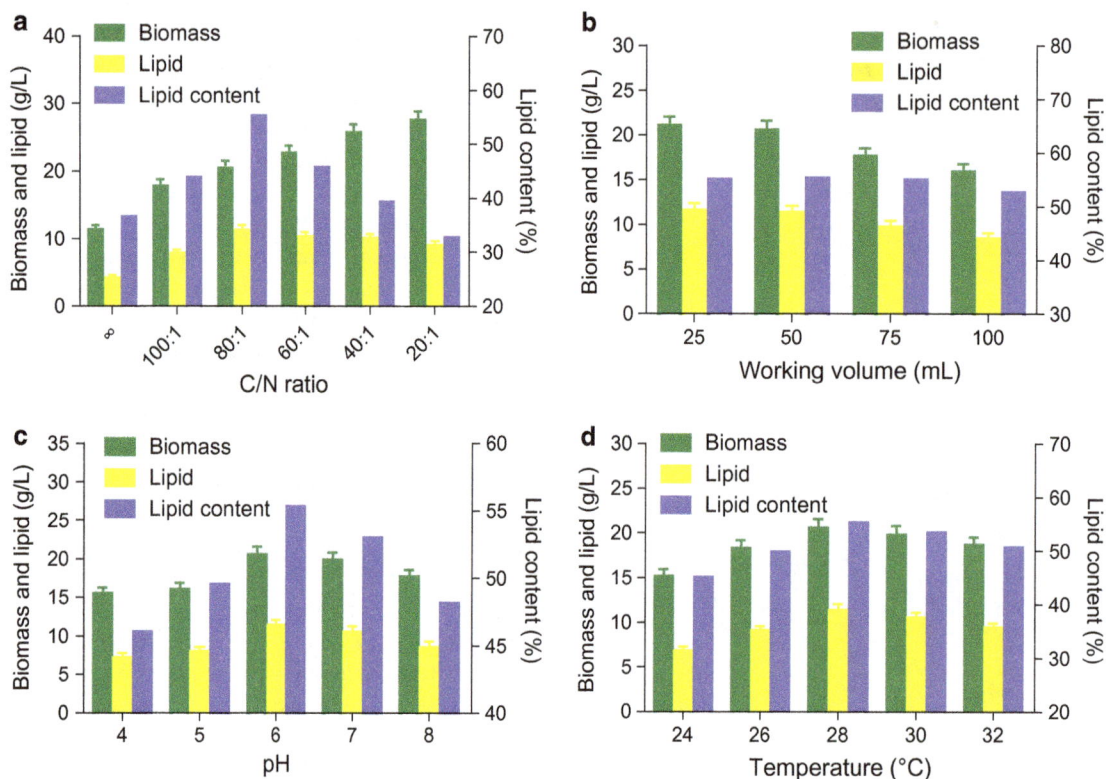

Fig. 3 a Effect of C/N molar ratio on SCO production by *T. cutaneum*. "∞" means no nitrogen source, i.e., ammonium sulfate, was added. **b** Influence of working volume (mL) on SCO production by *T. cutaneum*. **c** Effect of pH on SCO production by *T. cutaneum* in the enzymatic hydrolysate of SECS. **d** Effect of temperature on SCO production by *T. cutaneum*. The enzymatic hydrolysate, used for SCO production, contained 50.84 ± 2.37 g/L glucose and 6.65 ± 0.32 g/L xylose. All the results were obtained after 8 days fermentation. Data shown are means of at least three parallel samples (*n* ≥ 3) and *error bars* are standard deviations (mean ± SD)

decreased, the lipid content declined obviously because the lipid concentration decreased slightly but the cell biomass increased substantially. The results suggest that low C/N ratio is beneficial to cell growth but not to lipid accumulation, and that nitrogen source limitation rather than nitrogen source starvation facilitates SCO production. The best C/N ratio here is somewhat different from the work by Gao et al. [31] in which the C/N ratio about 50:1 was the best. This may be because we used different nitrogen sources and the type of nitrogen source affected SCO production [32]. In fact, the C/N ratio "∞" is not absolute because small quantity of organic nitrogen source exists in SECS enzymatic hydrolysate [31, 32], which was neglected in the calculation of the C/N ratio. That is the reason why *T. cutaneum* grew not so badly without addition of any nitrogen source. Therefore, the most suitable C/N ratio 80:1 was used into the subsequent experiments, where small quantity of organic nitrogen source in SECS was not taken into consideration and calculation.

Single cell oil production from monomeric sugars such as glucose and xylose is a process in great need of oxygen.

Thus, the working volume affects the SCO production in shaking flasks and the results are shown in Fig. 3b. The more the working volume is, the lower the cell biomass, lipid, and lipid content produced. This makes sense because the oxygen transfer and supply cannot meet the increasing demand with the working volume increasing. However, the working volumes 25 and 50 mL led to almost the same results in terms of cell biomass, lipid, and lipid content. Hence, 50 mL was used as the working volume in the subsequent SCO production in shaking flasks.

Figure 3c shows the influence of initial pH on SCO production by *T. cutaneum* and it was found that the most proper pH was 6.0. The initial pH values higher or lower than that had negative effect on SCO production. The optimal pH value here is the same as the pH in the fermentation for SCO production by *T. cutaneum* reported by Qi et al. [9] but different from the work reported by Gao et al. [31] and Liu et al. [33] in which the pH value 5.0 was used. Nevertheless, the difference is not so large. Additionally, from Fig. 3c, we can know that the effect of pH on SCO production *T. cutaneum* is not very obvious.

This indicates that *T. cutaneum* can tolerate wide range of pH. The pH resistance of *T. cutaneum* is probably innate owing to its exceptional habitats that range from industrial effluent to refinery waste. The initial pH value 6.0 was used in the subsequent experiments.

The effect of temperature on SCO production by *T. cutaneum* was studied and the results are shown in Fig. 3d. The optimal temperatures for *T. cutaneum* was 28 °C, which is the same as the report [9] but slightly lower than the reports [31, 33] in which 30 °C was used in the cultivation and fermentation. Actually, the difference between 28 and 30 °C was not that easy-to-see, though 28 °C was better which gave rise to the highest cell biomass, lipid, and lipid content. Accordingly, 28 °C was used in the following experiments.

Fermentation of SECS enzymatic hydrolysates

The enzymatic hydrolysates resulted from the different enzymatic hydrolysis processes (Fig. 2) were fermented by *T. cutaneum* to produce SCO. The results of SCO production are presented in Fig. 4. Figure 4a shows the time course of SCO production from the enzymatic hydrolysate resulted from the one-stage enzymatic hydrolysis of 100 g/L SECS, which contained 50.84 ± 2.37 g/L glucose and 6.65 ± 0.32 g/L xylose. *T. cutaneum* consumed almost all of glucose and xylose after 8 days of fermentation and produced 20.52 ± 1.09 g/L cell biomass and 11.35 ± 0.77 g/L lipid. The lipid content was 55.31%. The result is better than the work of Qi et al. [9], indicating that SECS enzymatic hydrolysate is the suitable substrate of *T. cutaneum* for SCO production which was also proven to be the suitable substrate of *Mortierella isabellina* for SCO production in our previous research [4]. Same as *M. isabellina*, *T. cutaneum* consumed glucose first and then xylose. This phenomenon is the same as the report by Gao et al. [31] but different from the report by Qi et al. [9] in which *T. cutaneum* seems to ferment glucose and xylose simultaneously.

When the solid loading of SECS increased from 100 to 300 g/L, the concentrations of glucose and xylose reached 108.65 ± 4.18 and 17.42 ± 0.85 g/L, respectively, after 72-h enzymatic hydrolysis, although the enzymatic hydrolysis yield was low. The high concentrations of fermentable sugars are beneficial to industrial applications such as bioethanol production because this could increase the concentrations of products and reduce the production cost [34]. Figure 4b shows the time course of SCO production by *T. cutaneum* in the SECS enzymatic hydrolysate containing 108.65 ± 4.18 g/L glucose and 17.42 ± 0.85 g/L xylose. It took more than 16 days for *T. cutaneum* to consume up all glucose and xylose. *T. cutaneum* produced 45.58 ± 2.95 g/L dry cell biomass

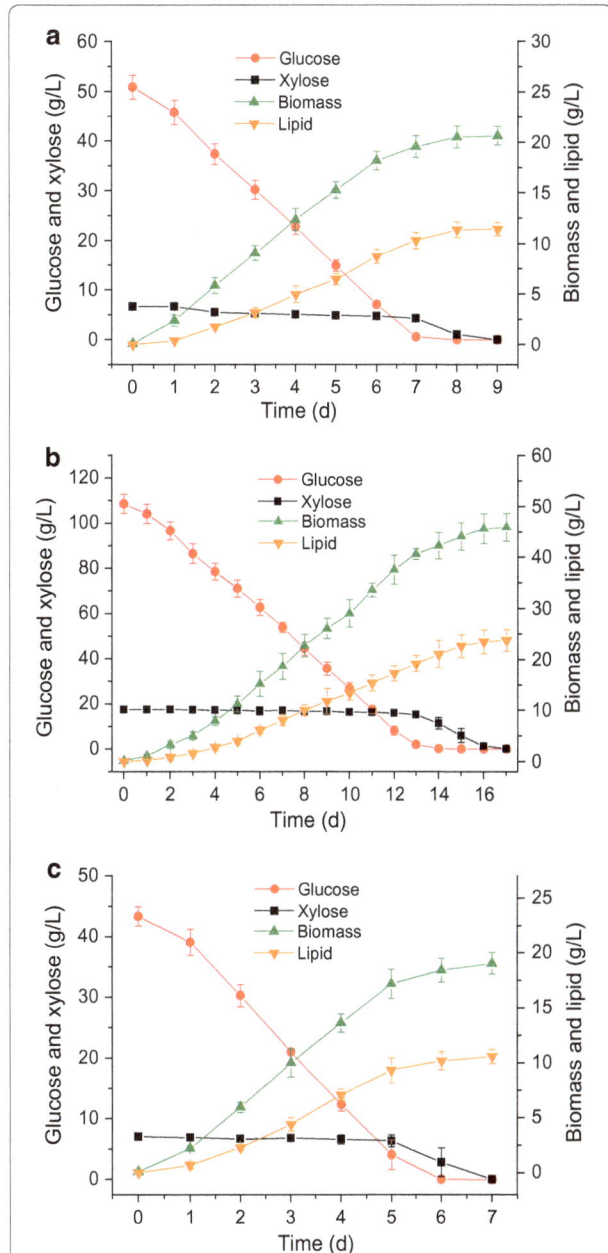

Fig. 4 Time courses of SCO production by *T. cutaneum* in the enzymatic hydrolysates containing 50.84 ± 2.37 g/L glucose and 6.65 ± 0.32 g/L xylose (**a**), 108.65 ± 4.18 g/L glucose and 17.42 ± 0.85 g/L xylose (**b**), and 43.31 ± 1.57 g/L glucose and 7.06 ± 0.39 g/L xylose (**c**), respectively. Data shown are means of at least three parallel samples ($n \geq 3$) and *error bars* are standard deviations (mean ± SD)

and 23.49 ± 2.33 g/L lipid after 16 days of fermentation. The fermentation rate and lipid productivity are as good as the fermentation of the SECS enzymatic hydrolysate containing 50.84 ± 2.37 g/L glucose and 6.65 ± 0.32 g/L xylose. However, too long fermentation period is not viable in industry because it has bigger risks of contaminations and needs much stricter fermentation process control, which means higher production cost.

Although three-stage enzymatic hydrolysis can improve the enzymatic hydrolysis efficiency, it has apparent shortcomings, one of which is lower concentrations of fermentable sugars. Surprisingly, this instead is an advantage in the SCO production by *M. isabellina* where high initial concentrations of sugars are unfavorable [4]. Here, we applied the three-stage enzymatic hydrolysis in the SCO production by *T. cutaneum* to investigate its influence on the whole bioprocess from SECS to SCO. The enzymatic hydrolysate resulted from the three-stage enzymatic hydrolysis of SECS, which contained 43.31 ± 1.57 g/L glucose and 7.06 ± 0.39 g/L xylose, were fermented by *T. cutaneum* for SCO production and the time course is shown in Fig. 4c. *T. cutaneum* exhausted all glucose and xylose within 7 days. After 7 days fermentation, 19.05 ± 0.98 g/L dry cell biomass and 10.60 ± 0.65 g/L lipid were produced. It had the shortest fermentation period and the highest lipid content, 55.64%. This indicates that *T. cutaneum*, like many other oleaginous microbes, prefers low substrate concentration when carrying out fermentation for SCO production. In addition, it was found in Table 1 (the data of biomass productivity) that the growth rate of *T. cutaneum* in lower sugar concentration is similar to that in higher sugar concentrations. Therefore, lower sugar concentration is better than higher sugar concentration for SCO production.

Trichosporon cutaneum spent about 8, 16, and 7 days in fermenting the enzymatic hydrolysate resulted from the one-stage enzymatic hydrolysis of 100 g SECS, the enzymatic hydrolysate resulted from the one-stage enzymatic hydrolysis of 300 g SECS, and the enzymatic hydrolysate resulted from the three-stage enzymatic hydrolysis of 300 g SECS, respectively. From the standpoint of application, the enzymatic hydrolysate resulted from the three-stage enzymatic hydrolysis of SECS containing lower concentrations of fermentable sugars was the most suitable substrate for *T. cutaneum* in the context of SCO production. Multi-stage enzymatic hydrolysis such as the three-stage one is able to enhance the efficiency of enzymatic saccharification, shortening the enzymatic hydrolysis period, lessening the enzyme dosage, and improving the enzymatic hydrolysis yield [4, 18, 19]. However, it has a very obvious disadvantage, it produces several-fold more hydrolysate and lowers the product concentration [18, 19], rendering it not a good option in many bioprocesses. Here, the disadvantage becomes the advantage. Therefore, the three-stage enzymatic hydrolysis was found to be applicable in the SCO production by *T. cutaneum*.

Table 1 Results of single cell oil (SCO) production by *T. cutaneum* from steam-exploded corn stover (SECS)

Bioprocess	*A*	*B*	*C*
SECS (g dry material)	100 + 11.51	300 + 41.45	300 + 27.63
Glucose (g/L)	50.84 ± 2.37	108.65 ± 4.18	43.31 ± 1.57
Xylose (g/L)	6.65 ± 0.32	17.42 ± 0.85	7.06 ± 0.39
Volume of enzymatic hydrolyzate (L)	1	1	3
Fermentation time (d)	8	16	7
Biomass (g/L dry cell biomass)	20.52 ± 1.09	45.58 ± 2.95	19.05 ± 0.98
Biomass yield (g/g glucose + xylose)	0.357	0.362	0.378
Biomass productivity (g/L/d)	2.565	2.849	2.721
Total yield of biomass (g/g SECS)	0.184	0.133	0.174
Lipid (g/L)	11.35 ± 0.77	23.49 ± 2.33	10.60 ± 0.65
Lipid yield (g/g glucose + xylose)	0.197	0.186	0.210
Lipid productivity (g/L/d)	1.419	1.468	1.514
Lipid content (%)	55.31	51.54	55.64
Total yield of lipid (g/g SECS)	0.102	0.069	0.097
Total time from SECS to SCO (h)	240	456	198
Total productivity of biomass (g/h)	0.086	0.100	0.289
Total productivity of lipid (g/h)	0.047	0.052	0.161
Enzyme input (FPIU/g lipid)	115.64	201.15	99.06
Handling capacity (g SECS/h)	0.465	0.749	1.655
Utilization ratio of SECS (%)	86.67 ± 4.06%	63.35 ± 2.53%	75.92 ± 2.96%

Comparison of different bioprocesses from SECS to SCO

Different bioprocesses from SECS to SCO were compared comprehensively and they were outlined in Fig. 5. The detailed information about these bioprocesses is listed in Table 1. The SECS used for the whole process is the sum of the SECS used as substrate for the enzymatic hydrolysis to produce fermentable sugars and the SECS used for on-site cellulase production. The total amounts of the SECS for the bioprocess A, B, and C were 111.51, 341.45, and 327.63 g (dry material), respectively. The bioprocess A produced 1 L SECS enzymatic hydrolysate containing 50.84 ± 2.37 g/L glucose and 6.65 ± 0.32 g/L xylose from 100 g SECS by one-stage enzymatic hydrolysis. The bioprocess B produced 1 L SECS enzymatic hydrolysate containing 108.65 ± 4.18 g/L glucose and 17.42 ± 0.85 g/L xylose from 300 g SECS by one-stage

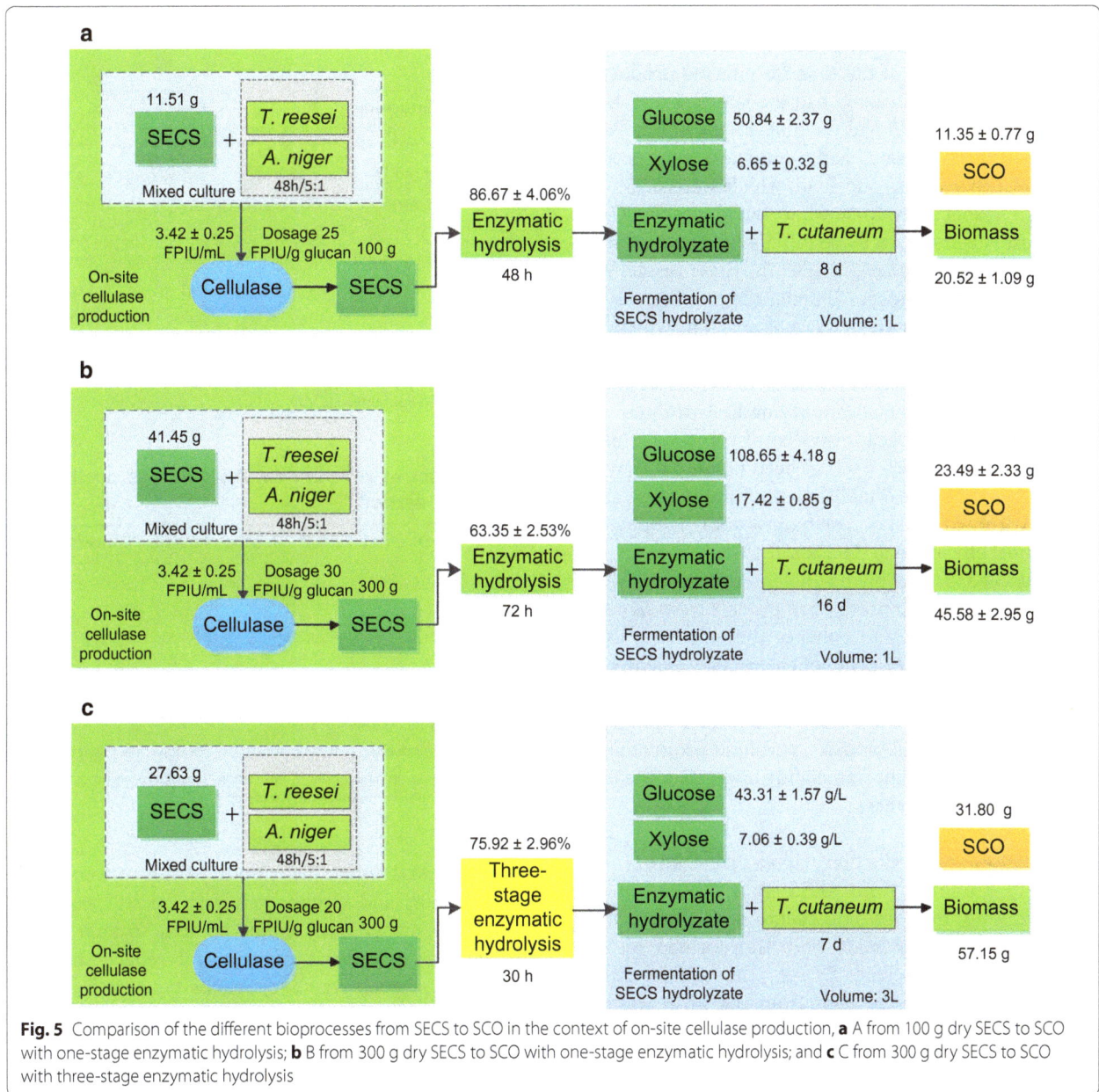

Fig. 5 Comparison of the different bioprocesses from SECS to SCO in the context of on-site cellulase production, **a** A from 100 g dry SECS to SCO with one-stage enzymatic hydrolysis; **b** B from 300 g dry SECS to SCO with one-stage enzymatic hydrolysis; and **c** C from 300 g dry SECS to SCO with three-stage enzymatic hydrolysis

enzymatic hydrolysis. The bioprocess C produced 3 L SECS enzymatic hydrolysate containing 43.31 ± 1.57 g/L glucose and 7.06 ± 0.39 g/L xylose from 300 g dry SECS by three-stage enzymatic hydrolysis.

These enzymatic hydrolysates of SECS were then fermented by *T. cutaneum* to produce SCO. The fermentation time for the bioprocess A, B, and C were 8, 16, and 7 days, respectively. The bioprocess C had the shortest fermentation period because it had the lowest starting sugar concentration. For the cell biomass and lipid, the bioprocess *C* is similar to the bioprocess *A*. The former one's cell biomass and lipid were

19.05 ± 0.98 and 10.60 ± 0.65 g/L, respectively, and the latter one's cell biomass and lipid were 20.52 ± 1.09 and 11.35 ± 0.77 g/L, respectively. In addition, the yields and productivities of the bioprocess *C* and *A* were close to each other (Table 1). They also had no big difference from those of the bioprocess *B*. This suggests that the different starting sugar concentrations just changed the fermentation time but made no big difference in other fermentation parameters.

Taking enzymatic hydrolysis and fermentation together into consideration, however, the bioprocess *B* had the lowest efficiency but the bioprocess *C* had the highest efficiency. The total time from SECS to SCO included the time for enzymatic hydrolysis and the time for SCO production, irrespective of the time for enzyme production or others. The bioprocess *C* had the shortest one 198 h while the bioprocess *B* had the longest one 456 h. Additionally, the bioprocess *C* had the highest total productivities of cell biomass and lipid, which were 0.289 and 0.161 g/h, respectively. Meanwhile, the bioprocess C had the lowest enzyme input 99.06 FPIU/g lipid and the highest handling capacity 1.655 g SECS/h. These results were achieved by the bioprocess C under the premise that the enzymatic hydrolysis yield was not so high, which was only 75.92 ± 2.96%. The results indicate that the efficiency of the bioprocess from SECS to SCO was greatly improved by the three-stage enzymatic hydrolysis. This work proved the application value of three-stage enzymatic hydrolysis, which had not been pointed out by the previous work on the establishment of multi-stage enzymatic hydrolysis [18, 19]. The bioprocess *C* is opined to be the most efficient one (Fig. 5c).

We succeeded in applying the three-stage enzymatic hydrolysis to the SCO production process from lignocellulose by *T. cutaneum* to enhance the efficiency, as we did before in the SCO production by *M. isabellina* [4]. As a whole, *T. cutaneum* outperformed *M. isabellina* in the SCO production from SECS. The former had higher fermentation rate, cell biomass, and lipid productivities, though the lipid content was slightly lower than the latter under same circumstances.

Self-metathesis for production of long-chain DCAs

The SCO produced from SECS by *T. cutaneum* was hydrolyzed to produce fatty acids and among them the unsaturated fatty acids were transformed by self-metathesis to long-chain DCAs. The self-metathesis of the unsaturated fatty acids from the SCO was carried out under a nitrogen atmosphere by using first- or second-generation Grubbs catalyst. The proportions of unsaturated fatty acids of the SCOs from the different bioprocesses are presented in Table 2. There is no difference in the composition of the fatty acids among the SCOs produced via the different bioprocesses established and compared previously. The unsaturated fatty acids from the SCOs produced via the bioprocess C were used as the substrate of self-metathesis for the production of DCAs.

The effect of catalyst on self-metathesis was investigated and the result is shown in Table 3. It was found that the second-generation Grubbs catalyst was more efficient than the first-generation Grubbs catalyst in catalyzing the self-metathesis. The catalyst dosage of 0.1 mol% was the most appropriate because the resulted conversion 81.15 ± 2.71% was acceptable. Ten times

Table 2 Fatty acid compositions (%) of single cell oil produced by *Trichosporon cutaneum* via the bioprocess *A*, *B*, and *C*

Fatty acid	Structure	A	B	C
Palmitic	C16:0	27.68 ± 1.53	27.55 ± 1.38	25.98 ± 1.05
Palmitoleic	C16:1Δ9	1.38 ± 0.09	1.27 ± 0.15	1.57 ± 0.12
Stearic	C18:0	10.07 ± 0.41	10.66 ± 0.54	10.18 ± 0.49
Oleic	C18:1Δ9	50.25 ± 2.14	49.22 ± 1.05	51.25 ± 1.82
Linoleic	C18:2Δ9,12	8.32 ± 0.38	8.09 ± 0.87	8.82 ± 0.43
γ-Linolenic	C18:3Δ6,9,12	0.63 ± 0.08	0.57 ± 0.06	0.69 ± 0.05
Others		1.67	2.64	1.51

The unit of the compositions of fatty acids is %

Data shown are means of at least three parallel samples ($n \geq 3$) and error bars are standard deviations (mean ± SD)

Table 3 Effect of different catalysts dosage on the self-metathesis reaction

Catalyst	Dosage (mol %)[a]	Reaction time (h)	Conversion[b] (%)
Grubbs 2nd	0.01	24	28.77 ± 0.76
Grubbs 2nd	0.1	1.5	81.15 ± 2.71
Grubbs 2nd	1	0.5	90.08 ± 3.25
Grubbs 1st	0.1	6	14.01 ± 0.49

Data shown are means of at least three parallel samples ($n \geq 3$) and error bars are standard deviations (mean ± SD)

[a] mol% is the unit defined as the molar quantity of catalyst per 100 g substrate

[b] Conversion (%) = (initial amount of fatty acids − residual amount of fatty acids) (g) × 100/initial amount of fatty acids (g)

more catalyst 1 mol% just improved the conversion by less than 10% and Grubbs catalysts are still expensive in market. Further work should be done to improve the efficiency of metathesis catalyst and reduce the cost. Some new Schrock-type and Grubbs-type catalysts have been invented [35, 36] but they are very expensive. Thus designing more efficient and cheaper catalyst is greatly needed because this decides the future of metathesis in DCAs production and other industrial applications.

The mechanism of the self-metathesis is illustrated in Fig. 6, which was deduced according to the theory of olefin metathesis. The double bonds "=" in the unsaturated fatty acids from the microbial lipids participated in the self-metathesis reaction, leading to new molecules with two carboxyl groups at both ends, i.e., DCAs [35, 36]. Moreover, the reaction produced alkene molecules which could be good feedstock for biofuels after some modifications to remove double bonds "=" [37, 38]. It is roughly estimated that half of the total amount of the unsaturated acids converted were transformed into long-chain DCAs and the other half were converted to byproducts. The products DCAs and the byproducts alkenes were identified by gas chromatography. So the mechanism

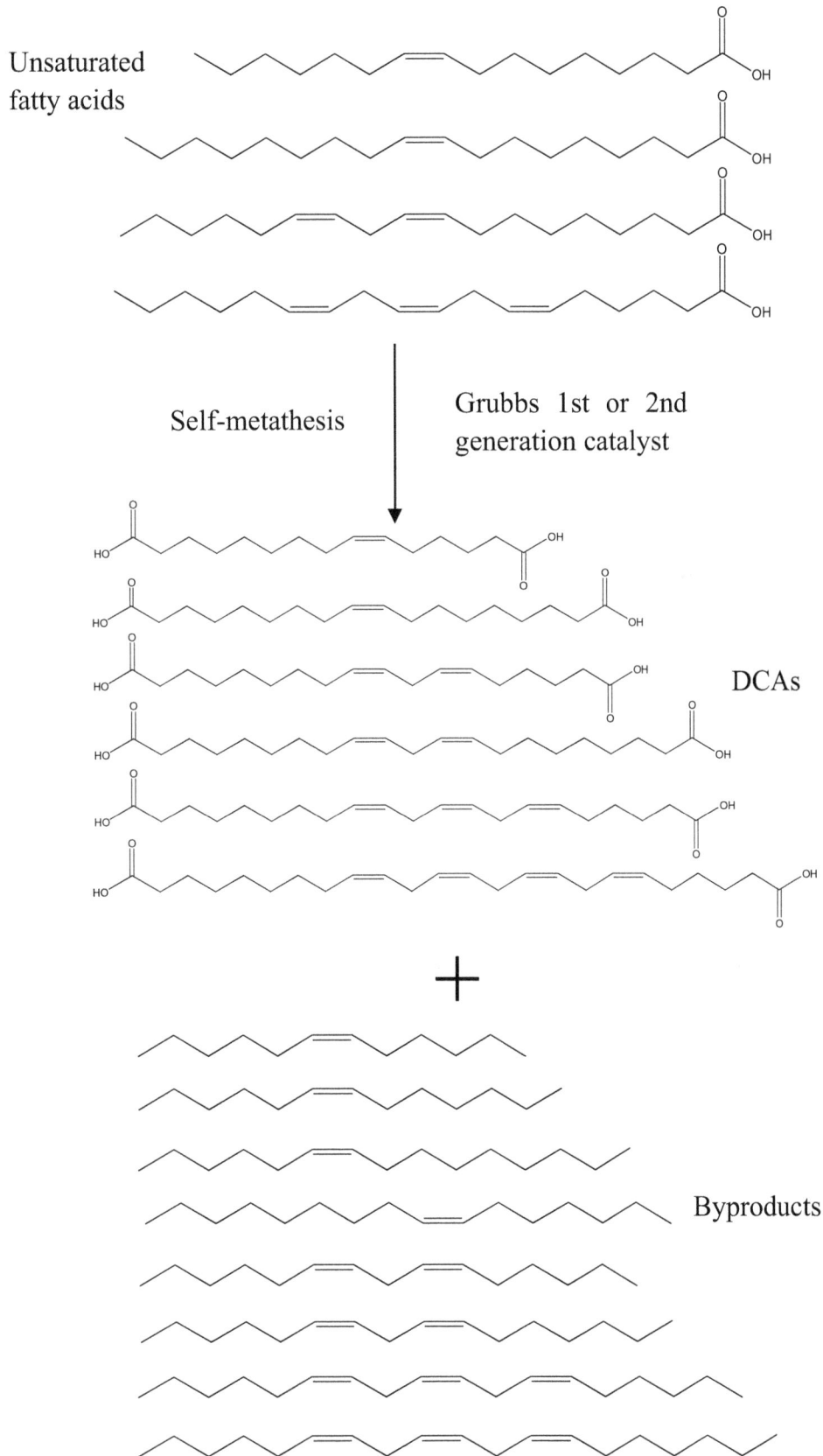

Fig. 6 The detail of the self-metathesis reaction

deduced was proven by this work. Ngo et al. employed self-metathesis to produce DCAs from plant oils [20, 21]. However, plant oils are not so renewable as SCO. Therefore, this work made substantial advancement in producing DCAs, establishing a bioprocess from corn stover to DCAs and improving the efficiency.

The overall process from corn stover to DCAs is outlined and illustrated in Fig. 7, where the mass balance of the whole process was estimated. As shown in Fig. 6, 409.54 g corn stover (dry material) formed 327.63 g SECS (dry material) by the pretreatment of steam explosion in which the recovery was assumed to be 80%. The recovery assumption was based on our previous work on SECS, which always fluctuated between 70 and 90% (data not shown). Of 327.63 g SECS, 27.63 g SECS was used as the inducer of cellulase production by the mixed culture of *T. reesei* and *A. niger* and 300 g SECS was used as the substrate of enzymatic hydrolysis for production of fermentable sugars. The three-stage

enzymatic hydrolysis produced 129.92 g glucose and 21.18 g xylose from 300 g SECS, the main fermentable sugars in the 3 L SECS enzymatic hydrolysate. Then the fermentable sugars were fermented and converted by *T. cutaneum* to yield 57.15 g dry cell biomass containing 31.80 g lipids. Subsequently, the lipids were extracted from *T. cutaneum* cells and hydrolyzed to form 23.80 g fatty acids, of which 14.83 g unsaturated fatty acids participated the self-metathesis reaction. After self-metathesis, 6.02 g long-chain DCAs and 6.02 g alkenes (Figs. 6, 7) were produced. In summary, 409.54 g corn stover could produce 6.02 g long-chain DCAs, 6.02 g alkenes, 8.97 g saturated fatty acids, 8.00 g glycerol, and 74.70 g lignin.

Figure 7 presents the outline of the process from corn stover to long-chain DCAs and the envisioned processes to make good use of the products and byproducts to produce value-added chemicals and fuels, strengthening the commercialization potential of the lignocellulose-based

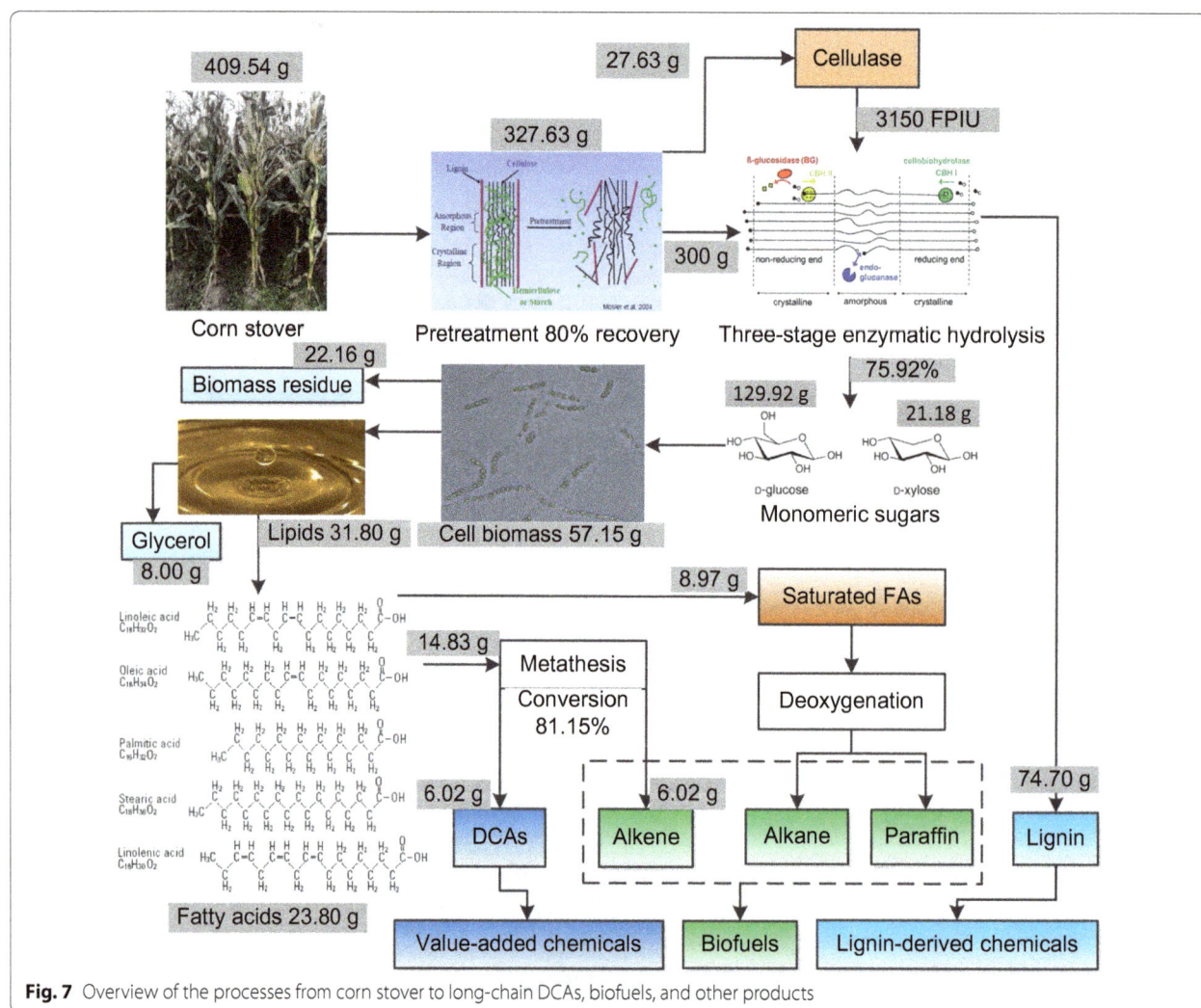

Fig. 7 Overview of the processes from corn stover to long-chain DCAs, biofuels, and other products

process. Long-chain DCAs are highly valuable platform chemicals as we aforementioned, which could be used as building blocks for a wide range of chemicals in industry [39, 40]. The saturated fatty acids, together with the byproducts of self-metathesis, could be transformed to biofuels [37, 38]. In addition, the side product lignin could also be transformed to valuable lignin-derived chemicals such as guaiacol and catechol [41]. This work provides a comprehensive way to utilize lignocellulose.

Here, we first report the whole process from corn stover to DCAs via a combination of the biological and chemical processes. Although some researchers reported the bioprocess from monomeric sugars to DCAs using engineered yeast [39], its controllability, efficiency, and versatility are not comparable with metathesis. Furthermore, no further advancement in that research has been reported. In short term, therefore, chemical routes are still dominant and irreplaceable because of their own merits. In long term, however, we look forward to combining the whole process from lignocellulose to DCAs or other chemicals into one step using lignocellulolytic microorganisms such as *T. reesei* and *Neurospora crassa* in the light of consolidated bioprocessing with the techniques of metabolic engineering and genome editing developing.

Conclusions

The mixed culture of *T. reesei* and *A. niger* was a good approach to enhance cellulase production and improve enzymatic hydrolysis of pretreated corn stover. In addition, the cellulase produced by the mixed culture outperformed other cellulases. The application value of the three-stage enzymatic hydrolysis was confirmed in the SCO production from SECS by *T. cutaneum*, which could improve the efficiency of the bioprocess. Then the unsaturated fatty acids from SCO were upgraded to long-chain DCAs by self-metathesis. A whole process from corn stover to long-chain DCAs via a combination of biological and chemical approaches was successfully established to comprehensively utilized lignocellulose to produce value-added chemicals. This work is an enlightening example of the comprehensive utilization of agricultural wastes.

Abbreviations

BGA: beta-glucosidase activity; CMC: carboxymethyl cellulose; C/N: ratio of carbon to nitrogen; DCA: α,ω-dicarboxylic acids; DCW: dry cell weight; FAME: fatty acid methyl esters; FPA: filter paper activity; GC: gas chromatography; GC–MS: gas chromatography–mass spectroscopy; HPLC: high-performance liquid chromatography; IU: international unit; IUPAC: International Union of Pure and Applied Chemistry; SCO: single cell oil; SECS: steam-exploded corn stover; YPD: yeast peptone dextrose.

Authors' contributions
CZ and HF conceived the ideas, designed the project, and performed the experiments. HF analyzed the data. HF and SC wrote the manuscript. All authors read and approved the final manuscript.

Author details
[1] College of Life Sciences, Northwest A&F University, 22 Xinong Road, Yangling 712100, Shaanxi, China. [2] National Engineering Laboratory for Cereal Fermentation Technology, Jiangnan University, 1800 Lihu Avenue, Wuxi 214122, Jiangsu, China.

Acknowledgements
Not applicable.

Competing interests
The authors declare that they have no competing interests.

Funding
This work is financially supported by the Fund for Doctoral Scientific Research (Z109021632). The authors also acknowledge the support from Jiangsu Youth Fund (BK20150130), the Start-up Fund for Talent Introduction (Z111021602) from Northwest A&F University, and the Fundamental Research Funds for the Central Universities (JUSRP115A17).

References
1. Fang H, Xia L. Cellulase production by recombinant *Trichoderma reesei* and its application in enzymatic hydrolysis of agricultural residues. Fuel. 2015;143:211–6.
2. Fang H, Xia L. Heterologous expression and production of *Trichoderma reesei* cellobiohydrolase II in *Pichia pastoris* and the application in the enzymatic hydrolysis of corn stover and rice straw. Biomass Bioenerg. 2015;78:99–109.
3. Fang X, Shen Y, Zhao J, Bao X, Qu Y. Status and prospect of lignocellulosic bioethanol production in China. Bioresour Technol. 2010;101:4814–9.
4. Fang H, Zhao C, Chen S. Single cell oil production by *Mortierella isabellina* from steam exploded corn stover degraded by three-stage enzymatic hydrolysis in the context of on-site enzyme production. Bioresour Technol. 2016;216:988–95.
5. Ageitos JM, Vallejo JA, Veiga-Crespo P, Villa TG. Oily yeasts as oleaginous cell factories. Appl Microbiol Biotechnol. 2011;90:1219–27.
6. Gao D, Zeng J, Zheng Y, Yu X, Chen S. Microbial lipid production from xylose by *Mortierella isabellina*. Bioresour Technol. 2013;133:315–21.
7. Zeng J, Zheng Y, Yu X, Yu L, Gao D, Chen S. Lignocellulosic biomass as a carbohydrate source for lipid production by *Mortierella isabellina*. Bioresour Technol. 2013;128:385–91.
8. Chen X, Li Z, Zhang X, Hu F, Ryu DD, Bao J. Screening of oleaginous yeast strains tolerant to lignocellulose degradation compounds. Appl Biochem Biotechnol. 2009;159:591–604.
9. Qi GX, Huang C, Chen XF, Xiong L, Wang C, Lin XQ, Shi SL, Yang D, Chen XD. Semi-pilot scale microbial oil production by *Trichosporon cutaneum* using medium containing corncob acid hydrolysate. Appl Biochem Biotechnol. 2016;179:625–32.
10. Wyman CE. What is (and is not) vital to advancing cellulosic ethanol. Trends Biotechnol. 2007;25:153–7.
11. Galbe M, Zacchi G. Pretreatment: the key to efficient utilization of lignocellulosic materials. Biomass Bioenerg. 2012;46:70–8.
12. Fang H, Zhao C, Song X-Y, Chen M, Chang Z, Chu J. Enhanced cellulolytic enzyme production by the synergism between *Trichoderma reesei* RUT-C30 and *Aspergillus niger* NL02 and by the addition of surfactants. Biotechnol Bioprocess Eng. 2013;18:390–8.
13. Fang H, Zhao C, Song XY. Optimization of enzymatic hydrolysis of steam-exploded corn stover by two approaches: response surface methodology

or using cellulase from mixed cultures of *Trichoderma reesei* RUT-C30 and *Aspergillus niger* NL02. Bioresour Technol. 2010;101:4111–9.

14. Alvira P, Tomas-Pejo E, Ballesteros M, Negro MJ. Pretreatment technologies for an efficient bioethanol production process based on enzymatic hydrolysis: a review. Bioresour Technol. 2010;101:4851–61.

15. Culbertson A, Jin M, da Costa Sousa L, Dale BE, Balan V. In-house cellulase production from AFEX™ pretreated corn stover using *Trichoderma reesei* RUT C-30. RSC Adv. 2013;3:25960.

16. Rana V, Eckard AD, Teller P, Ahring BK. On-site enzymes produced from *Trichoderma reesei* RUT-C30 and *Aspergillus saccharolyticus* for hydrolysis of wet exploded corn stover and loblolly pine. Bioresour Technol. 2014;154:282–9.

17. Zhang L, Liu Y, Niu X, Liu Y, Liao W. Effects of acid and alkali treated lignocellulosic materials on cellulase/xylanase production by *Trichoderma reesei* Rut C-30 and corresponding enzymatic hydrolysis. Biomass Bioenerg. 2012;37:16–24.

18. Yang J, Zhang X, Yong Q, Yu S. Three-stage enzymatic hydrolysis of steam-exploded corn stover at high substrate concentration. Bioresour Technol. 2011;102:4905–8.

19. Yang J, Zhang X, Yong Q, Yu S. Three-stage hydrolysis to enhance enzymatic saccharification of steam-exploded corn stover. Bioresour Technol. 2010;101:4930–5.

20. Ngo HL, Jones K, Foglia TA. Metathesis of unsaturated fatty acids: synthesis of long-chain unsaturated-alpha, omega-dicarboxylic acids. J Am Oil Chem Soc. 2006;83:629–34.

21. Ngo HL, Foglia TA. Synthesis of long chain unsaturated-α, ω-dicarboxylic acids from renewable materials via olefin metathesis. J Am Oil Chem Soc. 2007;84:777–84.

22. Ghose TK. Measurement of cellulase activities. Pure Appl Chem. 1987;59:257–68.

23. Miao C, Chakraborty M, Chen S. Impact of reaction conditions on the simultaneous production of polysaccharides and bio-oil from heterotrophically grown *Chlorella sorokiniana* by a unique sequential hydrothermal liquefaction process. Bioresour Technol. 2012;110:617–27.

24. Wang B, Xia L. High efficient expression of cellobiase gene from Aspergillus niger in the cells of *Trichoderma reesei*. Bioresour Technol. 2011;102:4568–72.

25. Zhang J, Zhong Y, Zhao X, Wang T. Development of the cellulolytic fungus *Trichoderma reesei* strain with enhanced beta-glucosidase and filter paper activity using strong artificial cellobiohydrolase 1 promoter. Bioresour Technol. 2010;101:9815–8.

26. Fang H, Xia L. High activity cellulase production by recombinant *Trichoderma reesei* ZU-02 with the enhanced cellobiohydrolase production. Bioresour Technol. 2013;144:693–7.

27. Juhász T, Szengyel Z, Réczey K, Siika-Aho M, Viikari L. Characterization of cellulases and hemicellulases produced by *Trichoderma reesei* on various carbon sources. Process Biochem. 2005;40:3519–25.

28. Ahamed A, Vermette P. Enhanced enzyme production from mixed cultures of *Trichoderma reesei* RUT-C30 and *Aspergillus niger* LMA grown as fed batch in a stirred tank bioreactor. Biochem Eng J. 2008;42:41–6.

29. Wen Z, Liao W, Chen S. Production of cellulase/β-glucosidase by the mixed fungi culture *Trichoderma reesei* and *Aspergillus phoenicis* on dairy manure. Process Biochem. 2005;40:3087–94.

30. Lu Y, Wang Y, Xu G, Chu J, Zhuang Y, Zhang S. Influence of high solid concentration on enzymatic hydrolysis and fermentation of steam-exploded corn stover biomass. Appl Biochem Biotechnol. 2010;160:360–9.

31. Gao Q, Cui Z, Zhang J, Bao J. Lipid fermentation of corncob residues hydrolysate by oleaginous yeast *Trichosporon cutaneum*. Bioresour Technol. 2014;152:552–6.

32. Xing D, Wang H, Pan A, Wang J, Xue D. Assimilation of corn fiber hydrolysates and lipid accumulation by *Mortierella isabellina*. Biomass Bioenerg. 2012;39:494–501.

33. Liu W, Wang Y, Yu Z, Bao J. Simultaneous saccharification and microbial lipid fermentation of corn stover by oleaginous yeast *Trichosporon cutaneum*. Bioresour Technol. 2012;118:13–8.

34. Liu ZH, Chen HZ. Simultaneous saccharification and co-fermentation for improving the xylose utilization of steam exploded corn stover at high solid loading. Bioresour Technol. 2016;201:15–26.

35. Deraedt C, d'Halluin M, Astruc D. Metathesis reactions: recent trends and challenges. Eur J Inorg Chem. 2013;2013(28):4881–908.

36. Rybak A, Fokou PA, Meier MAR. Metathesis as a versatile tool in oleochemistry. Eur J Lipid Sci Technol. 2008;110:797–804.

37. Bernas A, Myllyoja J, Salmi T, Murzin DY. Kinetics of linoleic acid hydrogenation on Pd/C catalyst. Appl Catal A. 2009;353:166–80.

38. Boda L, Onyestyák G, Solt H, Lónyi F, Valyon J, Thernesz A. Catalytic hydroconversion of tricaprylin and caprylic acid as model reaction for biofuel production from triglycerides. Appl Catal A. 2010;374:158–69.

39. Lu W, Ness JE, Xie W, Zhang X, Minshull J, Gross RA. Biosynthesis of monomers for plastics from renewable oils. J Am Chem Soc. 2010;132:15451–5.

40. Song JW, Jeon EY, Song DH, Jang HY, Bornscheuer UT, Oh DK, Park JB. Multistep enzymatic synthesis of long-chain alpha, omega-dicarboxylic and omega-hydroxycarboxylic acids from renewable fatty acids and plant oils. Angew Chem Int Ed Engl. 2013;52:2534–7.

41. Hicks JC. Advances in C–O Bond Transformations in Lignin-Derived Compounds for Biofuels Production. J Phys Chem Lett. 2011;2:2280–7.

14

Photoautotrophic production of polyhydroxyalkanoates in a synthetic mixed culture of *Synechococcus elongatus cscB* and *Pseudomonas putida cscAB*

Hannes Löwe, Karina Hobmeier, Manuel Moos, Andreas Kremling and Katharina Pflüger-Grau*

Abstract

Background: One of the major challenges for the present and future generations is to find suitable substitutes for the fossil resources we rely on today. Cyanobacterial carbohydrates have been discussed as an emerging renewable feedstock in industrial biotechnology for the production of fuels and chemicals, showing promising production rates when compared to crop-based feedstock. However, intrinsic capacities of cyanobacteria to produce biotechnological compounds are limited and yields are low.

Results: Here, we present an approach to circumvent these problems by employing a synthetic bacterial co-culture for the carbon-neutral production of polyhydroxyalkanoates (PHAs) from CO_2. The co-culture consists of two *bio-modules*: *Bio-module I*, in which the cyanobacterial strain *Synechococcus elongatus cscB* fixes CO_2, converts it to sucrose, and exports it into the culture supernatant; and *bio-module II*, where this sugar serves as C-source for *Pseudomonas putida cscAB* and is converted to PHAs that are accumulated in the cytoplasm. By applying a nitrogen-limited process, we achieved a maximal PHA production rate of 23.8 mg/(L day) and a maximal titer of 156 mg/L. We will discuss the present shortcomings of the process and show the potential for future improvement.

Conclusions: These results demonstrate the feasibility of mixed cultures of *S. elongatus cscB* and *P. putida cscAB* for PHA production, making room for the cornucopia of possible products that are described for *P. putida*. The construction of more efficient sucrose-utilizing *P. putida* phenotypes and the optimization of process conditions will increase yields and productivities and eventually close the gap in the contemporary process. In the long term, the co-culture may serve as a platform process, in which *P. putida* is used as a chassis for the implementation of synthetic metabolic pathways for biotechnological production of value-added products.

Keywords: Carbon neutral bioplastics, Polyhydroxyalkanoates (PHA), Synthetic co-culture, *Pseudomonas putida cscAB*, *Synechococcus elongatus cscB*, Cyanobacteria, CO_2 fixation

Background

For a long time, natural polymers like wood or wool have been used by humans to craft weapons and tools or to protect against the cold. This enhanced our ability to survive and allowed us to build cultures and to live in places that are hostile to our biology. From the nineteenth century, with the invention of modern polymer chemistry, many of these natural materials were complemented and/or replaced by modern plastics [1]. Plastics found application in all areas of our daily life. In 2015, the global plastics material production was estimated to reach 250 million tons per year [2] and most of the plastics produced were derived from petroleum. Because of the inevitable finiteness of fossil resources and the massive pollution caused by plastic wastes [3], contemporary research in this field is directed towards the exploration

*Correspondence: k.pflueger-grau@tum.de
Fachgebiet für Systembiotechnologie, Technische Universität München, Boltzmannstr 15, 85748 Garching, Germany

of renewable and biologically degradable sources of plastics. Advanced natural polymers like polyhydroxyalkanoates (PHA) that show thermoplastic, polypropylene-like properties, could be a valuable substitute for some applications. Under natural conditions, PHA, a linear polymer of 3-hydroxy fatty acids, serves as both energy and carbon storage in certain bacteria, among them *Pseudomonas putida*. In industrial production of PHA, substrate price is a key factor [4] as bio-based plastics have to compete with those made from fossil resources. Therefore, efforts are directed towards alternative feedstocks to reduce substrate costs in the overall process.

A newly discussed, potentially cheap source of substrates are carbohydrates produced by microalgae and cyanobacteria, for example, starch production by eukaryotic algae or sucrose production by recombinant cyanobacteria [5–8]. Compared to conventional crops, microalgae and cyanobacteria have the potential to reach higher areal yields and, additionally, their products do not interfere with the food markets. Further benefits include the ability to use salty or brackish water and bioreactors can be placed on non-arable land. The genetically engineered cyanobacterial strain *Synechococcus elongatus cscB* has recently been shown to export sucrose on a level comparable to sugar cane [6]. This was achieved by the introduction of only one heterologous gene encoding the sucrose permease CscB from *Escherichia coli* ATCC 700927. Under salt stress, *S. elongatus cscB* accumulates remarkable amounts of sucrose as a compatible solute, which are released into the surrounding medium by the activity of the heterologous permease CscB [6]. However, when these sugars are produced on a large scale, limitations will arise: These include the risk of contamination, as a carbon source is provided that can be used by heterotrophs, and economic aspects, e.g., the cost of sugar recovery from the fermentation broth [9, 10]. A recent approach to circumvent these problems is to convert the cyanobacterial feedstock directly into value-added products in a multispecies microbial factory in a so-called "one-pot" reaction [11]. By co-inoculation of both strains, the sugar-producing strain together with the product accumulating strain, several barriers can be overcome. Thus, the costs for sugar recovery from the cyanobacterial fermentation broth as well as the potential loss due to contamination are saved. This contributes to making the overall process economically more competitive. Furthermore, there are cases in which a positive effect of synthetic consortia on the productivity have been described [12, 13]. Along that line, two studies have been published recently that aimed to produce polyhydroxybutyrate (PHB) in a mixed culture of *S. elongatus cscB* with *Escherichia coli* or *Azotobacter vinelandii*, respectively [9, 14]. Even though the final titers reached were quite low

(around 1 mg/L), which might be a result of slow growth rates and suboptimal media composition, this shows the feasibility of the general approach and leaves room for improvement.

In this work, we aimed to produce biodegradable plastic from light and CO_2, tackling two of the major challenges of modern times: global warming and pollution by plastics. Strategies to combat these problems are the fixation of CO_2 to avoid its emission into the atmosphere and to find economically competitive processes for the production of plastic substitutes like PHA. Here, we present an approach in which the sucrose production of *S. elongatus cscB* is directly coupled to PHA accumulation by *P. putida cscAB* in a synthetic co-culture. Thus, CO_2 and sunlight are converted into carbon neutral bioplastics (Fig. 1). *P. putida* is known for its innate stress resistance and robustness and is therefore an excellent candidate as a mixed culture partner. The strain *P. putida cscAB* is genetically modified to be able to metabolize sucrose [15, 16], and is the strain of choice in this co-culture. We present the steps undertaken towards a functional mixed culture and the first improvements made to considerably increase the PHA content of the cells (to about 150 mg/L at the end).

Results and discussion

Sucrose production of *S. elongatus cscB* in BG-11+ medium

As *S. elongatus* accumulates sucrose as a compatible solute in response to the external salt concentration, we set out to identify the optimal NaCl concentration that would allow maximal sucrose excretion without severely inhibiting growth. Recently, it was shown that in normal BG-11 medium, *S. elongatus cscB* showed the highest sucrose production in the presence of 150 mM NaCl [6]. However, as the medium in this study was modified to meet the needs of both co-culture partners (see "Methods") and the cultivation conditions were different, the influence of NaCl on sucrose production and growth had to be assessed again in the newly defined medium in a 1.8 L photobioreactor at controlled pH. Therefore, *S. elongatus cscB* was cultivated in BG-11+ medium in the absence or presence of NaCl in concentrations ranging from 150 to 250 mM. Growth of the cells and excretion of sucrose were monitored by measuring the optical density and determining the concentrations of sucrose in the supernatants by HPLC (Fig. 2). The growth of *S. elongatus cscB* is clearly reduced by an increase in the external salt concentration, ranging from a biomass production rate of 0.236 g CDW/L d without NaCl to 0.059 g CDW/L d in the presence of 250 mM NaCl during light-limited, linear growth (Table 1). The highest sucrose productivity, however, was observed with 150 mM NaCl, which is in accordance with what was reported by Ducat

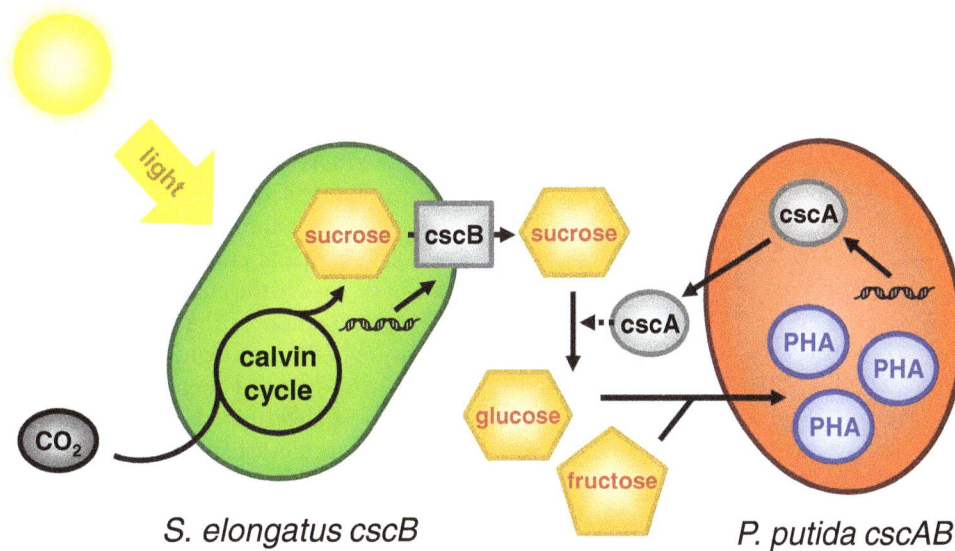

Fig. 1 Concept of the synthetic co-culture of *S. elongatus cscB* and *P. putida cscAB* for the production of PHA from CO_2 and light. CO_2 is fixed via the Calvin cycle to make sucrose, which in turn is secreted into the surrounding medium by the activity of the heterologous sucrose permease CscB. CscA produced by *P. putida cscAB* leaks out of the cell, where it splits sucrose extracellularly [16]. The monomers (glucose and fructose) are metabolized by *P. putida cscAB*, and polyhydroxyalkanoates (PHA) are accumulated in the cytoplasm [15]

et al. for normal BG-11 medium [6]. We reached a production rate of 0.346 ± 0.014 g/(L day) and a maximal titer of 2.63 g/L after 12 days (Table 1). By comparing the amount of sucrose produced under salt stress to the biomass production without NaCl, the carbon flux in the cyanobacteria is mirrored: At 150 mM NaCl, the produced mass of sucrose per unit time is higher than the amount of biomass produced without NaCl, indicating that a large fraction of the primary production is directed towards the synthesis of the compatible solute sucrose.

Growth of *P. putida cscAB* in modified BG-11 media

Next, we had to confirm that the metabolically engineered *P. putida cscAB* is able to grow in the modified BG-11 medium in the presence of 150 mM NaCl, the optimal salt concentration for sucrose production by *S. elongatus cscB*. Additionally, the influence of the changes in the medium composition had to be examined by comparing both BG-11 derived media. Therefore, *P. putida cscAB* was grown in BG-11[$-$NaCO$_3$, CaCl$_2$/100] and BG-11$^+$, in the presence of 150 mM NaCl and a mixture of glucose and fructose (each 1.5 g/L) as the carbon source (Fig. 3). We chose the monomers of sucrose as the carbon source to assay solely the effect of the medium composition on the growth of *P. putida cscAB* and not the influence of the efficiency of sucrose splitting. The strain grew well in both media with a growth

rate of 0.239 ± 0.001/h for BG-11[$-$NaCO$_3$, CaCl$_2$/100] and 0.305 ± 0.011/h for BG-11$^+$. The slight increase in the growth rate in BG-11$^+$ can most likely be attributed to the higher nutrient availability as the concentrations of potassium phosphate and magnesium sulfate were increased tenfold in this medium. This suggests that one or both of these nutrients are limiting factors in the original medium. The growth rate obtained in BG-11$^+$ was in the same range as the one determined in M9 medium [16], which is the standard minimal medium for *P. putida*. Thus, only minor adjustments in medium composition were necessary to achieve comparable growth of *P. putida cscAB* in BG-11 as well. This reflects the broad metabolic versatility and robustness of this bacterium. Other bacteria and eukaryotic culture partners seem to be less suited for co-cultivation in a photosynthetic consortium as they have higher nutritional demands. Recently, Hays et al. reported co-cultivation of three heterotrophs with *S. elongatus cscB*. However, in their process, the medium was also supplemented with ammonium salts and buffer [9]. The same seems to be true for oleaginous yeast as described by Li et al. [17]. As outlined above, supplementation of the medium with additives was not necessary with *P. putida cscAB* as co-cultivation partner. This underlines the versatility and suitability of *P. putida* as a chassis for industrial biotechnology [18].

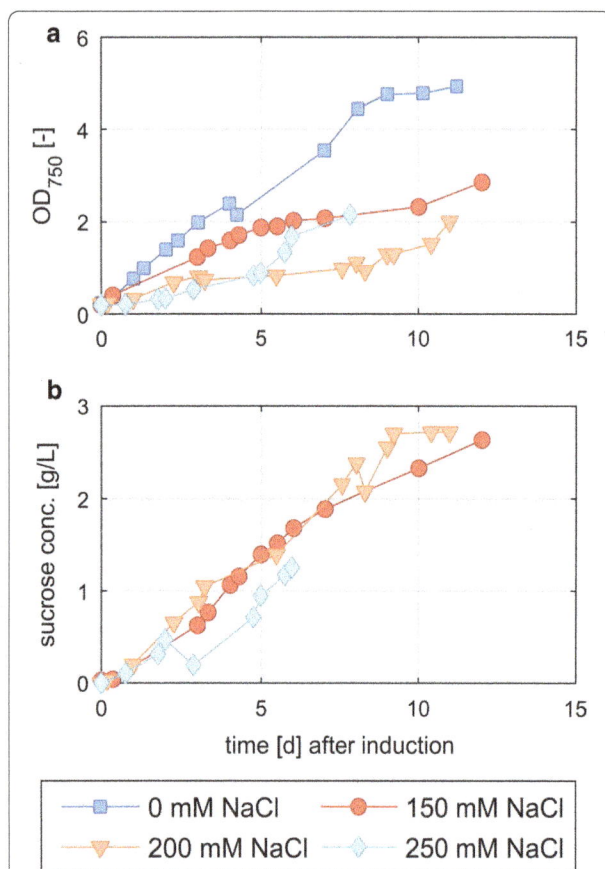

Fig. 2 Growth and sucrose concentration from fermentations of *S. elongatus cscB* in a photobioreactor in BG-11+ in the presence of different salt concentrations. Cells were grown in BG11+ medium with the NaCl concentration indicated, and growth and sucrose secretion were monitored. CscB production was induced with 0.1 mM IPTG at an OD of 0.1–0.2

Mixed culture of *S. elongatus cscB* and *P. putida cscAB*

Having two strains with complementary functions, one producing sucrose from CO_2, the other consuming sucrose, and a common medium allowing the growth of both, we set out to grow them simultaneously in the same cultivation vessel. Each strain represents a functional

Table 1 Maximal linear growth and sucrose production rates of *S. elongatus cscB* at different NaCl concentrations in BG-11+ medium

NaCl (mM)	Biomass production rate[a] [g_{CDW}/(L day)]	Sucrose production rate[a] [g/(L day)]
0	0.236 ± 0.00400	n.d.
150	0.134 ± 0.004	0.346 ± 0.014
200	0.116 ± 0.008	0.282 ± 0.012
250	0.059 ± 0.005	0.20 ± 0.03

[a] Standard deviations are regression errors, not derived from replicates

bio-module, which is linked to the other by sucrose transfer. To start the synthetic mixed culture, first *S. elongatus cscB* was inoculated in a photobioreactor in BG-11+ in the presence of 150 mM NaCl to promote sucrose accumulation right from the beginning. The heterotrophic organism was inoculated at least 1 day after induction of sucrose export with 0.1 mM IPTG to ensure that sucrose was readily available for *P. putida cscAB*. The total OD of the culture was monitored, i.e., the sum of *S. elongatus cscB* and *P. putida cscAB* cells, as well as the concentration of sucrose in the culture supernatant (Fig. 4). To have an approximation of the contribution of each strain to the overall OD, the colony forming units (CFUs) of each strain, and the cell counts of Nile red-stained *P. putida cscAB* were determined. Nile red predominantly stains lipophilic residues like PHAs [19] therefore, the number of cells reported represents only the proportion of *P. putida cscAB* cells that accumulated PHA in their cytoplasm. The overall OD of the mixed culture increased over a period of 7 days reaching a plateau of about OD = 2.3 at the end of the process (Fig. 4a). The growth behavior of *S. elongatus cscB* was similar to the process in pure culture at 150 mM NaCl (compare Figs. 2, 4a), hence there seems to be no severe negative effect from the presence of the commensal *P. putida cscAB*. Recently, co-cultivation of *S. elongatus cscB* with other organisms even showed a positive effect on the growth of the cyanobacterium [9]. Sucrose accumulated steadily up to 1.5 g/L in the culture supernatant until day 6. This is when *P. putida cscAB* reached a critical mass, and when sucrose started to be metabolized more rapidly than it was built. As the cell counts of *P. putida cscAB* increased, sucrose concentrations decreased, but no extracellular glucose or fructose accumulation was detected. Transient accumulation of the sugar monomers was observed in earlier studies, when *P. putida cscAB* was grown in M9 medium with sucrose as the sole carbon source [16]. It was attributed to the extracellular cleavage of sucrose by invertase CscA that was leaking out of the cells. Therefore, sucrose splitting seems to be the limiting factor for the growth of *P. putida cscAB*, as no sugar monomers were accumulated, and as with sucrose as C-source the growth rate was markedly reduced compared to a mixture of glucose and fructose in BG-11+ (compare to Fig. 3). Along the same line, pure cultures of *P. putida cscAB* in BG-11+ medium with sucrose as the sole carbon source showed inconsistent and very slow growth (data not shown). Thus, when sucrose is used as carbon source, the medium composition clearly has an influence on the growth of *P. putida cscAB*. However, the controlled process parameter in the photobioreactor and maybe the presence of *S. elongatus cscB* seem to have a stabilizing effect on *P. putida cscAB*, so that reliable growth in the co-culture was achieved.

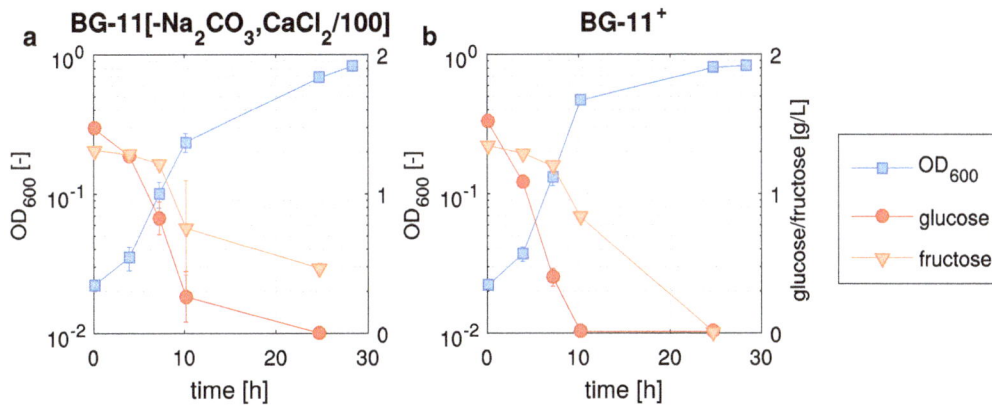

Fig. 3 Growth of *P. putida cscAB* in modified cyanobacterial BG-11 medium supplemented with glucose, fructose (each 1.5 g/L), and 150 mM NaCl. Experiments were conducted in 250 mL unbaffled shake flasks with an effective volume of 25 mL. Temperature was set to 30 °C, and agitation rate to 220 rpm. Shown are the means and standard deviations calculated from three biological replicates

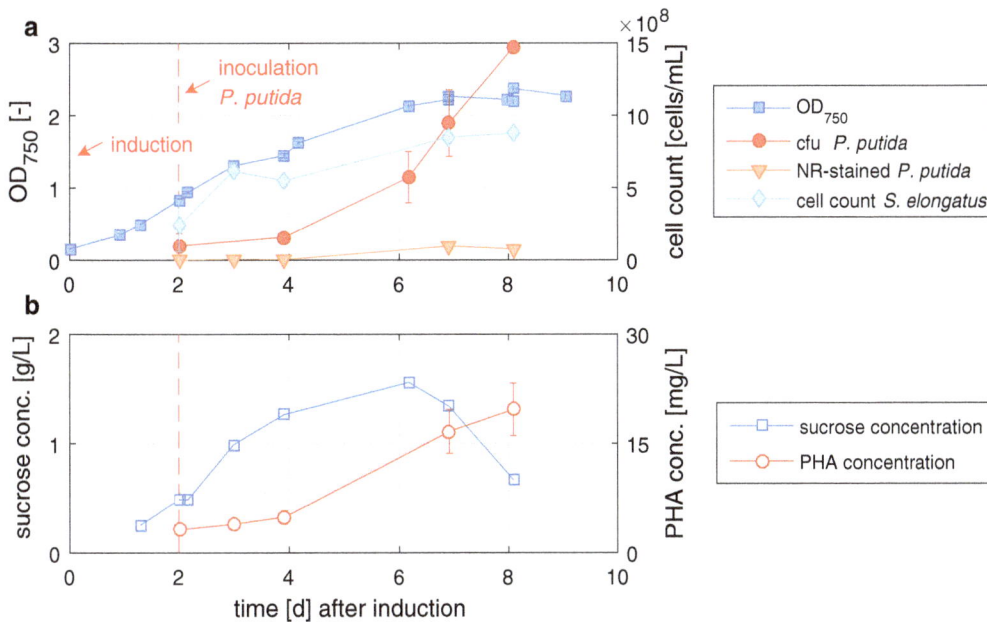

Fig. 4 Co-cultivation of *S. elongatus cscB* and *P. putida cscAB* in the photobioreactor. Optical density, Nile red-stained cell count of *P. putida cscAB*, cell count of *S. elongates cscB* and colony forming units (CFU) determined from plating are depicted over time in **a**. *Arrows* indicate the induction of sucrose export by the addition of 0.1 mM IPTG and inoculation of *P. putida cscAB*. Sucrose and PHA concentrations are plotted in **b**. Uncertainties in PHA concentrations were estimated from the propagation of errors of the PHA standards

Nevertheless, we assume that the growth of *P. putida cscAB* can be enhanced when sucrose is metabolized more efficiently. The sugar was detectable in the supernatant during the whole co-cultivation period, thus it was not consumed completely. This shortcoming will be tackled by the development of new sucrose splitting variants of *P. putida* with a higher splitting rate, which can be obtained for instance by active secretion of the invertase or screening for other sucrose permeases.

We also analyzed the PHA content of the cells, as *P. putida* accumulates these polymers to a small extent even under non-standard PHA accumulating conditions [20, 21]. *P. putida cscAB* produced PHA in the mixed culture at an approximate maximal production rate of 3.3 mg/ (L day) and a maximal titer of 19.7 mg/L was reached 6 days after inoculation with *P. putida cscAB* (Fig. 4b). It can be excluded that *S. elongatus cscB* is responsible for PHA formation since it lacks the corresponding genes for

PHA synthesis [22]. Moreover, the distribution pattern of 3-hydroxyalkanoic acids is typical for *P. putida* (Table 2), with 3-hydroxydecanoic acid being the most abundant monomer [23]. Apparently, PHA was accumulated in only a small fraction of the cells, which can be deduced when comparing the CFU/mL to the cell counts of Nile red-stained cells. On day 8, when the highest PHA titer was detected, only a very small fraction of the *P. putida cscAB* cells were stained with Nile red, i.e., accumulated PHA in their cytoplasm. To provide absolute numbers, further experiments, including counting cells with PHA granules under a microscope and comparing the numbers to counts gained by flow cytometry, are necessary to determine the efficiency of the staining method with Nile red. The flow cytograms, however, do provide qualitative information about the specific fluorescence and thus the PHA content of the stained cells, clearly showing the presence of PHA (see Additional file 1: Figure S2).

This way, the general process parameters for co-cultivation of *S. elongatus cscB* and *P. putida cscAB* for the production of PHAs from CO_2 and light were set and the feasibility of the process was confirmed. Without adding any other carbon source than CO_2 to the process, PHAs were produced by *P. putida*, however, yield and production rate were low.

Polyhydroxyalkanoate production in the co-culture under nitrogen-limiting conditions

One way to improve the PHA accumulation is to alter the C/N ratio of the growth conditions. Although not strictly necessary for PHA accumulation, nitrogen limitation in general increases the PHA content of the cells [24]. Therefore, we conducted the mixed culture experiment under nitrogen-limiting conditions. An initial batch phase was performed in which *S. elongatus cscB* was grown with a starting nitrate concentration of 48.0 mg/L and a nitrate feed of 9.2 ± 0.4 mg/day. Four days after induction of sucrose export with 0.1 mM IPTG and upon inoculation of the co-culture partner *P. putida cscAB*, the HNO_3-feed was increased to a rate of 46 ± 2 mg/day (NO_3^-). The nitrogen concentration was chosen so that it was just below the threshold that allows unlimited growth of *S. elongatus cscB*. Ideally, this should keep the steady-state concentration of nitrogen low enough to promote PHA formation by *P. putida cscAB*.

Table 2 Distribution of chain-lengths per mass fraction in PHA produced by *P. putida cscAB* in the mixed culture at maximal PHA concentration during the process

Chain-length (carbon number)	6	8	10	12	12:1
Mass fraction (%)	4.2	25.2	58.4	4.4	7.8

The results of this cultivation are illustrated in Fig. 5. After an initial period of 4 days, nitrate was no longer detectable in the photobioreactor, and is therefore, considered completely consumed by *S. elongatus cscB*. This was the time point of induction of *cscB* expression and is defined as day 0. The optical density started to rise, i.e., *S. elongatus cscB* started to grow, and on day 4, *P. putida cscAB* was inoculated. As expected, *S. elongatus cscB* grew linearly (dOD/dt = 0.105 ± 0.005/day during days 4–8, Fig. 5), although at a lower rate than under unlimited conditions (dOD/dt = 0.38 ± 0.03/day during days 1–4, Fig. 4). Sucrose was secreted constantly, reaching a maximal production rate of 0.316 g/(L day) and a maximal titer of 2.8 g/L 10 days after induction. The production rate reached was only marginally lower than the one obtained under non-limiting conditions (0.346 g/(L day), compared in Table 1). *P. putida cscAB* was apparently very stressed, which was manifested in non-reliable growth on LB agar plates when determining the CFUs (data not shown). Therefore, we could not use the cell count data from plating as an approximation of cells present in the co-culture. This behavior may be a result of the cumulative stress of a poor nitrogen source, nutrient limitation, low carbon availability because of low sucrose splitting rate, and possibly salt stress due to the presence of 150 mM NaCl in the medium. However, counting the Nile red-stained cells of *P. putida cscAB* by flow cytometry gave a qualitative estimate, and cell counts showed a steady course over time (Fig. 5a), which correlated nicely with the PHA concentration measured (Fig. 5b). The actual number of *P. putida cscAB* cells might be higher, assuming that only a fraction of the culture accumulates the polymer or are stained by the staining method. PHA accumulating cells as well as the PHA concentration increased until day 16, and then both decreased again. One possible explanation for the decrease might be cessation of sucrose metabolism and the consumption of intracellular reserves of PHAs, since it is possible that the actual number of *P. putida cscAB* cells continued to increase. PHA production reached a maximal production rate of 23.8 ± 6 mg/(L day) and a maximal titer of 156 ± 40 mg/L after 16 days. Thus, by applying nitrate limitation, we could increase the production rate about 7.3-fold.

To place these numbers in context: The achieved PHA level in the nitrogen-limited process exceeded the values reported for the mixed cultures of *S. elongatus cscB* with *E. coli* or *A. vinelandii* about a 150-fold [9, 14]. There are also efforts to produce PHB directly with recombinant cyanobacteria [25, 26], in which a high-cell dry weight fraction of PHB is already reached. However, there is no information about productivity available, and hence it cannot be determined how those strains compare to the process presented in this work.

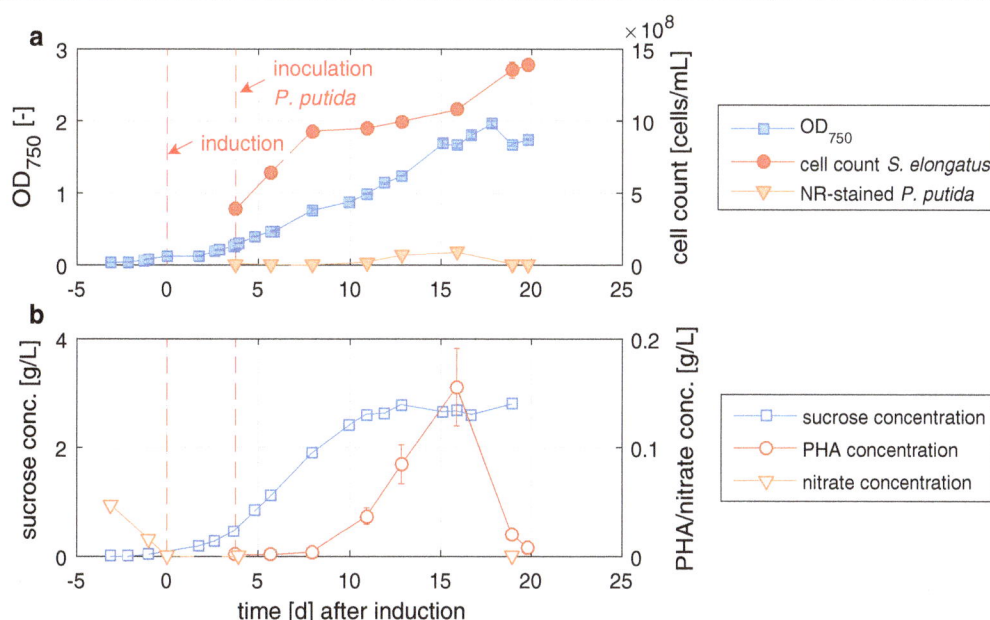

Fig. 5 Nitrate-limited co-cultivation of *S. elongatus cscB* and *P. putida cscAB*. Optical density, Nile red-stained cell count of *P. putida cscAB* and cell count of *S. elongatus cscB* are depicted over time in **a**. Sucrose, PHA and nitrate concentrations are plotted in **b**. Uncertainties in PHA concentrations were estimated from the propagation of errors of the PHA standards. Nitrate was fed at a constant rate of 46 ± 2 mg/day after inoculation with *P. putida cscAB*

A general challenge when working with phototrophic organisms is to reach the high-cell densities needed for efficient downstream processing. One problem is self-shading, which in the approach presented here could be solved by a compartmentalization of *S. elongatus cscB* and *P. putida cscAB*. In the process presented here, a major fraction of sucrose was left untouched by *P. putida cscAB*, which leaves room for the production of even more PHA. This might be facilitated by the construction of *P. putida* strains that split sucrose more efficiently. Furthermore, PHA production by *P. putida* can also be increased by metabolic engineering to direct more carbon flux to PHA formation [27] and thus reach higher PHA weight fractions of the cell dry weight, or by increasing the fraction of the population that actually accumulates PHAs. These attempts are currently under investigation in our laboratory.

Conclusion

In the work presented here, a mixed culture of *S. elongatus cscB* and *P. putida cscAB* was established in a lab-scale photobioreactor for the production of PHAs, a green plastic substitute. The concept of coupling a photosynthetic organism to a heterotrophic bacterium with the aim to produce an industrially relevant compound was successful and makes room for the implementation of a wide repertoire of bioproducts that can be

produced by *P. putida* [18, 28]. Only slight adjustments in the cyanobacterial growth medium composition were necessary to promote growth and PHA production by *P. putida cscAB*, showing the universality and robustness of this organism. When evaluating the productivity and yields of the co-culture, it has to be kept in mind that neither the process nor the organisms were optimized yet, which leaves great potential for optimization: One possible target is to increase the sucrose production by *S. elongatus cscB* through metabolic engineering, optimizing light influx or the use of higher photon flux densities. Indeed, under natural conditions at well-suited sites, photon flux densities are higher than the ones used in this work [29], which could result in an increased photosynthetic carbon fixation rate. Other potential targets include the improvement of sucrose consumption by *P. putida cscAB* or the more efficient channeling of the carbon flux towards PHA formation to achieve a higher weight fraction of PHA relative to the cellular dry weight. Here, we reached a maximal concentration of ~150 mg/L PHA, which is quite low, compared to pure cultures. However, this was achieved with a small number of cells compared to the cell densities reached in pure cultures, which are normally cultivated in fed-batch, high-cell density processes [23]. Nevertheless, it is crucial for efficient downstream processing to obtain a high concentration of PHA. One way to accomplish higher cell densities in coupled cultures is to separate the autotrophic and heterotrophic processes into different compartments [30] or to extend the process duration. In a possible future scenario, sunlight will be collected by a large area of photobioreactors and product formation by *P. putida* will be mainly limited by the sucrose production of *S. elongatus cscB*. For efficient downstream processing, *P. putida* could be either upconcentrated, e.g., by cultivation in a membrane bioreactor or hydrogel as suggested by Smith and Francis [14], or it could be auto-aggregated as described for other bacteria [31] and collected by decanting. Consequently, titers and production rates here cannot, and maybe do not have to, be compared to those of contemporary studies on PHA production in pure cultures. Future experiments with set-ups closer to a possible application scenario will elucidate the feasibility of this technology. In the future, the concept of the modular co-culture might also serve as a platform process with *S. elongatus cscB* fixing CO_2 and converting it to sucrose and *P. putida cscAB* serving as chassis for the implementation of synthetic pathways. This way, the product spectrum could be amplified tremendously.

In the long run, the applicability of commodities produced by microalgae will stand or fall depending on many factors: the development of cheaper and more efficient photobioreactors, the long-term dominance of cultures versus contaminations, the genetic streamlining of organisms, energy costs, the development of power-to-chemical processes and the price of oil, just to name a few. Whichever renewable technology will be feasible in the future, all research efforts in this area are important, as one fact is evident: availability of fossil resources will come to an end.

Methods
Bacterial strains and batch cultivation
Two organisms were used in the mixed culture: The autotrophic host, *S. elongatus cscB*, was kindly provided by Pamela Silver [6]. The heterotrophic commensal was *P. putida* EM178 *att*::miniTn7(eYFP) PP_3398::*cscAB*, a derivative of the prophage-free KT2440 strain EM178 [16]. For reasons of simplicity the strain will be called *P. putida cscAB*. Pre-cultures of *S. elongatus cscB* were grown in 100 mL shaking flasks with 40 mL of BG-11[−NaCO₃, CaCl₂/100] medium at 100 rpm, at 30 °C and a photon flux density of approximately 10–26 µmol/(m² s). A 30 W tubular fluorescent lamp was used for lighting, and air was the only source of CO_2. Experiments with *P. putida* alone were conducted in 250 mL flasks with 25 mL of BG-11[−NaCO₃, CaCl₂/100] or BG-11⁺ at 30 °C under shaking. Pre-cultures for *P. putida* experiments were grown in LB-medium overnight and washed once

in the cultivation medium of the main culture prior to inoculation.

Media
As a first step towards a functional co-culture between *S. elongatus cscB* and *P. putida cscAB*, a common growth medium for both organisms had to be defined, using original BG-11 (ATCC Medium 616) medium for blue-green algae [32] as starting point. The pH was shifted towards neutral pH, by omitting sodium carbonate, and $CaCl_2$ was reduced 100-fold to 3.4 µM, as it interfered with the growth of *P. putida* (see Additional file 1: Figure S1). This medium is referred to as BG-11[−NaCO₃, CaCl₂/100]. For its preparation four stock solutions were made, filter sterilized and stored at −20 °C in appropriate aliquots: solution 1 [100×]: 150 g/L NaNO₃, 3 g/L K₂HPO₄; solution 2 [1000×]: 75 g/L MgSO₄·7H₂O, 5 g/L citric acid, 6 g/L iron–ammonium citrate; 1.1 g/L disodium ethylen ediaminetetraacetate·2H₂O; solution 3 [1000×]: 0.36 g/L CaCl₂; trace element solution A5 [1000×]: 2.86 g/L H₃BO₃, 1.81 g/L MnCl₂·4H₂O, 0.222 g/L ZnSO₄·7H₂O, 0.39 g/L NaMoO₄·2H₂O, 0.079 g/L CuSO₄·5H₂O, 49.4 mg/L Co(NO₃)₂·6H₂O. The medium was prepared by adding each stock solution to autoclaved deionized water to reach 1× concentration (10 mL/L of solution 1, and 1 mL/L of solutions 2–4, respectively).

Additionally, based on this, a second medium was designed by increasing the phosphate and sulfate concentrations tenfold to exclude limitations that might occur when adding an additional microbe to the culture.

For this purpose, the components K₂HPO4 and MgSO₄·7H₂O were increased 10 times. They were added as separate, sterile stock solutions (300 g/L K₂HPO₄, 246.5 g/L MgSO₄·7H₂O). This enriched medium is designated herein BG-11⁺.

Cultivation in a photobioreactor
Liter-scale cultivations of *S. elongatus cscB* and mixed cultures were performed in a Labfors 5 Lux flat panel airlift photobioreactor (Infors AG, Switzerland) at a photon flux density of approximately 240 µmol/(m² s). The pH was controlled at 7.5 with 1 mol/L HNO₃ and an airflow of 2 L/min, enriched with 2% CO_2, was used as a carbon supply for autotrophic growth. The reactor was filled with 1.8 L of water containing the desired NaCl concentration, and autoclaved, and the medium ingredients were added sterilely through a septum from stock solutions. Twenty mL of a stationary *S. elongatus cscB* culture were used as inoculum and cells were grown until an optical density (OD₇₅₀) of 0.1–0.2 (equals a CDW of about 0.04–0.08 g/L) was reached; then 0.1 mmol/L IPTG was added to induce expression of *cscB* and thus sucrose export. For mixed culture cultivation, an over-night culture of *P.*

putida cscAB grown in BG-11$^+$ medium supplemented with 3 g/L glucose H_2O was then washed once with 5 mL reactor medium and added to the reactor.

In the case of the nitrate-limited process, the initial nitrate concentration was reduced to 50 mg/L NO_3^-. Upon inoculation with *S. elongatus cscB* a constant nitrate feed of 9.2 ± 0.5 mg/day was implemented by pumping a 0.03 mol/L HNO_3 solution into the reactor vessel. After inoculation with *P. putida cscAB*, the nitrate feed was increased to 46 ± 2 mg/day.

Samples were taken to measure the optical density, and culture supernatants were frozen and stored for subsequent HPLC analysis. At least every third day cells were diluted and the composition of mixed cultures was assessed by flow cytometry. The pellets of 5 mL of every sample were frozen at −80 °C for GC analysis. Growth of heterotrophic populations was additionally followed via plating in suitable dilutions on LB-medium agar plates.

Optical density, cell counting and determination of cell dry weight

The optical density of 200 µL culture was measured with an Infinity® microplate reader (Tecan, Austria) at wavelengths of 600 (*P. putida cscAB* cultivation) and 750 nm (*S. elongatus cscB* and mixed cultures). Sucrose, glucose, and fructose concentrations were determined by high performance liquid chromatography (HPLC) using an Agilent machine (Agilent 1100 series). Sugars were separated via a Shodex SH1011 column and a mobile phase of 0.5 mmol/L H_2SO_4 at a flow rate of 0.5 mL/min and a temperature of 30 °C. For flow cytometry, cultures were diluted to reach a cell count of about 800–1400 at a flow rate of 1–5 µL/s and injected into a CyFlow® instrument (Sysmex Partec GmbH, Germany) equipped with a laser (488 nm excitation wavelength). Fluorescence was measured at 536, 590, and 630 nm emission wavelengths. One mL of PHA-containing cells was centrifuged, stained with 10 µL Nile red solution (1 g/L in DMSO) and incubated for 5 min prior to dilution and followed by PHA-mediated Nile red fluorescence at 590 nm emission wavelength.

For cell dry weight measurement, an appropriate volume of cells was centrifuged at 8000×g for 10 min. The pellet was resuspended in phosphate buffered saline (PBS) and centrifuged again in a pre-dried and weighted 1.5 mL centrifuge tube. The supernatant was carefully discarded as completely as possible and the tube was dried at 60 °C for at least 3 days until the weight remained constant. The weight difference represented the dry weight. A correlation between optical density and cell dry weight was made for both organisms (data not shown) and cell dry weights of the other experiments were estimated from this correlation.

PHA determination

PHA content and composition were determined by gas chromatography (GC). A modified version of the propanylation protocols by Riis and Mai [33] and Furrer et al. [34] was applied. Samples were centrifuged at 17,000×g for 5 min and pellets were frozen and stored at −80 °C. To remove residual water, samples were freeze-dried for at least 3 days. The samples were dissolved in 2 mL of chloroform in an Ace® overpressure glass tube, and subsequently 1 mg of Poly-3-hydroxybutanoate (Sigma-Aldrich) and 0.2 mg of 3-methylbenzoic acid (Sigma-Aldrich) were added as internal standards. Two mL of an 80% (v/v) solution of 1-propanol and 37% HCl were pipetted into the mixture and the tube was sealed tightly. The bottom third of the tube was placed in an oil bath at 80 °C and mixed using a magnetic stirrer. After 16–24 h the tube was cooled to room temperature and 4 mL of bidistilled water were added. After shaking the tube vigorously, the tube was left at room temperature until the phases separated. The upper, aqueous phase was removed carefully, and the remaining liquid was dried with Na_2SO_4 and neutralized with Na_2CO_3. The remaining organic layer was transferred to a GC vial and injected into the GC machine (injection volume 1 µL). The samples were separated with a fused silica Stabilwax® column (Restek AG, Fuldabrueck, Germany) and measured with a flame ionization detector (detector temperature 245 °C). A temperature of 240 °C was set for the split/splitless injector (split ratio 1:10). Hydrogen gas was used as carrier gas at a flow rate of 3 mL/min. The different 3-hydroxyalkanoic acid esters were separated by applying a temperature gradient, starting at 80 °C (1 min), which then increased 5 °C every minute, stopping at 240 °C (hold time 5 min).

3-hydroxydecanoic acid (Sigma-Aldrich) and PHB were used as external standards. The response factors of the remaining 3-hydroxyalkanoic and 3-hydroxyalkenoic acids were inter- or extrapolated linearly from the two standards, according to Tan et al. [35].

Measurement of nitrate concentration

The nitrate/nitrite concentration of culture supernatants was determined using a colorimetric assay (Nitrite/Nitrate colorimetric method, Roche Diagnostics GmbH, Penzberg, Germany) in a microplate reader based on the enzymatic reduction of nitrate to nitrite. Nitrite reacts with a combination of dyes to form a diazo dye that can be measured at a wavelength of 540 nm and correlates linearly with the original nitrate/nitrite concentration.

Additional file

Additional file 1: Figure S1. Reduction of the $CaCl_2$ concentration promotes growth of *P. putida* EM178 in BG-11 [−NaCO$_3$] medium. Growth of *P. putida* EM178 with different concentrations of $CaCl_2$. Experiments were performed in 100 mL, unbaffled shake flasks filled with 10 mL of medium at 30 °C and an agitation rate of 220 rpm. Note that at 3.4 µM $CaCl_2$, the concentration chosen for BG11$^+$, no limitation in growth of *P. putida* EM178 is observed. **Figure S2.** Exemplary flow cytogram of Nile red-stained cells during nitrate-limited mixed culture of *P. putida cscAB* and *S. elongatus cscB* at the maximal concentration of PHA. The cells marked in the red circle only appeared upon staining with Nile red and are not found in the unstained control (data not shown). Cells below are unstained cells of both strains and undefined background.

Abbreviations
PHA: polyhydroxyalkanoate; PHB: polyhydroxyburate; CFU: colony forming units; GC: gas chromatopraphy; PBS: phosphate buffered saline; HPLC: high performance liquid chromatography; CDW: cellular dry weight; DMSO: dimethyl sulfoxide.

Authors' contributions
HL, AK, and KPG designed the experiments. HL, KH, and MM performed the experiments. HL and KPG wrote the manuscript. AK, KH, and MM carefully proofread the manuscript. All authors read and approved the final manuscript.

Acknowledgements
We would like to express our thanks to all supporters of this project, especially to Pamela Silver, Harvard Medical School, for providing us with *S. elongatus cscB*, Victor de Lorenzo, Centro Nacional de Biotecnología (CSIC), for the original *P. putida* strains and plasmids from the SEVA library, and the group of Thomas Brück, Technical University of Munich, especially Martina Haack for help with the GC analytics.

Competing interests
The authors declare that they have no competing interests.

Funding
No third party funding was obtained for this project.

References
1. Andrady AL, Neal MA. Applications and societal benefits of plastics. Phil Trans R Soc B Biol Sci. 2009;364:1977–84.
2. PlasticsEurope. Plastics- The facts 2014/2015. In: The unknown life of plastics. 2015. http://www.plasticseurope.org/plastics-industry/market-and-economics.aspx. Accessed 28 Mar 2017.
3. Jambeck JR, Geyer R, Wilcox C, Siegler TR, Perryman M, Andrady A, et al. Marine pollution. Plastic waste inputs from land into the ocean. Science. 2015;347:768–71.
4. Chen G-Q. A microbial polyhydroxyalkanoates (PHA) based bio- and materials industry. Chem Soc Rev. 2009;38:2434–46.
5. Brányiková I, Maršálková B, Doucha J, Brányik T, Bišová K, Zachleder V, et al. Microalgae—novel highly efficient starch producers. Biotechnol Bioeng. 2011;108:766–76.
6. Ducat DC, Avelar-Rivas JA, Way JC, Silver PA. Rerouting carbon flux to enhance photosynthetic productivity. Appl Environ Microbiol. 2012;78:2660–8.
7. Song K, Tan X, Liang Y, Lu X. The potential of *Synechococcus elongatus* UTEX 2973 for sugar feedstock production. Appl Microbiol Biotechnol. 2016;100:1–11.
8. Du W, Liang F, Duan Y, Tan X, Lu X. Exploring the photosynthetic production capacity of sucrose by cyanobacteria. Metab Eng. 2013;19:17–25.
9. Hays SG, Yan LLW, Silver PA, Ducat DC. Synthetic photosynthetic consortia define interactions leading to robustness and photoproduction. J Biol Eng. 2017;11:4.
10. Chisti Y. Constraints to commercialization of algal fuels. J Biotechnol. 2013;167:201–14.
11. Ortiz-Marquez JCF, Do Nascimento M, Zehr JP, Curatti L. Genetic engineering of multispecies microbial cell factories as an alternative for bioenergy production. Trends Biotechnol. 2013;31:521–9.
12. Patel VK, Sahoo NK, Patel AK, Rout PK, Naik SN, Kalra A. Exploring microalgae consortia for biomass production: a synthetic ecological engineering approach towards sustainable production of biofuel feedstock. In: Gupta SK, Malik A, Bux F, editors. Algal biofuels. Berlin: Springer; 2017. p. 109–26.
13. Abed R. Interaction between cyanobacteria and aerobic heterotrophic bacteria in the degradation of hydrocarbons. Int Biodeterior Biodegrad. 2010;64:58–64.
14. Smith MJ, Francis MB. A designed *A. vinelandii–S. elongatus* coculture for chemical photoproduction from air, water, phosphate, and trace metals. ACS Synth Biol. 2016;5:955–61.
15. Löwe H, Kremling A, Pflüger-Grau K. Bioplastik aus Licht und Luft—das Konzept einer synthetischen Ko-Kultur. Biospektrum. 2017;23:338–40.
16. Löwe H, Schmauder L, Hobmeier K, Kremling A, Pflüger-Grau K. Metabolic engineering to expand the substrate spectrum of *Pseudomonas putida* toward sucrose. Microbiologyopen. 2017;11:e00473.
17. Li T, Li C-T, Butler K, Hays SG, Guarnieri MT, Oyler GA, et al. Mimicking lichens: incorporation of yeast strains together with sucrose-secreting cyanobacteria improves survival, growth, ROS removal, and lipid production in a stable mutualistic co-culture production platform. Biotechnol Biofuels. 2017;10:55.
18. Nikel PI, Chavarría M, Danchin A, de Lorenzo V. From dirt to industrial applications: *Pseudomonas putida* as a synthetic biology chassis for hosting harsh biochemical reactions. Cur Opin Chem Biol. 2016;34:20–9.
19. Spiekermann P, Rehm BH, Kalscheuer R, Baumeister D, Steinbüchel A. A sensitive, viable-colony staining method using Nile red for direct screening of bacteria that accumulate polyhydroxyalkanoic acids and other lipid storage compounds. Arch Microbiol. 1999;171:73–80.
20. Follonier S, Panke S, Zinn M. A reduction in growth rate of *Pseudomonas putida* KT2442 counteracts productivity advances in medium-chain-length polyhydroxyalkanoate production from gluconate. Microb Cell Fact. 2011;10:25.
21. Poblete-Castro I, Escapa IF, Jäger C, Puchalka J, Lam CMC, Schomburg D, et al. The metabolic response of *P. putida* KT2442 producing high levels of polyhydroxyalkanoate under single- and multiple-nutrient-limited growth: highlights from a multi-level omics approach. Microb Cell Fact. 2012;11:34.
22. Beck C, Knoop H, Axmann IM, Steuer R. The diversity of cyanobacterial metabolism: genome analysis of multiple phototrophic microorganisms. BMC Genom. 2012;13:56.
23. Poblete-Castro I, Rodriguez AL, Lam CMC, Kessler W. Improved production of medium-chain-length polyhydroxyalkanoates in glucose-based fed-batch cultivations of metabolically engineered *Pseudomonas putida* strains. J Microbiol Biotechnol. 2014;24:59–69.
24. Prieto A, Escapa IF, Martínez V, Dinjaski N, Herencias C, de la Peña F, et al. A holistic view of polyhydroxyalkanoate metabolism in *Pseudomonas putida*. Environ Microbiol. 2016;18:341–57.
25. Bhati R, Samantaray S, Sharma L, Mallick N. Poly-β-hydroxybutyrate accumulation in cyanobacteria under photoautotrophy. Biotechnol J. 2010;5:1181–5.
26. Akiyama H, Okuhata H, Onizuka T, Kanai S, Hirano M, Tanaka S, et al. Antibiotics-free stable polyhydroxyalkanoate (PHA) production from carbon dioxide by recombinant *cyanobacteria*. Bioresour Technol. 2011;102:11039–42.
27. Acuña JMB-D, Bielecka A, Häussler S, Schobert M, Jahn M, Wittmann C, et al. Production of medium chain length polyhydroxyalkanoate in metabolic flux optimized *Pseudomonas putida*. Microb Cell Fact. 2014;13:88.
28. Loeschcke A, Thies S. *Pseudomonas putida*—a versatile host for the production of natural products. Appl Microbiol Biotechnol. 2015;99:6197–214.
29. Ge S, Smith RG, Jacovides CP, Kramer MG, Carruthers RI. Dynamics of photosynthetic photon flux density (PPFD) and estimates in coastal northern California. Theor Appl Climatol. 2011;105:107–18.

30. Smith MJ, Francis MB. Improving metabolite production in microbial co-cultures using a spatially constrained hydrogel. Biotechnol Bioeng. 2016;58:1711.

31. Nakajima M, Abe K, Ferri S, Sode K. Development of a light-regulated cell-recovery system for non-photosynthetic bacteria. Microb Cell Fact. 2016;15:31.

32. Stanier RY, Kunisawa R, Mandel M, Cohen-Bazire G. Purification and properties of unicellular blue-green algae (order Chroococcales). Bacteriol Rev. 1971;35:171–205.

33. Riis V, Mai W. Gas chromatographic determination of poly-β-hydroxybutyric acid in microbial biomass after hydrochloric acid propanolysis. J Chromatogr A. 1988;445:285–9.

34. Furrer P, Hany R, Rentsch D, Grubelnik A, Ruth K, Panke S, et al. Quantitative analysis of bacterial medium-chain-length poly([R]-3-hydroxyalkanoates) by gas chromatography. J Chromatogr A. 2007;1143:199–206.

35. Tan G-YA, Chen C-L, Ge L, Li L, Wang L, Zhao L, et al. Enhanced gas chromatography-mass spectrometry method for bacterial polyhydroxyalkanoates analysis. J Biosci Bioeng. 2014;117:379–82.

3ft2ddtßßfl3dlltxttft# 15

f# Heterologous co-expression of a yeast diacylglycerol acyltransferase (*ScDGA1*) and a plant oleosin (*AtOLEO3*) as an efficient tool for enhancing triacylglycerol accumulation in the marine diatom *Phaeodactylum tricornutum*

Nodumo Nokulunga Zulu[1,2], Jennifer Popko[1], Krzysztof Zienkiewicz[1], Pablo Tarazona[1], Cornelia Herrfurth[1] and Ivo Feussner[1,3,4]*

Abstract

Background: Microalgae are promising alternate and renewable sources for producing valuable products such as biofuel and essential fatty acids. Although this is the case, there are still challenges impeding on the effective commercial production of microalgal products. For instance, their product yield is still too low. Therefore, this study was oriented towards enhancing triacylglycerol (TAG) accumulation in the diatom *Phaeodactylum tricornutum* (strain Pt4). To achieve this, a type 2 acyl-CoA:diacylglycerol acyltransferase from yeast (*ScDGA1*) and the lipid droplet (LD) stabilizing oleosin protein 3 from *Arabidopsis thaliana* (*AtOLEO3*) were expressed in Pt4.

Results: The individual expression of *ScDGA1* and *AtOLEO3* in Pt4 resulted in a 2.3- and 1.4-fold increase in TAG levels, respectively, in comparison to the wild type. The co-expression of both, *ScDGA1* and *AtOLEO3*, was accompanied by a 3.6-fold increase in TAG content. On the cellular level, the lines co-expressing *ScDGA1* and *AtOLEO3* showed the presence of the larger and increased numbers of lipid droplets when compared to transformants expressing single genes and an empty vector. Under nitrogen stress, TAG productivity was further increased twofold in comparison to nitrogen-replete conditions. While TAG accumulation was enhanced in the analyzed transformants, the fatty acid composition remained unchanged neither in the total lipid nor in the TAG profile.

Conclusions: The co-expression of two genes was shown to be a more effective strategy for enhancing TAG accumulation in *P. tricornutum* strain Pt4 than a single gene strategy. For the first time in a diatom, a LD protein from a vascular plant, oleosin, was shown to have an impact on TAG accumulation and on LD organization.

Keywords: Biofuel, DGAT, Diatom, Lipid droplets, Microalgae, Neutral lipid biosynthesis

Background

In the past years, microalgal products have received a great deal of attention because of their potential in

resolving various problems encountered in environmental and health sectors [1, 2]. The fast growth rates and the ability to accumulate high yields of lipids have made microalgae better alternate sources for biofuel production in comparison to plants [3–6]. Furthermore, microalgae are capable of producing very high amounts of essential very long-chain polyunsaturated fatty acids (VLC-PUFAs) such as arachidonic acid (ARA),

*Correspondence: ifeussn@uni-goettingen.de
[1] Department of Plant Biochemistry, Albrecht-von-Haller-Institute for Plant Sciences, University of Goettingen, 37077 Goettingen, Germany Full list of author information is available at the end of the article

eicosapentaenoic acid (EPA), and docosahexaenoic acid (DHA) [3]. Despite the fact that microalgae have several advantages over plants as sources of biofuel and other highly valuable products, the microalgal industry is still not economically feasible [7]. One of the major bottlenecks of this industry is the trade-off between biomass production and lipid productivity as the strategies that are commonly used to enhance lipid accumulation impede biomass production [8]. Furthermore, product recovery from microalgal biomass is usually an expensive process [7]. Therefore, it is often preferred that microalgal products such as the essential VLC-PUFAs are deposited in triacylglycerols (TAGs), which are a preferred form of storage in terms of downstream processing [9].

TAGs can be synthesized via two different pathways, the acyl-CoA dependent de novo synthesis (also known as part of the Kennedy pathway) and the acyl-CoA independent pathway by degrading primarily existing organelle membranes [8]. In the Kennedy pathway, the last step of TAG synthesis involves the acylation of diacylglycerol (DAG) by an acyl-CoA:diacylglycerol acyltransferase (DGAT), whereas in the latter pathway DAG is acylated by the phospholipid:diacylglycerol acyltransferase (PDAT). Among DAGTs, the two mostly studied groups of these enzymes belong to type 1 and type 2. Although both types of DGATs perform similar functions, their sequences and preferences are different due to separate evolution [10]. In plants, both types of DGATs exist, and DGAT1 has been shown to be a major contributor of TAG accumulation in seeds [11, 12]. In contrast, the baker's yeast (*Saccharomyces cerevisiae*) only has a type 2 DGAT that is commonly known as *ScDGA1* [13]. *ScDGA1* has been previously shown to have a broad substrate specificity, in terms of the fatty acids it incorporates into TAGs [14]. This feature makes it a DGAT of interest in cases where the substrate pool is not well defined. Like plants and animals, microalgae have both types of DGATs. Remarkably, some microalgal species have multiple copies of DGAT2 genes [15]. For instance, *P. tricornutum* has 4 copies of DGAT2, whereas *Nannochloropsis oceanica* has 13 copies of DGATs, of which 12 encode DGAT2 [16]. Furthermore, DGAT2 from different species of microalgae exhibit diverse substrate specificities, which is reflected in the fatty acid composition of TAG among algal genera. Some microalgal species have high levels of VLC-PUFAs in their TAGs while others only have high amounts of unsaturated 16 carbon fatty acids. For example, *Ostreococcus tauri* accumulates high levels of DHA in TAGs [4, 14], whereas in *P. tricornutum* this VLC-PUFA is almost excluded from the TAGs [17].

Following their synthesis, TAGs are deposited in lipid droplets (LDs), which consist of a hydrophobic core that is surrounded by a monolayer of phospholipids with few embedded specific proteins [18]. LDs have a wide range of surface proteins, and in plants, oleosins have been found to be the most abundant proteins associated with the LDs in seeds [19]. Oleosins have been shown to stabilize LDs and prevent them from coalescing and degradation by providing steric hindrance and minimizing the access of TAG lipases, respectively, in seeds of flowering plants [18, 19]. Although oleosins are not present in microalgae, various other LD associated proteins have been characterized from various microalgal species [20, 21]. These include a major lipid droplet protein (MLDP) from *Chlamydomonas reinhardtii* [22], a lipid droplet surface protein (LDSP) from *Nannochlorospsis* sp. [23] and a stramenopile-type lipid droplet protein (StLDP) from *P. tricornutum* [20].

In this study, the diatom *P. tricornutum* was used as a model for engineering lipid metabolism. Diatoms are receiving a great deal of attention as a platform to produce biofuel [24], because they store TAGs as other microalgae as the major carbon source upon stress, i.e., when nitrogen becomes limiting. However, diatoms are better suited for biofuel production when compared to other classes of microalgae such as the chlorophytes because their fatty acid profiles are enriched with up to 80% of medium and long chain saturated and monounsaturated fatty acids of 14–16 carbon atoms [25]. In contrast, chlorophytes mostly have high levels of polyunsaturated long chain fatty acids, mostly C18 fatty acids. For biofuel production, medium chain saturated fatty acids are preferred because they increase the ignition quality of biofuels [25]. In addition to being good sources of biofuel, diatoms are also a source of the essential ω3 fatty acid EPA, which is not prevalent in chlorophytes [25]. In addition, they also produce pigments such as carotenoids, which have applications in the pharmaceutical, food, and cosmeceutical industries [26, 27].

Phaeodactylum tricornutum is one of the mostly studied diatoms, and in this study strain Pt4 was used as a model. Pt4 was isolated at the Island of Segelskär, Finland, which has a lower salt concentration in comparison to the regions of isolation for the other strains of *P. tricornutum* [28]. Therefore, Pt4 can be cultivated in industrial production systems close to biogas fermentation plants that are located far from coastal areas where salt water is not immediately available [17]. Although Pt4, accumulates low levels of TAGs in comparison to the other *P. tricornutum* strains [28], this opens up a platform for establishing higher TAG contents in a marine strain that can be cultivated at industrial scale without the risk of equipment corrosion caused by high salt content. In order to enhance TAG accumulation in Pt4, we have expressed a type 2 DGAT from *S. cerevisiae* (*ScDGA1*). We also pursued the possibility of further increasing

TAG productivity by blocking TAG degradation through the co-expression of *ScDGA1* and oleosin 3 from *A. thaliana* (*AtOLEO3*). We first assessed the individual expression of these genes in terms of TAG accumulation and then we compared this effect to data obtained when both, *ScDGA1* and *AtOLEO3*, are co-expressed in Pt4. Our results demonstrated that the co-expression of *ScDGA1* and *AtOLEO3* is a more effective approach towards enhancing TAG accumulation in Pt4 than the expression of single genes and demonstrates the potential of heterologous gene expression in microalgae as a tool for enhancing the TAG content.

Results

The expression of *ScDGA1* results in higher lipid accumulation

ScDGA1 has been previously demonstrated to have a broad substrate specificity, in terms of fatty acids it can incorporate into TAGs [14, 29]. For that reason, this type 2 DGAT was expressed in Pt4 towards achieving the aim of enhancing TAG accumulation in the background of an unknown substrate pool. The expression of *ScDGA1* in Pt4 was driven by the native fucoxanthin chlorophyll binding protein A (fcpA) promoter. This endogenous promoter is frequently used to drive stable gene expression in *P. tricornutum* and it is active during the photosynthetic phase of growth [30]. The resulting transformants were subjected to semi-quantitative PCR for confirming the expression of *ScDGA1* (Additional file 1: Figure S1). Of the eight lines that showed positive *ScDGA1* gene expression, four were randomly selected for further characterization. The assessment of growth profiles showed that there were no differences in growth patterns between the lines, empty vector control line and wild type. However, with the exception of line Pt4D1.1, which showed a lower cell concentration and chlorophyll a content (Fig. 1a, b). Regardless of its lower cell concentration, Pt4D1.1 accumulated the highest yields of total lipids and TAGs, reaching levels of 178 and 133 µg/mg, respectively, in comparison to the other lines and controls on day 7 (Fig. 1c). This corresponded to a twofold increase in total lipids and in TAGs in comparison to the wild type. For this reason, this line (Pt4D1.1) was selected for further analysis. It was also noted that in all Pt4 lines expressing the *ScDGA1* gene, there were no changes in fatty acid composition in total lipids or TAGs (Additional file 2: Figure S2).

The expression of *AtOLEO3* does not affect the growth of Pt4 but has a positive impact on TAG accumulation

Oleosins have been shown to specifically localize at the seed LDs of plants and are thought to stabilize and protect LDs from degradation [31]. Thus, since there are no oleosins in microalgae, we expressed the oleosin 3 gene from *A. thaliana* in order to test if its function to stabilize LDs is also conserved in other systems. To verify the localization of *AtOLEO3* in Pt4, the *AtOLEO3* encoding gene was tagged to YFP. The microscopic analysis showed that *AtOLEO3* specifically localized to the LDs when expressed in Pt4; this is indicated by the YFP signal surrounding the LDs (Fig. 2a). In a separate study, the randomly selected lines expressing non-tagged *AtOLEO3* (Pt4O3.1 and Pt4O3.2) were analyzed for growth rates and TAG accumulation. *AtOLEO3* gene expression was confirmed in these lines (Additional file 3: Figure S3). The obtained results showed that the lines grew in the same manner as the empty vector and wild type (Fig. 2b, c) but in terms of lipid accumulation, it was observed that line Pt4O3.1 accumulated more TAGs and total lipids, when compared to the empty vector and wild type. It reached 94.5 and 130.4 µg/mg, TAGs and total lipids, respectively, in comparison to the other line and controls at day 14 (Fig. 2d). This corresponded to a 1.4- and 1.2-fold increase in TAGs and total lipids respectively. Surprisingly, line Pt4O3.2 of the *AtOLEO3* transformants accumulated even less total lipids and TAGs when compared to the empty vector control and wild type. This could have been due to random integration of *AtOLEO3* at a critical site in the genome of Pt4. It was also noted that in both lines, the expression of *AtOLEO3* in Pt4 did not have any influence on the fatty acid composition neither of TAGs nor of total lipids (Additional file 4: Figure S4).

Co-expression of *ScDGA1* and *AtOLEO3* further increases TAG accumulation

The co-expression of a DGAT and a stabilizing LD protein had been shown to have a positive impact on TAG accumulation in the leaves of *A. thaliana* [32]. To our knowledge, the co-expression of DGAT and oleosin has not yet been demonstrated in microalgae. Thus, we tested the impact of co-expressing *ScDGA1* and *AtOLEO3* on growth rates and TAG levels in Pt4 under the control of the fcpA promoter. We selected two lines, (Pt4D1O3.1 and Pt4D1O3.2) showing positive gene expression for *ScDGA1* and *AtOLEO3* (Additional file 5: Figure S5), for analysis in terms of growth and lipid accumulation. The performance of these co-expressing lines was then assessed in parallel to line Pt4D1.1 (only expressing *ScDGA1*), in terms of growth rates and lipid accumulation (Fig. 3a, b). Similar to line Pt4D1.1, both co-expressing lines showed lower growth rates when compared to the empty vector line and the wild type (Fig. 3a, b). The highest total lipid and TAG yields were obtained from line Pt4D1O3.2, reaching 152 and 107 µg/mg, respectively (Fig. 3c); thus corresponding to a 2.2- and 3.6-fold increase in total lipids and TAGs, respectively. The

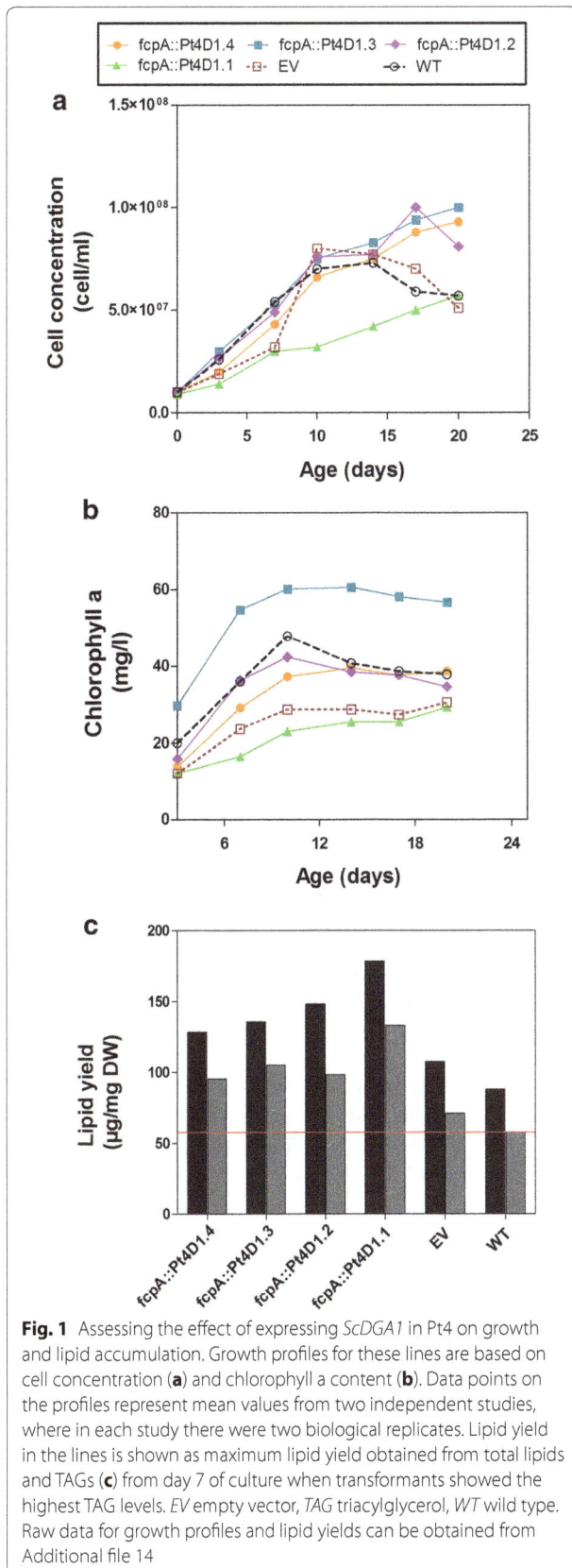

Fig. 1 Assessing the effect of expressing *ScDGA1* in Pt4 on growth and lipid accumulation. Growth profiles for these lines are based on cell concentration (**a**) and chlorophyll a content (**b**). Data points on the profiles represent mean values from two independent studies, where in each study there were two biological replicates. Lipid yield in the lines is shown as maximum lipid yield obtained from total lipids and TAGs (**c**) from day 7 of culture when transformants showed the highest TAG levels. *EV* empty vector, *TAG* triacylglycerol, *WT* wild type. Raw data for growth profiles and lipid yields can be obtained from Additional file 14

differences in TAG yields, especially between the double and single gene expressing lines, were also observed at the level of TAG productivities (Fig. 3d), where the best strategy would produce the highest TAG levels in a short period of time. As shown in Fig. 3d, all three transgenic lines displayed significantly higher levels of TAG productivities in comparison to the wild type; TAG productivities were increased by 2.2-, 3.5-, and 2.5-folds for Pt4D1O3.1, Pt4D1O3.2, and Pt4D1.1, respectively. Moreover, it was observed that line Pt4D1O3.2 (co-expressing *ScDGA1* and *AtOLEO3*) had a 1.4-fold higher TAG productivity (4.38 µg/mg/day) when compared to line Pt4D1.1 (3.1 µg/mg/day), which only expressed *ScDGA1* (Fig. 3d). Overall, these results strongly suggest that the co-expression of *ScDGA1* and *AtOLEO3* results in a further enhancement of TAG accumulation in Pt4, in comparison to when *ScDGA1* is expressed on its own.

In order to have a better understanding of the impact of *ScDGA1* and *AtOLEO3* genes on the formation of LDs, we analyzed the behavior of LDs in Pt4 lines co-expressing *ScDGA1* and *AtOLEO3* (lines Pt4D1O3.1 and Pt4D1O3.2) as well as lines individually expressing *ScDGA1* (Pt4D1.1) and *AtOLEO3* (Pt4O3.1), at cellular level (Fig. 4). This analysis showed that in comparison to the empty vector control, the individual expression of *ScDGA1* (Pt4D1.1) and *AtOLEO3* (Pt4O3.1) in Pt4 resulted in the formation of an increased number of LDs with an average diameter of 1.69 and 1.65 µm, respectively (Fig. 4a, b). In contrast, the co-expression of *ScDGA1* and *AtOLEO3* (Fig. 4a, for lines Pt4D1O3.1 and Pt4D1O3.2) was accompanied by the formation of prominent LDs, often of irregular shape. In these lines, the LDs occupied most of the cell volume and their average diameter reached 2.21 and 2.56 µm in lines Pt4D1O3.1 and Pt4D1O3.2, respectively (Fig. 4b). These large LDs in the lines co-expressing *ScDGA1* and *AtOLEO3* were however less numerous, when compared to single gene expressing lines (Fig. 4c).

Despite the significant increase in TAG yields, there were no changes observed in the fatty acid composition in response to the co-expression of *ScDGA1* and *AtO-LEO3* genes (Additional file 6: Figure S6). Similarly, there were no changes observed in the distribution of TAG molecular species (Additional file 7: Figure S7). TAG distribution remained similar between the transgenic lines and empty vector control. TAG species containing the saturated and monounsaturated fatty acids were still the major contributors in the TAG pool in all three lines (Pt4D1O3.1, Pt4D1O3.2, and Pt4D1.1). In addition, there were no new TAG species introduced by *ScDGA1*. Through the summing up of all detected TAG molecular species, we could validate our GC data, which showed

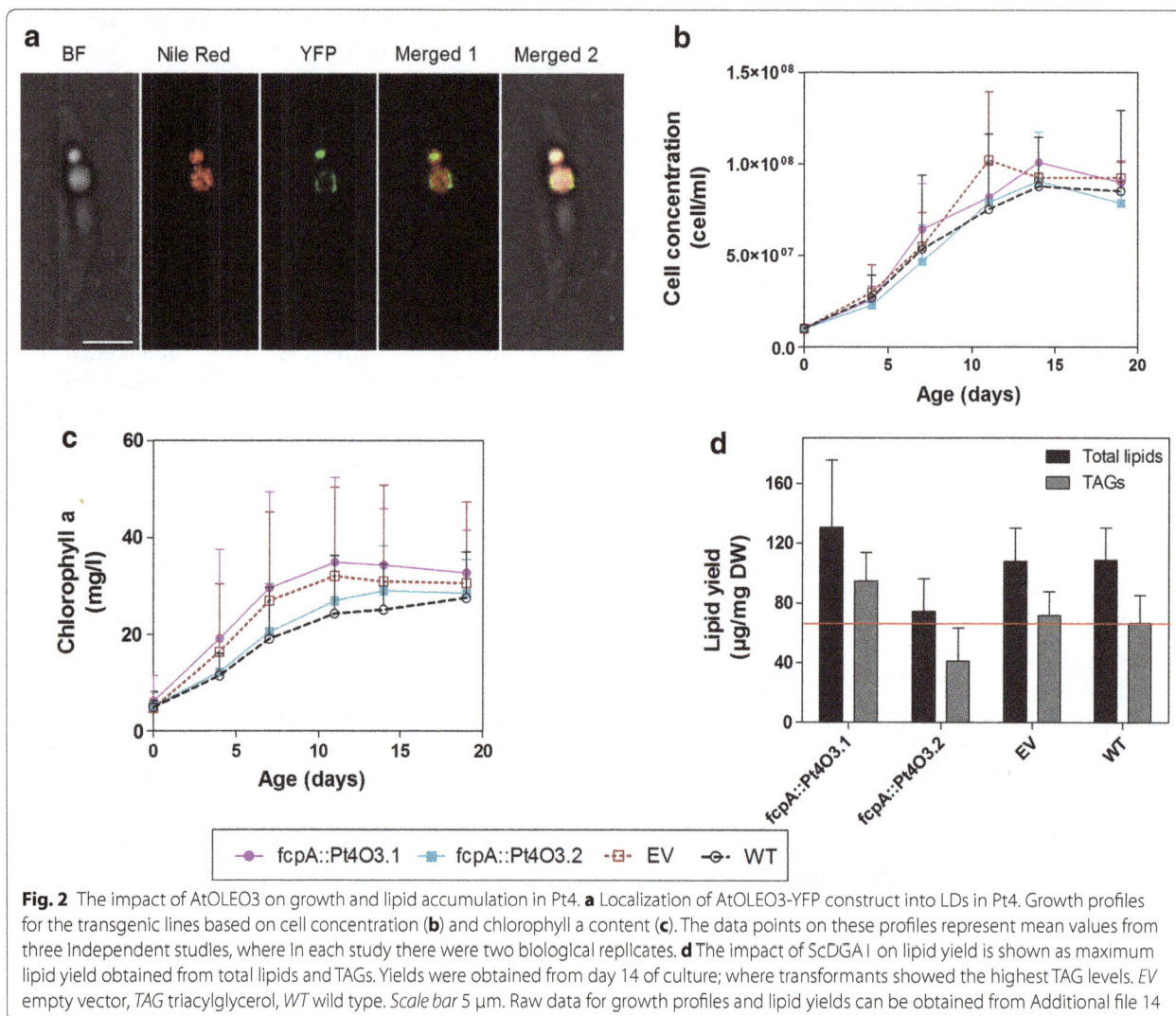

Fig. 2 The impact of AtOLEO3 on growth and lipid accumulation in Pt4. **a** Localization of AtOLEO3-YFP construct into LDs in Pt4. Growth profiles for the transgenic lines based on cell concentration (**b**) and chlorophyll a content (**c**). The data points on these profiles represent mean values from three independent studies, where in each study there were two biological replicates. **d** The impact of ScDGA1 on lipid yield is shown as maximum lipid yield obtained from total lipids and TAGs. Yields were obtained from day 14 of culture; where transformants showed the highest TAG levels. *EV* empty vector, *TAG* triacylglycerol, *WT* wild type. *Scale bar* 5 μm. Raw data for growth profiles and lipid yields can be obtained from Additional file 14

that line Pt4D1O3.2 had accumulated more TAGs than the other lines. In addition to analyzing TAG molecular species, the acyl-CoA pool was also assessed to check for any possible substrate limitations. This analysis was only carried out for the highest TAG accumulating line co-expressing *ScDGA1* and *AtOLEO3* (Pt4D1O3.2) and compared to the empty vector and wild type. Again, we found no significant differences in the acyl-CoA pool (Additional file 8: Figure S8).

Nitrogen starvation induces additional increase of TAG content in Pt4 co-expressing *ScDGA1* and *AtOLEO3*

Nitrogen (N) stress has been widely reported to influence TAG accumulation in microalgae [33]. Therefore, there was an interest in determining if TAG accumulation could be further enhanced in the obtained Pt4 lines, in response to N starvation. When exposed to N stress, lines Pt4D1O3.1 and Pt4D1O3.2 as well as Pt4D1.1

showed a further increment in TAG levels (Fig. 5). Interestingly, a similar pattern of TAG productivities was observed under both, N deplete and N replete conditions with line Pt4D1O3.2 showing the highest TAG productivity (6.9 μg/mg/day) in comparison to lines Pt4D1O3.1 (5.41 μg/mg/day) and Pt4D1.1 (5.06 μg/mg/day) (Fig. 5). Although it is clear that N stress further enhanced TAG productivities in the analyzed lines of Pt4, statistical analysis could not be performed due to the fact that the deviation between the samples exposed to N stress was high, possibly a consequence of increased sedimentation of Pt4 cells under these conditions.

Assessing the impact of the NR promoter in driving gene expression of *ScDGA1* and *AtOLEO3*

The nitrate reductase (NR) promoter is yet another endogenous promoter frequently used to drive gene expression in *P. tricornutum* [34]. Although this

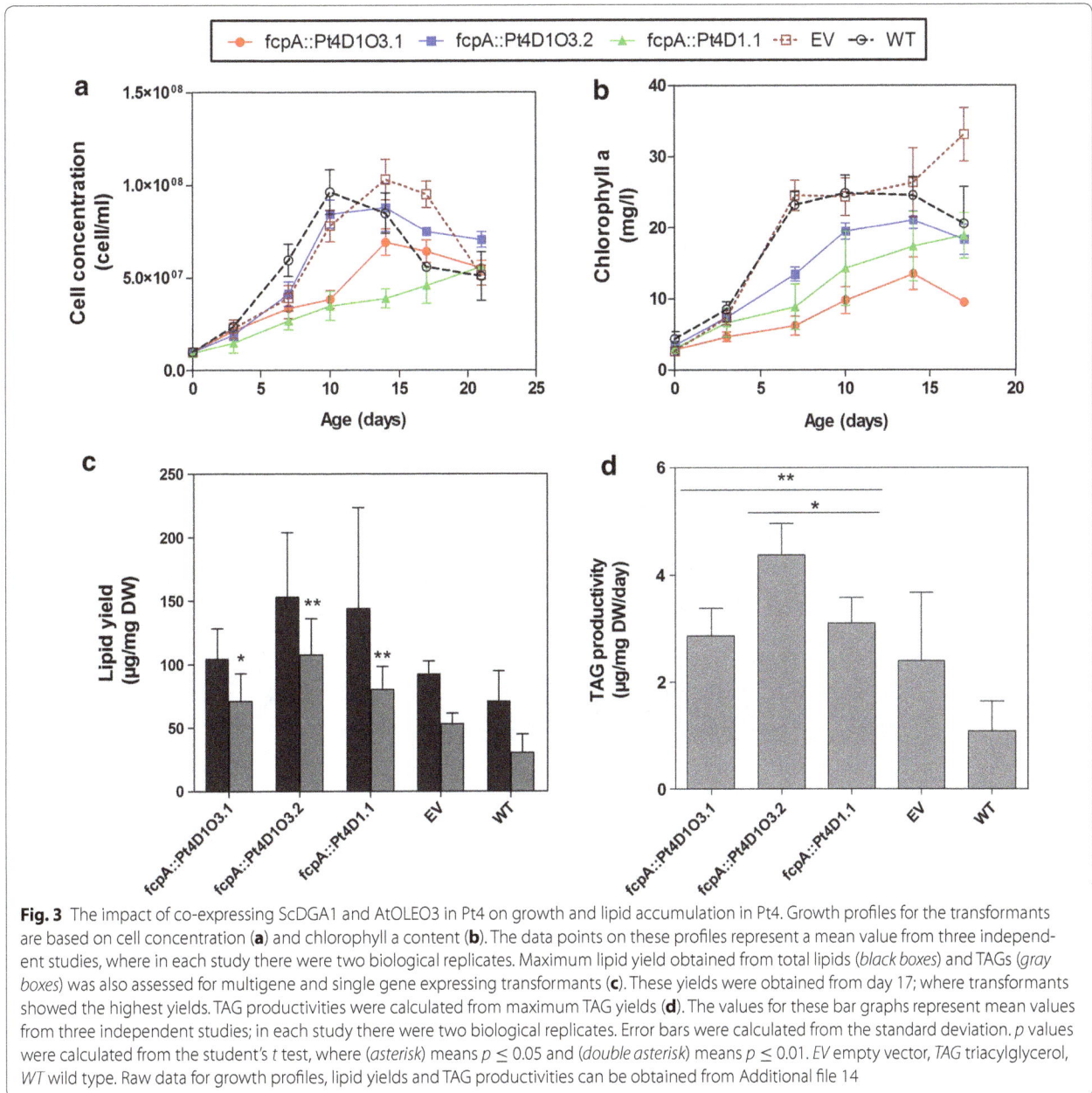

Fig. 3 The impact of co-expressing ScDGA1 and AtOLEO3 in Pt4 on growth and lipid accumulation in Pt4. Growth profiles for the transformants are based on cell concentration (**a**) and chlorophyll a content (**b**). The data points on these profiles represent a mean value from three independent studies, where in each study there were two biological replicates. Maximum lipid yield obtained from total lipids (*black boxes*) and TAGs (*gray boxes*) was also assessed for multigene and single gene expressing transformants (**c**). These yields were obtained from day 17; where transformants showed the highest yields. TAG productivities were calculated from maximum TAG yields (**d**). The values for these bar graphs represent mean values from three independent studies; in each study there were two biological replicates. Error bars were calculated from the standard deviation. p values were calculated from the student's t test, where (*asterisk*) means $p \leq 0.05$ and (*double asterisk*) means $p \leq 0.01$. *EV* empty vector, *TAG* triacylglycerol, *WT* wild type. Raw data for growth profiles, lipid yields and TAG productivities can be obtained from Additional file 14

promoter has recently been reported to be stronger than the fcpA promoter at driving stable gene expression in *P. tricornutum* [34], when we expressed YFP under both promoters (fcpA and NR) in Pt4, there appeared to be no differences in the YFP signal intensity (Additional file 9: Figure S9). Nonetheless, we also generated transformants expressing *ScDGA1* and *AOLEO3* under the NR promoter, in order to determine if there would be differences in TAG accumulation when compared to the lines expressing *ScDGA1* and *AOLEO3* genes under the control of the fcpA promoter. From the randomly selected

lines, three of them tested positive for gene expression (Additional file 10: Figure S10). Lines Pt4D1O3.14 and Pt4D1O3.40 were expressing *ScDGA1* under both promoters (fcpA and NR), such that there were two copies of *ScDGA1* in one construct and *AtOLEO3* was only expressed under the NR promoter. Line Pt4D1O3.35 was co-expressing *ScDGA1* and *AtOLEO3* under the NR promoter. There were no significant differences in growth behavior between the transgenic lines and the controls (Fig. 6a, b). The analysis of lipid yields showed that line Pt4D1O3.35 (*ScDGA1* and *AtOLEO3* under NR

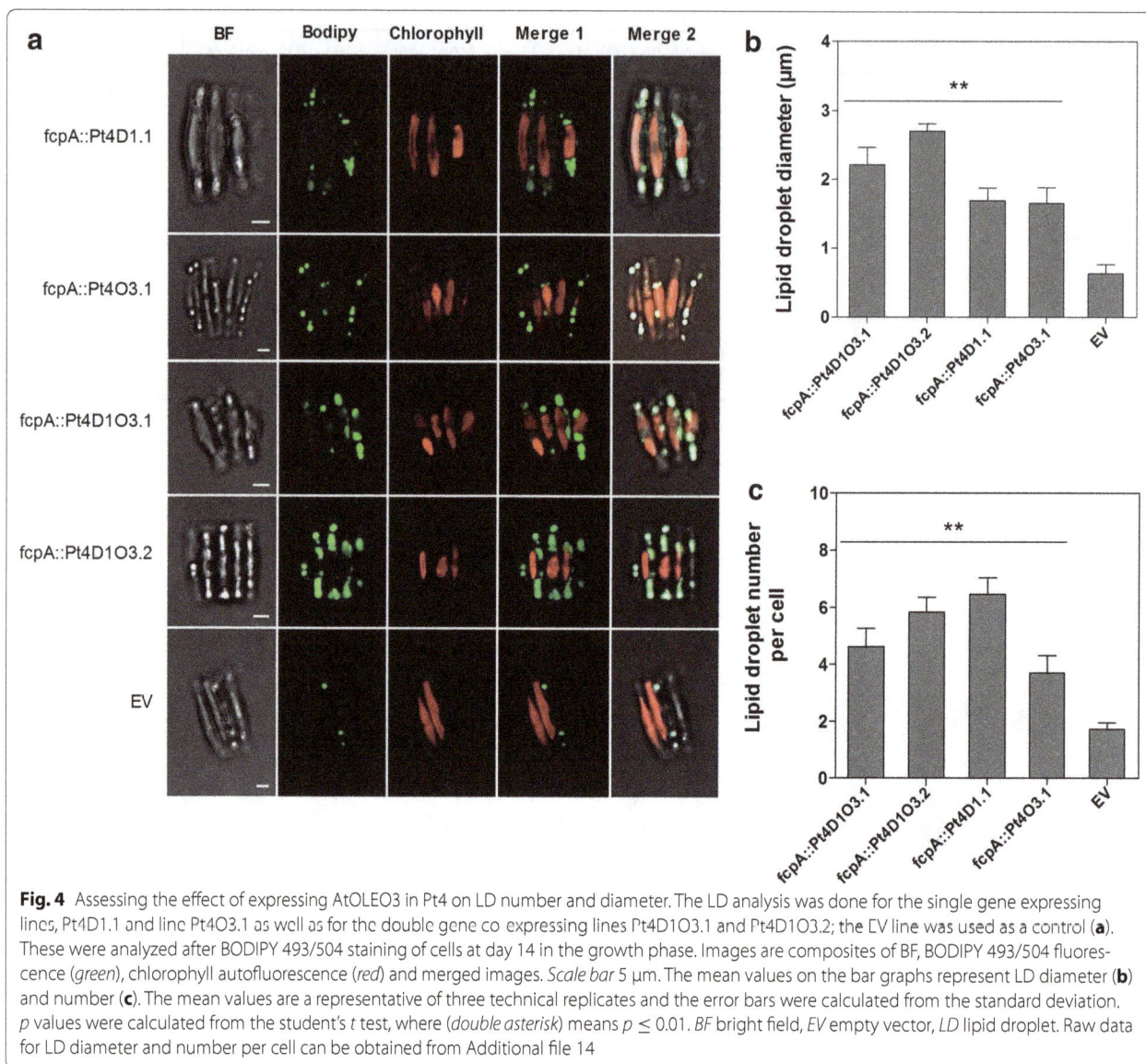

Fig. 4 Assessing the effect of expressing AtOLEO3 in Pt4 on LD number and diameter. The LD analysis was done for the single gene expressing lines, Pt4D1.1 and line Pt4O3.1 as well as for the double gene co-expressing lines Pt4D1O3.1 and Pt4D1O3.2; the EV line was used as a control (**a**). These were analyzed after BODIPY 493/504 staining of cells at day 14 in the growth phase. Images are composites of BF, BODIPY 493/504 fluorescence (*green*), chlorophyll autofluorescence (*red*) and merged images. *Scale bar* 5 μm. The mean values on the bar graphs represent LD diameter (**b**) and number (**c**). The mean values are a representative of three technical replicates and the error bars were calculated from the standard deviation. *p* values were calculated from the student's *t* test, where (*double asterisk*) means $p \leq 0.01$. *BF* bright field, *EV* empty vector, *LD* lipid droplet. Raw data for LD diameter and number per cell can be obtained from Additional file 14

promoter) accumulated more TAGs than the other two lines and the controls (Fig. 6c). Consequently, this effect was also visible when TAG productivities were compared since the line Pt4D1O3.35 had the highest TAG productivity of all analyzed lines (Fig. 6d). Noteworthy, the TAG productivity of line Pt4D1O3.35 obtained under the NR promoter (4.01 μg/mg/day) was only slightly lower than TAG productivity obtained under the fcpA promoter (4.4 μg/mg/day, as previously shown in Fig. 3d).

Discussion

The aim of this study was to improve TAG accumulation in *P. tricornutum*, strain Pt4, which naturally grows in less saline conditions in comparison to the other *P. tricornutum* strains [17, 28]. To achieve this aim, a gene

encoding an enzyme catalyzing the last and only specific metabolic step in TAG formation, DGAT2 from the baker's yeast (*ScDGA1*), together with a gene encoding the LD stabilizing protein oleosin 3 from *A. thaliana* (*AtOLEO3*) were expressed in Pt4, respectively. Oleosins in general have been shown to maximize TAG accumulation in plants, most likely by preventing the exposure of TAGs to TAG lipases [31]. *AtOLEO3* was selected because of its hampering effect on TAG degradation [32]. *ScDGA1* was selected to be expressed in Pt4 because of its broad substrate specificity [13, 14, 29, 35] and when expressed individually in Pt4, *ScDGA1* and *AtOLEO3* resulted in different responses in terms of growth rates and lipid accumulation. Individual expression of *ScDGA1* (line Pt4D1.1) accumulated the highest

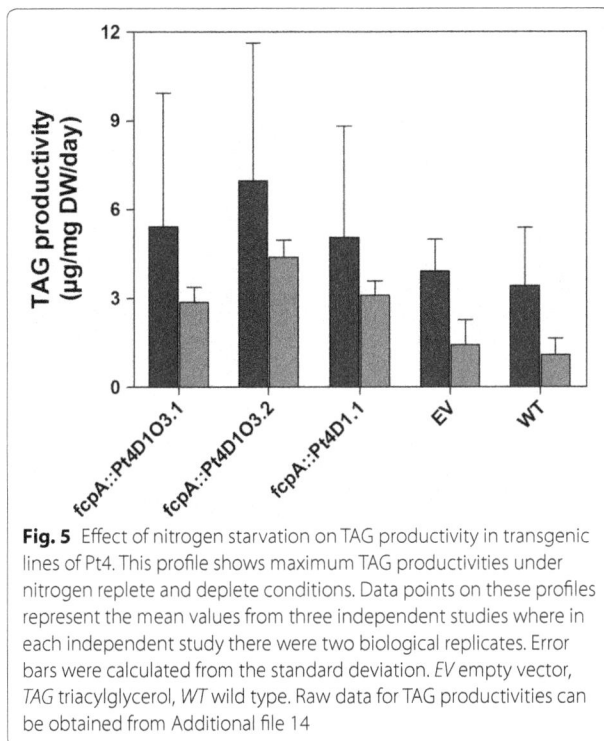

Fig. 5 Effect of nitrogen starvation on TAG productivity in transgenic lines of Pt4. This profile shows maximum TAG productivities under nitrogen replete and deplete conditions. Data points on these profiles represent the mean values from three independent studies where in each independent study there were two biological replicates. Error bars were calculated from the standard deviation. *EV* empty vector, *TAG* triacylglycerol, *WT* wild type. Raw data for TAG productivities can be obtained from Additional file 14

amounts of TAGs in comparison to the other *ScDGA1* lines and controls (Fig. 1c), reaching a 50% increase in TAG levels, when compared to the empty vector control. This promising increase in TAG accumulation was however accompanied by an impaired growth rate (Fig. 1a, b). Since this was the only line out of four independent events that showed this characteristics it is not clear if this was a result of *ScDGA1* expression or just a possible artifact caused by random integration of *ScDGA1* at a critical site in the genome of Pt4. However, it was previously observed that improving TAG synthesis in microalgae results in reduced growth rate [36], as the cells store more carbon instead of using it for growth. This was also the case when glycerol-3-phosphate dehydrogenase (G3PDH) was overexpressed in *P. tricornutum*, where this overexpression improved TAG accumulation but slowed down the growth rate [1]. Although the physiological nature of this conundrum still remains unknown, it seems likely that the changes in cellular carbon flux could potentially trigger yet unknown cellular energy stress-related pathways resulting in an impaired cell cycle and cell division. Contrary to this, the overexpression of the native DGAT2 in *P. tricornutum* was reported to increase neutral lipid levels without affecting biomass production [37]. Thus, it is possible that the observed negative correlation between oil accumulation and growth rates in Pt4 could simply be a result of heterologous gene expression, which affects the metabolism and developmental

program of the host cell. Interestingly, the expression of *AtOLEO3* in Pt4 (Pt4O3.1) resulted in an increased TAG accumulation and no differences in growth patterns were observed in comparison to the wild type and empty vector (Fig. 2b, c). Furthermore, total lipids and TAG levels were enhanced in this line in comparison to the wild type and empty vector (Fig. 2d). One of the main functions of oleosins in higher plants is to minimize TAG degradation. Thus, it is possible that *AtOLEO3* in Pt4 functions in a similar manner leading to effective TAG accumulation. This is in accordance with the reported overexpression of oleosin in rice kernels and *Arabidopsis* leaves, which both resulted in higher amounts of TAGs [31, 38]. Nonetheless, since gene expression and localization of *AtOLEO3* in LDs of Pt4 was confirmed (Additional file 3: Figure S3; Fig. 2a), we concluded that TAG accumulation in line Pt4O3.1 was a result of *AtOLEO3* expression.

For comparison purposes, the characterization of *ScDGA1* and *AtOLEO3* co-expressing lines (lines Pt4D1O3.1 and Pt4D1O3.2) was done in parallel with the *ScDGA1* expressing line (line Pt4D1.1). We showed that although growth was impaired in the co-expressing lines, TAG accumulation increased by 2.4- and 3.6-fold for Pt4D1O3.1 and Pt4D1O3.2, respectively, in comparison to the wild type (Fig. 3). This increase in TAG levels for both of these lines was higher than a previously reported study for *P. tricornutum* overexpressing the endogenous DGAT2 leading to approximately 1.4-fold increase in the neutral lipid content [37]. Furthermore, co-expressing *ScDGA1* and *AtOLEO3* in Pt4 resulted in significantly higher TAG productivities than when *ScDGA1* was individually expressed (Fig. 3c). In addition, the co-expression of these genes in Pt4 has resulted in a TAG content of 107 µg/mg, which is comparable to the TAG levels observed in the wild type *P. tricornutum* strain Pt1 [39]. In terms of total lipid content, the increase was estimated to be 1.5- and 2.2-folds for Pt4D1O3.1 and Pt4D1O3.2, respectively. Similarly, overexpression of endogenous DGAT2 enzyme in *P. tricornutum* was accompanied by a twofold increase of the total lipid content [40]. This clearly suggests that stacking two genes in an additive manner is more effective than expressing a single gene for improving lipid accumulation in *P. tricornutum*. Various other studies have demonstrated the efficiency of such a strategy in other species. For instance, co-expressing DGAT1 and oleosin in *A. thaliana* was more effective in enhancing TAG accumulation in comparison to when DGAT1 was expressed on its own [31]. This also remains in agreement with the role of oleosin in maximizing TAG accumulation also observed in our study. Moreover, in *P. tricornutum*, this double gene strategy has been shown to be effective in modifying the fatty acid composition as the co-expression of a desaturase and an elongase from

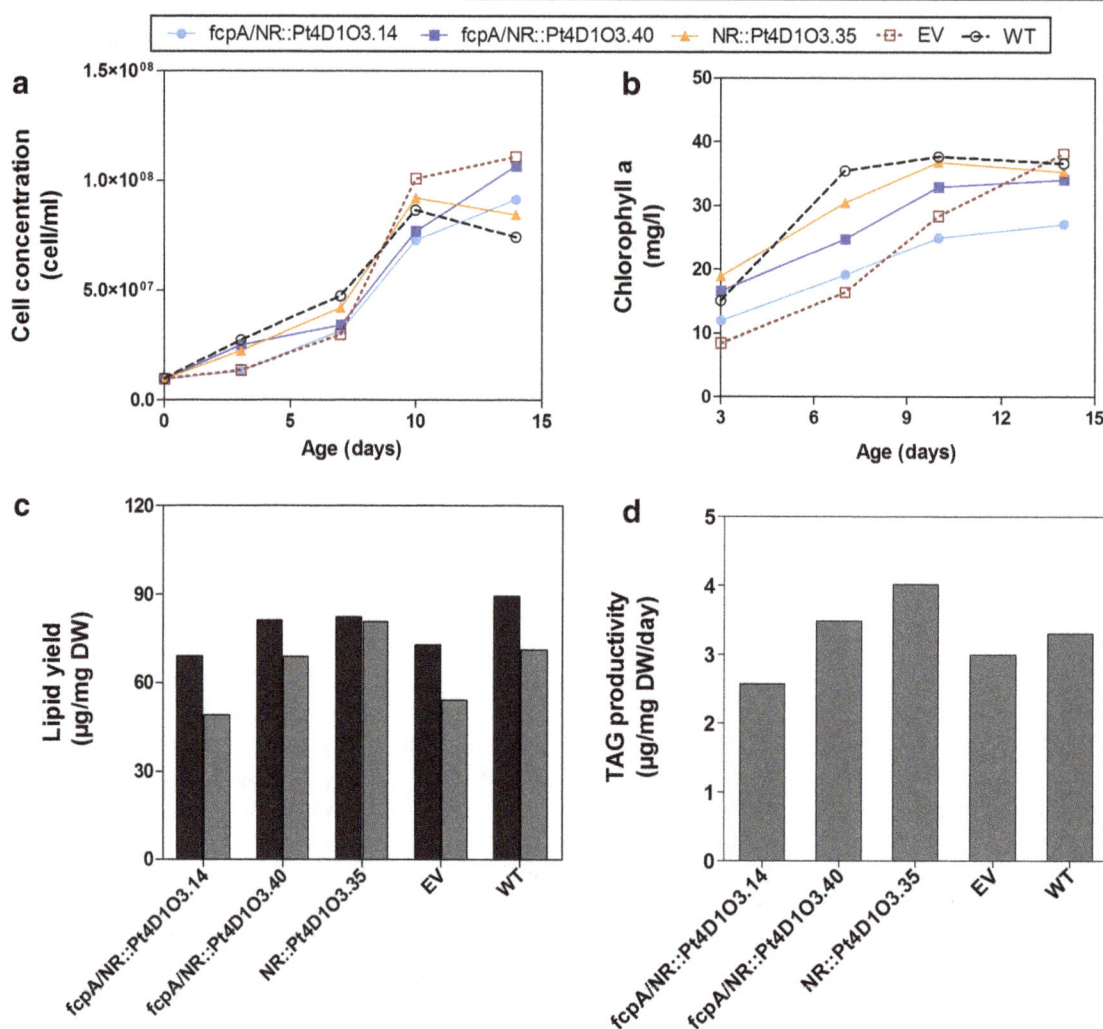

Fig. 6 The impact of the NR promoter on growth and lipid yield in lines co-expressing *ScDGA1* and *AtOLEO3*. Growth profiles are based on cell concentration (**a**) and chlorophyll a content (**b**). Maximum lipid yields (**c**) were obtained from day 17, a time point where the highest TAG yields were accumulated and TAG productivities (**d**) were calculated from maximum TAG yields. The bar graphs represent mean values of two independent studies, where in each study there were two biological replicates. *EV* empty vector, *NR* nitrate reductase, *TAG* triacylglycerol, *WT* wild type. Raw data for growth profiles, lipid yields, and TAG productivities can be obtained from Additional file 14

O. tauri had a significant impact on DHA levels in comparison to expressing the desaturase individually [41].

Assessing the impact of *ScDGA1* and *AtOLEO3* co-expression on LD formation in Pt4 showed that high TAG levels in lines Pt4D1O3.1 and Pt4D1O3.2 were accompanied by a formation of very large LDs. This effect was much less obvious in the other lines expressing either the single genes (Pt4D1.1 and Pt4O3.1) or an empty vector (Fig. 4). The formed LDs were less numerous in the co-expressing lines than in the single gene expressing lines but they filled up the majority of the cell volume (Fig. 4c, d). We propose that this is the result of a not yet optimized action of *ScDGA1* and *AtOLEO3*, where *ScDGA1* acts as an efficient TAG provider and oleosin functions

as a structural block involved in continuous TAG packaging into enlarging LDs and their prevention from being degraded by lipases. The problem may be solved by expressing the oleosin gene under a stronger promoter than the *ScDGA1* gene. However, similar results were obtained when DGAT1 was co-expressed with oleosin in the leaves of *A. thaliana* resulting in formation of larger LDs when compared to the wild type [31]. Overall, our results showed an evident positive correlation between TAG levels, LD number and sizes.

Since N stress has been commonly shown to enhance TAG accumulation in microalgae [33], we decided to expose the transformants, expressing the transgenes under the control of the fcpA promoter, to N limited conditions. Indeed, under N deprivation, all lines showed an increase in TAG productivities with line Pt4D1O3.2 still showing the highest TAG productivity when compared to other lines (Fig. 5). However, this increase in TAG was accompanied by an increased and early death of the cultured cells leading to an increased sediment of Pt4 cells under these conditions. Overall, this resulted in a significant decrease in biomass rendering this approach unprofitable for Pt4 cultures. Therefore, we suggest that a continuous culture that is continuously harvested may be a preferred strategy for efficient lipid production with Pt4 [24].

We also compared the impact of co-expressing *ScDGA1* and *AtOLEO3* in Pt4 not only under the control of fcpA promoter, but in addition by using the NR promoter. Both, the fcpA and the NR promoter seem to show no differences in TAG productivities under the conditions applied in this study (Fig. 6d). This is an interesting observation since they have been suggested to be active during different growth stages. However, this may be explained again by the assumption that a fast and exponential growth phase may be the primary factor that determines lipid productivity in Pt4.

The fatty acid profile in the TAGs of *P. tricornutum* is dominated by the saturated and monounsaturated fatty acids, with low amounts of VLC-PUFAs like EPA and DHA [17, 28, 41, 42]. In our study, the expression of *ScDGA1* did not result in any modifications in the fatty acid composition and the observed increase in TAG content was likely a result of elevated levels of saturated and monounsaturated fatty acids, which are the most valuable precursor fatty acids for biofuel production. Moreover, the expression of ScDGA1 neither changed the distribution of TAG species nor introduced new TAG molecular species to the TAG pool (Additional file 7: Figure S7). This is in agreement with previous reports [17, 41] and supports the notion that VLC-PUFAs are preferentially channeled into membrane lipids instead of being incorporated into TAGs. This strongly suggests that DGAT activity is not the limiting metabolic step leading to primarily saturated and monounsaturated TAG species [43].

Conclusions

To our knowledge, we have demonstrated for the first time in microalgae, and especially in a diatom (*P. tricornutum*) that expressing a TAG forming enzyme, yeast DGAT2, together with a plant LD stabilizing and protecting protein, oleosin, has a significant and additive effect on TAG accumulation. An increased TAG productivity was obtained when *ScDGA1* was co-expressed with *AtOLEO3*, in comparison to cultures that expressed *ScDGA1*

and *AtOLEO3* as single genes. Interestingly, the fatty acid composition and the profile of TAG molecular species were not altered through the expression of *ScDGA1*, suggesting that although DGAT activity is increasing TAG productivity it is not affecting its composition in this organism. In addition, we have provided first evidence that the expression of a LD protein from plant seeds may protect algal storage lipids from degradation and thereby improve TAG accumulation in microalgae as shown here for *P. tricornutum*. Although N deprivation of the cultures lead to a further increase in TAG productivity, it was accompanied by an increased sedimentation. Thus, rendering N limiting growth conditions unsuitable for optimal lipid production in Pt4. Overall, the approach used in our study was effective at enhancing TAGs species suitable for use in biofuel production. This approach contributes to a promising strategy for enhancing the TAG production in other hosts, including the commercial non-photosynthetic systems such as yeasts since they have much higher productivities than diatoms or microalgae.

Methods

Culture conditions for *Phaeodactylum tricornutum*

Phaeodactylum tricornutum strain Pt4 was purchased from the SAG culture collection of algae (Göttingen, Germany), with a designation number of SAG 1090-6. Pt4 was cultivated in batch cultures in Erlenmeyer flasks containing f/2 liquid medium [44] was used and its composition was modified by lowering the salt (NaCl) content from 1 to 0.7% (w/v). The cultures were incubated at 20 °C at a speed of 100 rpm, under a 16 h/8 h for day/night cycle, respectively. During the light phase, the cultures were illuminated at a light intensity of 200 µmol m^{-2} s^{-1}. Cultures in Erlenmeyer flasks were used as inocula for the inoculation of main cultures in 500 ml glass columns (Ochs Glasgerätebau, Bovenden, Germany). The starting cell concentration for all the experiments was always maintained at 1×10^7 cell/ml. In columns, air supplemented with 1% CO_2 was used to provide constant aeration to the cultures at a flow rate of 0.15 l/h. Under N deplete conditions, KNO_3 was omitted from f/2 media. The cells were harvested and washed in N deplete media prior to commencing with the N stress studies in order to ensure complete removal of N. Cell proliferation was assessed through cell concentration, which was measured under a microscope (BX51 Olympus microscope; software, Micro-Manager 1.4.22, Tokyo, Japan) by a Thoma cell counting chamber (Marienfeld, Germany). Chlorophyll a content was measured in order to assess cell proliferation according to a protocol described in [45].

Cloning of constructs into Pt4

The codon optimized *ScDGA1* and *AtOLEO3* full coding sequences (see Additional file 11) were cloned into Pt4 by the gateway cloning system (Thermo Fischer Scientific, Waltham, USA). Briefly, the *ScDGA1* and *AtOLEO3* genes first cloned into pEntry vectors using the *Spe*I and *Pac*I restriction sites with primers provided in Additional file 12: Table S1. Thereafter, the pEntry vectors were inserted into the pPha-ccdB destination vector, through the LR clonase reaction (Thermo Fischer Scientific, Waltham, USA). The pPha-ccdB destination vector had the pPhaT1 plasmid as a backbone and it consisted of *Sh ble* gene for selection in Zeocin containing f/2 media. Gene expression in Pt4 was driven by the fcpA and NR promoters. The pPha-NR promoter was obtained from Claudia Büchel, (Goethe Universität, Frankfurt am Main, Germany). Pt4 was transformed by particle bombardment according to a protocol described in [46], using a Bio-Rad Biolistic PDS-1000/He particle delivery system fitted with 1350 psi rupture disks. Following bombardment, cells were left to recover by incubation at 20 °C for 24 h in f/2 agar plates without a selection marker. Thereafter, they were transferred to f/2 agar plates containing zeocin (75 µg/ml). The surviving lines were transferred into fresh selection liquid f/2 media contained in microtitre plates for further characterization.

Analysis of transgenic Pt4

The selection criterion for transformants during the first screening of transgenic Pt4 was the expression of the transgene. This involved a random selection of transformants (each transformation round resulted in ~100 transformants of which 50 were tested for transgene integration, thereafter 10 transformants positive for transgene integration were tested for transgene expression. Transformants positive for transgene integration were then further checked for transgene expression, of which positive transformants were further processed for lipid yields. Genomic DNA was extracted using a Cetyltrimethyl Ammonium Bromide (CTAB) buffer (2% CTAB, 10 mM Tris–HCl pH 8, 20 mM EDTA pH 8, 1.4 M NaCl). The extracted genomic DNA then served as templates for amplification of *ScDGA*1 and *AtOLEO3*, using primers provided in Additional file 13: Table S2. Pt4 lines positive for gene integration were further analyzed for gene expression, which was done by semi-quantitative PCR. Briefly, RNA was extracted from fresh biomass using TRIZOL buffer according to a protocol described in [47]. The extracted RNA was treated with DNase I enzyme (Thermo Fischer Scientific, Waltham, USA) prior to cDNA synthesis, which was done by the RevertAid H-minus reverse transcriptase (Thermo Fischer Scientific, Waltham, USA), according to the

manufacturer's instructions. Standard PCR were then carried out to amplify the *ScDGA1* and *AtOLEO3* genes from the cDNA generated, using the primers provided in Additional file 13: Table S2. Lines that were positive for gene expression were bulked up for the analysis of growth and lipid accumulation.

Extraction and analysis of lipids from Pt4

Total lipids were extracted from lyophilized biomass with modifications according to a protocol described in [48]. Briefly, total lipids were extracted from 10 mg of lyophilized biomass using MTBE as the organic solvent. Prior to lipid extraction, internal standards were added; these were tri-15:0 TAG (Sigma-Aldrich, Munich, Germany) and Di-17:0 PC (Avanti Polar Lipids, Alabama, USA). TAGs were extracted from the total lipids by thin layer chromatography (TLC) using hexane:diethyl ether:acetic (80:20:1, v/v/v) acid as a mobile solvent. TAG bands were visualized by first spraying the TLC places with 0.2% (w/v) 8-anilinonaphthalene-1-sulfonic acid followed by exposure to UV light. TAG bands were scraped out from the TLC plates for further analysis. Total lipid and TAG samples were transesterified to generate fatty acid methyl esters (FAMEs) according to a protocol described in [49], however in the presence of an acid (H_2SO_4) instead of a base. FAMEs generated were subsequently analyzed by gas chromatography (Agilent GC 6890, Agilent Technologies, Waldbroon, Germany) coupled to a flame ionization detector (GC-FID). Samples were injected at temperature of 220 °C and a volume of 1 µl using a split mode injection. The separation of fatty acids was carried out in a DB-23 column (30 m × 0.25 mm with 0.25 µm coating thickness; Agilent Technologies, Waldbroon, Germany). Helium was used as a carrier gas and the temperature gradient used was 150 °C for 1 min, 150–200 °C at 4 °C/min, 200–250 °C at 20 °C/min, and 250 °C for 3 min. A FAME mix (C4–C24, Sigma-Aldrich, Munich, Germany) was used an external standard for fatty acid identification. The signals were integrated using the ChemStation Software (Agilent Technologies, Santa Clara, USA).

Analysis of TAG molecular species by UPLC-nano ESI-MS/MS

Total lipid extraction for this analysis was carried out in a similar manner as described above. The final lipid extracts were re-suspended in 0.8 ml tetrahydrofuran:methanol:water (4:4:1, v/v/v) prior to analysis. The separation of TAG species from the total extract was carried with a UPLC-ESI-MS/MS system (Waters Corp., Milford, USA) according to a protocol described in [50]. The system was equipped with the ACQUITY UPLC HSS T3 column (100 mm × 1 mm, 1.8 µm; Waters Corp., USA). Samples were injected at a volume of 2 µl, using the needle overfill mode. The flow rate was set at 0.10 ml/min with a separation temperature of 35 °C.

Extraction and analysis of acyl-CoAs

Acyl-CoAs were extracted from the highest TAG accumulating line (Pt4D1O3.2) grown under N replete conditions as well as from the empty vector and wild type according to a protocol described in [51], with some modifications. Briefly, 20 mg of lyophilized biomass was homogenized in 200 µl of freshly prepared extraction buffer [2-propanol, 2 ml; 50 mM KH_2PO_4 pH 7.2, 2 ml; acetic acid, 50 µl and 50 mg/ml BSA (fatty acid free), 80 µl]. The homogenate was mixed with 300 µl of saturated petroleum ether (saturated with 2-propanol: water (1:1, v/v). To force phase separation, samples were spun at 500 g or 2 min and the interphase was collected and washed with the saturated petroleum ether three times. Following the washing steps, the interphase fractions were mixed with 5 µl of saturated $(NH_4)_2SO_4$ and 600 µl of methanol:chloroform (2:1, v/v). Samples were then mixed thoroughly prior to incubation at room temperature for 20 min. Thereafter, they were centrifuged at 21,000g for 2 min. The supernatant was collected and dried under a stream of nitrogen and re-suspended in 300 µl of derivatization buffer [0.5 M chloracetaldehyde; 0.15 M citrate buffer, pH4, and 0.5% (w/v) SDS], and derivatized at 85 °C for 20 min. Thereafter, samples were cooled down and analyzed by high-performance liquid chromatography (HPLC), as described [51].

Assessment of the morphology and number of LDs

Cells were fixed in a mixture of 4% (w/v) paraformaldehyde in PBS (pH 7.4) overnight at 4 °C. After three washes in PBS buffer, cells were incubated with Bodipy 493/504 (Thermo Scientific, Grand Island, USA) at a final concentration of 10 µg/ml in PBS buffer for 1 h at room temperature. This was followed by the washing of cells three times in the same buffer and re-suspended in ProLong Gold anti-fade reagent (Thermo Fischer Scientific, Waltham, USA). Samples were then analyzed with a Zeiss LSM 510 META confocal laser scanning microscope (Carl Zeiss, Jena, Germany) using a 63× Plan-Apochromat 1.4 NA oil-immersion lens. An argon laser was used for Bodipy 493/504 and chlorophyll excitation and the emission spectra were collected in two separate channels, 500–515 nm and 630–670 nm, respectively. Z-series images were collected and processed with the LSM 5 Image Browser (Carl Zeiss, Jena, Germany). LD morphometrics were performed by the same software using 3D reconstruction confocal images of 70 cells for each analyzed line.

Additional files

Additional file 1: Figure S1. Semi-quantitative PCR for *ScDGA1* expressing lines. Screening for lines expressing the *ScDGA1* gene was done by semi-quantitative PCR where the cDNA was used as a template. cDNA extracted from the wild type was used as a negative control. Lines highlighted in red were selected for further analysis. M = marker, nc = negative control.

Additional file 2: Figure S2. The impact of *ScDGA1* on the FA composition in the total lipids (a) and TAGs (b) of Pt4. The FA composition is presented in relative amounts (% contribution of each FA). The mean values on the bars represent values from two independent studies, where in each study there were two biological replicates. This data is representative of samples taken on day 7 of culture, where maximum yields of lipids were accumulated. EV = empty vector, FA = fatty acid, TAG = triacylglycerol, WT = wild type. Raw data for fatty acid composition can be obtained from Additional file 14.

Additional file 3: Figure S3. Semi-quantitative PCR for *AtOLEO3* expressing lines. Screening for lines expressing the *AtOLEO3* gene was by semi-quantitative PCR where the cDNA was used as a template. cDNA extracted from the wild type was used as a negative control. Lines highlighted in red were selected for further analysis. M = marker, nc = negative control.

Additional file 4: Figure S4. The effect of *AtOLEO3* expression on the FA composition in the total lipids (a) and TAGs (b) of Pt4. The FA composition is presented in relative amounts (% contribution of each FA). Data points on the bars represent mean values from three independent studies, where in each study there were two biological replicates. Error bars were calculated from the standard deviation. This data is representative of samples taken on day 14 of culture, where maximum yields of lipids were accumulated. EV = empty vector, FA = fatty acid, TAG = triacylglycerol, WT = wild type. Raw data for fatty acid composition can be obtained from Additional file 14.

Additional file 5: Figure S5. Screening for lines co-expressing the *ScDGA1* and *AtOLEO3* genes. This screening was carried out by semi-quantitative PCR where the cDNA was used as a template. cDNA extracted from the wild type was used as a negative control. Lines highlighted in red were selected for further analysis. M = marker, nc = negative control.

Additional file 6: Figure S6. The impact of *ScDGA1* and *AtOLEO3* co-expression on the FA composition in the total lipids (a) and TAGs (b) of Pt4. The FA composition is presented in relative amounts (% contribution of each FA). Data points on the bars represent mean value from three independent studies, where in each study there were two biological replicates. Error bars were calculated from the standard deviation. This data is representative of samples taken on day 17. EV = empty vector, FA = fatty acid, TAG = triacylglycerol, WT = wild type. Raw data for fatty acid composition can be obtained from Additional file 14.

Additional file 7: Figure S7. The effect of ScDGA1 on the distribution of TAG molecular species. The distribution of TAG molecular species is presented in relative amounts (b) and the species were summed up to indicate the difference in TAG accumulation between transformants and controls (a). These bar graphs represent mean values of two biological replicates from a single experiment. The samples were taken from day 21 of culture, the last day in a growth curve. EV = empty vector, TAG = triacylglycerol, WT = wild type. Raw data for TAG molecular species can be obtained from Additional file 14.

Additional file 8: Figure S8. Acyl-CoA pool composition in the multigene expressing lines and controls. The acyl-CoAs were extracted from the highest TAG accumulating line (Pt4D1O3.2), empty vector control (EV) and from the wild type strain of Pt4 (WT) on the last day (21st) of the growth curves. Error bars were calculated from the standard deviation, with three technical replicates from a single experiments. EV = empty vector, WT = wild type. Raw data for acyl-CoA pool composition can be obtained from Additional file 14.

Additional file 9: Figure S9. Comparison of the fcpA and NR promoters in driving YFP expression in Pt4. In both lines, the YFP signal is present in the cytosol. The EV line was used as a negative control and it showed no

YFP fluorescence. Images are composites of BF, YFP fluorescence (green), chlorophyll autofluorescence (red) and merged images. Scale bar = 5 μm. BF = bright field, EV = empty vector, NR = nitrate reductase, YFP = yellow fluorescent protein.

Additional file 10: Figure S10. Semi-quantitative PCR for *ScDGA1* and *AtOLEO3* co-expressing lines under the fcpA promoter and NR promoter. Screening for lines co-expressing the *ScDGA1* and *AtOLEO3* genes was by semi-quantitative PCR where the cDNA was used as a template. cDNA extracted from the wild type was used as a negative control. M = marker, nc = negative control, NR = nitrate reductase.

Additional file 11. Nucleotide sequences for *ScDGA1* and *AtOLEO3* used in this study.

Additional file 12: Table S1. Restriction site primers used for the cloning of *ScDGA1* and AtOLEO3.

Additional file 13: Table S2. Primers used for assessing gene integration and expression of *ScDGA1* and *AtOLEO3* genes in Pt4.

Additional file 14. Raw data used to generate the figures. Data used to generate Fig. 1 and Additional file 2: Figure S2 is presented under tab, ScDGA1. Raw data for Fig. 2 and Additional file 4: Figure S4 is presented under tab, AtOLEO3. Raw data for Fig. 3 and Additional file 6: Figure S6 is presented under tab, ScDGA1 + AtOLEO3. Raw data for Fig. 4 is presented under tab LD morphology. Raw data for Fig. 5 is presented under tab, ScDGA1 + AtOLEO3 N deplete. Raw data for Fig. 6 is presented under tab ScDGA1 + AtOLEO3 NR promoter. Raw data for Additional file 7: Figure S7 is presented under tab, TAG molecular species. Raw data for Additional file 8: Figure S8 is presented under tab Acyl-CoA pool.

Abbreviations
AtOLEO3: oleosin 3 from *Arabidopsis thaliana*; CLSM: confocal laser scanning microscopy; DGAT: acyl-CoA:diacylglycerol acyltransferase; FA: fatty acid; FAME: fatty acid methyl ester; GC-FID: gas chromatography-flame ionization detector; HPLC: high-performance liquid chromatography; LD: lipid droplet; ScDGA1: acyl-CoA:diacylglycerol acyltransferase 1 from *Saccharomyces cerevisiae*; TAG: triacylglycerol; UPLC: ultra-performance liquid chromatography; VLC-PUFA: very long-chain polyunsaturated fatty acid.

Authors' contributions
NZ performed the experiments, analyzed data, and wrote the first draft of the manuscript. JP, KZ, PT, and CH performed experiments and analyzed data. IF supervised the study, designed experiments, analyzed data, coordinated the write up, and editing of this article. All authors read and approved the final manuscript.

Author details
[1] Department of Plant Biochemistry, Albrecht-von-Haller-Institute for Plant Sciences, University of Goettingen, 37077 Goettingen, Germany. [2] Novagreen Projektmanagement GmbH, 49377 Vechta, Germany. [3] Department of Plant Biochemistry, Goettingen Center for Molecular Biosciences (GZMB), University of Goettingen, 37077 Goettingen, Germany. [4] Department of Plant Biochemistry, International Center for Advanced Studies of Energy Conversion (ICASEC), University of Goettingen, 37077 Goettingen, Germany.

Acknowledgements
The authors would like to thank Sabine Freitag for technical assistance with the GC analysis. The authors also thank Prof. Claudia Büchel (Goethe Universität, Frankfurt, Germany) for providing the NR promoter. This work was supported by the People Programme (Marie Curie Actions) of the European Union's Seventh Framework Programme FP7/2007-2013/under REA Grant Agreement Numbers 317184 and 627266.

Competing interests
The authors declare that they have no competing interests.

Funding
The research leading to these results has received funding from the People Programme (Marie Curie Actions) of the European Union's Seventh Framework Programme FP7 under REA Grant Agreement Nos. 317184 (PHOTO.COMM) and 627266 (AlgaeOilSynth) for NZ and KZ, respectively. JP received funding for initiating the work through the European Union's Seventh Framework Programme FP7 under Grant Agreement KBBE-4-GIAVAP (GIAVAP). This material reflects only the authors' views and the European Union is not liable for any use that may be made of the information therein. IF acknowledges funding from the German Research Council (DFG, INST 186/822-1).

References
1. Yao Y, Lu Y, Peng K-T, Huang T, Niu Y-F, Xie W-H, Yang W-D, Liu J-S, Li H-Y. Glycerol and neutral lipid production in the oleaginous marine diatom *Phaeodactylum tricornutum* promoted by overexpression of glycerol-3-phosphate dehydrogenase. Biotechnol Biofuels. 2014;7(1):110.
2. Ruiz J, Olivieri G, de Vree J, Bosma R, Willems P, Reith JH, Eppink MHM, Kleinegris DMM, Wijffels RH, Barbosa MJ. Towards industrial products from microalgae. Energy Environ Sci. 2016;9(10):3036–43.
3. Hamilton ML, Warwick J, Terry A, Allen MJ, Napier JA, Sayanova O. Towards the industrial production of omega-3 long chain polyunsaturated fatty acids from a genetically modified diatom *Phaeodactylum tricornutum*. PLoS ONE. 2015;10(12):e0144054.
4. Khozin-Goldberg I, Leu S, Boussiba S. Microalgae as a source for VLC-PUFA production. In: Nakamura Y, Li-Beisson Y, editors. Lipids in plant and algae development. Cham: Springer International Publishing; 2016. p. 471–510.
5. Li-Beisson Y, Shorrosh B, Beisson F, Andersson MX, Arondel V, Bates PD, Baud S, Bird D, DeBono A, Durrett TP et al. Acyl-lipid metabolism. In: The arabidopsis book. The American Society of Plant Biologists; 2013. p. e0161.
6. Hannon M, Gimpel J, Tran M, Rasala B, Mayfield S. Biofuels from algae: challenges and potential. Biofuels. 2010;1(5):763–84.
7. Medipally SR, Yusoff FM, Banerjee S, Shariff M. Microalgae as sustainable renewable energy feedstock for biofuel production. BioMed Res Int. 2015;2015:13.
8. Du Z-Y, Benning C. Triacylglycerol accumulation in photosynthetic cells in plants and algae. In: Nakamura Y, Li-Beisson Y, editors. Lipids in Plant and Algae Development. Cham: Springer International Publishing; 2016. p. 179–205.
9. Dong T, Knoshaug EP, Pienkos PT, Laurens LML. Lipid recovery from wet oleaginous microbial biomass for biofuel production: a critical review. Appl Energy. 2016;177:879–95.
10. Cao H. Structure-function analysis of diacylglycerol acyltransferase sequences from 70 organisms. BMC Res Notes. 2011;4(1):249.
11. Chapman KD, Ohlrogge JB. Compartmentation of triacylglycerol accumulation in plants. J Biol Chem. 2012;287(4):2288–94.
12. Cagliari A, Margis R, dos Santos Maraschin F, Turchetto-Zolet AC, Loss G, Margis-Pinheiro M. Biosynthesis of triacylglycerols (TAGs) in plants and algae. Int J Plant Biol. 2011;2(1):10.
13. Sorger D, Daum G. Triacylglycerol biosynthesis in yeast. Appl Microbiol Biotechnol. 2003;61(4):289–99.
14. Wagner M, Hoppe K, Czabany T, Heilmann M, Daum G, Feussner I, Fulda M. Identification and characterization of an acyl-CoA:diacylglycerol acyltransferase 2 (DGAT2) gene from the microalga *O. tauri*. Plant Physiol Biochem. 2010;48(6):407–16.
15. Zienkiewicz K, Du Z-Y, Ma W, Vollheyde K, Benning C. -induced neutral lipid biosynthesis in microalgae—molecular, cellular and physiological insights. Biochim Biophys Acta. 2016;1861(9, Part B):1269–81.
16. Zienkiewicz K, Zienkiewicz A, Poliner E, Du Z-Y, Vollheyde K, Herrfurth C, Marmon S, Farré EM, Feussner I, Benning C. Nannochloropsis, a rich source of diacylglycerol acyltransferases for engineering of triacylglycerol content in different hosts. Biotechnol Biofuels. 2017;10(1):8.
17. Popko J, Herrfurth C, Feussner K, Ischebeck T, Iven T, Haslam R, Hamilton M, Sayanova O, Napier J, Khozin-Goldberg I, et al. Metabolome analysis reveals betaine lipids as major source for triglyceride formation, and the accumulation of sedoheptulose during nitrogen-starvation of *Phaeodactylum tricornutum*. PLoS ONE. 2016;11(10):e0164673.

18. Chapman KD, Dyer JM, Mullen RT. Biogenesis and functions of lipid droplets in plants. J Lipid Res. 2012;53(2):215–26.

19. Huang M-D, Huang AHC. Bioinformatics reveal five lineages of oleosins and the mechanism of lineage evolution related to structure/function from green algae to seed plants. Plant Physiol. 2015;169(1):453–70.

20. Yoneda K, Yoshida M, Suzuki I, Watanabe MM. Identification of a major lipid droplet protein in a marine diatom *Phaeodactylum tricornutum*. Plant Cell Physiol. 2016;57(2):397–406.

21. Huang N-L, Huang M-D, Chen T-LL, Huang AHC. Oleosin of subcellular lipid droplets evolved in green algae. Plant Physiol. 2013;161(4):1862–74.

22. Moellering ER, Benning C. RNA interference silencing of a major lipid droplet protein affects lipid droplet size in *Chlamydomonas reinhardtii*. Eukaryot Cell. 2010;9(1):97–106.

23. Vieler A, Brubaker SB, Vick B, Benning C. A lipid droplet protein of Nannochloropsis with functions partially analogous to plant oleosins. Plant Physiol. 2012;158(4):1562–9.

24. Vinayak V, Manoylov KM, Gateau H, Blanckaert V, Hérault J, Pencréac'h G, Marchand J, Gordon R, Schoefs B. Diatom milking: a review and new approaches. Mar Drugs. 2015;13(5):2629–65.

25. Merz CR, Main KL. Microalgae (diatom) production-the aquaculture and biofuel nexus. In: 2014 Oceans-St John's. St John's: IEEE; 2014. p. 1–10.

26. Fu W, Wichuk K, Brynjólfsson S. Developing diatoms for value-added products: challenges and opportunities. New Biotechnol. 2015;32(6):547–51.

27. Vílchez C, Forján E, Cuaresma M, Bédmar F, Garbayo I, Vega JM. Marine carotenoids: biological functions and commercial applications. Mar Drugs. 2011;9(3):319.

28. Abida H, Dolch L-J, Mei C, Villanova V, Conte M, Block MA, Finazzi G, Bastien O, Tirichine L, Bowler C, et al. Membrane glycerolipid remodeling triggered by nitrogen and phosphorus starvation in *Phaeodactylum tricornutum*. Plant Physiol. 2015;167:118–36.

29. Guihéneuf F, Leu S, Zarka A, Khozin-Goldberg I, Khalilov I, Boussiba S. Cloning and molecular characterization of a novel acyl-CoA:diacylglycerol acyltransferase 1-like gene (PtDGAT1) from the diatom *Phaeodactylum tricornutum*. FEBS J. 2011;278(19):3651–66.

30. Zaslavskaia LA, Lippmeier JC, Kroth PG, Grossman AR, Apt KE. Transformation of the diatom *Phaeodactylum tricornutum* (Bacillariophyceae) with a variety of selectable marker and reporter genes. J Phycol. 2000;36(2):379–86.

31. Winichayakul S, Scott RW, Roldan M, Hatier J-HB, Livingston S, Cookson R, Curran AC, Roberts NJ. In vivo packaging of triacylglycerols enhances Arabidopsis leaf biomass and energy density. Plant Physiol. 2013;162(2):626–39.

32. Vanhercke T, El Tahchy A, Liu Q, Zhou X-R, Shrestha P, Divi UK, Ral J-P, Mansour MP, Nichols PD, James CN, et al. Metabolic engineering of biomass for high energy density: oilseed-like triacylglycerol yields from plant leaves. Plant Biotechnol J. 2014;12(2):231–9.

33. Li-Beisson Y, Nakamura Y, Harwood J. Lipids: from chemical structures, biosynthesis, and analyses to industrial applications. In: Nakamura Y, Li-Beisson Y, editors. Lipids in plant and algae development. Cham: Springer International Publishing; 2016. p. 1–18.

34. Chu L, Ewe D, Río Bártulos C, Kroth PG, Gruber A. Rapid induction of GFP expression by the nitrate reductase promoter in the diatom *Phaeodactylum tricornutum*. PeerJ. 2016;4:e2344.

35. Kamisaka Y, Kimura K, Uemura H, Yamaoka M. Overexpression of the active diacylglycerol acyltransferase variant transforms *Saccharomyces cerevisiae* into an oleaginous yeast. Appl Microbiol Biotechnol. 2013;97(16):7345–55.

36. Tan KWM, Lee YK. The dilemma for lipid productivity in green microalgae: importance of substrate provision in improving oil yield without sacrificing growth. Biotechnol Biofuels. 2016;9(1):255.

37. Niu Y-F, Zhang M-H, Li D-W, Yang W-D, Liu J-S, Bai W-B, Li H-Y. Improvement of neutral lipid and polyunsaturated fatty acid biosynthesis by overexpressing a type 2 diacylglycerol acyltransferase in marine diatom *Phaeodactylum tricornutum*. Mar Drugs. 2013;11(11):4558–69.

38. Wu Y-Y, Chou Y-R, Wang C-S, Tseng T-H, Chen L-J, Tzen JTC. Different effects on triacylglycerol packaging to oil bodies in transgenic rice seeds by specifically eliminating one of their two oleosin isoforms. Plant Physiol Biochem. 2010;48(2–3):81–9.

39. Lu Y, Wang X, Balamurugan S, Yang W-D, Liu J-S, Dong H-P, Li H-Y. Identification of a putative seipin ortholog involved in lipid accumulation in marine microalga *Phaeodactylum tricornutum*. J Appl Phycol. 2017. doi:10.1007/s10811-10017-11173-10818.

40. Dinamarca J, Levitan O, Kumaraswamy GK, Lun DS, Falkowski PG. Overexpression of a diacylglycerol acyltransferase gene in *Phaeodactylum tricornutum* directs carbon towards lipid biosynthesis. J Phycol. 2017;53(2):405–14.

41. Hamilton ML, Haslam RP, Napier JA, Sayanova O. Metabolic engineering of *Phaeodactylum tricornutum* for the enhanced accumulation of omega-3 long chain polyunsaturated fatty acids. Metabol Eng. 2014;22:3–9.

42. Mühlroth A, Li K, Røkke G, Winge P, Olsen Y, Hohmann-Marriott M, Vadstein O, Bones A. Pathways of lipid metabolism in marine algae, co-expression network, bottlenecks and candidate genes for enhanced production of EPA and DHA in species of Chromista. Mar Drugs. 2013;11(11):4662.

43. Bigogno C, Khozin-Goldberg I, Adlerstein D, Cohen Z. Biosynthesis of arachidonic acid in the oleaginous microalga *Parietochloris incisa* (*Chlorophyceae*): radiolabeling studies. Lipids. 2002;37(2):209–16.

44. Guillard RR, Ryther JH. Studies of marine planktonic diatoms. I. *Cyclotella nana* Hustedt, and *Detonula confervacea* (cleve) Gran. Can J Microbiol. 1962;8:229–39.

45. Khozin-Goldberg I, Shrestha P, Cohen Z. Mobilization of arachidonyl moieties from triacylglycerols into chloroplastic lipids following recovery from nitrogen starvation of the microalga *Parietochloris incisa*. Biochim Biophys Acta. 2005;1738(1–3):63–71.

46. Falciatore A, Casotti R, Leblanc C, Abrescia C, Bowler C. Transformation of nonselectable reporter genes in marine diatoms. Mar Biotechnol. 1999;1(3):239–51.

47. Chomczynski P, Sacchi N. Single-step method of RNA isolation by acid guanidinium thiocyanate-phenol-chloroform extraction. Anal Biochem. 1987;162(1):156–9.

48. Matyash V, Liebisch G, Kurzchalia TV, Shevchenko A, Schwudke D. Lipid extraction by methyl-tert-butyl ether for high-throughput lipidomics. J Lipid Res. 2008;49(5):1137–46.

49. Hornung E, Korfei M, Pernstich C, Struss A, Kindl H, Fulda M, Feussner I. Specific formation of arachidonic acid and eicosapentaenoic acid by a front-end D^5-desaturase from *Phytophthora megasperma*. Biochim Biophys Acta. 2005;1686(3):181–9.

50. Tarazona P, Feussner K, Feussner I. An enhanced plant lipidomics method based on multiplexed liquid chromatography–mass spectrometry reveals additional insights into cold- and drought-induced membrane remodeling. Plant J. 2015;84(3):621–33.

51. Larson TR, Graham IA. Technical Advance: a novel technique for the sensitive quantification of acyl CoA esters from plant tissues. Plant J. 2001;25(1):115–25.

PERMISSIONS

All chapters in this book were first published in BFB, by BioMed Central; hereby published with permission under the Creative Commons Attribution License or equivalent. Every chapter published in this book has been scrutinized by our experts. Their significance has been extensively debated. The topics covered herein carry significant findings which will fuel the growth of the discipline. They may even be implemented as practical applications or may be referred to as a beginning point for another development.

The contributors of this book come from diverse backgrounds, making this book a truly international effort. This book will bring forth new frontiers with its revolutionizing research information and detailed analysis of the nascent developments around the world.

We would like to thank all the contributing authors for lending their expertise to make the book truly unique. They have played a crucial role in the development of this book. Without their invaluable contributions this book wouldn't have been possible. They have made vital efforts to compile up to date information on the varied aspects of this subject to make this book a valuable addition to the collection of many professionals and students.

This book was conceptualized with the vision of imparting up-to-date information and advanced data in this field. To ensure the same, a matchless editorial board was set up. Every individual on the board went through rigorous rounds of assessment to prove their worth. After which they invested a large part of their time researching and compiling the most relevant data for our readers.

The editorial board has been involved in producing this book since its inception. They have spent rigorous hours researching and exploring the diverse topics which have resulted in the successful publishing of this book. They have passed on their knowledge of decades through this book. To expedite this challenging task, the publisher supported the team at every step. A small team of assistant editors was also appointed to further simplify the editing procedure and attain best results for the readers.

Apart from the editorial board, the designing team has also invested a significant amount of their time in understanding the subject and creating the most relevant covers. They scrutinized every image to scout for the most suitable representation of the subject and create an appropriate cover for the book.

The publishing team has been an ardent support to the editorial, designing and production team. Their endless efforts to recruit the best for this project, has resulted in the accomplishment of this book. They are a veteran in the field of academics and their pool of knowledge is as vast as their experience in printing. Their expertise and guidance has proved useful at every step. Their uncompromising quality standards have made this book an exceptional effort. Their encouragement from time to time has been an inspiration for everyone.

The publisher and the editorial board hope that this book will prove to be a valuable piece of knowledge for researchers, students, practitioners and scholars across the globe.

LIST OF CONTRIBUTORS

Ajaya K. Biswal, Li Tan and Melani A. Atmodjo
Department of Biochemistry and Molecular Biology, University of Georgia, Athens, GA 30602, USA.
Complex Carbohydrate Research Center, University of Georgia, 315 Riverbend Rd., Athens, GA 30602-4712, USA
DOE-BioEnergy Science Center (BESC), Oak Ridge 37831, TN, USA

Jaclyn DeMartini
DOE-BioEnergy Science Center (BESC), Oak Ridge 37831, TN, USA.
Center for Environmental Research and Technology (CE-CERT) and Department of Chemical and Environmental Engineering, University of California Riverside, Riverside 92507, CA, USA
DuPont Industrial Biosciences, Palo Alto, CA 94304, USA

Ivana Gelineo-Albersheim
Complex Carbohydrate Research Center, University of Georgia, 315 Riverbend Rd., Athens, GA 30602-4712, USA
DOE-BioEnergy Science Center (BESC), Oak Ridge 37831, TN, USA

Kimberly Hunt
Complex Carbohydrate Research Center, University of Georgia, 315 Riverbend Rd., Athens, GA 30602-4712, USA
DOE-BioEnergy Science Center (BESC), Oak Ridge 37831, TN, USA
South Georgia State College, Douglas, GA 31533, USA

Ian M. Black
Complex Carbohydrate Research Center, University of Georgia, 315 Riverbend Rd., Athens, GA 30602-4712, USA

Sushree S. Mohanty and David Ryno
Complex Carbohydrate Research Center, University of Georgia, 315 Riverbend Rd., Athens, GA 30602-4712, USA
DOE-BioEnergy Science Center (BESC), Oak Ridge 37831, TN, USA

Charles E. Wyman
DOE-BioEnergy Science Center (BESC), Oak Ridge 37831, TN, USA
Center for Environmental Research and Technology (CE-CERT) and Department of Chemical and Environmental Engineering, University of California Riverside, Riverside 92507, CA, USA

Debra Mohnen
Department of Biochemistry and Molecular Biology, University of Georgia, Athens, GA 30602, USA
Complex Carbohydrate Research Center, University of Georgia, 315 Riverbend Rd., Athens, GA 30602-4712, USA
DOE-BioEnergy Science Center (BESC), Oak Ridge 37831, TN, USA

Canhui Lu
State Key Laboratory of Polymer Materials Engineering, Polymer Research Institute of Sichuan University, Chengdu 610065, China

Dong Tian
State Key Laboratory of Polymer Materials Engineering, Polymer Research Institute of Sichuan University, Chengdu 610065, China
State Key Laboratory of Bioreactor Engineering, East China University of Science and Technology, 130 Meilong Road, Shanghai 200237, China
Forest Products Biotechnology/ Bioenergy Group, Department of Wood Science, Faculty of Forestry, University of British Columbia, 2424 Main Mall, Vancouver, BC V6T 1Z4, Canada

Jinguang Hu
State Key Laboratory of Bioreactor Engineering, East China University of Science and Technology, 130 Meilong Road, Shanghai 200237, China
Forest Products Biotechnology/ Bioenergy Group, Department of Wood Science, Faculty of Forestry, University of British Columbia, 2424 Main Mall, Vancouver, BC V6T 1Z4, Canada

Jie Bao
State Key Laboratory of Bioreactor Engineering, East China University of Science and Technology, 130 Meilong Road, Shanghai 200237, China

Richard P. Chandra and Jack N. Saddler
Forest Products Biotechnology/ Bioenergy Group, Department of Wood Science, Faculty of Forestry, University of British Columbia, 2424 Main Mall, Vancouver, BC V6T 1Z4, Canada

Feng Li, Yuanxiu Li, Xiaofei Li, Changji Yin, Xingjuan An, Xiaoli Chen, Yao Tian and Hao Song
Key Laboratory of Systems Bioengineering (Ministry of Education), School of Chemical Engineering and Technology, Tianjin University, Tianjin 300072, China
SynBio Research Platform, Collaborative Innovation Centre of Chemical Science and Engineering, Tianjin University, Tianjin 300072, China

Liming Sun
Petrochemical Research Institute, PetroChina Company Limited, Beijing 102206, People's Republic of China

Farzad Taheripour, Xin Zhao and Wallace E. Tyner
Department of Agricultural Economics, Purdue University, West Lafayette, USA

Ling Chen, Yanping Jiang, Gen Zou and Zhihua Zhou
CAS-Key Laboratory of Synthetic Biology, CAS Center for Excellence in Molecular Plant Sciences, Institute of Plant Physiology and Ecology, Chinese Academy of Science, Fenglin Rd 300, Shanghai 200032, China

Rui Liu
CAS-Key Laboratory of Synthetic Biology, CAS Center for Excellence in Molecular Plant Sciences, Institute of Plant Physiology and Ecology, Chinese Academy of Science, Fenglin Rd 300, Shanghai 200032, China
University of Chinese Academy of Sciences, Beijing 100049, China

Jie Wang, Michael Chae and David C. Bressler
Department of Agricultural, Food and Nutritional Science, University of Alberta, Edmonton T6G 2P5, Canada

Dominic Sauvageau
Department of Chemical and Materials Engineering, University of Alberta, Edmonton T6G 1H9, Canada

Daniel Jaeger and Jan H. Mussgnug
Algae Biotechnology and Bioenergy, Faculty of Biology, Center for Biotechnology (CeBiTec), Bielefeld University, 33615 Bielefeld, Germany

Anika Winkler and Jörn Kalinowski
Microbial Genomics and Biotechnology, Center for Biotechnology (CeBiTec), Bielefeld University, 33615 Bielefeld, Germany

Alexander Goesmann
Bioinformatics and Systems Biology, Justus-Liebig-Universität, 35392 Gießen, Germany

Olaf Kruse
Algae Biotechnology and Bioenergy, Faculty of Biology, Center for Biotechnology (CeBiTec), Bielefeld University, 33615 Bielefeld, Germany
AlgaeBiotechnology and Bioenergy, Faculty of Biology, Center for Biotechnology (CeBiTec), Bielefeld University, Universitaetsstrasse 27, 33615 Bielefeld, Germany

Xiaoqing Wang, Davinia Salvachúa, Violeta Sànchez i Nogué, William E. Michener, Adam D. Bratis and Gregg T. Beckham
National Bioenergy Center, National Renewable Energy Laboratory, Golden, CO 80401, USA

John R. Dorgan
Chemical and Biological Engineering Department, Colorado School of Mines, Golden, CO 80401, USA

Yu Matsuoka
Department of Bioscience and Bioinformatics, Kyushu Institute of Technology, 680-4 Kawazu, Iizuka, Fukuoka 820-8502, Japan

Hiroyuki Kurata
Department of Bioscience and Bioinformatics, Kyushu Institute of Technology, 680-4 Kawazu, Iizuka, Fukuoka 820-8502, Japan
Biomedical Informatics R&D Center, Kyushu Institute of Technology, 680-4 Kawazu, Iizuka, Fukuoka 820-8502, Japan

Man Zhou, Tao Wang, Lina Gao, Huijun Yin and Xin Lü
College of Food Science and Engineering, Northwest A&F University, Yangling, Shaanxi Province, China

Peng Guo and Cheng Cai
College of Information Engineering, Northwest A&F University, Yangling, Shaanxi Province, China

Jie Gu
College of Natural Resources and Environment, Northwest A&F University, Yangling, Shaanxi Province, China

Vera Novy and Bernd Nidetzky
Institute of Biotechnology and Biochemical Engineering, Graz University of Technology, Graz, Austria

Ruifei Wang, Johan O. Westman and Carl Johan Franzén
Division of Industrial Biotechnology, Department of Biology and Biological Engineering, Chalmers University of Technology, Gothenburg, Sweden

Kannikar Charoensuk
Division of Product Development and Management Technology, Faculty of Agro-Industrial Technology, Rajamangala University of Technology Tawan-ok, Chanthaburi Campus, Chanthaburi 22100, Thailand

Tomoko Sakurada and Masayuki Murata
Life Science, Graduate School of Science and Technology for Innovation, Yamaguchi University, Ube 755-8505, Japan

Amina Tokiyama
Department of Biological Chemistry, Faculty of Agriculture, Yamaguchi University, 1677-1 Yoshida, Yamaguchi 753-8515, Japan

Tomoyuki Kosaka and Mamoru Yamada
Life Science, Graduate School of Science and Technology for Innovation, Yamaguchi University, Ube 755-8505, Japan
Department of Biological Chemistry, Faculty of Agriculture, Yamaguchi University, 1677-1 Yoshida, Yamaguchi 753-8515, Japan
Research Center for Thermotolerant Microbial Resources, Yamaguchi University, Yamaguchi 753-8315, Japan

Pornthap Thanonkeo
Department of Biotechnology, Faculty of Technology, Khon Kaen University, Khon Kaen 40002, Thailand

Chen Zhao and Shaolin Chen
College of Life Sciences, Northwest A&F University, 22 Xinong Road, Yangling 712100, Shaanxi, China

Hao Fang
College of Life Sciences, Northwest A&F University, 22 Xinong Road, Yangling 712100, Shaanxi, China

National Engineering Laboratory for Cereal Fermentation Technology, Jiangnan University, 1800 Lihu Avenue, Wuxi 214122, Jiangsu, China

Hannes Löwe, Karina Hobmeier, Manuel Moos, Andreas Kremling and Katharina Pflüger-Grau
Fachgebiet für Systembiotechnologie, Technische Universität München, Boltzmannstr 15, 85748 Garching, Germany

Nodumo Nokulunga Zulu
Department of Plant Biochemistry, Albrecht-von-Haller-Institute for Plant Sciences, University of Goettingen, 37077 Goettingen, Germany
Novagreen Projektmanagement GmbH, 49377 Vechta, Germany

Jennifer Popko, Krzysztof Zienkiewicz, Pablo Tarazona and Cornelia Herrfurth
Department of Plant Biochemistry, Albrecht-von-Haller-Institute for Plant Sciences, University of Goettingen, 37077 Goettingen, Germany

Ivo Feussner
Department of Plant Biochemistry, Albrecht-von-Haller-Institute for Plant Sciences, University of Goettingen, 37077 Goettingen, Germany
Department of Plant Biochemistry, Goettingen Center for Molecular Biosciences (GZMB), University of Goettingen, 37077 Goettingen, Germany
Department of Plant Biochemistry, International Center for Advanced Studies of Energy Conversion (ICASEC), University of Goettingen, 37077 Goettingen, Germany

Index